Managing and Mining Uncertain Data

ADVANCES IN DATABASE SYSTEMS

Volume 35

Series Editors

Ahmed K. Elmagarmid
Purdue University
West Lafayette, IN 47907

Amit P. Sheth
Wright State University
Dayton, Ohio 45435

Other books in the Series:

PRIVACY-PRESERVING DATA MINING: ***Models and Algorithms*** *by Charu C. Aggarwal and Philip S. Yu*; ISBN: 978-0-387-70991-8

SEQUENCE DATA MINING *by Guozhu Dong, Jian Pei*; ISBN: 978-0-387-69936-3

DATA STREAMS: ***Models and Algorithms***, *edited by Charu C. Aggarwal*; ISBN: 978-0-387-28759-1

SIMILARITY SEARCH: ***The Metric Space Approach***, *P. Zezula, G. Amato, V. Dohnal, M. Batko*; ISBN: 0-387-29146-6

STREAM DATA MANAGEMENT, *Nauman Chaudhry, Kevin Shaw, Mahdi Abdelguerfi*; ISBN: 0-387-24393-3

FUZZY DATABASE MODELING WITH XML, *Zongmin Ma*; ISBN: 0-387-24248-1

MINING SEQUENTIAL PATTERNS FROM LARGE DATA SETS, *Wei Wang and Jiong Yang*; ISBN: 0-387-24246-5

ADVANCED SIGNATURE INDEXING FOR MULTIMEDIA AND WEB APPLICATIONS, *Yannis Manolopoulos, Alexandros Nanopoulos, Eleni Tousidou*; ISBN: 1-4020-7425-5

ADVANCES IN DIGITAL GOVERNMENT: ***Technology, Human Factors, and Policy,*** *edited by William J. McIver, Jr. and Ahmed K. Elmagarmid*; ISBN: 1-4020-7067-5

INFORMATION AND DATABASE QUALITY, *Mario Piattini, Coral Calero and Marcela Genero*; ISBN: 0-7923-7599-8

DATA QUALITY, *Richard Y. Wang, Mostapha Ziad, Yang W. Lee*: ISBN: 0-7923-7215-8

THE FRACTAL STRUCTURE OF DATA REFERENCE: ***Applications to the Memory Hierarchy,*** *Bruce McNutt*; ISBN: 0-7923-7945-4

SEMANTIC MODELS FOR MULTIMEDIA DATABASE SEARCHING AND BROWSING, *Shu-Ching Chen, R.L. Kashyap, and Arif Ghafoor*; ISBN:0-7923-7888-1

INFORMATION BROKERING ACROSS HETEROGENEOUS DIGITAL DATA: ***A Metadata-based Approach,*** *VipulKashyap, AmitSheth*; ISBN:0-7923-7883-0

For a complete listing of books in this series, go to http://www.springer.com

Managing and Mining Uncertain Data

Edited by

Charu C. Aggarwal
IBM T.J. Watson Research Center
USA

Editor
Charu C. Aggarwal
IBM Thomas J. Watson Research Center
19 Skyline Drive
Hawthorne, NY 10532
charu@us.ibm.com

Series Editors
Ahmed K. Elmagarmid
Purdue University
West Lafayette, IN 47907

Amit P. Sheth
Wright State University
Dayton, Ohio 45435

ISBN 978-1-4419-3517-5 e-ISBN 978-0-387-09690-2
DOI 10.1007/978-0-387-09690-2

Printed on acid-free paper

springer.com

Preface

Uncertain data management has seen a revival in interest in recent years because of a number of new fields which utilize this kind of data. For example, in fields such as privacy-preserving data mining, additional errors may be added to data in order to mask the identity of the records. Often the data may be imputed using statistical methods such as forecasting. In such cases, the data is uncertain in nature. Such data sets may often be probabilistic in nature. In other cases, databases may show *existential uncertainty* in which one or more records may be present or absent from the data set. Such data sets lead to a number of unique challenges in processing and managing the underlying data.

The field of uncertain data management has been studied in the traditional database literature, but the field has seen a revival in recent years because of new ways of collecting data. The field of uncertain data management presents a number of challenges in terms of collecting, modeling, representing, querying, indexing and mining the data. We further note that many of these issues are inter-related and cannot easily be addressed independently. While many of these issues have been addressed in recent research, the research in this area is often quite varied in its scope. For example, even the underlying assumptions of uncertainty are different across different papers. It is often difficult for researchers and students to find a single place containing a coherent discussion on the topic.

This book is designed to provide a coherent treatment of the topic of uncertain data management by providing surveys of the key topics in this field. The book is structured as an edited volume containing surveys by prominent researchers in the field. The choice of chapters is carefully designed, so that the overall content of the uncertain data management and mining field is covered reasonably well. Each chapter contains the key research content on a particular topic, along with possible research directions. This includes a broad overview of the topic, the different models and systems for uncertain data, discussions on database issues for managing uncertain data, and mining issues with uncertain data. Two of the most prominent systems for uncertain data have also been described in the book in order to provide an idea how real uncertain data management systems might work. The idea is to structurally organize the topic, and provide insights which are not easily available otherwise. It is hoped that this structural organization and survey approach will be a great help to students, researchers, and practitioners in the field of uncertain data management and mining.

Contents

List of Figures

List of Tables

Chapter 1

AN INTRODUCTION TO UNCERTAIN DATA ALGORITHMS AND APPLICATIONS

Charu C. Aggarwal
IBM T. J. Watson Research Center
Hawthorne, NY 10532
charu@us.ibm.com

Abstract

In recent years, uncertain data has become ubiquitous because of new technologies for collecting data which can only measure and collect the data in an imprecise way. Furthermore, many technologies such as privacy-preserving data mining create data which is inherently uncertain in nature. As a result there is a need for tools and techniques for mining and managing uncertain data. This chapter discusses the broad outline of the book and the methods used for various uncertain data applications.

1. Introduction

In recent years many new techniques for collecting data have resulted in an increase in the availability of uncertain data. While many applications lead to data which contains errors, we refer to *uncertain data sets* as those in which the level of uncertainty can be quantified in some way. Some examples of applications which create uncertain data are as follows:

- Many scientific measurement techniques are inherently imprecise. In such cases, the level of uncertainty may be derived from the errors in the underlying instrumentation.

- Many new hardware technologies such as sensors generate data which is imprecise. In such cases, the error in the sensor network readings can be modeled, and the resulting data can be modeled as imprecise data.

- In many applications such as the tracking of mobile objects, the *future trajectory* of the objects is modeled by forecasting techniques. Small errors in current readings can get magnified over the forecast into the distant future of the trajectory. This is frequently encountered in cosmological applications when one models the probability of encounters with Near-Earth-Objects (NEOs). Errors in forecasting are also encountered in non-spatial applications such as electronic commerce.

- In many applications such as privacy-preserving data mining, the data is modified by adding perturbations to it. In such cases, the format of the output [5] is exactly the same as that of uncertain data.

A detailed survey of uncertain data mining and management algorithms may be found in [2]. In this book, we discuss techniques for mining and managing uncertain data. The broad areas covered in the book are as follows:

- **Modeling and System Design for Uncertain Data:** The nature of complexity captured by the uncertain data representation relies on the model used in order to capture it. The most general model for uncertain data is the *possible worlds model*[1], which tries to capture all the possible states of a database which are consistent with a given schema. The generality of the underlying scheme provides the power of the model. On the other hand, it is often difficult to leverage a very general representation for application purposes. In practice, a variety of simplifying assumptions (independence of tuples or independence of attributes) are used in order to model the behavior of the uncertain data. On the other hand, more sophisticated techniques such as probabilistic graphical models can be used in order to model complex dependencies. This is a natural tradeoff between representation power and utility. Furthermore, the design of the system used for representing, querying and manipulating uncertain data critically depends upon the model used for representation.

- **Management of Uncertain Data:** The process of managing uncertain data is much more complicated than that for traditional databases. This is because the uncertainty information needs to be represented in a form which is easy to process and query. Different models for uncertain data provide different tradeoffs between usability and expressiveness. Clearly, the best model to use depends upon the application at hand. Furthermore, effective query languages need to be designed for uncertain data and index structures need to be constructed. Most data management operations such as indexing, join processing or query processing need to be fundamentally re-designed.

- **Mining Uncertain Data:** The uncertainty information in the data is useful information which can be leveraged in order to improve the quality of the underlying results. For example, in a classification application, a feature with greater uncertainty may not be as important as one which has a lower amount of uncertainty. Many traditional applications such as classification, clustering, and frequent pattern mining may need to re-designed in order to take the uncertainty into account.

This chapter is organized as follows. In the next section, we will discuss the broad areas of work in the topic of uncertain data. Each of these areas is represented by a chapter in the book. The next section will discuss a summary of the material discussed in the chapter and its relationship to other chapters in the book. Section 3 contains the conclusions.

2. Algorithms for Uncertain Data

This section will provide a chapter-by-chapter overview of the different topics which are discussed in this book. The aim is to cover the modeling, management and mining topics fairly comprehensively. The key algorithms in the field are described fairly comprehensively in the different chapters and the relevant pointers are provided. The key topics discussed in the book are as follows:

Models for Uncertain Data. A clear challenge for uncertain data management is underlying data representation and modeling [13, 16, 20]. This is because the underlying representation in the database defines the power of the different approaches which can be used. Chapter 2 provides a clear discussion of the several models which are used for uncertain data management. A related issue is the representation in relational databases, and its relationship with the query language which is finally used. Chapter 3 also discusses the issue of relational modeling of uncertain data, though with a greater emphasis on relational modeling and query languages. While chapter 2 discusses the formal definitions of different kinds of models, chapter 3 discusses some of the more common and simplified models which are used in the literature. The chapter also discusses the implications of using different kinds of models from the relational algebra perspective.

Probabilistic Graphical Models. Probabilistic Graphical Models are a popular and versatile class of models which have significantly greater expressive power because of their graphical structure. They allow us to intuitively capture and reason about complex interactions between the uncertainties of different data items. Chapter 4 discusses a number of common graphical models such as Bayesian Networks and Markov Networks. The chapter discusses the application of these models to the representation of uncertainty. The chapter also discusses how queries can be effectively evaluated on uncertain data with the use of graphical models.

Systems for Uncertain Data. We present two well known systems for uncertain data. These are the *Trio* and *MayBMS* systems. These chapters will provide a better idea of how uncertain data management systems work in terms of database manipulation and querying. The *Trio* system is described in chapter 5, whereas the *MayBMS* system is discussed in chapter 6. Both these chapters provide a fairly comprehensive study of the different kinds of systems and techniques used in conjunction with these systems.

Data Integration. Uncertain data is often collected from disparate data sources. This leads to issues involving database integration. Chapter 7 discusses issues involved in database integration of uncertain data. The most important issue with uncertain data is to use schema mappings in order to match the uncertain data from disparate sources.

Query Estimation and Summarization of Uncertain Data Streams. The problem of querying is one of the most fundamental database operations. Query estimation is a closely related problem which is often required for a number of database operations. A closely related problem is that of resolving *aggregate queries* with the use of probabilistic techniques such as sketches. Important statistical measures of streams such as the quantiles, minimum, maximum, sum, count, repeat-rate, average, and the number of distinct items are useful in a variety of database scenarios. Chapter 8 discusses the issue of sketching probabilistic data streams, and how the synopsis may be used for estimating the above measures.

Join Processing of Uncertain Data. The problem of join processing is challenging in the context of uncertain data, because the join-attribute is probabilistic in nature. Therefore, the join operation needs to be redefined in the context of probabilistic data. Chapter 9 discusses the problem of join processing of uncertain data. An important aspect of join processing algorithms is that the uncertainty model significantly affects the nature of join processing. The chapter discusses different kinds of join methods such as the use of *confidence-based join methods*, *similarity joins* and *spatial joins*.

Indexing Uncertain Data. The problem of indexing uncertain data is especially challenging because the diffuse probabilistic nature of the data can reduce the effectiveness of index structures. Furthermore, the challenges for indexing can be quite different, depending upon whether the data is discrete, continuous, spatio-temporal, or how the probabilistic function is defined [8, 9, 12, 22, 23]. Chapter 10 provides a comprehensive overview of the problem of indexing uncertain data. This chapter discusses the problem of indexing both continuous and discrete data. Chapter 11 further discusses the problem of indexing uncertain data in the context of spatiotemporal data. Chapters 10 and 11 provide a fairly comprehensive survey of the different kinds of techniques which are often used for indexing and retrieval of uncertain data.

Probabilistic XML Data. XML data poses a number of special challenges in the context of uncertainty because of the structural nature of the underlying data. Chapter 12 discusses uncertain models for probabilistic XML data. The chapter also describes algebraic techniques for manipulating XML data. This includes probabilistic aggregate operations and the query language for XML data (known as PXML). The chapter discusses both special cases for probability distributions as well as arbitrary probability distributions for representing probabilistic XML data.

Clustering Uncertain Data. Data mining problems are significantly influenced by the uncertainty in the underlying data, since we can leverage the uncertainty in order to improve the quality of the underlying results. Clustering is one of the most comprehensively studied problems in the uncertain data mining literature. Recently, techniques have been designed for clustering uncertain data. These include the *UMicro* algorithm, the UK-means algorithms, the FDBSCAN, and FOPTICS algorithms [6, 18, 19, 21]. Recently, some approximation algorithms [7] have also been developed for clustering uncertain data. Chapter 13 discusses a comprehensive overview of the different algorithms for clustering uncertain data.

General Transformations for Uncertain Data Mining. A natural approach to uncertain data management techniques is to use general transformations [3] which can create *intermediate representations* which adjust for the uncertainty. These intermediate representations can then be leveraged in order to improve the quality of the underlying results. Chapter 14 discusses such an approach with the use of density based transforms. The idea is to create a probability density representation of the data which takes the uncertainty into account during the transformation process. The chapter discusses two applications of this approach to the problems of classification and outlier detection. We note that the approach can be used for any data mining problem, as long as a method can be found to use intermediate density transformations for data mining purposes.

Frequent Pattern Mining. Chapter 15 surveys a number of different approaches for frequent pattern mining of uncertain data. In the case of transactional data, items are assumed to have *existential probabilities* [4, 10, 11], which characterize the likelihood of presence in a given transaction. This includes Apriori-style algorithms, candidate generate-and-test algorithms, pattern growth algorithms and hyper-structure based algorithms. The chapter examines the uniqueness of the tradeoffs involved for pattern mining algorithms in the uncertain case. The chapter compares many of these algorithms for the challenging case of high existential probabilities, and shows that the behavior is quite different from deterministic algorithms. Most of the literature [10, 11] studies the case of low existential probabilities. The chapter suggests that the behavior is quite different for the case of high-existential probabilities. This is because many of the pruning techniques designed for the case of low existential probabilities do not work well for the case when these probabilities are high.

Applications to Biomedical Domain. We provide one application chapter in order to provide a flavor of the application of uncertain DBMS techniques to a real application. The particular application picked in this case is that of biomedical images. Chapter 16 is a discussion of the application of uncertain data management techniques to the biomedical domain. The chapter is particular interesting in that it discusses the application of many techniques discussed in this book (such as indexing and join processing) to an application domain. While the chapter discusses the biological image domain, the primary goal is to present an example of the application of many of the discussed techniques to a particular application.

3. Conclusions

In this chapter, we introduced the problem of uncertain data mining, and discussed an overview of the different facets of this area covered by this book. Uncertain data management promises to be a new and exciting field for practitioners, students and researchers. It is hoped that this book is able to provide a broad overview of this topic, and how it relates to a variety of data mining and management applications. This book discusses both data management and data mining issues. In addition, the book discusses an application domain for the field of uncertain data. Aside from the topics discussed in the book, some of the open areas for research in the topic of uncertain data are as follows:

- **Managing and Mining Techniques under General Models:** Most of the uncertain data mining and management algorithms use a variety of simplifying assumptions in order to allow effective design of the underlying algorithms. Examples of such simplifying assumptions could imply tuple or attribute independence. In more general scenarios, one may want to use more complicated schemas to represent uncertain databases. Some models such as probabilistic graphical models [15] provide greater expressivity in capturing such cases. However, database management and mining techniques become more complicated under such models.

Most of the current techniques in the literature do not use such general models. Therefore, the use of such models for developing DBMS techniques may be a fruitful future area of research.

- **Synergy between Uncertain Data Acquisition and Usage:** The utility of the field can increase further only if a concerted effort is made to standardize the uncertainty in the data to the models used for the general management and mining techniques. For example, the output of both the privacy-preserving publishing and the sensor data collection fields are typically uncertain data. In recent years, some advances have been made [5, 14] in order to design models for data acquisition and creation, which naturally pipeline onto useful uncertain representations. A lot more work remains to be done in a variety of scientific fields in order to facilitate model based acquisition and creation of uncertain data.

Acknowledgements

Research was sponsored in part by the US Army Research laboratory and the UK ministry of Defense under Agreement Number W911NF-06-3-0001. The views and conclusions contained in this document are those of the author and should not be interpreted as representing the official policies of the US Government, the US Army Research Laboratory, the UK Ministry of Defense, or the UK Government. The US and UK governments are authorized to reproduce and distribute reprints for Government purposes.

References

[1] S. Abiteboul, P. C. Kanellakis, G. Grahne. "On the Representation and Querying of Sets of Possible Worlds." in *Theoretical Computer Science*, 78(1): 158-187 (1991)

[2] C.C. Aggarwal, P. S. Yu. " A Survey of Uncertain Data Algorithms and Applications," in *IEEE Transactions on Knowledge and Data Engineering*, to appear, 2009.

[3] C. C. Aggarwal, "On Density Based Transforms for Uncertain Data Mining," in *ICDE Conference Proceedings*, 2007.

[4] C. C. Aggarwal, Y. Li, J. Wang, J. Wang. "Frequent Pattern Mining with Uncertain Data." *IBM Research Report*, 2008.

[5] C. C. Aggarwal, "On Unifying Privacy and Uncertain Data Models," in*ICDE Conference Proceedings*, 2008.

[6] C. C. Aggarwal and P. S. Yu, "A Framework for Clustering Uncertain Data Streams," in *ICDE Conference*, 2008.

[7] G. Cormode, and A. McGregor, "Approximation algorithms for clustering uncertain data," in *PODS Conference*, pp. 191-200, 2008.

[8] R. Cheng, Y. Xia, S. Prabhakar, R. Shah, and J. Vitter, "Efficient Indexing Methods for Probabilistic Threshold Queries over Uncertain Data," in *VLDB Conference Proceedings*, 2004.

[9] R. Cheng, D. Kalashnikov, S. Prabhaker: "Evaluating Probabilistic Queries over Imprecise Data" in *SIGMOD Conference*, 2003.

[10] C.-K. Chui, B. Kao, E. Hung. "Mining Frequent Itemsets from Uncertain Data." *PAKDD Conference*, 2007.

[11] C.-K. Chui, B. Kao. "Decremental Approach for Mining Frequent Itemsets from Uncertain Data." *PAKDD Conference*, 2008.

[12] D. Pfozer, C. Jensen. Capturing the uncertainty of moving object representations. in *SSDM Conference*, 1999.

[13] A. Das Sarma, O. Benjelloun, A. Halevy, and J. Widom, "Working Models for Uncertain Data," in *ICDE Conference Proceedings*, 2006.

[14] A. Deshpande, C. Guestrin, S. Madden, J. M. Hellerstein, W. Hong. "Model-Driven Data Acquisition in Sensor Networks." in *VLDB Conference*, 2004.

[15] A. Deshpande, S. Sarawagi. "Probabilistic Graphical Models and their Role in Databases." in *VLDB Conference*, 2007.

[16] H. Garcia-Molina, and D. Porter, "The Management of Probabilistic Data," in *IEEE Transactions on Knowledge and Data Engineering*, vol. 4, pp. 487–501, 1992.

[17] B. Kanagal, A. Deshpande, "Online Filtering, Smoothing and Probabilistic Modeling of Streaming data," in *ICDE Conference*, 2008.

[18] H.-P. Kriegel, and M. Pfeifle, "Density-Based Clustering of Uncertain Data," in *ACM KDD Conference Proceedings*, 2005.

[19] H.-P. Kriegel, and M. Pfeifle, "Hierarchical Density Based Clustering of Uncertain Data," in *ICDM Conference*, 2005.

[20] L. V. S. Lakshmanan, N. Leone, R. Ross, and V. S. Subrahmanian, "ProbView: A Flexible Probabilistic Database System," in *ACM Transactions on Database Systems*, vol. 22, no. 3, pp. 419–469, 1997.

[21] W. Ngai, B. Kao, C. Chui, R. Cheng, M. Chau, and K. Y. Yip, "Efficient Clustering of Uncertain Data," in *ICDM Conference Proceedings*, 2006.

[22] S. Singh, C. Mayfield, S. Prabhakar, R. Shah, S. Hambrusch. "Indexing Uncertain Categorical Data", in *ICDE Conference*, 2007.

[23] Y. Tao, R. Cheng, X. Xiao, W. Ngai, B. Kao, S. Prabhakar. "Indexing Multi-dimensional Uncertain Data with Arbitrary Probabality Density Functions", in *VLDB Conference*, 2005.

Chapter 2

MODELS FOR INCOMPLETE AND PROBABILISTIC INFORMATION

Todd J. Green
Department of Computer and Information Science
University of Pennsylvania
tjgreen@cis.upenn.edu

Abstract We discuss, compare and relate some old and some new models for incomplete and probabilistic databases. We characterize the expressive power of c-tables over infinite domains and we introduce a new kind of result, algebraic completion, for studying less expressive models. By viewing probabilistic models as incompleteness models with additional probability information, we define completeness and closure under query languages of general probabilistic database models and we introduce a new such model, probabilistic c-tables, that is shown to be complete and closed under the relational algebra. We also identify fundamental connections between query answering with incomplete and probabilistic databases and data provenance. We show that the calculations for incomplete databases, probabilistic databases, bag semantics, lineage, and why-provenance are particular cases of the same general algorithms involving semi-rings. This further suggests a comprehensive provenance representation that uses semi-rings of polynomials. Finally, we show that for positive Boolean c-tables, containment of positive relational queries is the same as for standard set semantics.

Keywords: Incomplete databases, probabilistic databases, provenance, lineage, semi-rings

1. Introduction

This chapter provides a survey of models for incomplete and probabilistic information from the perspective of two recent papers that the author has written with Val Tannen [28] and Grigoris Karvounarakis and Val Tannen [27]. All the concepts and technical developments that are not attributed specifically to another publication originate in these two papers.

The representation of incomplete information in databases has been an important research topic for a long time, see the references in [25], in Ch.19 of [2], in [43], in [48, 36], as well as the recent [45, 42, 41, 4]. Moreover, this work is closely related to recently active research topics such as inconsistent databases and repairs [5], answering queries using views [1], data exchange [20], and data provenance [9, 8]. The classic reference on incomplete databases remains [30] with the fundamental concept of c-table and its restrictions to simpler tables with variables. The most important result of [30] is the query answering algorithm that defines an algebra on c-tables that corresponds exactly to the usual relational algebra ($\mathcal{RA}$). A recent paper [41] has defined a hierarchy of incomplete database models based on finite sets of choices and optional inclusion. We shall give below **comparisons** between the models [41] and the tables with variables from [30].

Two criteria have been provided for comparisons among all these models: [30, 41] discuss *closure* under relational algebra operations, while [41] also emphasizes *completeness*, specifically the ability to represent all finite incomplete databases. We point out that the latter is not appropriate for tables with variables over an infinite domain, and we describe another criterion, **$\mathcal{RA}$-completeness**, that fully characterizes the expressive power of c-tables.

We outline a method for the study of models that are not complete. Namely, we consider combining existing models with queries in various fragments of relational algebra. We then ask how big these fragments need to be to obtain a combined model that is complete. We give a number of such **algebraic completion** results.

Early on, probabilistic models of databases were studied less intensively than incompleteness models, with some notable exceptions [10, 6, 39, 34, 17]. Essential progress was made independently in three papers [22, 33, 47] that were published at about the same time. [22, 47] assume a model in which tuples are taken independently in a relation with given probabilities. [33] assumes a model with a separate distribution for each attribute in each tuple. All three papers attacked the problem of calculating the probability of tuples occurring in query answers. They solved the problem by developing more general models in which rows are **annotated** with additional information ("event expressions," "paths," "traces"), and they noted the similarity with the conditions in c-tables.

We go beyond the problem of individual tuples in query answers by defining **closure** under a query language for probabilistic models. Then we describe **probabilistic c-tables** which add *to the c-tables themselves* probability distributions for the values taken by their variables. Here is an example of such a representation that captures the set of instances in which Alice is taking a course that is Math with probability 0.3; Physics (0.3); or Chemistry (0.4), while Bob takes the same course as Alice, provided that course is Physics or

Chemistry and Theo takes Math with probability 0.85:

Student	*Course*	*Condition*
Alice	x	
Bob	x	$x = \text{phys} \vee x = \text{chem}$
Theo	math	$t = 1$

$$x = \begin{cases} \text{math} & : 0.3 \\ \text{phys} & : 0.3 \\ \text{chem} & : 0.4 \end{cases}$$

$$t = \begin{cases} 0: & 0.15 \\ 1: & 0.85 \end{cases}$$

The concept of probabilistic c-table allows us to solve the closure problem by using the same algebra on c-tables defined in [30].

We also give a **completeness** result by showing that probabilistic Boolean c-tables (all variables are two-valued and can appear only in the conditions, not in the tuples) can represent *any* probabilistic database.

An important conceptual point is that, at least for the models we consider, the probabilistic database models can be seen, as **probabilistic counterparts** of incomplete database models. In an incompleteness model a tuple or an attribute value in a tuple may or may not be in the database. In its probabilistic counterpart, these are seen as elementary events with an assigned probability. For example, the models used in [22, 33, 47] are probabilistic counterparts of the two simplest incompleteness models discussed in [41]. As another example, the model used in [17] can be seen as the probabilistic counterpart of an incompleteness model one in which tuples sharing the same key have an exclusive-or relationship.

A consequence of this observation is that, in particular, query answering for probabilistic c-tables will allow us to solve the problem of calculating probabilities about query answers for any model that can be defined as a probabilistic counterpart of the incompleteness models considered in [30, 41].

Besides the models for incomplete and probabilistic information, several other forms of **annotated relations** have appeared in various contexts in the literature. Query answering in these settings involves generalizing $\mathcal{RA}$ to perform corresponding operations on the annotations.

In data warehousing, [14] and [15] compute lineages for tuples in the output of queries, in effect generalizing $\mathcal{RA}$ to computations on relations annotated with sets of contributing tuples. For curated databases, [9] proposes decorating output tuples with their why-provenance, essentially the set of sets of contributing tuples. Finally, $\mathcal{RA}$ on bag semantics can be viewed as a generalization to annotated relations, where a tuple's annotation is a number representing its multiplicity.

We observe that in all of these cases, the calculations with annotations are strikingly similar. This suggests looking for an algebraic structure on annotations that captures the above as particular cases. It turns out that the right structure to use for this purpose is that of **commutative semi-rings**. In fact, one can show that the laws of commutative semi-rings are *forced* by certain expected identities in $\mathcal{RA}$. Having identified commutative semi-rings as the right algebraic structure, we argue that a symbolic representation of semi-ring calculations is just what is needed to record, document, and track $\mathcal{RA}$ querying from input to output for applications which require rich **provenance** information. It is a standard philosophy in algebra that such symbolic representations form *the most general* such structure. In the case of commutative semi-rings, just as for rings, the symbolic representation is that of polynomials. This strongly suggests using polynomials to capture provenance.

The rest of this chapter is organized as follows:

- We develop the basic notions of **representation systems** for **incomplete information databases**, and we give several examples (Section 2).

- We define two measures of expressive power for representation systems, **$\mathcal{RA}$-Completeness** and **finite completeness**. $\mathcal{RA}$-com-pleteness characterizes the expressiveness of c-tables, and finite completeness the expressiveness of a restricted system which we call finite c-tables (Section 3).

- We examine the related notion of **closure** of representation systems under relational operations (Section 4).

- We define the notion of **algebraic completion**, and we give a number of results showing, for various representation systems not closed under the full relational algebra, that "closing" them under (certain fragments of) the relational algebra yields expressively complete representation systems (Section 5).

- We develop the basic notions of **probabilistic representation systems** (Section 6) and present **probabilistic counterparts** of various representation systems for incomplete databases (Sections 7 and 8).

- We observe patterns in the calculations used in incomplete and probabilistic databases, bag semantics, and why-provenance which motivate the more general study of **annotated relations** (Section 9).

- We define **K-relations**, in which tuples are annotated (tagged) with elements from K. We define a generalized positive algebra on K-relations and argue that K must be a **commutative semi-ring** (Section 10).

- For **provenance semi-rings** we use polynomials with integer coefficients, and we show that positive algebra semantics for any commutative semi-rings **factors** through the provenance semantics (Section 11).

- We consider **query containment** w.r.t. K-relation semantics and we show that for unions of conjunctive queries and when K is a **distributive lattice**, query containment is the same as that given by standard set semantics (Section 12).

2. Incomplete Information and Representation Systems

Our starting point is suggested by the work surveyed in [25], in Ch. 19 of [2], and in [43]. A database that provides incomplete information consists of a *set of possible instances*. At one end of this spectrum we have the conventional single instances, which provide "complete information." At the other end we have the set of *all* allowable instances which provides "no information" at all, or "zero information."

We adopt the formalism of relational databases over a fixed countably infinite domain $\mathbb{D}$. We use the unnamed form of the relational algebra. To simplify the notation we will work with relational schemas that consist of a single relation name of arity n. Everything we say can be easily reformulated for arbitrary relational schemas. We shall need a notation for the set of *all* (conventional) instances of this schema, i.e., all the finite n-ary relations over $\mathbb{D}$:

$$\mathcal{N} := \{I \mid I \subseteq \mathbb{D}^n,\ I \text{ finite}\}$$

DEFINITION 2.1 *An* **incomplete(-information) database** *(***i-database** *for short),* $\mathcal{I}$*, is a set of conventional instances, i.e., a subset* $\mathcal{I} \subseteq \mathcal{N}$*.*

The usual relational databases correspond to the cases when $\mathcal{I} = \{I\}$. The **no-information** or **zero-information database** consists of *all* the relations: $\mathcal{N}$.

Conventional relational instances are finite. However, because $\mathbb{D}$ is infinite incomplete databases are in general infinite. Hence the interest in finite, syntactical, representations for incomplete information.

DEFINITION 2.2 *A* **representation system** *consists of a set (usually a syntactically defined "language") whose elements we call tables, and a function Mod that associates to each table* T *an incomplete database Mod*(T)*.*

The notation corresponds to the fact that T can be seen as a logical assertion such that the conventional instances in *Mod*(T) are in fact the *models* of T (see also [38, 44]).

The classical reference [30] considers three representation systems: **Codd tables**, v**-tables**, and c**-tables**. v-tables are conventional instances in which variables can appear in addition to constants from $\mathbb{D}$. If T is a v-table then[1]

$$Mod(T) := \{\nu(T) \mid \nu : Var(T) \rightarrow \mathbb{D} \text{ is a valuation for the variables of } T\}$$

Codd tables are v-tables in which all the variables are distinct. They correspond roughly to the current use of nulls in SQL, while v-tables model "labeled" or "marked" nulls. c-tables are v-tables in which each tuple is annotated with a *condition* — a Boolean combination of equalities involving variables and constants. The tuple condition is tested for each valuation ν and the tuple is discarded from $\nu(T)$ if the condition is not satisfied.

EXAMPLE 2.3 *A v-table and its possible worlds.*

$$R := \begin{array}{|ccc|}\hline 1 & 2 & x \\ 3 & x & y \\ z & 4 & 5 \\ \hline\end{array} \qquad Mod(R) = \left\{ \begin{array}{|ccc|}\hline 1 & 2 & 1 \\ 3 & 1 & 1 \\ 1 & 4 & 5 \\ \hline\end{array}, \begin{array}{|ccc|}\hline 1 & 2 & 2 \\ 3 & 2 & 1 \\ 1 & 4 & 5 \\ \hline\end{array}, \ldots, \begin{array}{|ccc|}\hline 1 & 2 & 77 \\ 3 & 77 & 89 \\ 97 & 4 & 5 \\ \hline\end{array}, \ldots \right\}$$

EXAMPLE 2.4 *A c-table and its possible worlds.*

$$S := \begin{array}{|ccc|c|}\hline 1 & 2 & x & \\ 3 & x & y & x = y \land z \neq 2 \\ z & 4 & 5 & x \neq 1 \lor x \neq y \\ \hline\end{array}$$

$$Mod(S) = \left\{ \begin{array}{|ccc|}\hline 1 & 2 & 1 \\ 3 & 1 & 1 \\ \hline\end{array}, \begin{array}{|ccc|}\hline 1 & 2 & 2 \\ 1 & 4 & 5 \\ \hline\end{array}, \ldots, \begin{array}{|ccc|}\hline 1 & 2 & 77 \\ 97 & 4 & 5 \\ \hline\end{array}, \ldots \right\}$$

Several other representation systems have been proposed in a recent paper [41]. We illustrate here three of them and we discuss several others later. A **?-table** is a conventional instance in which tuples are optionally labeled with "?," meaning that the tuple may be missing. An **or-set-table** looks like a conventional instance but or-set values [31, 37] are allowed. An or-set value $\langle 1, 2, 3 \rangle$ signifies that exactly one of 1, 2, or 3 is the "actual" (but unknown) value. Clearly, the two ideas can be combined yielding another representation systems that we might (awkwardly) call **or-set-?-tables.**[2]

EXAMPLE 2.5 *An or-set-?-table and its possible worlds.*

$$T := \begin{array}{|ccc|c|}\hline 1 & 2 & \langle 1,2 \rangle & \\ 3 & \langle 1,2 \rangle & \langle 3,4 \rangle & \\ \langle 4,5 \rangle & 4 & 5 & ? \\ \hline\end{array} \qquad Mod(T) = \left\{ \begin{array}{|ccc|}\hline 1 & 2 & 1 \\ 3 & 1 & 3 \\ 4 & 4 & 5 \\ \hline\end{array}, \begin{array}{|ccc|}\hline 1 & 2 & 1 \\ 3 & 1 & 3 \\ \hline\end{array}, \ldots, \begin{array}{|ccc|}\hline 1 & 2 & 2 \\ 3 & 2 & 4 \\ \hline\end{array} \right\}$$

3. $\mathcal{RA}$-Completeness and Finite Completeness

"Completeness" of expressive power is the first obvious question to ask about representation systems. This brings up a fundamental difference between the representation systems of [30] and those of [41]. The presence of

[1] We follow [2, 41] and use the *closed-world assumption (CWA)*. [30] uses the *open-world assumption (OWA)*, but their results hold for CWA as well.
[2] In [41] these three systems are denoted by $\mathcal{R}_?$, $\mathcal{R}^A$ and $\mathcal{R}^A_?$.

variables in a table T and the fact that $\mathbb{D}$ is infinite means that *Mod*(T) may be infinite. For the tables considered in [41], *Mod*(T) is always finite.

[41] defines completeness as the ability of a representation system to represent "all" possible incomplete databases. For the kind of tables considered in [41] the question makes sense. But in the case of the tables with variables in [30] this is hopeless for trivial reasons. Indeed, in such systems there are only countably many tables while there are uncountably many incomplete databases (the subsets of $\mathcal{N}$, which is infinite). We will discuss separately below *finite completeness* for systems that only represent finite databases. Meanwhile, we will develop a different yardstick for the expressive power of tables with variables that range over an infinite domain.

c-tables and their restrictions (v-tables and Codd tables) have an inherent limitation: the cardinality of the instances in *Mod*(T) is at most the cardinality of T. For example, the zero-information database $\mathcal{N}$ *cannot* be represented with c-tables. It also follows that among the incomplete databases that are representable by c-tables the "minimal"-information ones are those consisting for some m of all instances of cardinality up to m (which are in fact representable by Codd tables with m rows). Among these, we make special use of the ones of cardinality 1:

$$\mathcal{Z}_k := \{\{t\} \mid t \in \mathbb{D}^k\}.$$

Hence, $\mathcal{Z}_k$ consists of *all* the one-tuple relations of arity k. Note that $\mathcal{Z}_k = \mathit{Mod}(Z_k)$ where Z_k is the Codd table consisting of a single row of k distinct variables.

DEFINITION 3.1 *An incomplete database $\mathcal{I}$ is $\mathcal{RA}$-**definable** if there exists a relational algebra query q such that $\mathcal{I} = q(\mathcal{Z}_k)$, where k is the arity of the input relation name in q.*

THEOREM 3.2 *If $\mathcal{I}$ is an incomplete database representable by a c-table T, i.e., $\mathcal{I} = \mathit{Mod}(T)$, then $\mathcal{I}$ is $\mathcal{RA}$-definable.*

Proof: Let T be a c-table, and let $\{x_1, \ldots, x_k\}$ denote the variables in T. We want to show that there exists a query q in $\mathcal{RA}$ such that $q(\mathit{Mod}(Z_k)) = \mathit{Mod}(T)$. Let n be the arity of T. For every tuple $t = (a_1, \ldots, a_n)$ in T with condition $T(t)$, let $\{x_{i_1}, \ldots, x_{i_j}\}$ be the variables in $T(t)$ which do not appear in t. For $1 \le i \le n$, define C_i to be the singleton $\{c\}$, if $a_i = c$ for some constant c, or $\pi_j(Z_k)$, if $a_i = x_j$ for some variable x_j. For $1 \le j \le k$, define C_{n+j} to be the expression $\pi_{i_j}(Z_k)$, where x_j is the jth variable in $T(t)$ which does not appear in t. Define q to be the query

$$q := \bigcup_{t \in T} \pi_{1,\ldots,n}(\sigma_{\psi(t)}(C_1 \times \cdots \times C_{n+k})),$$

where $\psi(t)$ is obtained from $T(t)$ by replacing each occurrence of a variable x_i with the index j of the term C_j in which x_i appears. To see that $q(Mod(Z_k)) = Mod(T)$, since Z_k is a c-table, we can use Theorem 4.2 and check that, in fact, $\bar{q}(Z_k) = T$ where $\bar{q}$ is the translation of q into the c-tables algebra (see the proof of Theorem 4.2). Note that we only need the *SPJU* fragment of $\mathcal{RA}$. ∎

EXAMPLE 3.3 *The c-table from Example 2.4 is definable as* $Mod(S) = q(\mathcal{Z}_3)$ *where q is the following query with input relation name V of arity 3:* $q(V) := \pi_{123}(\{1\} \times \{2\} \times V) \cup \pi_{123}(\sigma_{2=3,4\neq\text{'}2\text{'}}(\{3\} \times V)) \cup \pi_{512}(\sigma_{3\neq\text{'}1\text{'},3\neq 4}(\{4\} \times \{5\} \times V))$.

REMARK 3.4 *It turns out that the i-databases representable by c-tables are also definable via $\mathcal{RA}$ starting from the absolute zero-information instance, $\mathcal{N}$. Indeed, it can be shown (Proposition 15.1) that for each k there exists an $\mathcal{RA}$ query q such that $\mathcal{Z}_k = q(\mathcal{N})$. From there we can apply Theorem 3.2. The class of incomplete databases $\{\mathcal{I} \mid \exists q \in \mathcal{RA} \text{ s.t. } \mathcal{I} = q(\mathcal{N})\}$ is strictly larger than that representable by c-tables, but it is still countable hence strictly smaller than that of all incomplete databases. Its connections with FO-definability in finite model theory might be interesting to investigate.*

Hence, c-tables are in some sense "no more powerful" than the relational algebra. But are they "as powerful"? This justifies the following:

DEFINITION 3.5 *A representation system is* $\mathcal{RA}$-**complete** *if it can represent any $\mathcal{RA}$-definable i-database.*

Since Z_k is itself a c-table the following is an immediate corollary of the fundamental result of [30] (see Theorem 4.2 below). It also states that the converse of Theorem 3.2 holds.

THEOREM 3.6 *c-tables are $\mathcal{RA}$-complete.*

This result is similar in nature to Corollary 3.1 in [25]. However, the exact technical connection, if any, is unclear, since Corollary 3.1 in [25] relies on the certain answers semantics for queries.

We now turn to the kind of completeness considered in [41].

DEFINITION 3.7 *A representation system is* **finitely complete** *if it can represent any finite i-database.*

The finite incompleteness of ?-tables, or-set-tables, or-set-?-tables and other systems is discussed in [41] where a finitely complete representation system $\mathcal{R}^A_{\text{prop}}$ is also given (we repeat the definition in the Appendix). Is finite completeness a reasonable question for c-tables, v-tables, and Codd tables? In general, for such tables *Mod*(T) is infinite (all that is needed is a tuple with at least one variable and with an infinitely satisfiable condition). To facilitate comparison with the systems in [41] we define *finite-domain* versions of tables with variables.

DEFINITION 3.8 *A* **finite-domain** *c-table (v-table, Codd table) consists of a c-table (v-table, Codd table) T together with a finite $dom(x) \subset \mathbb{D}$ for each variable x that occurs in T.*

Note that finite-domain Codd tables are equivalent to or-set tables. Indeed, to obtain an or-set table from a Codd table, one can see $\text{dom}(x)$ as an or-set and substitute it for x in the table. Conversely, to obtain a Codd table from an or-set table, one can substitute a fresh variable x for each or-set and define $\text{dom}(x)$ as the contents of the or-set.

In light of this connection, finite-domain v-tables can be thought of as a kind of "correlated" or-set tables. Finite-domain v-tables are strictly more expressive than finite Codd tables. Indeed, every finite Codd table is also a finite v-table. But, the set of instances represented by e.g. the finite v-table $\{(1, x), (x, 1)\}$ where $\text{dom}(x) = \{1, 2\}$ cannot be represented by any finite Codd table. Finite-domain v-tables are themselves finitely incomplete. For example, the i-database $\{\{(1, 2)\}, \{(2, 1)\}\}$ cannot be represented by any finite v-table.

It is easy to see that finite-domain c-tables are finitely complete and hence equivalent to [41]'s $\mathcal{R}^A_{\text{prop}}$ in terms of expressive power. In fact, this is true even for the fragment of finite-domain c-tables which we will call *Boolean c-tables*, where the variables take only Boolean values and are only allowed to appear in conditions (never as attribute values).

THEOREM 3.9 *Boolean c-tables are finitely complete (hence finite-domain c-tables are also finitely complete).*

Proof: Let $\mathcal{I} = \{I_1, \ldots, I_m\}$ be a finite i-database. Construct a Boolean c-table T such that $Mod(T) = \mathcal{I}$ as follows. Let $\ell := \lceil \lg m \rceil$. For $1 \leq i < m$, put all the tuples from I_i into T with condition φ_i, defined

$$\varphi_i := \bigwedge_j \neg x_j \wedge \bigwedge_k x_k,$$

where the first conjunction is over all $1 \leq j \leq \ell$ such that jth digit in the ℓ-digit binary representation of $i - 1$ is 0, and the second conjunction is over all $1 \leq k \leq \ell$ such that the kth digit in the ℓ-digit binary representation of $i - 1$ is 1. Finally, put all the tuples from I_m into T with condition $\varphi_m \vee \cdots \vee \varphi_{2^\ell}$. ∎

Although Boolean c-tables are complete there are clear advantages to using variables in tuples also, chief among them being *compactness* of representation.

EXAMPLE 3.10 *Consider the finite v-table* $\{(x_1, x_2, \ldots, x_m)\}$ *where* $dom(x_1) = dom(x_2) = \cdots = dom(x_m) = \{1, 2, \ldots, n\}$. *The equivalent Boolean c-table has* n^m *tuples.*

If we additionally restrict Boolean c-tables to allow conditions to contain only true or a single variable which appears in no other condition, then we obtain a representation system which is equivalent to ?-tables.

Since finite c-tables and $\mathcal{R}^A_{\text{prop}}$ are each finitely complete there is an obvious naïve algorithm to translate back and forth between them: list all the instances the one represents, then use the construction from the proof of finite completeness for the other. Finding a more practical "syntactic" algorithm is an interesting open question.

4. Closure Under Relational Operations

DEFINITION 4.1 *A representation system is* **closed** *under a query language if for any query q and any table T there is a table T′ that represents q(Mod(T)).*

(For notational simplicity we consider only queries with one input relation name, but everything generalizes smoothly to multiple relation names.)

This definition is from [41]. In [2], a *strong* representation system is defined in the same way, with the significant addition that T' should be *computable* from T and q. It is not hard to show, using general recursion-theoretic principles, that there exist representation systems (even ones that only represent finite i-databases) which are closed as above but not strong in the sense of [2]. However, the concrete systems studied so far are either not closed or if they are closed then the proof provides also the algorithm required by the definition of strong systems. Hence, we see no need to insist upon the distinction.

THEOREM 4.2 ([30]) *c-tables, finite-domain c-tables, and Boolean c-tables are closed under the relational algebra.*

Proof: (Sketch.) We repeat here the essentials of the proof, including most of the definition of the c-table algebra. For each operation u of the relational algebra [30] defines its operation on the c-table conditions as follows. For projection, if V is a list of indexes, the condition for a tuple t in the output is given by

$$\bar{\pi}_V(T)(t) := \bigvee_{t' \in T \text{ s.t. } \pi_V(t') = t} T(t')$$

where $T(t')$ denotes the condition associated with t' in T. For selection, we have

$$\bar{\sigma}_P(T)(t) := T(t) \wedge P(t)$$

where $P(t)$ denotes the result of evaluating the selection predicate P on the values in t (for a Boolean c-table, this will always be true or false, while for

c-tables and finite-domain c-tables, this will be in general a Boolean formula on constants and variables). For cross product and union, we have

$$(T_1 \bar{\times} T_2)(t) \;:=\; T_1(t) \wedge T_2(t)$$
$$(T_1 \bar{\cup} T_2)(t) \;:=\; T_1(t) \vee T_2(t)$$

Difference and intersection are handled similarly. By replacing u's by $\bar{u}$ we translate any relational algebra expression q into a c-table algebra expression $\bar{q}$ and it can be shown that

LEMMA 4.3 *For all valuations* ν, $\nu(\bar{q}(T)) = q(\nu(T))$.

From this, $Mod(\bar{q}(T)) = q(Mod(T))$ follows immediately. ∎

In Section 10, we shall see a generalization of the (positive) c-table algebra and Lemma 4.3 in the context of annotated relations.

5. Algebraic Completion

None of the incomplete representation systems we have seen so far is closed under the full relational algebra. Nor are two more representation systems considered in [41], $\mathcal{R}_{\text{sets}}$ and $\mathcal{R}_{\oplus\equiv}$ (we repeat their definitions in the Appendix).

PROPOSITION 5.1 ([30, 41]) *Codd tables and v-tables are not closed under e.g. selection. Or-set tables and finite v-tables are also not closed under e.g. selection. ?-tables, $\mathcal{R}_{sets}$, and $\mathcal{R}_{\oplus\equiv}$ are not closed under e.g. join.*

We have seen that "closing" minimal-information one-row Codd tables (see before Definition 3.5) $\{Z_1, Z_2, \ldots\}$, by relational algebra queries yields equivalence with the c-tables. In this spirit, we will investigate "how much" of the relational algebra would be needed to complete the other representation systems considered. We call this kind of result *algebraic completion.*

DEFINITION 5.2 *If $(\mathcal{T}, Mod)$ is a representation system and $\mathcal{L}$ is a query language, then the representation system* obtained by closing $\mathcal{T}$ under $\mathcal{L}$ *is the set of tables $\{(T, q) \mid T \in \mathcal{T}, q \in \mathcal{L}\}$ with the function $Mod : \mathcal{T} \times \mathcal{L} \to \mathcal{N}$ defined by $Mod(T, q) := q(Mod(T))$.*

We are now ready to state the results regarding algebraic completion.

THEOREM 5.3 ($\mathcal{RA}$-COMPLETION)

1 *The representation system obtained by closing Codd tables under $SPJU$ queries is $\mathcal{RA}$-complete.*

2 *The representation system obtained by closing v-tables under SP queries is $\mathcal{RA}$-complete.*

Proof: (Sketch.) For each case we show that given a arbitrary c-table T one can construct a table S and a query q of the required type such that $\bar{q}(S) = T$. Case 1 is a trivial corollary of Theorem 3.2. The details for Case 2 are in the Appendix. ■

Note that in general there may be a "gap" between the language for which closure fails for a representation system and the language required for completion. For example, Codd tables are not closed under selection, but at the same time closing Codd tables under selection does not yield an $\mathcal{RA}$-complete representation system. (To see this, consider the incomplete database represented by the v-table $\{(x, 1), (x, 2)\}$. Intuitively, selection alone is not powerful enough to yield this incomplete database from a Codd table, as, selection operates on one tuple at a time and cannot correlate two un-correlated tuples.) On the other hand, it is possible that some of the results we present here may be able to be "tightened" to hold for smaller query languages, or else proved to be "tight" already. This is an issue which may be worth examining in the future.

We give now a set of analogous completion results for the finite case.

THEOREM 5.4 (FINITE-COMPLETION)

1 *The representation system obtained by closing or-set-tables under PJ queries is finitely complete.*

2 *The representation system obtained by closing finite v-tables under PJ or S^+P queries is finitely complete.*

3 *The representation system obtained by closing $\mathcal{R}_{sets}$ under PJ or PU queries is finitely complete.*

4 *The representation system obtained by closing $\mathcal{R}_{\oplus\equiv}$ under S^+PJ queries is finitely complete.*

Proof:(Sketch.) In each case, given an arbitrary finite incomplete data-base, we construct a table and query of the required type which yields the incomplete database. The details are in the Appendix. ■

Note that there is a gap between the $\mathcal{RA}$-completion result for Codd tables, which requires $SPJU$ queries, and the finite-completion result for finite Codd tables, which requires only PJ queries. A partial explanation is that proof of the latter result relies essentially on the finiteness of the i-database.

More generally, if a representation system can represent arbitrarily-large i-databases, then closing it under $\mathcal{RA}$ yields a finitely complete representation system, as the following theorem makes precise (see Appendix for proof).

THEOREM 5.5 (GENERAL FINITE-COMPLETION) *Let $\mathcal{T}$ be a representation system such that for all $n \geq 1$ there exists a table T in $\mathcal{T}$ such that*

$|Mod(T)| \geq n$. *Then the representation system obtained by closing T under $\mathcal{RA}$ is finitely-complete.*

COROLLARY 5.6 *The representation system obtained by closing ?-tables under $\mathcal{RA}$ queries is finitely complete.*

6. Probabilistic Databases and Representation Systems

Finiteness assumption For the entire discussion of probabilistic database models we will assume that *the domain of values $\mathbb{D}$ is finite.* Infinite domains of values are certainly interesting in practice; for some examples see [33, 45, 41]. Moreover, in the case of incomplete databases we have seen that they allow for interesting distinctions.[3] However, finite probability spaces are much simpler than infinite ones and we will take advantage of this simplicity. The issues related to probabilistic databases over infinite domains are nonetheless interesting and worth pursuing in the future.

We wish to model probabilistic information using a probability space whose possible outcomes are all the conventional instances. Recall that for simplicity we assume a schema consisting of just one relation of arity n. The finiteness of $\mathbb{D}$ implies that there are only finitely many instances, $I \subseteq \mathbb{D}^n$.

By **finite probability space** we mean a probability space (see e.g. [18]) $(\Omega, \mathcal{F}, \Pr[\,])$ in which the set of outcomes Ω is *finite* and the σ-field of events $\mathcal{F}$ consists of *all* subsets of Ω. We shall use the equivalent formulation of pairs (Ω, p) where Ω is the finite set of outcomes and where the *outcome probability assignment* $p : \Omega \to [0, 1]$ satisfies $\sum_{\omega \in \Omega} p(\omega) = 1$. Indeed, we take $\Pr[A] = \sum_{\omega \in A} p(\omega)$.

DEFINITION 6.1 *A* **probabilistic(-information) database** *(sometimes called in this paper a* **p-database***) is a finite probability space whose outcomes are all the conventional instances, i.e., a pair $(\mathcal{N}, p)$ where $\sum_{I \in \mathcal{N}} p(I) = 1$.*

Demanding the direct specification of such probabilistic databases is unrealistic because there are 2^N possible instances, where $N := |\mathbb{D}|^n$, and we would need that many (minus one) probability values. Thus, as in the case of incomplete databases we define **probabilistic representation systems** consisting of "probabilistic tables" (prob. tables for short) and a function *Mod* that associates to each prob. table T a probabilistic database *Mod*(T). Similarly, we define **completeness** (finite completeness is the only kind we have in our setting).

To define closure under a query language we face the following problem. Given a probabilistic database $(\mathcal{N}, p)$ and a query q (with just one input relation name), how do we define the probability assignment for the instances in $q(\mathcal{N})$?

[3]Note however that the results remain true if $\mathbb{D}$ is finite; we just require an infinite supply of *variables*.

It turns out that this is a common construction in probability theory: image spaces.

DEFINITION 6.2 *Let (Ω, p) be a finite probability space and let $f : \Omega \to \Omega'$ where Ω' is some finite set. The* **image** *of (Ω, p) under f is the finite probability space (Ω', p') where* [4] $p'(\omega') := \sum_{f(\omega)=\omega'} p(\omega)$.

Again we consider as query languages the relational algebra and its sublanguages defined by subsets of operations.

DEFINITION 6.3 *A probabilistic representation system is* **closed** *under a query language if for any query q and any prob. table T there exists a prob. table T' that represents $q(Mod(T))$, the image space of $Mod(T)$ under q.*

7. Probabilistic ?-Tables and Probabilistic Or-Set Tables

Probabilistic ?-tables (p-?-tables for short) are commonly used for probabilistic models of databases [47, 22, 23, 16] (they are called the "independent tuples" representation in [42]). Such tables are the probabilistic counterpart of ?-tables where each "?" is replaced by a probability value. Example 7.4 below shows such a table. The tuples not explicitly shown are assumed tagged with probability 0. Therefore, we define a p-?-table as a mapping that associates to each $t \in \mathbb{D}^n$ a probability value p_t. In order to represent a probabilistic database, papers using this model typically include a statement like "every tuple t is in the outcome instance with probability p_t, independently from the other tuples" and then a statement like

$$\Pr[I] = \Big(\prod_{t \in I} p_t\Big)\Big(\prod_{t \notin I} (1 - p_t)\Big).$$

In fact, to give a rigorous semantics, one needs to define the events $E_t \subseteq \mathcal{N}$, $E_t := \{I \mid t \in I\}$ and then to prove the following.

PROPOSITION 7.1 *There exists a unique probabilistic database such that the events E_t are jointly independent and $\Pr[E_t] = p_t$.*

This defines p-?-tables as a probabilistic representation system. We shall however provide an equivalent but more perspicuous definition. We shall need here another common construction from probability theory: product spaces.

DEFINITION 7.2 *Let $(\Omega_1, p_1), \ldots, (\Omega_n, p_n)$ be finite probability spaces. Their* **product** *is the space $(\Omega_1 \times \cdots \times \Omega_n, p)$ where*[5] *we have:*

$$p(\omega_1, \ldots, \omega_n) := p_1(\omega_1) \cdots p_n(\omega_n)$$

[4]It is easy to check that the $p'(\omega')$'s do actually add up to 1.

[5]Again, it is easy to check that the outcome probability assignments add up to 1.

This definition corresponds to the intuition that the n systems or phenomena that are modeled by the spaces $(\Omega_1, p_1), \ldots, (\Omega_n, p_n)$ behave without "interfering" with each other. The following formal statements summarize this intuition.

PROPOSITION 7.3 *Consider the product of the spaces* $(\Omega_1, p_1), \ldots, (\Omega_n, p_n)$. *Let* $A_1 \subseteq \Omega_1, \ldots, A_n \subseteq \Omega_n$.

1 We have $\Pr[A_1 \times \cdots \times A_n] = \Pr[A_1] \cdots \Pr[A_n]$.

2 The events $A_1 \times \Omega_2 \times \cdots \times \Omega_n, \Omega_1 \times A_2 \times \cdots \times \Omega_n, \ldots, \Omega_1 \times \Omega_2 \times \cdots \times A_n$ *are jointly independent in the product space.*

Turning back to p-?-tables, for each tuple $t \in \mathbb{D}^n$ consider the finite probability space $B_t := (\{\mathsf{true}, \mathsf{false}\}, p)$ where $p(\mathsf{true}) := p_t$ and $p(\mathsf{false}) = 1 - p_t$. Now consider the product space

$$P := \prod_{t \in \mathbb{D}^n} B_t$$

We can think of its set of outcomes (abusing notation, we will call this set P also) as the set of functions from $\mathbb{D}^n$ to $\{\mathsf{true}, \mathsf{false}\}$, in other words, predicates on $\mathbb{D}^n$. There is an obvious function $f : P \to \mathcal{N}$ that associates to each predicate the set of tuples it maps to true.

All this gives us a p-database, namely the image of P under f. It remains to show that it satisfies the properties in Proposition 7.1. Indeed, since f is a bijection, this probabilistic database is in fact *isomorphic* to P. In P the events that are in bijection with the E_t's are the Cartesian product in which there is exactly one component $\{\mathsf{true}\}$ and the rest are $\{\mathsf{true}, \mathsf{false}\}$. The desired properties then follow from Proposition 7.3.

We define now another simple probabilistic representation system called **probabilistic or-set-tables** (p-or-set-tables for short). These are the probabilistic counterpart of or-set-tables where the attribute values are, instead of or-sets, finite probability spaces whose outcomes are the values in the or-set. p-or-set-tables correspond to a simplified version of the ProbView model presented in [33], in which plain probability values are used instead of confidence intervals.

EXAMPLE 7.4 *A p-or-set-table S, and a p-?-table T.*

$S :=$

1	$\langle 2:0.3, 3:0.7 \rangle$
4	5
$\langle 6:0.5, 7:0.5 \rangle$	$\langle 8:0.1, 9:0.9 \rangle$

$T :=$

1	2	0.4
3	4	0.3
5	6	1.0

A p-or-set-table determines an instance by choosing an outcome in each of the spaces that appear as attribute values, *independently*. Recall that or-set tables are equivalent to finite-domain Codd tables. Similarly, a p-or-set-table corresponds to a Codd table T plus for each variable x in T a finite probability space $\mathrm{dom}(x)$ whose outcomes are in $\mathbb{D}$. This yields a p-database, again by image space construction, as shown more generally for c-tables next in Section 8.

Query answering The papers [22, 47, 33] have considered, independently, the problem of calculating the probability of tuples appearing in query answers. This does *not* mean that in general $q(\mathit{Mod}(T))$ can be represented by another tuple table when T is some p-?-table and $q \in \mathcal{RA}$ (neither does this hold for p-or-set-tables). This follows from Proposition 5.1. Indeed, if the probabilistic counterpart of an incompleteness representation system $\mathcal{T}$ is closed, then so is $\mathcal{T}$. Hence the lifting of the results in Proposition 5.1 and other similar results.

Each of the papers [22, 47, 33] recognizes the problem of query answering and solves it by developing a more general model in which rows contain additional information *similar in spirit* to the conditions that appear in c-tables (in fact [22]'s model is essentially what we call probabilistic Boolean c-tables, see next section). It turns out that one can actually use a probabilistic counterpart to c-tables themselves together with the algebra on c-tables given in [30] to achieve the same effect.

8. Probabilistic c-tables

DEFINITION 8.1 *A* **probabilistic** c-**table** *(pc*-**tables** *for short) consists of a c-table T together with a* finite probability space *$dom(x)$ (whose outcomes are values in $\mathbb{D}$) for each variable x that occurs in T.*

To get a probabilistic representation system consider the product space

$$V := \prod_{x \in \mathit{Var}(T)} \mathrm{dom}(x)$$

The outcomes of this space are in fact the *valuations* for the c-table T! Hence we can define the function $g : V \to \mathcal{N}, g(\nu) := \nu(T)$ and then define $\mathit{Mod}(T)$ as the image of V under g.

Similarly, we can talk about Boolean pc-tables, pv-tables and probabilistic Codd tables (the latter related to [33], see previous section). Moreover, the p-?-tables correspond to restricted Boolean pc-tables, just like ?-tables.

THEOREM 8.2 *Boolean pc-tables are complete (hence pc-tables are also complete).*

Proof: Let $I_1, \ldots, I_k$ denote the instances with non-zero probability in an arbitrary probabilistic database, and let $p_1, \ldots, p_k$ denote their probabilities. Construct a probabilistic Boolean c-table T as follows. For $1 \leq i \leq k-1$, put the tuples from I_i in T with condition $\neg x_1 \wedge \cdots \wedge \neg x_{i-1} \wedge x_i$. Put the tuples from I_k in T with condition $\neg x_1 \wedge \cdots \wedge \neg x_{k-1}$. For $1 \leq i \leq k-1$,

A B C	
$a\ b\ c$	b_1
$d\ b\ e$	b_2
$f\ g\ e$	b_3

(a)

A C	
$a\ c$	$(b_1 \wedge b_1) \vee (b_1 \wedge b_1)$
$a\ e$	$b_1 \wedge b_2$
$d\ c$	$b_1 \wedge b_2$
$d\ e$	$(b_2 \wedge b_2) \vee (b_2 \wedge b_2) \vee (b_2 \wedge b_3)$
$f\ e$	$(b_3 \wedge b_3) \vee (b_3 \wedge b_3) \vee (b_2 \wedge b_3)$

(b)

A C	
$a\ c$	b_1
$a\ e$	$b_1 \wedge b_2$
$d\ c$	$b_1 \wedge b_2$
$d\ e$	b_2
$f\ e$	b_3

(c)

Figure 2.1. Boolean c-tables example

set $\Pr[x_i = \textsf{true}] := p_i/(1 - \sum_{j=1}^{i-1} p_j)$. It is straightforward to check that this yields a table such that $\Pr[I_i] = p_i$. ■

The previous theorem was independently observed in [42] and [28].

THEOREM 8.3 *pc-tables (and Boolean pc-tables) are closed under the relational algebra.*

Proof:(Sketch.) For any pc-table T and any $\mathcal{RA}$ query q we show that the probability space $q(Mod(T))$ (the image of $Mod(T)$ under q) is in fact the same as the space $Mod(\bar{q}(T))$. The proof of Theorem 4.2 already shows that the outcomes of the two spaces are the same. The fact that the probabilities assigned to each outcome are the same follows from Lemma 4.3. ■

The proof of this theorem gives in fact an algorithm for constructing the answer as a p-database itself, represented by a pc-table. In particular this will work for the models of [22, 33, 47] or for models we might invent by adding probabilistic information to v-tables or to the representation systems considered in [41]. The interesting result of [16] about the applicability of an "extensional" algorithm to calculating answer tuple probabilities can be seen also as characterizing the conjunctive queries q which for any p-?-table T are such that the c-table $\bar{q}(T)$ is in fact equivalent to some p-?-table.

9. Queries on Annotated Relations

In this section we compare the calculations involved in query answering in incomplete and probabilistic databases with those for two other important examples. We observe similarities between them which will motivate the general study of annotated relations.

As a first example, consider the Boolean c-table in Figure 2.1(a), and the following $\mathcal{RA}$ query, which computes the union of two self-joins:

$$q(R) := \pi_{\mathrm{AC}}(\pi_{\mathrm{AB}}R \bowtie \pi_{\mathrm{BC}}R \cup \pi_{\mathrm{AC}}R \bowtie \pi_{\mathrm{BC}}R)$$

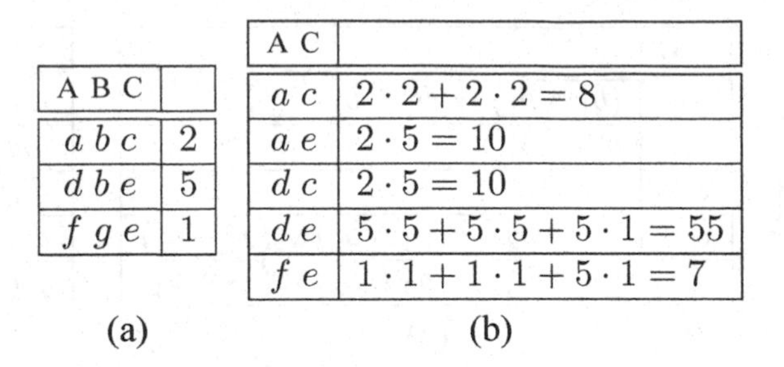

(a)

A B C	
$a\ b\ c$	2
$d\ b\ e$	5
$f\ g\ e$	1

(b)

A C	
$a\ c$	$2 \cdot 2 + 2 \cdot 2 = 8$
$a\ e$	$2 \cdot 5 = 10$
$d\ c$	$2 \cdot 5 = 10$
$d\ e$	$5 \cdot 5 + 5 \cdot 5 + 5 \cdot 1 = 55$
$f\ e$	$1 \cdot 1 + 1 \cdot 1 + 5 \cdot 1 = 7$

Figure 2.2. Bag semantics example

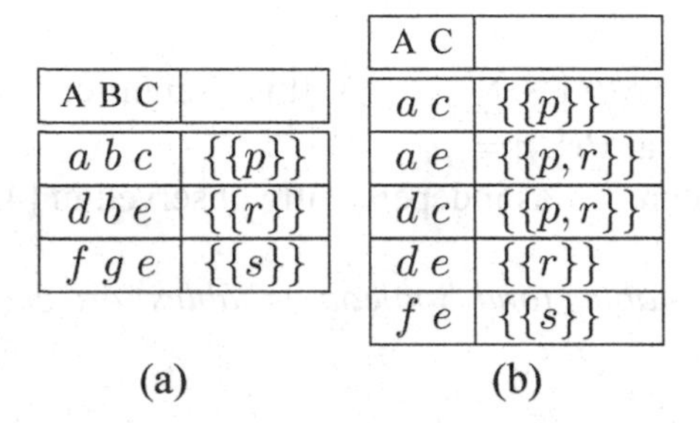

(a)

A B C	
$a\ b\ c$	$\{\{p\}\}$
$d\ b\ e$	$\{\{r\}\}$
$f\ g\ e$	$\{\{s\}\}$

(b)

A C	
$a\ c$	$\{\{p\}\}$
$a\ e$	$\{\{p, r\}\}$
$d\ c$	$\{\{p, r\}\}$
$d\ e$	$\{\{r\}\}$
$f\ e$	$\{\{s\}\}$

Figure 2.3. Minimal witness why-provenance example

The Imielinski-Lipski algorithm (cf. Theorem 4.2) produces the Boolean c-table shown in Figure 2.1(b), which can be simplified to the one shown in Figure 2.1(c). The annotations on the tuples of this c-table are such that it correctly represents the possible worlds of the query answer:

$$Mod(q(R)) = q(Mod(R))$$

Another kind of table with annotations is a *multiset* or *bag*. In this case, the annotations are natural numbers which represent the multiplicity of the tuple in the multiset. (A tuple not listed in the table has multiplicity 0.) Query answering on such tables involves calculating not just the tuples in the output, but also their multiplicities.

For example, consider the multiset shown in Figure 2.2(a). Then $q(R)$, where q is the same query from before, is the multiset shown in Figure 2.2(b). Note that for projection and union we add multiplicities while for join we multiply them. There is a striking similarity between the arithmetic calculations we do here for multisets, and the Boolean calculations for the c-table.

A third example involves the *minimal witness why-provenance* proposed in [9] for tracking the processing of scientific data. Here source tuples are annotated with their own tuple ids, and answering queries involves calculating the set of sets of ids of source tuples which "contribute together" for a given output tuple. The minimal witness why-provenance W for an output tuple t is required to be minimal in the sense that for any A, B in W neither is a subset of the other.

Figure 2.3(a) shows an example of a source table, where t_1, t_2, t_3 are tuple ids. Considering again the same query q as above, the algorithm of [9] produces the table with why-provenance annotations shown in Figure 2.3(b). Note again the similarity between this table and the example earlier with Boolean *c*-tables.

10. *K*-Relations

In this section we unify the examples above by considering generalized relations in which the tuples are annotated (*tagged*) with information of various kinds. Then, we will define a generalization of the positive relational algebra ($\mathcal{RA}^+$) to such tagged-tuple relations. The examples in Section 9 will turn out to be particular cases.

We use here the named perspective [2] of the relational model in which tuples are functions $t : U \to \mathbb{D}$ with U a finite set of attributes and $\mathbb{D}$ a domain of values. We fix the domain $\mathbb{D}$ for the time being and we denote the set of all such U-tuples by U-Tup. (Usual) relations over U are subsets of U-Tup.

A notationally convenient way of working with tagged-tuple relations is to model tagging by a function on all possible tuples, with those tuples not considered to be "in" the relation tagged with a special value. For example, the usual set-theoretic relations correspond to functions that map U-Tup to $\mathbb{B} = \{\mathsf{true}, \mathsf{false}\}$ with the tuples in the relation tagged by true and those not in the relation tagged by false.

DEFINITION 10.1 *Let K be a set containing a distinguished element 0. A* K**-relation** *over a finite set of attributes U is a function $R : U\text{-}\mathsf{Tup} \to K$ such that its* **support** *defined by $\mathsf{supp}(R) := \{t \mid R(t) \neq 0\}$ is finite.*

In generalizing $\mathcal{RA}^+$ we will need to assume more structure on the set of tags. To deal with selection we assume that the set K contains two distinct values $0 \neq 1$ which denote "out of" and "in" the relation, respectively. To deal with union and projection and therefore to combine different tags of the same tuple into one tag we assume that K is equipped with a binary operation "+". To deal with natural join (hence intersection and selection) and therefore to combine the tags of joinable tuples we assume that K is equipped with another binary operation "·".

DEFINITION 10.2 *Let $(K, +, \cdot, 0, 1)$ be an algebraic structure with two binary operations and two distinguished elements. The operations of the* **positive algebra** *are defined as follows:*

empty relation *For any set of attributes* U*, there is* $\emptyset : U\text{-Tup} \rightarrow K$ *such that* $\emptyset(t) = 0$.

union *If* $R_1, R_2 : U\text{-Tup} \rightarrow K$ *then* $R_1 \cup R_2 : U\text{-Tup} \rightarrow K$ *is defined by*

$$(R_1 \cup R_2)(t) \ := \ R_1(t) + R_2(t)$$

projection *If* $R : U\text{-Tup} \rightarrow K$ *and* $V \subseteq U$ *then* $\pi_V R : V\text{-Tup} \rightarrow K$ *is defined by*

$$(\pi_V R)(t) \ := \sum_{t=t' \text{ on } V \text{ and } R(t') \neq 0} R(t')$$

(here $t = t'$ on V *means* t' *is a* U*-tuple whose restriction to* V *is the same as the* V*-tuple* t*; note also that the sum is finite since* R *has finite support)*

selection *If* $R : U\text{-Tup} \rightarrow K$ *and the selection predicate* **P** *maps each* U*-tuple to either* 0 *or* 1 *then* $\sigma_{\mathbf{P}} R : U\text{-Tup} \rightarrow K$ *is defined by*

$$(\sigma_{\mathbf{P}} R)(t) \ := \ R(t) \cdot \mathbf{P}(t)$$

Which $\{0, 1\}$*-valued functions are used as selection predicates is left unspecified, except that we assume that* false*—the constantly* 0 *predicate, and* true*—the constantly* 1 *predicate, are always available.*

natural join *If* $R_i : U_i\text{-Tup} \rightarrow K \ \ i = 1, 2$ *then* $R_1 \bowtie R_2$ *is the* K*-relation over* $U_1 \cup U_2$ *defined by*

$$(R_1 \bowtie R_2)(t) \ := \ R_1(t_1) \cdot R_2(t_2)$$

where $t_1 = t$ on U_1 *and* $t_2 = t$ on U_2 *(recall that* t *is a* $U_1 \cup U_2$*-tuple).*

renaming *If* $R : U\text{-Tup} \rightarrow K$ *and* $\beta : U \rightarrow U'$ *is a bijection then* $\rho_\beta R$ *is a* K*-relation over* U' *defined by*

$$(\rho_\beta R)(t) \ := \ R(t \circ \beta)$$

PROPOSITION 10.3 *The operation of* $\mathcal{RA}^+$ *preserve the finiteness of supports therefore they map* K*-relations to* K*-relations. Hence, Definition 10.2 gives us an algebra on* K*-relations.*

This definition generalizes the definitions of $\mathcal{RA}^+$ for the motivating examples we saw. Indeed, for $(\mathbb{B}, \vee, \wedge, \text{false}, \text{true})$ we obtain the usual $\mathcal{RA}^+$ with set semantics. For $(\mathbb{N}, +, \cdot, 0, 1)$ it is $\mathcal{RA}^+$ with bag semantics.

For the Imielinski-Lipski algebra on c-tables we consider the set of Boolean expressions over some set B of variables which are *positive*, i.e., they involve only disjunction, conjunction, and constants for true and false. Then we identify those expressions that yield the same truth-value for all Boolean assignments of the variables in B.[6] Denoting by $\mathsf{PosBool}(B)$ the result and applying Definition 10.2 to the structure $(\mathsf{PosBool}(B), \vee, \wedge, \mathsf{false}, \mathsf{true})$ produces exactly the Imielinski-Lipski algebra.

These three structures are examples of *commutative semi-rings*, i.e., algebraic structures $(K, +, \cdot, 0, 1)$ such that $(K, +, 0)$ and $(K, \cdot, 1)$ are commutative monoids, $\cdot$ is distributive over $+$ and $\forall a,\ 0\cdot a = a\cdot 0 = 0$. Further evidence for requiring K to form such a semi-ring is given by

PROPOSITION 10.4 *The following $\mathcal{RA}$ identities:*

- *union is associative, commutative and has identity $\emptyset$;*
- *join is associative, commutative and distributive over union;*
- *projections and selections commute with each other as well as with unions and joins (when applicable);*
- $\sigma_{\mathsf{false}}(R) = \emptyset$ *and* $\sigma_{\mathsf{true}}(R) = R$.

hold for the positive algebra on K-relations if and only if $(K, +, \cdot, 0, 1)$ is a commutative semi-ring.

Glaringly absent from the list of relational identities are the idempotence of union and of (self-)join. Indeed, these fail for the bag semantics, an important particular case of the general treatment presented here.

Any function $h : K \to K'$ can be used to transform K-relations to K'-relations simply by applying h to each tag (note that the support may shrink but never increase). Abusing the notation a bit we denote the resulting transformation from K-relations to K'-relations also by h. The $\mathcal{RA}$ operations we have defined work nicely with semi-ring structures:

PROPOSITION 10.5 *Let $h : K \to K'$ and assume that K, K' are commutative semi-rings. The transformation given by h from K-relations to K'-relations commutes with any $\mathcal{RA}^+$ query (for queries of one argument) $q(h(R)) = h(q(R))$ if and only if h is a semi-ring homomorphism.*

Proposition 10.5 has some useful applications. For example, for Boolean c-tables and semi-ring homomorphisms $\mathsf{Eval}_\nu : \mathsf{PosBool}(B) \to \mathbb{B}$ corresponding to valuations of the variables $\nu : B \to \mathbb{B}$, Proposition 10.5 generalizes Lemma 4.3 and can be used to establish the closure of $\mathsf{PosBool}(B)$-annotated relations (in the sense of Section 4) under $\mathcal{RA}^+$ queries.

[6] in order to permit simplifications; it turns out that this is the same as transforming using the axioms of *distributive lattices* [13]

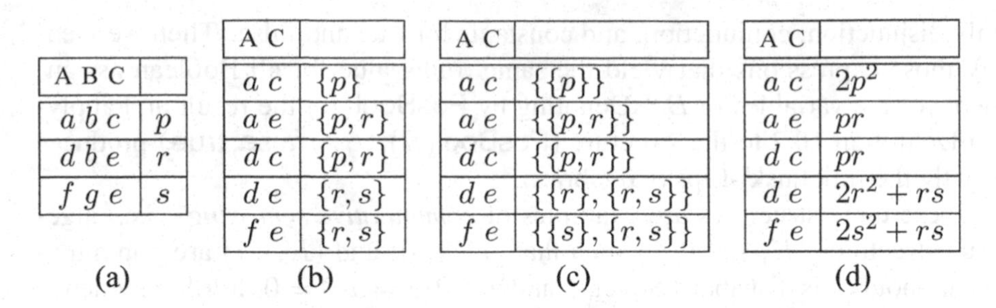

(a)

A B C	
$a\ b\ c$	p
$d\ b\ e$	r
$f\ g\ e$	s

(b)

A C	
$a\ c$	$\{p\}$
$a\ e$	$\{p, r\}$
$d\ c$	$\{p, r\}$
$d\ e$	$\{r, s\}$
$f\ e$	$\{r, s\}$

(c)

A C	
$a\ c$	$\{\{p\}\}$
$a\ e$	$\{\{p, r\}\}$
$d\ c$	$\{\{p, r\}\}$
$d\ e$	$\{\{r\}, \{r, s\}\}$
$f\ e$	$\{\{s\}, \{r, s\}\}$

(d)

A C	
$a\ c$	$2p^2$
$a\ e$	pr
$d\ c$	pr
$d\ e$	$2r^2 + rs$
$f\ e$	$2s^2 + rs$

Figure 2.4. Lineage, why-provenance, and provenance polynomials

11. Polynomials for Provenance

Lineage was defined in [14, 15] as a way of relating the tuples in a query output to the tuples in the query input that "contribute" to them. The lineage of a tuple t in a query output is in fact the *set* of all contributing input tuples.

Computing the lineage for queries in $\mathcal{RA}^+$ turns out to be exactly Definition 10.2 for the semi-ring $(\mathcal{P}(X) \cup \{\bot\}, +, \cdot, \bot, \emptyset)$ where X consists of the ids of the tuples in the input instance, $\bot + S = S + \bot = S$, $S \cdot \bot = \bot \cdot S = \bot$, and $S + T = S \cdot T = S \cup T$ if $S, T \neq \bot$[7]

For example, we consider the same tuples as in relation R used in the examples of Section 9 but now we tag them with their own ids p,r,s, as shown in Figure 2.4(a). The resulting R can be seen as a $\mathcal{P}(\{p, r, s\})$-relation by replacing p with $\{p\}$, etc. Applying the query q from Section 9 to R we obtain according to Definition 10.2 the $\mathcal{P}(\{p, r, s\})$-relation shown in Figure 2.4(b).

A related but finer-grained notion of provenance, called why-provenance, was defined in [9].[8] The why-provenance of a tuple t in a query output is the *set* of *sets* of input tuples which contribute together to produce t. The lineage of t can be obtained by flattening the why-provenance of t.

As with lineage, computing the why-provenance for queries in $\mathcal{RA}^+$ can be done [8] using Definition 10.2, this time for the semi-ring $(\mathcal{P}(\mathcal{P}(X)), \cup, \uplus, \emptyset, \{\emptyset\})$ where X is the set of tuple ids for the input instance and $A \uplus B$ is the pairwise union of A and B, i.e., $A \uplus B := \{a \cup b : a \in A, b \in B\}$. For example, the R in Figure 2.4(a) can be seen as a why-provenance relation by replacing p with $\{\{p\}\}$, etc. Applying the query q from Section 9 to R we obtain according to Definition 10.2 the why-provenance relation shown in Figure 2.4(c).

[7] This definition for lineage, due to [8], corrects the one which appeared in [27].

[8] The distinction between lineage and why-provenance, which went unnoticed in [9] and [27], was pointed out in [8].

Finally, to return to the third example of Section 9, the minimal witness why-provenance can be computed [8] using a semi-ring whose domain is $\mathsf{irr}(\mathcal{P}(X))$, the set of *irredundant* subsets of $\mathcal{P}(X)$, i.e., W is in $\mathsf{irr}(\mathcal{P}(X))$ if for any A, B in W neither is a subset of the other. We can associate with any $W \in \mathcal{P}(X)$ a unique irredundant subset of W by repeatedly looking for elements A, B such that $A \subset B$ and deleting B from W. Then we define a semi-ring $(\mathsf{irr}(\mathcal{P}(X)), +, \cdot, 0, 1)$ as follows:

$$\begin{array}{rclrcl} I + J & := & \mathsf{irr}(I \cup J) & I \cdot J & := & \mathsf{irr}(I \uplus J) \\ 0 & := & \emptyset & 1 & := & \{\emptyset\} \end{array}$$

The table in Figure 2.3(b) is obtained by applying the query q from Section 9 to R of Figure 2.3(a) according to Definition 10.2 for the minimal why-provenance semi-ring. Note that this is a well-known semi-ring: the construction above is the construction for the free distributive lattice generated by the set X. Moreover, it is isomorphic $\mathsf{PosBool}(X)$! This explains the similarity between the calculations in Figure 2.1 and Figure 2.3.

These examples illustrate the limitations of lineage and why-provenance (also recognized in [12]). For example, in the query result in Figure 2.4(b) (f, e) and (d, e) have the same lineage, the input tuples with id r and s. However, the query can also calculate (f, e) from s alone and (d, e) from r alone. In a provenance application in which one of r or s is perhaps less trusted or less usable than the other the effect can be different on (f, e) than on (d, e) and this cannot be detected by lineage. Meanwhile, with why-provenance we do see that (f, e) can be calculated from s alone and (d, e) from r alone, but we have lost information about *multiplicities* (the number of times a tuple was used in a self-join, or the number of derivations of an output tuple in which a given set of tuples is involved) which may be needed to calculate a more refined notion of trust. It seems that we need to know not just *which* input tuples contribute but also exactly *how* they contribute.[9]

On the other hand, by using the different operations of the semi-ring, Definition 10.2 appears to fully "document" how an output tuple is produced. To record the documentation as tuple tags we need to use a semi-ring of symbolic expressions. In the case of semi-rings, like in ring theory, these are the *polynomials*.

DEFINITION 11.1 *Let X be the set of tuple ids of a (usual) database instance I. The* **positive algebra provenance semi-ring** *for I is the semi-ring of polynomials with variables (a.k.a. indeterminates) from X and coefficients from $\mathbb{N}$,*

[9]In contrast to why-provenance, the notion of provenance we describe could justifiably be called **how-provenance**.

with the operations defined as usual[10]: $(\mathbb{N}[X], +, \cdot, 0, 1)$.

Example of provenance computation. Start again from the relation R in Figure 2.4(a) in which tuples are tagged with their own id. R can be seen as an $\mathbb{N}[p, r, s]$-relation. Applying to R the query q from Section 9 and doing the calculations in the provenance semi-ring we obtain the $\mathbb{N}[p, r, s]$-relation shown in Figure 2.4(c). The provenance of (f, e) is $2s^2 + rs$ which can be "read" as follows: (f, e) is computed by q in three different ways; two of them use the input tuple s *twice*; the third uses input tuples r and s. We also see that the provenance of (d, e) is different and we see *how* it is different! □

The following standard property of polynomials captures the intuition that $\mathbb{N}[X]$ is as "general" as any semi-ring:

PROPOSITION 11.2 *Let K be a commutative semi-ring and X a set of variables. For any valuation $v : X \to K$ there exists a unique homomorphism of semi-rings*

$$\mathsf{Eval}_v : \mathbb{N}[X] \to K$$

such that for the one-variable monomials we have $\mathsf{Eval}_v(x) = v(x)$.

As the notation suggests, $\mathsf{Eval}_v(P)$ evaluates the polynomial P in K given a valuation for its variables. In calculations with the integer coefficients, na where $n \in \mathbb{N}$ and $a \in K$ is the sum in K of n copies of a. Note that $\mathbb{N}$ is embedded in K by mapping n to the sum of n copies of 1_K.

Using the Eval notation, for any $P \in \mathbb{N}[x_1, \ldots, x_n]$ and any K the **polynomial function** $f_P : K^n \to K$ is given by:

$$f_P(a_1, \ldots, a_n) := \mathsf{Eval}_v(P) \quad v(x_i) = a_i, i = 1..n$$

Putting together Propositions 10.5 and 11.2 we obtain the following conceptually important fact that says, informally, that the semantics of $\mathcal{RA}^+$ on K-relations for any semi-ring K **factors** through the semantics of the same in provenance semi-rings.

THEOREM 11.3 *Let K be a commutative semi-ring, let R be a K-relation, and let X be the set of tuple ids of the tuples in* $\mathsf{supp}(R)$. *There is an obvious valuation $v : X \to K$ that associates to a tuple id the tag of that tuple in R.*

We associate to R an "abstractly tagged" version, denoted $\bar{R}$, which is an $X \cup \{0\}$-relation. $\bar{R}$ is such that $\mathsf{supp}(\bar{R}) = \mathsf{supp}(R)$ *and the tuples in*

[10]These are polynomials in commutative variables so their operations are the same as in middle-school algebra, except that subtraction is not allowed.

$\mathsf{supp}(\bar{R})$ *are tagged by their own tuple id. Note that as an* $X \cup \{0\}$*-relation,* $\bar{R}$ *is a particular kind of* $\mathbb{N}[X]$*-relation.*

Then, for any $\mathcal{RA}^+$ *query* q *we have*[11]

$$q(R) = \mathsf{Eval}_v(q(\bar{R}))$$

To illustrate an instance of this theorem, consider the provenance polynomial $2r^2 + rs$ of the tuple (d, e) in Figure 2.4(c). Evaluating it in $\mathbb{N}$ for $p = 2, r = 5, s = 1$ we get 55 which is indeed the multiplicity of (d, e) in Figure 2.2(a).

12. Query Containment

Here we present some results about query containment w.r.t. the general semantics in K-relations.

DEFINITION 12.1 *Let* K *be a naturally ordered commutative semi-ring and let* q_1, q_2 *be two queries defined on* K*-relations. We define containment with respect to* K*-relations semantics by*

$$q_1 \sqsubseteq_K q_2 \overset{\text{def}}{\Leftrightarrow} \forall R\, \forall t\; q_1(R)(t) \leq q_2(R)(t)$$

When K is $\mathbb{B}$ and $\mathbb{N}$ we get the usual notions of query containment with respect to set and bag semantics.

Some simple facts follow immediately. For example if $h : K \to K'$ is a semi-ring homomorphism such that $h(x) \leq h(y) \Rightarrow x \leq y$ and q_1, q_2 are $\mathcal{RA}^+$ queries it follows from Prop. 10.5 that $q_1 \sqsubseteq_{K'} q_2 \Rightarrow q_1 \sqsubseteq_K q_2$. If instead h is a surjective homomorphism then $q_1 \sqsubseteq_K q_2 \Rightarrow q_1 \sqsubseteq_{K'} q_2$.

The following result allows us to use the decidability of containment of unions of conjunctive queries [11, 40].

THEOREM 12.2 *If* K *is a* distributive lattice *then for any* q_1, q_2 *unions of conjunctive queries we have*

$$q_1 \sqsubseteq_K q_2 \text{ iff } q_1 \sqsubseteq_{\mathbb{B}} q_2$$

Proof:(sketch) One direction follows because $\mathbb{B}$ can be homomorphically embedded in K. For the other direction we use the existence of query body homomorphisms to establish mappings between monomials of provenance polynomials. Then we apply the factorization theorem (11.3) and the idempotence and absorption laws of K. ■

Therefore, if K is a distributive lattice for (unions of) conjunctive queries containment with respect to K-relation semantics is decidable by the same

[11] To simplify notation we have stated this theorem for queries of one argument but the generalization is immediate.

procedure as for standard set semantics. $\mathsf{PosBool}(B)$ is a distributive lattice, as is the semi-ring $([0,1], \max, \min, 0, 1)$ which is related to *fuzzy sets* [46] and could be referred to as the *fuzzy semi-ring*. A theorem similar to the one above is shown in [32] but the class of algebraic structures used there does not include $\mathsf{PosBool}(B)$ (although it does include the fuzzy semi-ring).

13. Related Work

Lineage and why-provenance were introduced in [14, 15, 9], (the last paper uses a tree data model) but the relationship with [30] was not noticed. The papers on probabilistic databases [22, 47, 33] note the similarities with [30] but do not attempt a generalization.

Two recent papers on provenance, although independent of our work, have a closer relationship to the approach outlined here. Indeed, [12] identifies the limitations of why-provenance and proposes *route-provenance* which is also related to derivation trees. The issue of infinite routes in recursive programs is avoided by considering only *minimal* ones. [7] proposes a notion of lineage of tuples for a type of incomplete databases but does not consider recursive queries. It turns out that we can also describe the lineage in [7] by means of a special commutative semi-ring.

The first attempt at a general theory of relations with annotations appears to be [32] where axiomatized *label systems* are introduced in order to study containment.

14. Conclusion and Further Work

The results on algebraic completion may not be as tight as they can be. Ideally, we would like to be able show that for each representation system we consider, the fragment of $\mathcal{RA}$ we use is minimal in the sense that closing the representation system under a more restricted fragment does not obtain a complete representation system.

We did not consider c-tables with global conditions [24] nor did we describe the exact connection to logical databases [38, 44]. Even more importantly, we did not consider complexity issues as in [3]. All of the above are important topics for further work, especially the complexity issues and the related issues of *succinctness/compactness* of the table representations.

As we see, in pc-tables the probability distribution is on the values taken by the variables that occur in the table. The variables are assumed independent here. This is a lot more flexible (as the example shows) than independent tuples, but still debatable. Consequently, to try to make pc-tables even more flexible, it would be worthwhile to investigate models in which the assumption that the variables take values independently is relaxed by using conditional probability distributions [21].

It would be interesting to connect this work to the extensive literature on *disjunctive databases*, see e.g., [35], and to the work on probabilistic object-oriented databases [19].

Probabilistic modeling is by no means the only way to model uncertainty in information systems. In particular it would be interesting to investigate *possibilistic* models [29] for databases, perhaps following again, as we did here, the parallel with incompleteness.

Query answering on annotated relations can be extended beyond $\mathcal{RA}^+$ to recursive Datalog programs, using semi-rings of formal power series (see [27] for details). These formal power series, which can be represented finitely using a system of equations, are the foundation of trust policies and incremental update propagation algorithms in the ORCHESTRA collaborative data sharing system [26].

Beyond the technical results, the approach surveyed above can be regarded also as arguing that various forms of K-relations, even multisets, provide coarser forms of provenance while the polynomial annotations are, by virtue of their "universality" (as illustrated by the factorization theorem) the most general form of annotation possible with the boundaries of semi-ring structures. This might be a perspective worth using when, in the future, we search for provenance structures for data models other than relational.

Acknowledgments

The author is grateful to Grigoris Karvounarakis and Val Tannen, his co-authors of the papers [28, 27] on which this survey is based.

References

[1] S. Abiteboul and O. M. Duschka. Complexity of answering queries using materialized views. In *PODS*, pages 254–263, 1998.

[2] S. Abiteboul, R. Hull, and V. Vianu. *Foundations of Databases*. Addison–Wesley, Reading, MA, 1995.

[3] S. Abiteboul, P. Kanellakis, and G. Grahne. On the representation and querying of sets of possible worlds. *Theor. Comput. Sci.*, 78(1):159–187, 1991.

[4] L. Antova, T. Jansen, C. Koch, and D. Olteanu. Fast and simple relational processing of uncertain data. In *ICDE*, 2008.

[5] M. Arenas, L. E. Bertossi, and J. Chomicki. Answer sets for consistent query answering in inconsistent databases. *TPLP*, 3(4-5):393–424, 2003.

[6] D. Barbara, H. Garcia-Molina, and D. Porter. A probabilistic relational data model. In *EDBT*, pages 60–74, New York, NY, USA, 1990.

[7] O. Benjelloun, A. D. Sarma, A. Y. Halevy, and J. Widom. ULDBs: Databases with uncertainty and lineage. In *VLDB*, 2006.

[8] P. Buneman, J. Cheney, W.-C. Tan, and S. Vansummeren. Curated databases. In *PODS*, pages 1–12, 2008.

[9] P. Buneman, S. Khanna, and W.-C. Tan. Why and where: A characterization of data provenance. In *ICDT*, 2001.

[10] R. Cavallo and M. Pittarelli. The theory of probabilistic databases. In *VLDB*, pages 71–81, 1987.

[11] A. K. Chandra and P. M. Merlin. Optimal implementation of conjunctive queries in relational data bases. In *STOC*, 1977.

[12] L. Chiticariu and W.-C. Tan. Debugging schema mappings with routes. In *VLDB*, 2006.

[13] P. Crawley and R. P. Dilworth. *Algebraic Theory of Lattices*. Prentice Hall, 1973.

[14] Y. Cui. *Lineage Tracing in Data Warehouses*. PhD thesis, Stanford University, 2001.

[15] Y. Cui, J. Widom, and J. L. Wiener. Tracing the lineage of view data in a warehousing environment. *TODS*, 25(2), 2000.

[16] N. Dalvi and D. Suciu. Efficient query evaluation on probabilistic databases. In *VLDB*, pages 864–875, 2004.

[17] D. Dey and S. Sarkar. A probabilistic relational model and algebra. *ACM TODS*, 21(3):339–369, 1996.

[18] R. Durrett. *Probability: Theory and Examples*. Duxbury Press, 3rd edition, 2004.

[19] T. Eiter, J. J. Lu, T. Lukasiewicz, and V. S. Subrahmanian. Probabilistic object bases. *ACM Trans. Database Syst.*, 26(3):264–312, 2001.

[20] R. Fagin, P. G. Kolaitis, R. J. Miller, and L. Popa. Data exchange: Semantics and query answering. In *ICDT*, pages 207–224, London, UK, 2003. Springer-Verlag.

[21] N. Friedman, L. Getoor, D. Koller, and A. Pfeffer. Learning probabilistic relational models with structural uncertainty. In *Proc. ICML*, 2001.

[22] N. Fuhr and T. Rölleke. A probabilistic relational algebra for the integration of information retrieval and database systems. *TOIS*, 14(1), 1997.

[23] E. Grädel, Y. Gurevich, and C. Hirch. The complexity of query reliability. In *PODS*, pages 227–234, 1998.

[24] G. Grahne. Horn tables - an efficient tool for handling incomplete information in databases. In *PODS*, pages 75–82. ACM Press, 1989.

[25] G. Grahne. *The Problem of Incomplete Information in Relational Databases*, volume 554 of *Lecture Notes in Computer Science*. Springer-Verlag, Berlin, 1991.

[26] T. J. Green, G. Karvounarakis, Z. G. Ives, and V. Tannen. Update exchange with mappings and provenance. In *VLDB*, 2007.

[27] T. J. Green, G. Karvounarakis, and V. Tannen. Provenance semi-rings. In *PODS*, 2007.

[28] T. J. Green and V. Tannen. Models for incomplete and probabilistic information. In *EDBT Workshops*, 2006.

[29] J. Y. Halpern. *Reasoning About Uncertainty*. MIT Press, Cambridge, MA, 2003.

[30] T. Imieliński and W. Lipski, Jr. Incomplete information in relational databases. *J. ACM*, 31(4):761–791, 1984.

[31] T. Imieliński, S. A. Naqvi, and K. V. Vadaparty. Incomplete objects — a data model for design and planning applications. In *SIGMOD*, pages 288–297, 1991.

[32] Y. E. Ioannidis and R. Ramakrishnan. Containment of conjunctive queries: beyond relations as sets. *TODS*, 20(3), 1995.

[33] L. V. S. Lakshmanan, N. Leone, R. Ross, and V. S. Subrahmanian. Probview: a flexible probabilistic database system. *ACM TODS*, 22(3):419–469, 1997.

[34] L. V. S. Lakshmanan and F. Sadri. Probabilistic deductive databases. In *ILPS*, pages 254–268, Cambridge, MA, USA, 1994. MIT Press.

[35] N. Leone, F. Scarcello, and V. S. Subrahmanian. Optimal models of disjunctive logic programs: Semantics, complexity, and computation. *IEEE Trans. Knowl. Data Eng.*, 16(4):487–503, 2004.

[36] L. Libkin. *Aspects of Partial Information in Databases*. PhD thesis, University of Pennsylvania, 1994.

[37] L. Libkin and L. Wong. Semantic representations and query languages for or-sets. *J. Computer and System Sci.*, 52(1):125–142, 1996.

[38] R. Reiter. A sound and sometimes complete query evaluation algorithm for relational databases with null values. *J. ACM*, 33(2):349–370, 1986.

[39] F. Sadri. Modeling uncertainty in databases. In *ICDE*, pages 122–131. IEEE Computer Society, 1991.

[40] Y. Sagiv and M. Yannakakis. Equivalences among relational expressions with the union and difference operators. *J. ACM*, 27(4), 1980.

[41] A. D. Sarma, O. Benjelloun, A. Halevy, and J. Widom. Working models for uncertain data. In *ICDE*, 2006.

[42] D. Suciu and N. Dalvi. Foundations of probabilistic answers to queries (tutorial). In *SIGMOD*, pages 963–963, New York, NY, USA, 2005. ACM Press.

[43] R. van der Meyden. Logical approaches to incomplete information: A survey. In J. Chomicki and G. Saake, editors, *Logics for Databases and Information Systems*. Kluwer Academic Publishers, Boston, 1998.

[44] M. Y. Vardi. Querying logical databases. *JCSS*, 33(2):142–160, 1986.

[45] J. Widom. Trio: A system for integrated management of data, accuracy, and lineage. In *CIDR*, Jan. 2005.

[46] L. A. Zadeh. Fuzzy sets. *Inf. Control*, 8(3), 1965.

[47] E. Zimányi. Query evaluation in probabilistic databases. *Theoretical Computer Science*, 171(1–2):179–219, 1997.

[48] E. Zimányi and A. Pirotte. Imperfect information in relational databases. In *Uncertainty Management in Information Systems*, pages 35–88. Kluwer, 1996.

15. Appendix

PROPOSITION 15.1 *There exists a relational query q such that* $q(\mathcal{N}) = \mathcal{Z}_n$.

Proof: Define sub-query q' to be the relational query

$$q'(V) := V - \pi_\ell(\sigma_{\ell \neq r}(V \times V)),$$

where ℓ is short for $1, \ldots, n$ and $\ell \neq r$ is short for $1 \neq n+1 \vee \cdots \vee n \neq 2n$. Note that q' yields V if V consists of a single tuple and $\emptyset$ otherwise. Now define q to be the relational query

$$q(V) := q'(V) \cup (\{t\} - \pi_\ell(\{t\} \times q'(V))),$$

where t is a tuple chosen arbitrarily from $\mathbb{D}^n$. It is clear that $q(\mathcal{N}) = \mathcal{Z}_n$. ■

DEFINITION 15.2 *A table in the representation system* $\mathcal{R}_{sets}$ *is a multiset of sets of tuples, or* blocks, *each such block optionally labeled with a '?'. If* T *is an* $\mathcal{R}_{sets}$ *table, then Mod*(T) *is the set of instances obtained by choosing one tuple from each block not labeled with a '?', and at most one tuple from each block labeled with a '?'.*

DEFINITION 15.3 *A table in the representation system* $\mathcal{R}_{\oplus\equiv}$ *is a multiset of tuples* $\{t_1, \ldots, t_m\}$ *and a conjunction of logical assertions of the form* $i \oplus j$ *(meaning* t_i *or* t_j *must be present in an instance, but not both) or* $i \equiv j$ *(meaning* t_i *is present in an instance iff* t_j *is present in the instance). If* T *is an* $\mathcal{R}_{\oplus\equiv}$ *table then Mod*(T) *consists of all subsets of the tuples satisfying the conjunction of assertions.*

DEFINITION 15.4 *A table in the representation system* $\mathcal{R}^A_{prop}$ *is a multiset of or-set tuples* $\{t_1, \ldots, t_m\}$ *and a Boolean formula on the variables* $\{t_1, \ldots, t_m\}$. *If* T *is an* $\mathcal{R}^A_{prop}$ *table then Mod*(T) *consists of all subsets of the tuples satisfying the Boolean assertion, where the variable* t_i *has value* true *iff the tuple* t_i *is present in the subset.*

Theorem 5.3 ($\mathcal{RA}$-Completion).

1. *The representation system obtained by closing Codd tables under* $SPJU$ *queries is* $\mathcal{RA}$*-complete.*

2. *The representation system obtained by closing v-tables under* SP *queries is* $\mathcal{RA}$*-complete.*

Proof: In each case we show that given an arbitrary c-table T, one can construct a table S and a query q such that $\bar{q}(S) = T$.

1 Trivial corollary of Theorem 3.2.

2 Let k be the arity of T. Let $\{t_1, \ldots, t_m\}$ be an enumeration of the tuples of T, and let $\{x_1, \ldots, x_n\}$ be an enumeration of the variables which appear in T. Construct a v-table S with arity $k + n + 1$ as follows. For every tuple t_i in T, put exactly one tuple t'_i in S, where t'_i agrees with t_i on the first k columns, the $k + 1$st column contains the constant i, and the last m columns contain the variables $x_1, \ldots, x_m$. Now let q be the SP query defined

$$q := \pi_{1,\ldots,k}(\sigma_{\bigvee_{i=1}^{m} k+1=`i' \wedge \psi_i}(S))$$

where ψ_i is obtained from the condition $T(t_i)$ of tuple t_i by replacing variable names with their corresponding indexes in S.

■

Theorem 5.4 (Finite Completion).

1 The representation system obtained by closing or-set-tables under PJ queries is finitely complete.

2 The representation system obtained by closing finite v-tables under PJ or S^+P queries is finitely complete.

3 The representation system obtained by closing $\mathcal{R}_{sets}$ under PJ or PU queries is finitely complete.

4 The representation system obtained by closing $\mathcal{R}_{\oplus\equiv}$ under S^+PJ queries is finitely complete.

Proof: Fix an arbitrary finite incomplete database $\mathcal{I} = \{I_1, \ldots, I_n\}$ of arity k. It suffices to show in each case that one can construct a table T in the given representation system and a query q in the given language such that $q(Mod(T)) = \mathcal{I}$.

1 We construct a pair of or-set-tables S and T as follows. (They can be combined together into a single table, but we keep them separate to simplify the presentation.) For each instance I_i in $\mathcal{I}$, we put all the tuples of I_i in S, appending an extra column containing value i. Let T be the or-set-table of arity 1 containing a single tuple whose single value is the or-set $\langle 1, 2, \ldots, n\rangle$. Now let q be the S^+PJ query defined:

$$q := \pi_{1,\ldots,k}\sigma_{k+1=k+2}(S \times T).$$

2 Completion for PJ follows from Case 1 and the fact that finite v-tables are strictly more expressive than or-set tables. For S^+P, take the finite

v-table representing the cross product of S and T in the construction from Case 1, and let q be the obvious S^+P query.

3 Completion for PJ follows from Case 1 and the fact (shown in [41]) that or-set-tables are strictly less expressive than $\mathcal{R}_{\text{sets}}$. Thus we just need show the construction for PU. We construct an $\mathcal{R}_{\text{sets}}$ table T as follows. Let m be the cardinality of the largest instance in $\mathcal{I}$. Then T will have arity km and will consist of a single block of tuples. For every instance I_i in $\mathcal{I}$, we put one tuple in T which has every tuple from I_i arranged in a row. (If the cardinality of I_i is less than m, we pad the remainder with arbitrary tuples from I_i.) Now let q be the PU query defined as follows:

$$q := \bigcup_{i=0}^{m-1} \pi_{ki,\ldots,ki+k-1}(T)$$

4 We construct a pair of $\mathcal{R}_{\oplus\equiv}$-tables S and T as follows. (S can be encoded as a special tuple in T, but we keep it separate to simplify the presentation.) Let $m = \lceil \lg n \rceil$. T is constructed as in Case 2. S is a binary table containing, for each i, $1 \leq i \leq m$, a pair of tuples $(0, i)$ and $(1, i)$ with an exclusive-or constraint between them. Let sub-query q' be defined

$$q' := \prod_{i=1}^{m} \pi_1(\sigma_{2=`i'}(S))$$

The S^+PJ query q is defined as in Case 2, but using this definition of q'.

■

Theorem 5.5 (General Finite Completion). *Let $\mathcal{T}$ be a representation system such that for all $n \geq 1$ there exists a table T in $\mathcal{T}$ such that $|Mod(T)| \geq n$. Then the representation system obtained by closing $\mathcal{T}$ under $\mathcal{RA}$ is finitely-complete.* **Proof:** Let $\mathcal{T}$ be a representation system such that for all $n \geq 1$ there is a table T in $\mathcal{T}$ such that $|Mod(T)| \geq n$. Let $\mathcal{I} = \{I_1, ..., I_k\}$ be an arbitrary non-empty finite set of instances of arity m. Let T be a table in $\mathcal{T}$ such that $Mod(T) = \{J_1, \ldots, J_\ell\}$, with $\ell \geq k$. Define $\mathcal{RA}$ query q to be

$$q(V) := \bigcup_{1 \leq i \leq k-1} I_i \times q_i(V) \cup \bigcup_{k \leq i \leq \ell} I_k \times q_i(V),$$

where I_i is the query which constructs instance I_i and $q_i(V)$ is the Boolean query which returns true iff V is identical to I_i (which can be done in $\mathcal{RA}$). Then $q(Mod(T)) = \mathcal{I}$. ■

Chapter 3

RELATIONAL MODELS AND ALGEBRA FOR UNCERTAIN DATA

Sumit Sarkar
University of Texas at Dallas
School of Management
sumit@utdallas.edu

Debabrata Dey
University of Washington, Seattle
Michael G. Foster School of Business
ddey@u.washington.edu

Abstract Uncertainty in data values is pervasive in all real-world environments and have received a lot of attention in the literature. Over the last decade or so, several extensions to the relational model have been proposed to address the issue of how data uncertainty can be modeled in a comprehensive manner. This chapter provides a summary of the major extensions. We discuss the strengths and weaknesses of these models and show the underlying similarities and differences.

Keywords: Data uncertainty, incomplete information, probabilistic database, relational algebra, query language

1. Introduction

Database systems are widely used as a part of information systems in a variety of applications. The environment that a database attempts to represent is often very uncertain; this translates into uncertainty in terms of the data values. There could be several reasons for uncertainty in data items. The actual value of a data item may be unknown or not yet realized. For example, one may need to store the uncertainty associated with future stock prices in a securities database [1]; this information would be useful in developing a portfolio of investments with specified characteristics. Uncertainty in data items may also arise from consolidation or summarization of data [2]. For example, the results of market surveys are often expressed in a consolidated manner in which the details of individual consumer preferences are summarized. Such information is of considerable importance in designing new products. Another well-documented source of data uncertainty is data heterogeneity [8, 16]. When two heterogeneous databases show different values for the same data item, its actual value is not known with certainty. This has become an important concern in developing corporate data warehouses which consolidate data from multiple heterogeneous data sources. When the values for common data items are not consistent it may not be easy to establish which values are correct. How should such data be stored? One option is to store only values of those data items that are consistent. Of course, this would lead to ignoring a large number of data items for which we have some information. The other alternative is to record the different possible values for such data items, recognizing that there is some uncertainty associated with those values. By recording such data, and recognizing explicitly the uncertainty associated with those values, the user would be able to decide (perhaps using a cost-benefit analysis) when it is appropriate to use such data. The above examples illustrate the need to represent uncertainty in data models.

The relational data model has become the dominant model for handling data for a majority of applications. The relational model provides a range of advantages, such as access flexibility, logical and physical data independence, data integrity, reduced (and controlled) data redundancy, and enhanced programmer productivity. Unfortunately, the standard relational model does not have a comprehensive way to handle incomplete and uncertain data. As discussed earlier, such data, however, exist everywhere in the real world. Having no means to model these data, the relational model ignores all uncertain data and focuses primarily on values that are known for sure; uncertain data items are represented using "null" values, which are special symbols often employed to represent the fact that the value is either unknown or undefined [4, 15]. Consequently, relational databases do not yield satisfactory results in many real-world situations. Since the standard relational model cannot represent the inherent uncertain nature of the data, it cannot be used directly. This has led to several efforts to extend the relational model to handle uncertain data.

In order to store uncertain data, one needs to specify the nature of uncertainty that is being considered. It is well-documented that there are two types of uncertainties in the real world: uncertainty due to *vagueness* and uncertainty due to *ambiguity* [13]. Uncertainty due to vagueness is associated with the difficulty of making sharp or precise distinctions in the real world. For example, subjective terms such as tall, far, and heavy are vague. These cases can be modeled reasonably well with the help of tools such as fuzzy set theory. Uncertainty due to ambiguity, on the other hand, is associated with situations in which the choices among several precise alternatives cannot be perfectly resolved. Such situations are better modeled using the probability measure [13]. Probability theory has a rich theoretical basis, is easy to interpret, and empirically testable. Therefore, we restrict our discussion here to models that use probability theory and its associated calculus.

To summarize, we consider extensions of the relational model that generalize the standard relational model to allow the handling of uncertain data. The use of the relational framework enables the use of the powerful relational algebra operations to the extent possible. We focus on those extensions that use probability measures in representing uncertainty. The choice of probability theory allows the use of established probability calculus in manipulating uncertain data in the relations, and helps extend and redefine the relational algebra operations. Such models are referred to as probabilistic relational data models.

There have been several alternative modeling frameworks presented in the literature. We discuss the seminal probabilistic relational models in this chapter. These models can be viewed as belonging to one of the following approaches:

- Point-valued probability measures are assigned to every tuple in a relation. The probability measure indicates the joint probability of all the attribute values in that tuple. This includes the work of Cavallo and Pittarelli [3], Pittarelli [17], Dey and Sarkar [5], and Fuhr and Rölleke [11], among others.

- Point-valued probability measures are assigned to attribute values or sets of attribute values, resulting in a nested relational model. The model proposed by Barbara et al [1] uses this approach.

- Interval-valued probability measures are assigned to sets of attribute values, which is also a nested relational model. The choice of interval-valued probabilities helps capture the error in measurement approaches, and allows for a generalized set of combination strategies. Lakshmanan et al [14] present a model using this approach.

- Interval-valued probability measures are assigned to complex values, which are treated as tuples. The model proposed by Eiter et al [9] uses this approach.

We provide the main theoretical underpinnings of each of the above extensions in the next section. In Section 3, we provide a detailed discussion of the extended algebra proposed in one of these models (the one developed by Dey and Sarkar [5]). This forms the basis for comparing the operations suggested in the different approaches, as discussed in Section 4. Section 5 provides directions for future research.

2. Different Probabilistic Data Models

We discuss extant research on probabilistic data models. Each model is associated with a representation scheme for uncertain data, and an associated algebra specifying how the data in the relations should be manipulated for relational and set theoretic operations. We discuss in this section the important representation schemes proposed by different sets of researchers, and discuss the assumptions underlying these representations.

Our discussion is organized in the following manner. We first discuss those models that assign *point-valued* probability measures to each *tuple* in a relation. Next, we discuss a model that assigns *point-valued* probability measures at the *attribute* level. After that, we describe a model that assigns probability *intervals* at the *attribute* level. Finally, we discuss a model that assigns probability *intervals* at the *tuple* level. Other important aspects of the models are discussed along with the models.

2.1 Point-Valued Probability Measures Assigned to Each Tuple

By assigning point-valued probability measures to each tuple in a relation, the models in this category adhere to the first Normal Form (1NF). This makes the implementation of such models straightforward, and the probabilistic operations defined on such models more in line with traditional relational operations.

Cavallo and Pittarelli [3], Pittarelli [17]. Cavallo and Pittarelli are generally credited with providing the first probabilistic data model. They attempted to generalize the relational model by replacing the characteristic function of a relation with a probability distribution function, so that facts about which one is uncertain can also be handled. The probability assigned to each tuple indicates the joint probability of all the attribute values in the tuple. The tables shown in Figure 3.1 illustrate their model for a database that stores information on employees in an organization [17]. The first table stores information about the department to which an employee belongs. The second table stores information about the quality of an employee and the bonus the employee may be eligible for. The third table captures the expected sales generated by employees in the coming year.

An important requirement in their model is that the total probability assigned to all tuples in a relation should be one. This implies that tuples are disjoint; in other words, the set of attribute-values in a tuple is mutually ex-

Table 1

Employee	Department	$p_1(t)$
Jon Smith	Toy	0.5
Fred Jones	Houseware	0.5

Table 2

Employee	Quality	Bonus	$p_2(t)$
Jon Smith	Great	Yes	0.2
Jon Smith	Good	Yes	0.25
Jon Smith	Fair	No	0.05
Fred Jones	Good	Yes	0.5

Table 3

Employee	Sales	$p_3(t)$
Jon Smith	30–34 K	0.15
Jon Smith	35–39K	0.35
Fred Jones	20–24 K	0.25
Fred Jones	25–29K	0.25

Figure 3.1. Probabilistic Database with Employee Information (Reproduced from [17])

clusive of the sets of attribute-values for every other tuple in a relation. This constraint is reasonable when a relation is used to store data about one uncertain event, where the event is characterized by one or more attributes. However, it is not very convenient to store data on multiple entity instances, which is typically the case for traditional relations. As a result, even though it may be known with certainty that Jon Smith works in the Toy department and Fred Jones works in the Houseware department, Table 1 shows the respective probabilities to be equal to 0.5 each. The known probability for the department affiliation of each employee in this example can be recovered by multiplying each probability by the number of tuples [17]. An alternative approach to using their representation would be to use a separate relation for every distinct entity (or relationship) instance that is captured in a relation. Of course, this may result in an unmanageably large number of tables.

Dey and Sarkar [5]. Dey and Sarkar present a probabilistic relational model that does away with the restriction that the sum of probabilities of tuples in a relation must equal one. Instead, they consider keys to identify object instances, and use integrity constraints to ensure that the probabilities associated with different tuples representing the same object instance should not exceed one. If an object is known to exist with certainty, the probability stamps associated with the corresponding key value sum up to one exactly. Their model also allows storing data about objects whose very existence may be uncertain. The probability stamps associated with the key value of such an object sum up to less than one. Figure 3.2 illustrates an Employee relation in their model.

EMP#	ssn	lName	fName	rank	salary	dept	pS
3025	086-63-0763	Lyons	James	clerk	15K	toy	0.2
3025	086-63-0763	Lyons	James	cashier	20K	shoe	0.6
3025	086-63-0763	Lyons	James	cashier	15K	auto	0.2
6723	089-83-0789	Kivari	Jack	clerk	18K	toy	0.4
6723	089-83-0789	Kivari	Jack	cashier	20K	auto	0.4
6879	098-84-1234	Peters	Julia	clerk	25K	toy	0.3
6879	098-84-1234	Peters	Julia	clerk	27K	toy	0.1
6879	098-84-1234	Peters	Julia	cashier	25K	shoe	0.6

Figure 3.2. A Probabilistic Relation Employee (Reproduced from [5])

In this table, the primary key is EMP#, and the last column pS denotes the probability associated with each row of the relation. The pS column for the first row has the value 0.2; it means that there is a probability of 0.2 that there exists an employee with the following associated values: 3025 for EMP#, 086-63-0763 for ssn, Lyons for lName, James for fName, clerk for rank, 15K for salary, and toy for dept. All other rows are interpreted in a similar fashion. The probability stamp of a tuple is, therefore, the joint probability of the given realizations of all the attributes (in that tuple) taken together. Probabilities of individual attributes can be derived by appropriately marginalizing the distribution. For example, the first three rows indicate that it is known with certainty that (i) there exists an employee with EMP# 3025, and (ii) the ssn, lName and fName for this employee are 086-63-0763, Lyons and James, respectively. Similarly, the probability of an employee having EMP# = 3025 and rank = “cashier" is 0.8 (from the second and the third rows). Also, in the example shown in Figure 3.2, the probability masses associated with EMP# 6723 add up to 0.8; this means that the existence of an employee with EMP# 6723 is not certain and has a probability of 0.8.

This model assumes that the tuples with the same key value are disjoint, and each such tuple refers to mutually exclusive states of the world. At the same time, tuples with different key values are assumed to be independent of each other. Further, attributes in different relations are assumed to be independent, conditional on the key values. Dependent attributes are stored with their full joint distributions in the same relation.

Their model is unisorted, with the only valid object being a relation. The algebra described on their model is shown to be a consistent extension of traditional relational algebra, and reduces to the latter when there is no uncertainty associated with attribute values.

Fuhr and Rölleke [11]. Fuhr and Rölleke were motivated to develop a probabilistic relational model in order to integrate a database with information retrieval systems. The driving force behind their model is to extend the relational model in such a way that it can handle probabilistic weights required for performing information retrieval. In document indexing, terms are assigned weights with respect to the documents in which they occur, and these weights are taken into account in retrieving documents, where the probability of relevance of a document with respect to a query is estimated as a function of the indexing weights of the terms involved (for example, [10, 18]).

Analogous to [5], they too consider a model that is closed with respect to the operations defined, and assign probabilities to tuples as a whole. An important aspect of their approach is to associate each tuple of a probabilistic relation with a probabilistic event. A probabilistic relation corresponds to an ordinary relation where the membership of a single tuple in this relation is affected by a probabilistic event. If the event is true, the tuple belongs to the relation; otherwise it does not belong to the relation. For each event, the probability of being true must be provided.

Events are considered to be of two types: basic and complex. Complex events are Boolean combinations of basic events. Tuples in base relations are associated with basic events, while tuples in derived relations are associated with complex events. Event keys are used as identifiers for tuples in a relation. Figure 3.3 illustrates probability relations in their framework [11].

The example shows two relations, DocTerm and DocAu. The relation DocTerm stores weighted index terms for some documents, while DocAu provides the probability that an author is competent in the subjects described in a document. In these examples, event keys are represented as a combination of the relation name and the attribute values; e.g., DT(1,IR) is the event key for the first tuple in the relation DocTerm. It is suggested that actual implementations use internal IDŠs.

All basic events are assumed to be independent of each other. However, in order to handle imprecise attributes with disjoint values, they modify the independence assumption by introducing the notion of a *disjointness key*. The disjointness key is used as a kind of integrity constraint in their framework, and is analogous to that of a primary key in [5]. Since complex events are combinations of basic events, they do not have to be independent of other events (basic or complex). The authors identify conditions under which operations on complex events are probabilistically meaningful. Attributes in different relations are implicitly assumed to be independent of each other.

DocTerm	β	DocNo	Term
DT(1,IR)	0.9	1	IR
DT(2,DB)	0.7	2	DB
DT(3,IR)	0.8	3	IR
DT(3,DB)	0.5	3	DB
DT(4,AI)	0.8	4	AI

DocAu	β	DocNo	AName
DA(1,Bauer)	0.9	1	Bauer
DA(2,Bauer)	0.3	2	Bauer
DA(2,Meier)	0.9	2	Meier
DA(2,Schmidt)	0.8	2	Schmidt
DA(3,Schmidt)	0.7	3	Schmidt
DA(4,Koch)	0.9	4	Koch
DA(4,Bauer)	0.6	4	Bauer

Figure 3.3. Relations DocTerm and DocAu (Reproduced from [11])

2.2 Point-Valued Probability Measures Assigned to Attributes and Attribute Sets

Barbara et al [1]. This model extends the relational model by assigning probabilities to values of attributes. Relations have keys which are assumed to be deterministic. Non-key attributes describe the properties of entities, and may be deterministic or stochastic in nature. Figure 3.4 illustrates an example relation in this model.

Key	Independent Deterministic	Interdependent Stochastic	Independent Stochastic
EMPLOYEE	DEPARTMENT	QUALITY BONUS	SALES
Jon Smith	Toy	0.4 [Great Yes] 0.5 [Good Yes] 0.1 [Fair No]	0.3 [30–34K] 0.7 [35–39K]
Fred Jones	Houseware	1.0 [Good Yes]	0.5 [20–24K] 0.5 [25–29K]

Figure 3.4. Example Probabilistic Relation (Reproduced from [1])

The example relation stores information on two entities, "Jon Smith" and "Fred Jones." Since key values are deterministic, the two entities exist with certainty. The attribute DEPARTMENT is also deterministic in this example. Therefore, it is certain that Jon Smith works in the Toy department, and Fred Jones in the Houseware department. The attributes QUALITY and BONUS are probabilistic and jointly distributed. The interpretation is that QUALITY and BONUS are random variables whose outcome (jointly) depends on the EMPLOYEE under consideration. In this example, the probability that Jon Smith has a great quality and will receive a bonus is 0.4. The last attribute, SALES, describes the expected sales in the coming year by the employee and is assumed to be probabilistic but independent of the other non-key attributes.

In this model, each stochastic attribute is handled as a discrete probability distribution function. This means that the probabilities for each attribute in a tuple must add up to 1.0. However, to account for situations where the full distribution is not known or is difficult to specify exactly, an important feature of this model is the inclusion of *missing probabilities*. The example in Figure 3.5 illustrates how missing probabilities can be represented in this model.

Key	Independent Deterministic	Interdependent Stochastic	Independent Stochastic
EMPLOYEE	DEPARTMENT	QUALITY BONUS	SALES
Jon Smith	Toy	0.3 [Great Yes] 0.4 [Good Yes] 0.2 [Fair *] 0.1 [* *]	0.3 [30–34K] 0.5 [35–39K] 0.2 [*]

Figure 3.5. Example Probabilistic Relation with Missing Probabilities (Reproduced from [1])

In the example shown in Figure 3.5, a probability of 0.2 has not been assigned to a particular sales range. While the authors assume that this missing probability is distributed over all ranges in the domain, they do not make any assumptions as to how it is distributed. Since the missing probability may or may not be in the range $30–34K, the probability that the sales will be $30–34K next year is interpreted to lie between 0.3 and 0.3+0.2. In other words, the probability 0.3 associated with the sales range $30–34K is a lower bound. Similarly, 0.5 is a lower bound for the probability associated with $35–39K. The missing probability for the joint distribution over the attributes QUALITY and BONUS is interpreted similarly. A probability of 0.1 is distributed in an undetermined way over all possible quality and bonus pairs, while 0.2 is distributed only over pairs that have a "Fair" quality component. Thus, the probability that Smith is rated as "Great" and gets a bonus is between 0.3 and 0.3+0.1.

The incorporation of missing probabilities in a probabilistic model is one of the important contributions of this work. It allows the model to capture uncertainty in data values as well as in the probabilities. It facilitates inserting data into a probabilistic relation, as it is not necessary to have all the information about a distribution before a tuple can be entered. The authors go on to show how missing probabilities can arise during relational operations, even when the base relations have no missing probability. It also makes it possible to eliminate uninteresting information when displaying relations. For example, a user may only be interested in seeing values with probability greater than 0.5; the rest can be ignored.

Attributes in a relation are implicitly assumed to be independent of attributes in other relations, conditioned on the key values.

2.3 Interval-Valued Probability Measures Assigned to Attribute-Values

This line of research attaches a probabilistic *interval* to each value from a subset of possible values of an imprecise attribute. The relational algebra operations are then generalized to combine such probabilities in a suitable manner. A variety of strategies are considered regarding combinations of probability intervals.

Lakshmanan et al [14]. A goal of this model is to provide a unified framework to capture various strategies for conjunction, disjunction, and negation of stochastic events captured in a database. The different strategies are developed to handle different assumptions regarding the underlying events, such as independence, mutual exclusion, ignorance, positive correlation, etc. In order to accommodate the different strategies, this model works with probability intervals instead of point-valued probabilities. This is because, depending on the assumptions made regarding the underlying events, probability intervals can arise when deriving probabilities for complex events. Interval-valued probabilities are also considered useful and appropriate when there is noise in the process of measuring the desired probability parameters, and it is important to capture the margin of error in the probability of an event.

The model associates probabilities with individual level elements (i.e., attribute values), although the authors note that the element-level probabilities can be converted into a representation that associates probabilities with whole tuples. Figure 3.6 provides an example of a probabilistic relation called Target.

The example relation has three attributes, Location, Object, and Band. The relation includes three tuples. The first tuple has one value for the attribute Location ("site1") with the probability interval [1,1] associated with it, implying there is no uncertainty associated with that value. The tuple also has one value associated with the attribute Object ("radar_type1"), again with the prob-

LOC	OBJ	Band
site1 $h_1(site1)$=[1,1]	radar_type1 $h_2(radar\_type1)$=[1,1]	750, 800 $h_3(750)$=[0.4,0.7] $h_3(800)$=[0.5,0.9]
site2 $h_4(site2)$=[1,1]	{radar_type1,radar_type2} $h_5(radar\_type1)$=[0.8,0.9] $h_5(radar\_type2)$=[0.8,0.3][a]	700 $h_6(700)$=[1,1]
site3 $h_7(site3)$=[1,1]	{radar_type1,radar_type2} $h_8(radar\_type1)$=[0.4,0.7] $h_8(radar\_type2)$=[0.5,0.6]	700, 750 $h_9(700)$=[0.6,0.9] $h_9(750)$=[0,0.4]

[a]Clearly, there is a typographical error in [14], since the upper bound cannot be less than the lower bound.

Figure 3.6. A Probabilistic Relation Target with Three Attributes (Reproduced from [14])

ability interval [1,1]. It has two values associated with the attribute Band, the value 750 with the probability interval [0.4,0.7], and 800 with the probability interval [0.5,0.9]. The intervals provide the lower and upper bounds for the probabilities associated with each value.

The model associates each tuple with a possible world, and interprets a probabilistic tuple as an assignment of probabilities to the various worlds associated with that tuple (in other words, a world represents a possible realization of a probabilistic tuple). Thus, there are two worlds associated with the first tuple in the example relation:

$$w_1 = (\text{site1, radar\_type1, 750}),$$
$$w_2 = (\text{site1, radar\_type1, 800}).$$

A world-id is used to identify each possible world, and is associated with tuples in base relations. In order to identify complex events that result from Boolean combinations of probabilistic tuples, the model proposes the use of annotated tuples. Such Boolean combinations of possible worlds are referred to as paths, and annotated tuples use such paths to identify probabilistic tuples that appear in derived relations, i.e., views. The algebra defined on their model manipulates data, probability bounds, as well as paths. The paths keep track of the navigation history for tuples appearing in views, and are expected to be maintained internally. The paths help encode interdependencies between attributes, and hence between tuples, and enable enforcing integrity constraints in relations. As mentioned earlier, an important consideration behind the proposed operations is that they be amenable to a wide range of possible assumptions about tuples. The authors propose a set of postulates that combination strategies should satisfy, and define the operations accordingly.

2.4 Interval-valued Probability Measures Assigned to Tuples

Eiter et al [9]. Eiter et al extend the probabilistic model in [14] by incorporating complex values (also called derived events) in relations. Probabilistic intervals are assigned to tuples in their entirety. Tuples in base relations are associated with basic events. Derived tuples are associated with general events, than include conjunctions and disjunctions of basic events. Every complex value v is associated with a probability interval $[l, u]$, and an event e. The interval $[l, u]$ represents the likelihood that v belongs to a relation of the database, where e records information about how the value is derived. Figure 3.7 provides an example relation with probabilistic complex values.

v		l	u	e
patient John	diseases {lung cancer, tuberculosis}	0.7	0.9	$e_1 \vee e_2$
patient Jack	diseases {leprosy}	0.5	0.7	e_3

Figure 3.7. A Probabilistic Complex Value Relation (Reproduced from [9])

The example relation holds data about patients and their diseases. The first tuple in the relation shows that the probability that a patient John suffers from both lung cancer as well as tuberculosis lies between 0.7 and 0.9. Similarly, the second tuple shows that the probability that a patient Jack suffers from leprosy is between 0.5 and 0.7. The complex event $(e_1 \vee e_2)$ is associated with the first tuple, and is assumed to be derived by combining tuples corresponding to basic events.

As in [14], basic events are not assumed to be pairwise independent or mutually exclusive. The probability range of a complex event is computed using whatever dependence information is available about the basic events. The combination strategies refine the ones presented in [14].

3. Probabilistic Relational Algebra

In this section, we provide the details of the probabilistic relational model and algebra proposed in [5].

3.1 Basic Definitions

Let $\mathcal{N} = \{1, 2, \ldots, n\}$ be an arbitrary set of integers. A *relation scheme* R is a set of *attribute names* $\{A_1, A_2, \ldots, A_n\}$, one of which may be a probability stamp pS. Corresponding to each attribute name A_i, $i \in \mathcal{N}$, is a set

D_i called the *domain* of A_i. If $A_i = pS$, then $D_i = (0, 1]$. The *multiset* $\boldsymbol{D} = \{D_1, D_2, \ldots, D_n\}$ is called the domain of R. A *tuple* x over R is a function from R to $\boldsymbol{D}$ ($x : R \to \boldsymbol{D}$), such that $x(A_i) \in D_i$, $i \in \mathcal{N}$. In other words, a tuple x over R can be viewed as a set of attribute name-value pairs: $x = \{\langle A_i, v_i \rangle | \forall i \in \mathcal{N}(A_i \in R \wedge v_i \in D_i)\}$. *Restriction of a tuple* x over S, $S \subset R$, written $x(S)$, is the sub-tuple containing values for attribute names in S only, i.e., $x(S) = \{\langle A, v \rangle \in x | A \in S\}$. The formal interpretation of a tuple is as follows: a tuple x over R represents one's belief about attributes (in R) of a real world object. If $pS \in R$, then a probability of $x(pS) > 0$ is assigned to the fact that an object has the values $x(R - \{pS\})$ for the corresponding attributes. In other words, the attribute pS represents the joint distribution of all the attributes taken together:

$$x(pS) = \Pr\left[R - \{pS\} = x(R - \{pS\})\right].$$

If $pS \notin R$, i.e., if the relation scheme R is deterministic, then every tuple on R is assigned a probability of one, and is not explicitly written; in that case, it is implicitly assumed that $x(pS) = 1$.

Two tuples x and y on relation scheme R are *value-equivalent* (written $x \simeq y$) if and only if, for all $A \in R$, $(A \neq pS) \Rightarrow (y(A) = x(A))$. Value-equivalent tuples are not allowed in a relation; they must be *coalesced*. Two types of *coalescence* operations on value-equivalent tuples are defined:

1 The *coalescence-PLUS* operation is used in the definition of the projection operation. Coalescence-PLUS (denoted by $\oplus$) on two value-equivalent tuples x and y is defined as:

$$z = x \oplus y \Leftrightarrow (x \simeq y) \wedge (z \simeq x) \wedge \Big(z(pS) = \min\{1, x(pS) + y(pS)\}\Big).$$

2 The *coalescence-MAX* operation is used in the definition of the union operation. Coalescence-MAX (denoted by $\odot$) on two value-equivalent tuples x and y is defined as:

$$z = x \odot y \Leftrightarrow (x \simeq y) \wedge (z \simeq x) \wedge \Big(z(pS) = \max\{x(pS), y(pS)\}\Big).$$

The idea of value-equivalent tuples and coalescence operations need not be confined to just two tuples. Given m tuples $x_1, x_2, \ldots, x_m$, all of which are on the same relation scheme, they are said to be value-equivalent if $x_i \simeq x_j$ for all i, j; $1 \leq i, j \leq m$. Coalescence-PLUS, for example, on all these m value-equivalent tuples will recursively coalesce all the tuples pair-wise, i.e.,

$$\bigoplus_{i=1}^{m} x_i = (\ldots((x_1 \oplus x_2) \oplus x_3) \oplus \ldots \oplus x_{m-1}) \oplus x_m.$$

Let R be a relation scheme. A *relation* r on the scheme R is a finite collection of tuples x on R such that no two tuples in r are value-equivalent.

3.2 Primary and Foreign Keys

In the relational model, every tuple in a relation represents a unique object (i.e., an entity or a relationship) from the real world; a *superkey* is a set of attributes that uniquely identifies a tuple, and hence an object. A superkey, in that sense, is an *object surrogate*, one that uniquely identifies every object. A *candidate key* is a minimal superkey, minimal in the sense that no attribute can be dropped without sacrificing the property of uniqueness. For each relation, only one candidate key is chosen as the *primary key* of that relation.

In the probabilistic extension, where every tuple has a probability stamp that represents the joint probability of occurrence of the attribute values in that tuple, each tuple cannot stand for a unique object. Associated with every object there may be several tuples representing the complete joint distribution of its attributes. This suggests that one must retain the *object surrogate* interpretation of the primary key (i.e., unique identifier of real world objects) and discard the notion of the primary key as a unique identifier of tuples.

The term *foreign key* retains the usual meaning in this model. In other words, a foreign key of a relation scheme R is a set of attributes $F \subset R$ that *refers* to a primary key K of some relation scheme S. Attributes in F and K may have different names, but they relate to the same real world property of an object and come from the same domain. If r and s are relations on schemes R and S respectively, we call r the *referring* (or, *referencing*) relation and s the *referred* (or, *referenced*) relation. This is written symbolically as: $r.F \longrightarrow s.K$. The possibility that r and s are the same relation is not excluded. Primary and foreign keys are useful in enforcing important integrity constraints on probabilistic relations.

Intra-Relational Integrity Constraints: Let r be any relation on scheme R with primary key K. The following intra-relational constraints are imposed on r:

1 The total probability associated with a primary key value must be no more than one. In other words, for all $x \in r$,

$$\sum_{\substack{y \in r \\ y(K)=x(K)}} y(pS) \leq 1.$$

2 For all $x \in r$, no part of $x(K)$ can be null.

3 For all $x \in r$, if $pS \in R$, then $x(pS) \in (0,1]$ and $x(pS)$ is not null.

Referential Integrity Constraints: Let r and s be two relations on schemes R and S respectively. Let K_R and K_S be the primary keys of R and

S, and let $r.F \longrightarrow s.K_S$ for some $F \subset R$. The following referential constraints are imposed on r and s:

1 For all $x \in r$, if there exists an attribute $A \in F$ such that $x(A)$ is null, then for all other attributes $B \in F$, $x(B)$ is also null. This ensures that the foreign key value of a tuple is not partially null.

2 For all $x \in r$, either $x(F)$ is null (fully), or there exists $y \in s$ such that

$$\sum_{\substack{z \in r \\ z(K_R F) = x(K_R F)}} z(pS) \leq \sum_{\substack{z \in s \\ z(K_S) = y(K_S)}} z(pS),$$

where $K_R F$ is a shorthand for $K_R \cup F$.

3.3 Relational Operations

Based on the above definitions, we now present the algebraic operations for this model. In addition to the basic relational operations, a special operation called *conditionalization* is introduced. This is useful in answering queries involving non-key attributes in relations.

Union. Let r and s be relations on the same scheme R. Then the union of these two relations is defined as:

$$\begin{aligned} r \cup s \quad = \quad & \Big\{ x(R) \Big| \Big((x \in r) \wedge \big(\forall y \in s (y \not\simeq x) \big) \Big) \vee \\ & \Big((x \in s) \wedge \big(\forall y \in r (y \not\simeq x) \big) \Big) \vee \\ & \Big(\exists y \in r \ \exists z \in s (x = y \odot z) \Big) \Big\}. \end{aligned}$$

This operation is a straightforward extension of the deterministic counterpart, with the added restriction that if there are value-equivalent tuples in the participating relations then the higher probability stamp is included in the result. It can be easily verified that union is *commutative*, *associative*, and *idempotent*. An alternative definition of the union operation may be obtained by replacing $\odot$ with $\oplus$ in the above definition; in that case, however, the union operation would not be idempotent.

Difference. Let r and s be as above. Then the difference of these two relations is given by:

$$r - s = \Big\{x(R) \,\Big|\, \Big((x \in r) \wedge \Big(\forall y \in s(y \not\simeq x)\Big)\Big) \vee \Big(\exists y \in r\ \exists z \in s\Big((x \simeq y \simeq z) \wedge \Big(y(pS) > z(pS)\Big) \wedge \Big(x(pS) = y(pS) - z(pS)\Big)\Big)\Big)\Big\}.$$

In this operation, if there are value-equivalent tuples in the participating relations, then the difference in the probability stamps is included in the result provided the difference is positive.

Projection. Let r be a relation on scheme R, and let $S \subset R$. The projection of r onto S is defined as:

$$\Pi_S(r) = \left\{ x(S) \,\middle|\, x = \bigoplus_{\substack{y \in r \\ y(S) \simeq x}} y(S) \right\}.$$

This operation provides the marginal distribution for a subset of attributes. The result is meaningful if a candidate key is included in the projection list, as the marginalization is then conducted separately for each object captured in the relation.

Selection. Let r be a relation on scheme R. Let Θ be a set of comparators over domains of attribute names in R. Let P be a predicate (called the selection predicate) formed by attributes in R, comparators in Θ, constants in the domain of A for all $A \in R$, and logical connectives. The selection on r for P, written $\sigma_P(r)$, is the set $\{x \in r | P(x)\}$. This operation is defined at the tuple level, and is identical to its counterpart in traditional relation algebra.

Natural Join. Let r and s be any two relations on schemes R and S respectively, and let $R' = R - \{pS\}$ and $S' = S - \{pS\}$. The natural join of r and s is defined as:

$$r \bowtie s = \Big\{x(R \cup S) \,\Big|\, \exists y \in r\ \exists z \in s\Big(\Big(x(R') = y(R')\Big) \wedge \Big(x(S') = z(S')\Big) \wedge \Big(x(pS) = y(pS) z(pS)\Big)\Big)\Big\}.$$

Note that the attributes in R and S should be independent for the natural join operation to yield meaningful results. It can be easily verified that the natural join is *commutative* and *associative*, but it is not *idempotent*.

Rename. The rename operation (ρ) is used to change the names of some attributes of a relation. Let r be a relation on scheme R, and let A and B be attributes satisfying $A \in R$, $B \notin R$. Let A and B have the same domain, and let $S = (R - \{A\}) \cup \{B\}$. Then r with A renamed to B is given by:

$$\rho_{A \leftarrow B}(r) = \left\{ y(S) \;\middle|\; \exists x \in r \left(\Big(y(S - B) = x(R - A) \Big) \wedge \Big(y(B) = x(A) \Big) \right) \right\}.$$

If pS is renamed, it loses its special meaning and behaves like just another user-defined attribute.

Conditionalization. Let r be a relation on scheme R, and $S \subset R - \{pS\}$. The conditionalization of r on S is given by:

$$\Upsilon_S(r) = \left\{ x(R) \;\middle|\; \exists y \in r \left((x \simeq y) \wedge \left(x(pS) = \frac{y(pS)}{\eta_{S,r}(y)} \right) \right) \right\},$$

where $\eta_{S,r}(x)$ is a function defined on a tuple $x \in r$ if $pS \in R$, and is given by:

$$\eta_{S,r}(x) = \min \left\{ 1, \quad \sum_{\substack{y \in r \\ y(S) = x(S)}} y(pS) \right\}.$$

The conditionalization operation on S revises the probability stamp associated with each tuple by changing the marginal probability of the values for attributes in S to unity. In other words, after conditionalization, the relation can be interpreted as the joint conditional distribution of all attributes in $(R - S - \{pS\})$ given the values of attributes in S. As a result, this operation is useful, for example, in answering queries about non-key attributes of a relation for a given key value, or, before performing the join operation to obtain meaningful results. Note that, for the conditional probabilities to be meaningful, it may be necessary to include a candidate key as part of S.

Other relational operations such as intersection and Cartesian product can be expressed in terms of the above basic operations:

- *Intersection.* Let r and s be relations on the same scheme R. Then the intersection of these two relations is given by:

$$r \cap s \;=\; \Big\{x(R) \;\Big|\; \exists y \in r \; \exists z \in s \Big((x \simeq y \simeq z) \wedge \Big(x(pS) = \min\{y(pS), z(pS)\}\Big)\Big)\Big\}.$$

  It can be easily verified that $r \cap s = r - (r - s)$.

- *Cartesian Product.* The Cartesian product of two relations is a special case of a natural join [Codd 1990], where the relations do not have any common attribute name (with the possible exception of pS). Let R and S be two relation schemes satisfying $(R \cap S) - \{pS\} = \emptyset$. Let r and s be relations on the schemes R and S, respectively. The Cartesian product of r and s is a relation on scheme $(R \cup S)$ given by: $r \times s = r \bowtie s$.

- *Theta-join.* Let R, S, r and s be as above. Let Θ be a set of comparators over domains of attributes in $(R \cup S)$. Let P be any predicate formed by attributes in $(R \cup S)$, comparators in Θ, constants in the domain of A for all $A \in (R \cup S)$, and logical connectives. The theta-join between r and s is given by: $r \bowtie_P s = \sigma_P(r \bowtie s)$.

- *Alpha-cut.* The alpha-cut operation selects only those tuples from a relation that have a probability of α or more. Let r be a relation on scheme R. Let $R' = R - \{pS\}$. Then alpha-cut of r, denoted $\Phi_\alpha(r)$, is $\{x(R') | (x \in r) \wedge (x(pS) \geq \alpha)\}$. It is easy to verify that $\Phi_\alpha(r) = \Pi_{R'}(\sigma_{pS \geq \alpha}(r))$.

3.4 Relational Algebra

Assume that $\boldsymbol{U}$ is a set of attribute names, called the *universe*. $\boldsymbol{U}$ may have the probability stamp pS as only one of its element. Let $\mathcal{D}$ be a set of domains, and let **dom** be a total function from $\boldsymbol{U}$ to $\mathcal{D}$. Let $\boldsymbol{R} = \{R_1, R_2, \ldots, R_p\}$ denote a set of distinct relation schemes, where $R_i \subset \boldsymbol{U}$, for $1 \leq i \leq p$. Let $d = \{r_1, r_2 \ldots, r_p\}$ be a set of relations, such that r_i is a relation on R_i, $1 \leq i \leq p$. Θ denotes a set of comparators over domains in $\mathcal{D}$. The *relational algebra* over $\boldsymbol{U}$, $\mathcal{D}$, **dom**, $\boldsymbol{R}$, d and Θ is the 7-tuple $\mathcal{R} = (\boldsymbol{U}, \mathcal{D}, \mathbf{dom}, \boldsymbol{R}, d, \Theta, O)$, where O is the set of operators union, difference, natural join, projection, selection, rename and consolidation using attributes in $\boldsymbol{U}$ and comparators in Θ, and logical connectives. An *algebraic expression* over $\mathcal{R}$ is any expression formed legally (according to the restrictions on the operators) from the relations in d and constant relations over schemes in $\boldsymbol{U}$, using the operators in O.

The *relational algebraic expressions* and their *schemes* over $\mathcal{R}$ are defined recursively (according to the restrictions on the operators) as follows:

1 Let $Q = \{C_1, C_2, \ldots, C_k\} \subset \boldsymbol{U}$ be any relational scheme, and let $c_i \in \text{dom}(C_i)$, $1 \leq i \leq k$. Then $\{\langle c_1 : C_1, c_2 : C_2, \ldots, c_k : C_k \rangle\}$ is a relational expression over scheme Q called a constant.

2 Each $r_i \in d$ is a relational expression over the scheme R_i, $1 \leq i \leq p$.

3 If E_1 and E_2 are relational expressions over the same scheme Q, then so are the following: (i) $E_1 \cup E_2$, (ii) $E_1 - E_2$, and (iii) $\sigma_P(E_1)$, where P is a selection predicate.

4 If E is a relational expression over the scheme Q, and $S \subset Q$, then $\Pi_S(E)$ is a relational expression over the scheme S.

5 If E_1 and E_2 are relational expressions over schemes Q_1 and Q_2, then so is $E_1 \bowtie E_2$ over the scheme $Q_1 \cup Q_2$.

6 If E is a relational expression over Q, and A and B are attributes with the same domain, then $\rho_{A \leftarrow B}(E)$ is a relational expression over $(Q - \{A\}) \cup \{B\}$.

7 If E is a relational expression over Q, then so is $\Upsilon_S(E)$, for all $S \subset (Q - \{pS\})$.

Dey and Sarkar [5] show that this algebra is closed, is a consistent extension of the traditional relational algebra, and reduces to the latter.

3.5 Incomplete Distribution and Null Values

We now turn our attention to the case where the joint probability distribution of the attributes of an object is partially specified. For example, it is possible that the existence of an employee is certain (i.e., the marginal probability of the key EMP# is one), but the marginal distribution of the salary of that employee is not completely specified. This scenario is illustrated in the relation shown in Figure 3.8, where the existence of an employee with EMP#=6879 is known with certainty; the marginal distribution of rank is completely specified for this employee, but the marginal distribution for salary and department information is not completely available.

The EMPLOYEE relation in Figure 3.8 models this type of incompleteness with the help of a *null value* "∗." It means that a portion of the probability mass is associated with a value that is unknown. For example, out of a total of 1.0, only 0.3 is associated with a known value of salary for EMP#=6879; remaining 0.7 is given to the null value.

Interpretation of Partial Distribution. An important question is the interpretation of the probability stamp when the joint probability distribution is not fully specified. How one interprets the probability stamp has to do with the

EMP#	rank	salary	dept	pS
3025	clerk	15K	toy	0.2
3025	cashier	20K	shoe	0.6
3025	cashier	15K	auto	0.2
6723	clerk	18K	toy	0.4
6723	cashier	20K	auto	0.4
6723	$*$	$*$	$*$	0.1
6879	clerk	25K	toy	0.3
6879	clerk	$*$	toy	0.1
6879	cashier	$*$	$*$	0.6

Figure 3.8. EMPLOYEE: A Probabilistic Relation with Null Values

interpretation given to the portion of the total probability mass (associated with a key value) that is not specified, called the *missing probability* in [1]. There are two possible interpretations that may be given to the missing probability. The first is that the missing probability is associated with realizations of those values of attributes that are not already included in the relation. Thus, in Figure 3.8, the missing probability of 0.1 for EMP# 6723 could be distributed in any manner over those joint realizations for rank, salary and department that are not already included in the table. With this interpretation, the probability stamps for tuples that do appear in the relation are construed as point estimates of the conditional probabilities for given values of the attributes. Therefore, the probability that EMP#=6723, rank="clerk," salary=18K and dept="toy" is interpreted to be 0.4. Similarly, the probability that EMP#=6879 and dept="toy" is 0.4.

The second interpretation for the missing probabilities is that they could be distributed over the entire set of realizations of the attributes, including the ones that already appear in the relation. In that case, the uncertainty associated with the attribute values for tuples that appear in the relation are represented by probability intervals, and not point estimates. The probability stamp associated with a tuple is then the lower bound of the probability interval for that tuple (as in [1]). Consider the previous example of EMP# 6723; this key value has a missing probability of 0.1. Since this probability mass could be assigned to any value, including those that have already appeared, the probability that EMP#=6723, rank="clerk," salary=18K and dept="toy" lies in the interval $[0.4, 0.5]$. Similarly, the probability that EMP#=6879 and dept="toy" lies in the interval $[0.4, 1.0]$. When the distribution is completely specified, the interval clearly reduces to a point.

Extended Relational Operations. The basic algebraic operations can be extended to incorporate the null values as possible attribute values. An important feature of this extension is that the semantics associated with each of the above two interpretations is preserved as a result of all the basic relational operations, i.e., the extended operations can handle both interpretations of missing probabilities. Consequently, depending on their preference, users can represent uncertainties regarding attribute values either as point estimates or as intervals. The result of the relational operations will be consistent with the user's interpretation of the original tables. First, a few definitions are necessary.

Let x be any tuple on scheme R. If $A \in R$ and $x(A)$ is not null, x is called *definite* on A, written $x(A)\downarrow$. For $S \subset R$, $x(S)\downarrow$ if $x(A)\downarrow$ for all $A \in S$. A tuple x is said to *subsume* a tuple y, both on scheme R, written $x \geq y$, if for all $A \in R$, $y(A)\downarrow$ implies $x(A) = y(A)$.

Now, the concept of value-equivalent tuples must be redefined for the ones that might have null values. Let R be a relation scheme and let $R' = R - \{pS\}$. For any two tuples x and y on R,

$$(x \simeq y) \;\Leftrightarrow\; (x(R') \geq y(R')) \wedge (y(R') \geq x(R')).$$

Again, value-equivalent tuples are not allowed to co-exist in a relation; they must be coalesced. The coalescence-PLUS and the coalescence-MAX operations as defined earlier also work for this extension.

As far as the basic relational operations are concerned, the previous definitions of the union, difference, projection, and rename operations can be used with the extended definition of value-equivalent tuples. Thus, only the selection, natural join and conditionalization operations have to be redefined.

- *Selection.* Let R, r, Θ, P be as in the earlier definition of the *selection* operation. Let $S \subset R$ be the set of attributes involved in P. Then, $\sigma_P(r) = \{x \in r | x(S)\downarrow \wedge P(x)\}$. In other words, tuples with null values for attributes involved in the selection predicate are not considered.

- *Natural Join.* Let r and s be any two relations on schemes R and S respectively. Let $Q = R \cap S$, $R' = R - \{pS\}$ and $S' = S - \{pS\}$. Then,

$$r \bowtie s \;=\; \left\{ x(R \cup S) \;\middle|\; \exists y \in r\, \exists z \in s \Big(y(Q)\downarrow \wedge z(Q)\downarrow \wedge \left(x(R') = y(R')\right) \wedge \left(x(S') = z(S')\right) \wedge \left(x(pS) = y(pS)z(pS)\right)\Big)\right\}.$$

  In other words, join operation matches tuples on non-null attribute values only.

- *Conditionalization.* Let r be a relation on scheme R, and $S \subset R-\{pS\}$. The conditionalization of r on S is given by:

$$\Upsilon_S(r) = \left\{ x(R) \;\middle|\; \exists y \in r \left(y(S) \downarrow \wedge (x \simeq y) \wedge \left(x(pS) = \frac{y(pS)}{\eta_{S,r}(y)} \right) \right) \right\},$$

where $\eta_{S,r}(y)$ is as before. Again, tuples with null values for attributes in S are excluded in performing the conditionalization operation.

Finally, [5] introduces a new operation called the *N-th moment*. This operation allows one to obtain interesting aggregate properties of different attribute names based on the original distribution of those attribute names represented in the form of a relation. The N-th moment of a probability distribution is traditionally defined in the following manner: let ψ be a random variable with domain Ψ and probability density function $f_\psi(x)$, $x \in \Psi$; its N-th moment, $\mu_N(\psi)$, is then defined as:

$$\mu_N(\psi) = \mathrm{E}[\psi^N] \;\; = \int\limits_{x \in \Psi} x^N f_\psi(x) dx.$$

Moments of a distribution are useful in obtaining aggregate properties of a distribution such as mean, standard deviation, skewness and kurtosis. For example, the standard deviation of the random variable ψ can be easily obtained from its first and second moments:

$$\mathrm{Stdev}(\psi) = \sqrt{\mu_2(\psi) - (\mu_1(\psi))^2}.$$

These aggregate properties are not only useful in understanding the overall nature of a distribution, but also in comparing two different distributions. This is why moments are a very important tool in statistical analysis. The N-th moment operation helps to form an overall opinion about the nature of real world objects, as well as allows various statistical analysis to be performed on the stored data.

- *N-th Moment.* Let r be a relation on scheme R. Let $R' = R - \{pS\}$ and $S \subset R'$. The N-th moment of r given S, written $\mu_{S,N}(r)$, is defined as:

$$\begin{aligned} \mu_{S,N}(r) \;\; = \;\; & \left\{ x(R') \;\middle|\; \exists y \in r \left(y(S) \downarrow \wedge \left(x(S) = y(S) \right) \wedge \right.\right. \\ & \left.\left. \left(\forall A \in (R' - S) \left(x(A) = m_{S,r,N}(y, A) \right) \right) \right) \right\}, \end{aligned}$$

where,

$$m_{S,r,N}(x,A) = \begin{cases} \dfrac{\sum\limits_{\substack{y\in r \\ y(A)\downarrow \\ y(S)=x(S)}} (y(A))^N y(pS)}{\sum\limits_{\substack{y\in r \\ y(A)\downarrow \\ y(S)=x(S)}} y(pS)}, & \text{if } pS\in R \text{ and } A\in R' \text{ is numeric,} \\ \Omega, & \text{otherwise.} \end{cases}$$

A few comments are in order about the *N-th moment* operation. First, Ω is a special type of null value generated as a result of this operation on non-numeric attributes. Second, this operation is really a family of operations, because one gets a different operation for each positive integer N. For example, to obtain the expected value of different attributes, one can use the *first moment*, i.e., $N = 1$. If the first moment operation is applied on the EMPLOYEE relation shown in Figure 3.8 with S={EMP#}, one would obtain the expected value of all other attributes given EMP#; this is illustrated in Figure 3.9. Third, it is

EMP#	rank	salary	dept
3025	Ω	18K	Ω
6723	Ω	19K	Ω
6879	Ω	25K	Ω

Figure 3.9. EMPLOYEE Relation after First Moment Operation

possible to define other operations—such as *standard deviation*, *skewness* and *kurtosis*—based on the above class of operations. Finally, as can be seen from the definition, null values $(*)$ are not considered in calculating moments. In other words, only the explicitly specified part of the distribution is considered in calculation of moments.

4. Algebraic Implications of the Different Representations and Associated Assumptions

In Section 2, we discussed the main differences across the various models in terms of how the uncertainty in data is represented by the models, and the underlying assumptions for each approach. Having discussed the model from [5] in considerable detail in Section 3, we now highlight the key differences in the algebra proposed for each of the different approaches, and discuss where each approach provides different functionalities to users. To keep this discussion brief and easy to follow, we consider the model in [5] as the benchmark and contrast the differences.

4.1 Models Assigning Point-Valued Probability Measures at the Tuple Level

We first examine the unique features of the model proposed by Cavallo and Pittarelli [3], and subsequently refined in [17]. In their work, they focus on the operations *projection*, *selection*, and *join*. Since, in their model, a relation can be viewed as storing the distribution of an uncertain event, their projection operation provides the marginal distribution of the projected attributes. Thus, the operation in [5] is similar to theirs, except that [5] allows data on multiple objects (events) to be stored in a relation, leading to marginalization only within tuples that share the primary key value. The selection operations in [3] and [5] are also similar, with the difference that [3] requires that the probabilities associated with selected tuples are normalized to add up to one (in keeping with their requirement about a relation). Their join operation is motivated by the idea that, to the extent possible, the full distribution across the collective set of attributes can be reconstructed from the projections of the joined relation into the schema for the participating relations. Using this motivation, they propose that the maximum entropy distribution be obtained for the tuples in the joined relation, while preserving the marginal probability distributions associated with the participating relations. If the non-common attributes of a participating relation is conditionally independent of the non-common attributes of the other participating relation given the common attributes, then their operation calculates the distribution for the tuples in the resulting relation by using this conditionalization explicitly. The operation provided in [5] is analogous, where it is implicitly assumed that the conditional independence property holds (if this property does not hold, then the participating relations are considered to be ill formed, as they cannot capture the uncertainty associated with the full set of attributes in the two relations). Neither [3] nor [17] provide a formal discussion of the *union* and *difference* operations, although they note that updates to the database can be viewed as revising the distribution associated with the tuples in a manner consistent with incrementing the relative frequencies of observations that lead to the update process.

Fuhr and Rölleke [11] extend the traditional relational algebra for the five basic operations. In order to account for the distinction they make in their independence assumptions regarding relations with basic events and those with complex events, their algebra is described separately for each of them. We first discuss their operations defined for basic events, and then contrast them with operations for complex events. Their approach assumes that tuples corresponding to basic events are independent of each other, and their *selection* and *natural join* operations are identical to those presented in [5]. The other three operations work differently, because value-equivalent tuples are involved. Thus, for the *projection* operation, they compute the probability of a tuple that is part of multiple tuples in the original relation by forming the disjunction of the events associated with the original tuples. Implicit in their approach is the assumption that the primary key attribute is not included in the projected relation (in which case marginalization would be the appropriate approach). The *union* operation likewise computes the probability of a value-equivalent tuple in the result by considering the disjunction of the corresponding tuples in the participating relation. Their *difference* operation returns only those tuples that appear in the first relation and do not have value-equivalent tuples in the second one.

When considering operations on complex events, Fuhr and Rölleke [11] propose that the appropriate Boolean expression implied by the operation be taken into account when calculating the probabilities for tuples in the resulting relation. As mentioned earlier, to facilitate this, they explicitly record the complex events corresponding to tuples in derived relations. When further operations are performed on these derived relations, the event expressions of relevant tuples are examined to determine if the tuples can be considered independent. If that is the case, the operations remain unchanged. When that is not the case, the process requires transforming the Boolean expression for each tuple in the result into its equivalent disjunctive normal form and then computing the desired probability.

4.2 Models Assigning Point-Valued Probability Measures at the Attribute Level

Barbara et al [1] focus primarily on the operations *projection*, *selection*, and *join*, and then present a set of new operations that do not have counterparts in conventional relational algebra. We do not include the non-conventional operations in this discussion. In their model relations must have deterministic keys. As a result, the projection operation requires that the key attribute(s) be included in the projection. Further, since probabilities are stored at the attribute-level, and there exists only one tuple for any key value, the operation cannot lead to value-equivalent tuples. When a projection includes a subset of dependent stochastic attributes, the marginal probabilities are returned for the projected attributes. If wildcards denoting missing probabilities are involved, then the wild-cards are treated as just another attribute-value. Two types of conditions are provided for the selection operation: the certainty condition and the possibility condition. These two types of conditions exploit the semantics of the missing probabilities. A query with a certainty condition selects tuples that are guaranteed to meet the selection criteria regardless of how the missing probabilities may be assigned. A query with a possibility condition selects tuples for which there exists some feasible assignment of the missing probabilities that would lead that tuple to meet the selection criteria. When the selection condition involves a subset of stochastically dependent non-key attributes, it involves an implicit projection operation. Their natural join operation requires that the common attribute(s) must be the key to one of the relations. Since attributes in a relation are assumed to be conditionally independent of attributes in other relations given the key value, the probability distribution for stochastic attribute values in the result of the join are obtained by multiplying the probabilities associated with the participating attributes in the two relations.

4.3 Models Assigning Interval-Valued Probability Measures at the Attribute Level

In their model, Lakshmanan et al [14] define the operations on the annotated representations of their probabilistic relations. The annotated representation includes, in addition to the possible attribute values, the upper and lower bounds for the probability associated with the set of specified attribute values, and the path (which is a Boolean expression involving world-ids). As mentioned earlier, the operations manipulate the attribute values, the probability intervals, as well as the paths.

The *selection* operation in their algebra is practically identical to that in traditional relational algebra, with the bounds and the path associated with each tuple included in the result (which is also annotated). Their *projection* operation does not eliminate value-equivalent tuples. As a result, value-equivalent tuples may appear in the result; these tuples are distinguished by their associated paths and the probability bounds.

Their *Cartesian product* and *join* operations are not straightforward extensions of the classical relational algebra operations. Their definitions incorporate several possible strategies for combining probabilistic tuples that involve conjunctions and disjunctions of events. Thus, in concatenating tuple t_1 from relation R_1 with tuple t_2 from relation R_2, the probability interval associated with the result depends on whatever relationship is known among tuples t_1 and t_2. For instance, if the tuples are assumed to be independent, the probability interval for the result is different from the scenario where the tuples are assumed to be positively correlated. In the former scenario, the bounds for the result are obtained as the products of the respective bounds on the participating tuples. In the latter scenario, the bounds correspond to the minimum values of the bounds of the participating tuples. With such scenarios in mind, they define a generic concatenation operation on tuples, with the restriction that the user should specify a probabilistic strategy that satisfies several proposed postulates on the structure and semantics of computing concatenations of tuples (i.e., conjunctions of events). The path information also plays a role if necessary; e.g., if the participating tuples correspond to inconsistent states of the world (one event is the negation of the other), then the resulting tuple is not included. When performing a Cartesian product, the user is provided the flexibility to specify which strategy to use.

To handle value-equivalent tuples, they propose an operation called compaction, which is intuitively similar to the coalescence operations defined in [5]. The compaction operation uses a disjunctive combination strategy for evaluating the probability intervals and paths for the resulting tuple. They propose a generic disjunction strategy to accommodate different assumptions regarding the value-equivalent tuples.

Their *union* operation is analogous to the traditional union operation, again with the paths and probability bounds used to distinguish value-equivalent tuples. Their *difference* operation, on the other hand, explicitly takes into account value-equivalent tuples. The basic intuition behind calculating the probability bounds involves taking the conjunction of the event associated with the tuple in the first relation and the negation of the event associated with the value-equivalent tuple in the second relation. This is complicated by the fact that multiple value-equivalent tuples may appear in both the participating relations, each associated with a different path. Since one of these paths may be a subset of another path, special checks are developed that determine which tuples are subsumed in this manner. The final output includes tuples associated with paths that are not subsumed.

The actual computation of the probability intervals depend on the assumptions being specified by the user. For some assumptions, the values can be easily obtained. In other situations, linear programs are used to compute the tightest possible bounds, given the available information.

4.4 Models Assigning Interval-Valued Probability Measures to Tuples

Eiter et al [9] generalize the annotated representation of the interval-valued model presented in [14] by allowing complex values. In their model, each tuple is assigned an event, which is analogous to a path in the model of Lakshmanan et al [14]. An important distinction is that the combination strategies in their algebra are based on axioms of probability theory, instead of the postulates for combination functions suggested in [14]. In their conjunction and disjunction strategies, they incorporate additional types of dependence information such as left implication, right implication, exhaustion, and antivalence, which have not been considered in prior works. They are able to identify combinations of probabilistic pairs of events and dependence information that are unsatisfiable, which enables them to refine the combination strategies presented by Lakshmanan et al. In certain cases, their approach is able to obtain tighter bounds on the probability intervals.

While the work by Eiter et al [9] is motivated by Lakshmanan et al [14], the resulting representations they consider are very similar to the ones in [5], with the difference that interval-valued probabilities are considered. Their operations are, of course, generalized to allow for the different combination strategies, which lead to several possible ways to compute the probability intervals for tuples in derived relations.

4.5 Some Observations on the Independence Assumption Across Tuples

The various probabilistic relational models can also be viewed as belonging to one of two groups. In the first group, it is assumed (explicitly or implicitly) that tuples corresponding to different objects are independent. In the other, no such assumptions are made. The former include the models presented by Cavallo and Pittarelli [3], Pittarelli [17], Dey and Sarkar [5], and Barbara et al [1]. The models proposed by Fuhr and Rölleke [11], Lakshmanan et al [14], and Eiter et al [9] belong to the latter group. An outcome of relaxing this assumption is that, in order to compute the probabilities (point-valued or interval-valued) associated with tuples in the result of operations, it is necessary to keep track of additional information regarding how the tuple was derived. Fuhr and Rölleke use information in the form of complex events, Lakshmanan et al use paths, and Eiter et al use derived events to store these kinds of information along with the relations. In addition, they assume that additional knowledge about these dependencies is available in order to make the right probabilistic inferences when computing the result. Consequently, the models in the latter group, while providing a more generalized set of options, also impose additional requirements for operational consideration. First, they require that users be able to specify how the dependencies should be considered for tuples within each relation, as well as for tuples across all the relations. Second, as Fuhr and Rölleke observe, even when dependence models may be more appropriate, the additional parameters needed would often not be available to compute the desired probabilities. They go on to note that complete probability assignments will not be feasible for a database with a reasonable number of events, if every possible dependency is to be accurately captured.

5. Concluding Remarks

Although relational databases enjoy a very wide-spread popularity in modern information systems, they lack the power to model uncertainty in data items. In this chapter, we provide a summary of the major extensions that attempt to overcome this limitation. We discuss the strengths and weaknesses of these models and show the underlying similarities and differences.

Before any of these models can be implemented, a more complete framework needs to be developed that deals with the issues of table structure and normal forms, belief revision, and a non-procedural query language. For the model proposed by Dey and Sarkar [5], these issues have been addressed in a series of follow-up articles [6–8]. To the best of our knowledge, such follow-up work has not been undertaken for the other extensions. Future research could examine these issues for the other models.

Another issue of practical significance is how to obtain the probability distributions for representing the uncertainty associated with data items. Clearly, one comprehensive scheme that works in all situations is unlikely to emerge. Therefore, context-driven schemes need to be devised. In a recent article, Jiang et al [12] examine this issue in the context of heterogeneous data sources. Future research need to examine other contexts that lead to data uncertainty and develop appropriate schemes for those contexts as well.

References

[1] D. Barbara, H. Garcia-Molina, and D. Porter. The Management of Probabilistic Data. *IEEE Transactions on Knowledge and Data Engineering*, 4(5):487–502, October 1992.

[2] J. Bischoff and T. Alexander. *Data Warehouse: Practical Advice from the Experts*, Prentice-Hall, 1997.

[3] R. Cavallo and M. Pittarelli. The Theory of Probabilistic Databases. *Proceedings of the 13th VLDB Conference*, pp. 71–81, Brighton, September 1–4 1987.

[4] C.J. Date. *Relational Database: Selected Writings*, Addison-Wesley, 1986.

[5] D. Dey and S. Sarkar. A Probabilistic Relational Model and Algebra. *ACM Transactions on Database Systems*, 21(3):339–369, September 1996.

[6] D. Dey and S. Sarkar. PSQL: A Query Language for Probabilistic Relational Data. *Data and Knowledge Engineering*, 28(1):107–120, October 1998.

[7] D. Dey and S. Sarkar. Modifications of Uncertain Data: A Bayesian Framework for Belief Revision. *Information Systems Research*, 11(1):1–16, March 2000.

[8] D. Dey and S. Sarkar. Generalized Normal Forms for Probabilistic Relational Data. *IEEE Transactions on Knowledge and Data Engineering*, 14(3):485–497, May/June 2002.

[9] T. Eiter, T. Lukasiewicz, and M. Walter. A Data Model and Algebra for Probabilistic Complex Values. *Annals of Mathematics and Artificial Intelligence*, 33(2–4):205–252, December 2001.

[10] N. Fuhr and C. Buckley. A Probabilistic Learning Approach for Document Indexing. *ACM Transactions on Information Systems*, 9(3):223–248, July 1991.

[11] N. Fuhr and T. Rölleke. A Probabilistic Relational Algebra for the Integration of Information Retrieval and Database Systems. *ACM Transactions on Information Systems*, 15(1):32–66, January 1997.

[12] Z. Jiang, S. Sarkar, P. De, and D. Dey. A Framework for Reconciling Attribute Values from Multiple Data Sources. *Management Science*, 53(12):1946–1963, December 2007.

[13] G.J. Klir T.A. and Folger. *Fuzzy Sets, Uncertainty, and Information*, Prentice Hall, Englewood Cliffs, NJ, 1988.

[14] L. Lakshmanan, N. Leone, R. Ross, and V.S. Subrahmanian. Probview: A Flexible Probabilistic Database System. *ACM Transactions on Database Systems*, 22(3):419–469, September 1997.

[15] D. Maier. *The Theory of Relational Databases*, Computer Science Press, 1983.

[16] A. Motro. Accommodating Imprecision in Database Systems: Issues and Solutions. *ACM SIGMOD Record*, 19(4):69–74, December 1990.

[17] M. Pittarelli. An Algebra for Probabilistic Databases, *IEEE Transactions on Knowledge and Data Engineering*, 6(2):293–303, April 1994.

[18] G. Salton and C. Buckley. Term Weighting Approaches in Automatic Text Retrieval. *Information Processing & Management*, 24(5):513–523, 1988.

Chapter 4

GRAPHICAL MODELS FOR UNCERTAIN DATA

Amol Deshpande
University of Maryland, College Park, MD
amol@cs.umd.edu

Lise Getoor
University of Maryland, College Park, MD
getoor@cs.umd.edu

Prithviraj Sen
University of Maryland, College Park, MD
sen@cs.umd.edu

Abstract Graphical models are a popular and well-studied framework for compact representation of a joint probability distribution over a large number of interdependent variables, and for efficient reasoning about such a distribution. They have been proven useful in a wide range of domains from natural language processing to computer vision to bioinformatics. In this chapter, we present an approach to using graphical models for managing and querying large-scale uncertain databases. We present a unified framework based on the concepts from graphical models that can model not only tuple-level and attribute-level uncertainties, but can also handle arbitrary correlations that may be present among the data; our framework can also naturally capture *shared correlations* where the same uncertainties and correlations occur repeatedly in the data. We develop an efficient strategy for query evaluation over such probabilistic databases by casting the query processing problem as an *inference* problem in an appropriately constructed graphical model, and present optimizations specific to probabilistic databases that enable efficient query evaluation. We conclude the chapter with a discussion of related and future work on these topics.

Keywords: Graphical models; probabilistic databases; inference; first-order probabilistic models.

1. Introduction

An increasing number of real-world applications are demanding support for managing, storing, and querying uncertain data in relational database systems. Examples include data integration [14], sensor network applications [22], information extraction systems [34], mobile object tracking systems [11] and others. Traditional relational database management systems are not suited for storing or querying uncertain data, or for reasoning about the uncertainty itself – commonly desired in these applications. As a result, numerous approaches have been proposed to handle uncertainty in databases over the years [32, 10, 24, 26, 4, 39, 11, 22, 14, 8, 6]. However, most of these approaches make simplistic and restrictive assumptions concerning the types of uncertainties that can be represented. In particular, many of the proposed models can only capture and reason about tuple-level existence uncertainties, and cannot be easily extended to handle uncertain attribute values which occur naturally in many domains. Second, they make highly restrictive independence assumptions and cannot easily model correlations among the tuples or attribute values.

Consider a simple car advertisement database (Figure 4.1) containing information regarding pre-owned cars for sale, culled from various sources on the Internet. By its very nature, the data in such a database contains various types of uncertainties that interact in complex ways. First off, we may have uncertainty about the validity of a tuple – older ads are likely to correspond to cars that have already been sold. We may represent such uncertainty by associating an *existence probability* (denoted $prob_e$) with each tuple. Second, many of the attribute values may not be known precisely. In some cases, we may have an explicit probability distribution over an attribute value instead (e.g. the *SellerID* attribute for Ad 103 in Figure 4.1(a)). More typically, we may have a joint probability distribution over the attributes, and the uncertainty in the attribute values for a specific tuple may be computed using the known attribute values for that tuple. Figure 4.1(d) shows such a joint probability distribution over the attributes *make, model* and *mpg*; this can then be used to compute a distribution over the *mpg* attribute for a specific tuple (given the tuple's *make* and/or *model* information). Finally, the data may exhibit complex attribute-level or tuple-level correlations. For instance, since the ads 101 and 102 are both entered by the same seller, their validity is expected to be highly correlated; such a correlation may be represented using a joint probability distribution as shown in Figure 4.1(c).

Many other application domains naturally produce correlated data as well [52]. For instance, data integration may result in relations containing duplicate tuples that refer to the same *entity*; such tuples must be modeled as *mutually exclusive* [10, 1]. Real-world datasets such as the Christmas Bird Count [16] naturally contain complex correlations among tuples. Data generated by sen-

<table>
<tr><th>Ad</th><th colspan="2">SellerID</th><th>Date</th><th colspan="2">Type</th><th>Model</th><th>mpg</th><th>Price</th><th>prob_e</th></tr>
<tr><td>101</td><td colspan="2">201</td><td>1/1</td><td colspan="2">Sedan</td><td>Civic(EX)</td><td>?</td><td>$6000</td><td>0.5</td></tr>
<tr><td>102</td><td colspan="2">201</td><td>1/10</td><td colspan="2">Sedan</td><td>Civic(DX)</td><td>?</td><td>$4000</td><td>0.45</td></tr>
<tr><td rowspan="3">103</td><td>-</td><td>prob</td><td rowspan="3">1/15</td><td>-</td><td>prob</td><td rowspan="3">Civic</td><td rowspan="3">?</td><td rowspan="3">$12000</td><td rowspan="3">0.8</td></tr>
<tr><td>201</td><td>0.6</td><td>Sedan</td><td>0.3</td></tr>
<tr><td>202</td><td>0.4</td><td>Hybrid</td><td>0.7</td></tr>
<tr><td>104</td><td colspan="2">202</td><td>1/1</td><td colspan="2">Hybrid</td><td>Civic</td><td>?</td><td>$20000</td><td>0.2</td></tr>
<tr><td>105</td><td colspan="2">202</td><td>1/1</td><td colspan="2">Hybrid</td><td>Civic</td><td>?</td><td>$20000</td><td>0.2</td></tr>
</table>

(a) Advertisements

SellerID	Reputation
201	Shady
202	Good

(b) Sellers

Ad 101	Ad 102	*prob*
valid	valid	0.4
valid	invalid	0.1
invalid	valid	0.05
invalid	invalid	0.45

(c)

Type	Model	mpg	*prob*
Sedan	Civic(EX)	26	0.2
		28	0.6
		30	0.2
	Civic(DX)	32	0.1
		35	0.7
		37	0.2
	Civic	28	0.4
		35	0.6
Hybrid	Civic	45	0.4
		50	0.6

(d)

Figure 4.1. (a,b) A simple car advertisement database with two relations, one containing uncertain data; (c) A joint probability function (*factor*) that represents the correlation between the validity of two of the ads ($prob_e$ for the corresponding tuples in the *Advertisements* table can be computed from this); (d) A *shared* factor that captures the correlations between several attributes in *Advertisements* – this can be used to obtain a probability distribution over missing attribute values for any tuple.

sor networks is typically highly correlated, both in time and space [22]. Finally, data generated through the application of a machine learning technique (e.g. classification labels) typically exhibits complex correlation patterns. Furthermore, the problem of handling correlations among tuples arises naturally during query evaluation *even when one assumes that the base data tuples are independent*. In other words, the independence assumption is not closed under the relational operators, specifically *join* [26, 14].

In this chapter, we present a framework built on the foundations of probabilistic graphical models that allows us to uniformly handle uncertainties and correlations in the data, while keeping the basic probabilistic framework simple and intuitive. The salient features of our proposed framework are as follows:

- Our framework enables us to uniformly represent both tuple-level and attribute-level uncertainties and correlations through the use of *conditional probability distributions* and *joint probability factors*. Our proposed model is based on the commonly-used *possible world semantics* [26, 14], and as a result, every relational algebra query has precise and clear semantics on uncertain data.

- Our framework can represent and exploit recurring correlation patterns (called *shared factors*) that are common in many application domains and are also manifested during the query evaluation process itself (Figure 4.1(d) shows one such shared factor).

- We show how to cast query evaluation on probabilistic databases as an *inference* problem in probabilistic graphical models and develop techniques for efficiently constructing such models during query processing. This equivalence not only aids in our understanding of query evaluation on uncertain databases, but also enables transparent technology transfer by allowing us to draw upon the prior work on inference in the probabilistic reasoning community. In fact several of the novel inference algorithms we develop for query evaluation over probabilistic databases are of interest to the probabilistic reasoning community as well.

Our focus in this chapter is on management of large-scale uncertain data using probabilistic graphical models. We differentiate this from the dual problem of casting *inference in probabilistic graphical models* as *query evaluation* in an appropriately designed database (where the conditional probability distributions are stored as database relations) [9]. We revisit this issue in Section 5, along with several other topics such as probabilistic relational models and the relationship between our approach and other probabilistic query evaluation approaches.

The rest of the paper is organized as follows. We begin with a brief overview of graphical models (Section 2); we focus on representation and inference, and refer the reader to several texts on machine learning [44, 13, 35, 27] for learning and other advanced issues. We then present our framework for representing uncertain data using these concepts (Section 3). Next we develop an approach to cast query processing in probabilistic databases as an inference problem, and present several techniques for efficient inference (Section 4). We conclude with a discussion of related topics such as probabilistic relational models, safe plans, and lineage-based approaches (Section 5).

2. Graphical Models: Overview

Probabilistic graphical models (PGMs) comprise a powerful class of approaches that enable us to compactly represent and efficiently reason about very large joint probability distributions [44, 13]. They provide a principled approach to dealing with the uncertainty in many application domains through the use of probability theory, while effectively coping with the computational and representational complexity through the use of graph theory. They have been proven useful in a wide range of domains including natural language processing, computer vision, social networks, bioinformatics, code design, sensor networks, and unstructured data integration to name a few. Techniques from graphical models literature have also been applied to many topics directly of interest to the database community including information extraction, sensor data analysis, imprecise data representation and querying, selectivity estimation for query optimization, and data privacy.

At a high level, our goal is to efficiently represent and operate upon a joint distribution Pr over a set of random variables $\mathcal{X} = \{X_1, \ldots, X_n\}$. Even if these variables are binary-valued, a naive representation of the joint distribution requires the specification of 2^n numbers (the probabilities of the 2^n different assignments to the variables), which would be infeasible except for very small n. Fortunately, most real-world application domains exhibit a high degree of structure in this joint distribution that allows us to factor the representation of the distribution into modular components. More specifically, PGMs exploit *conditional independences* among the variables:

DEFINITION 2.1 *Let* $\mathbf{X}$, $\mathbf{Y}$, *and* $\mathbf{Z}$ *be sets of random variables.* $\mathbf{X}$ *is* conditionally independent *of* $\mathbf{Y}$ *given* $\mathbf{Z}$ *(denoted* $\mathbf{X} \perp \mathbf{Y} | \mathbf{Z}$*) in distribution* Pr *if:*

$$\Pr(\mathbf{X} = \mathbf{x}, \mathbf{Y} = \mathbf{y} | \mathbf{Z} = \mathbf{z}) = \Pr(\mathbf{X} = \mathbf{x} | \mathbf{Z} = \mathbf{z}) \Pr(\mathbf{Y} = \mathbf{y} | \mathbf{Z} = \mathbf{z})$$

for all values $\mathbf{x} \in dom(\mathbf{X})$, $\mathbf{y} \in dom(\mathbf{Y})$ *and* $\mathbf{z} \in dom(\mathbf{Z})$.

A graphical model consists of two components: (1) A graph whose nodes are the random variables and whose edges connect variables that interact directly; variables that are not directly connected are conditionally independent given some combination of the other variables. (2) A set of small functions called *factors* each over a subset of the random variables.

DEFINITION 2.2 *A* factor $f(\mathbf{X})$ *is a function over a (small) set of random variables* $\mathbf{X} = \{X_1, \ldots, X_k\}$ *such that* $f(\mathbf{x}) \geq 0 \ \forall \mathbf{x} \in dom(X_1) \times \ldots \times dom(X_k)$.

The set of factors that can be associated with a graphical model is constrained by the nature (undirected vs directed) and the structure of the graph as we will see later. Note that it is not required that $f(\mathbf{x})$ be ≤ 1; in other words, factors are not required to be (but can be) probability distributions.

DEFINITION 2.3 *A* probabilistic graphical model *(PGM)* $\mathcal{P} = \langle \mathcal{F}, \mathcal{X} \rangle$ *defines a joint distribution over the set of random variables* $\mathcal{X}$ *via a set of factors* $\mathcal{F}$, *each defined over a subset of* $\mathcal{X}$. *Given a complete joint assignment* $\mathbf{x} \in dom(X_1) \times \cdots \times dom(X_n)$ *to the variables in* $\mathcal{X}$, *the joint distribution is defined by:*

$$\Pr(\mathbf{x}) = \frac{1}{\mathcal{Z}} \prod_{f \in \mathcal{F}} f(\mathbf{x}_f)$$

where $\mathbf{x}_f$ *denotes the assignments restricted to the arguments of* f *and* $\mathcal{Z} = \sum_{\mathbf{x}'} \prod_{f \in \mathcal{F}} f(\mathbf{x}'_f)$ *is a normalization constant.*

The power of graphical models comes from the graphical representation of factors that makes it easy to understand, reason about, and operate upon them. Depending on the nature of the interactions between the variables, there are two popular classes of graphical models, *Bayesian networks (directed models)*, and *Markov networks (undirected models)*. These differ in the family of probability distributions they can represent, the set of factorizations they allow, and the way in which the interactions are quantified along the edges. We discuss these briefly in turn.

2.1 Directed Graphical Models: Bayesian Networks

Directed graphical models, popularly known as Bayesian networks, are typically used to represent causal or asymmetric interactions amongst a set of random variables. A directed edge from variable X_i to variable X_j in the graph (which must be *acyclic*) is used to indicate that X_i directly influences X_j. A canonical set of conditional independences encoded by a directed graphical model is obtained as follows: a node X_j is independent of all its non-descendants given the values of its parents. In other words, if X_i is not a descendant or a parent of X_j, then $X_i \perp X_j | parents(X_j)$. The rest of the conditional independences encoded by the model can be derived from these.

The probability distribution that a directed graphical model represents can be factorized as follows:

$$\Pr(X_1, \ldots, X_n) = \prod_{i=1}^{n} \Pr(X_i | \text{parents}(X_i))$$

In other words, each of the factors associated with a Bayesian network is a conditional probability distribution (CPD) over a node given its parents in the graph.

Figure 4.2 shows a simple example Bayesian network that models the *location, age, degree, experience*, and *income* of a person. In this application domain, we might model the *location* to be independent from the rest of the variables (as captured by not having any edges to or from the corresponding node in the graph). For simplicity, we also model the *age* and *degree* to be independent from each other if no other information about the person is known. Although *income* is influenced by *degree, age*, and *experience*, in most cases, the influence from *age* will be indirect, and will disappear given the *experience* of the person; in other words, once the *experience* level of a person is known, the *age* does not provide any additional information about the *income*. This is modeled by not having any direct edge from *age* to *income*. The figure also shows the factors that will be associated with such a Bayesian network

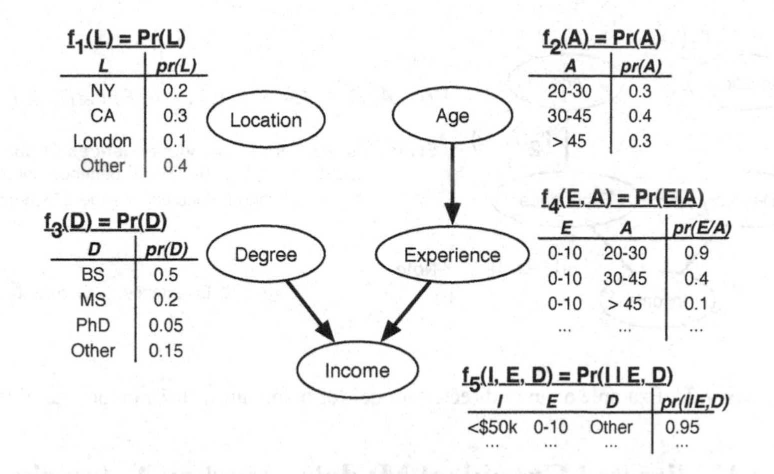

Figure 4.2. Example of a directed model for a domain with 5 random variables

(one CPD each corresponding to each node), and the expression for the joint probability distribution as a product of the factors.

A domain expert typically chooses the edges to be added to the model, although the graph could also be learned from a training dataset. A sparse graph with few edges leads to more compact representation and (typically) more efficient inference, but a denser graph might be required to capture all the interactions between the variables faithfully.

The compactness of representing a joint probability distribution using a Bayesian network is evident from the above example. If each of the variables has domain of size 10, the size of the joint pdf will be 10^5, whereas the number of probabilities required to store the factors as shown in the figure is only about 1000, an order of magnitude reduction.

Since Bayesian networks are easy to design, interpret and reason about, they are extensively used in practice. Some popular examples of Bayesian networks include Hidden Markov Models [47, 56], Kalman Filters [37, 57], and QMR networks [40, 33].

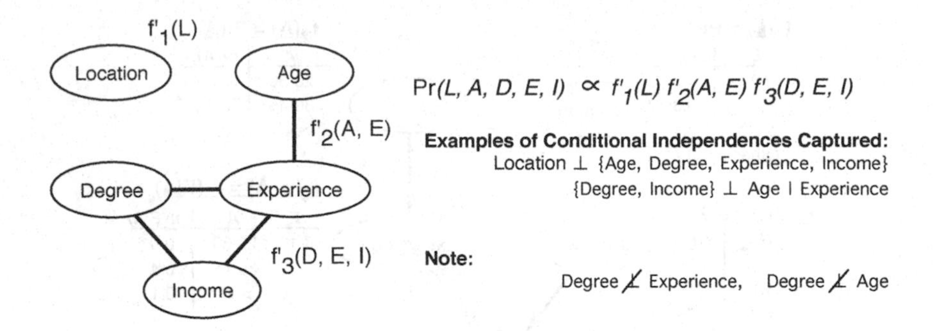

Figure 4.3. Example of an undirected model for a domain with 5 random variables

2.2 Undirected Graphical Models: Markov Networks

Undirected graphical models, or Markov Networks, are useful for representing distributions over variables where there is no natural directionality to the influence of one variable over another and where the interactions are more symmetric. Examples include the interactions between atoms in a molecular structure, the dependencies between the labels of pixels of an image, or the interactions between environmental properties sensed by geographically co-located sensors [22]. Markov networks are sometimes preferred over Bayesian networks because they provide a simpler model of independences between variables.

The probability distribution represented by a Markov network factorizes in a somewhat less intuitive manner than Bayesian networks; in many cases, the factors may only indicate the relative compatibility of different assignments to the variables, but may not have any straightforward probabilistic interpretation. Let G be the undirected graph over the random variables $\mathcal{X} = \{X_1, \ldots, X_n\}$ corresponding to a Markov network, and let $\mathcal{C}$ denote the set of cliques (complete subgraphs) of G. Then the probability distribution represented by the Markov network factorizes as follows:

$$\Pr(X_1, \ldots, X_n) = \frac{1}{Z} \prod_{C \in \mathcal{C}} f_C(X_C)$$

where $f_C(X_C)$ are the factors (also called *potential functions*) each over a complete subgraph of G. $Z = \sum_X \prod_{C \in \mathcal{C}} f_C(X_C)$ is the normalization constant.

Figure 4.3 shows an example Markov network over the same set of random variables as above. The maximal complete subgraphs of the network are $\{Location\}, \{Degree, Experience, Income\}, \{Age, Experience\}$ and factors may be defined over any of these sets of random variables, or their subsets.

The conditional independences captured by a Markov network are determined as follows: if a set of nodes $\mathbf{X}$ separates sets of nodes $\mathbf{Y}$ and $\mathbf{Z}$ (i.e., if by removing the nodes in $\mathbf{X}$, there are no paths between a node in $\mathbf{Y}$ and a node in $\mathbf{Z}$), then $\mathbf{Y}$ and $\mathbf{Z}$ are conditionally independent given $\mathbf{X}$. Figure 4.3 also shows the conditional independences captured by our example network.

An important subclass of undirected models is the class of *decomposable models* [20]. In a decomposable model, the graph is constrained to be *chordal* (*triangulated*) and the factors are the joint probability distributions over the maximal cliques of the graph. These types of models have many desirable properties such as closed product form factorizations that are easy to compute and reason about [21]. Further, these bear many similarities to the notion of *acyclic database schemas* [5].

2.3 Inference Queries

Next we consider the main types of tasks (queries) that are commonly performed over the model. The most common query type is the *conditional probability query*, $\Pr(\mathbf{Y} \mid \mathbf{E} = \mathbf{e})$. Such a query consists of two parts: (1) the *evidence*, a subset $\mathbf{E}$ of random variables in the network, and an instantiation $\mathbf{e}$ to these variables; and (2) the *query*, a subset $\mathbf{Y}$ of random variables in the network. Our task is to compute $\Pr(\mathbf{Y} \mid \mathbf{E} = \mathbf{e}) = \frac{\Pr(\mathbf{Y},\mathbf{e})}{\Pr(\mathbf{e})}$, i.e., the probability distribution over the values $\mathbf{y}$ of $\mathbf{Y}$, conditioned on the fact that $\mathbf{E} = \mathbf{e}$.

A special case of conditional probability queries is simply *marginal computation queries*, where we are asked to compute the marginal probability distribution $\Pr(\mathbf{Y})$ over a subset of variables $\mathbf{Y}$.

Another type of query that often arises, called *maximum a posteriori (MAP)*, is finding the most probable assignment to some subset of variables. As with conditional probability queries, we are usually given evidence $\mathbf{E} = \mathbf{e}$, and a set of query variables, $\mathbf{Y}$. In this case, however, our goal is to compute the *most likely assignment* to $\mathbf{Y}$ given the evidence $\mathbf{E} = \mathbf{e}$, i.e.:

$$argmax_{\mathbf{y}} \Pr(\mathbf{y}, \mathbf{e})$$

where, in general, $argmax_x f(x)$ represents the value of x for which $f(x)$ is maximal. Note that there might be more than one assignment that has the highest posterior probability. In this case, we can either decide that the MAP task is to return the set of possible assignments, or to return an arbitrary member of that set.

A special variant of this class of queries is the *most probable explanation (MPE)* queries. An MPE query tries to find the most likely assignment to all of the (non-evidence) variables, i.e., $\mathbf{Y} = \mathbf{X} - \mathbf{E}$. MPE queries are somewhat easier than MAP queries, which are much harder to answer than the other tasks; this is because MAP queries contain both summations and maximizations, thus combining the elements of both conditional probabilities queries and MPE queries.

The simplest way to use the graphical model to answer any of these queries is: (1) generate the joint probability distribution over all the variables, (2) condition it using the evidence (generating another joint pdf), and then (3) sum over the unneeded variables (in the case of a conditional probability query) or search for the most likely entry (in the case of an MPE query). For example, consider the example shown in Figure 4.2, and lets say we want to compute the marginal probability distribution corresponding to $income$ (I). This distribution can be obtained from the full joint distribution by summing out the rest of the variables:

$$\begin{aligned} \Pr(I) &= \Sigma_{L,A,D,E} \Pr(I, L, A, D, E) \\ &= \Sigma_{L,A,D,E}\ f_1(L) f_2(A) f_3(D) f_4(E, A) f_5(I, E, D) \end{aligned}$$

However, this approach is not very satisfactory and is likely to be infeasible in most cases, since it results in an exponential space and computational blowup that the graphical model representation was designed to avoid. In general, the exact computation of either of the inference tasks is #P-complete. However, many graphical models that arise in practice have certain properties that allow efficient probabilistic computation [59]. More specifically, the problem can be solved in polynomial time for graphical models with bounded tree-width [50].

Variable elimination (VE) [59, 19], also known as *bucket elimination*, is an exact inference algorithm that has the ability to exploit this structure. Intuitively variable elimination specifies the order in which the variables are summed out (eliminated) from the above expression; eliminating a variable requires multiplying all factors that contain the variable, and then summing out the variable. Say we chose the order: L, A, D, E, then the computation is as follows (the expression evaluated in each step is underlined, and its result is bold-faced in the next step):

$$\begin{aligned} \Pr(I) &= \Sigma_{L,A,D,E} f_1(L) f_2(A) f_3(D) f_4(E, A) f_5(I, E, D) \\ &= \Sigma_E(\Sigma_D f_5(I, E, D) f_3(D)\ (\Sigma_A f_2(A) f_4(E, A)\ \underline{(\Sigma_L f_1(L))})) \\ &= \Sigma_E(\Sigma_D f_5(I, E, D) f_3(D)\ \underline{(\Sigma_A f_2(A) f_4(E, A))}) \\ &= \Sigma_E\underline{(\Sigma_D f_5(I, E, D) f_3(D))}\, \mathbf{g_1(E)} \\ &= \underline{\Sigma_E\, \mathbf{g_2(I, E)} g_1(E)} \\ &= \mathbf{g_3(I)} \end{aligned}$$

The order in which the variables are summed out is known as the *elimination order*, and the cost of running VE depends on the choice of the elimination order. Even though finding the optimal ordering is NP-hard [2] (this is closely related to the problem of finding the optimal triangulation of a graph), good heuristics are available [7, 18].

Another popular algorithm for exact inference is the *junction tree* algorithm [13, 30]. A junction tree is an efficient data structure for reusing work for several inference queries on the same graph. Once a junction tree is constructed, we can provide exact answers to inference queries over any subset of variables in the same clique by running the sum-product message passing or belief propagation algorithms. The message passing algorithm runs in time that is linear in the number of cliques in the tree and exponential in the size of the largest clique in the tree (which is same as the tree-width of the model).

However, many real-life graphical models give rise to graphs with large tree-widths, and the design of good approximation schemes in such cases is an active topic of research in the statistics and probabilistic reasoning communities. The most commonly used techniques include methods based on belief propagation (e.g. loopy belief propagation [42]), sampling-based techniques (e.g. Gibbs sampling, particle filters [3, 38]) and variational approximation methods [36] to name a few. We refer the reader to [35] for further details.

3. Representing Uncertainty using Graphical Models

We are now ready to define a probabilistic database in terms of a PGM. The basic idea is to use random variables to depict the uncertain attribute values and factors to represent the uncertainty and the correlations. Let R denote a probabilistic relation or simply, relation, and let $attr(R)$ denote the set of attributes of R. A relation R consists of a set of probabilistic tuples or simply, tuples, each of which is a mapping from $attr(R)$ to random variables. Let $t.a$ denote the random variable corresponding to tuple $t \in R$ and attribute $a \in attr(R)$. Besides mapping each attribute to a random variable, every tuple t is also associated with a boolean-valued random variable which captures the existence uncertainty of t and we denote this by $t.e$.

DEFINITION 3.1 *A* probabilistic database *or simply, a* database, $\mathcal{D}$ *is a pair* $\langle \mathcal{R}, \mathcal{P} \rangle$ *where* $\mathcal{R}$ *is a set of relations and* $\mathcal{P}$ *denotes a PGM defined over the set of random variables associated with the tuples in* $\mathcal{R}$.

Figure 4.4(a) shows a small two-relation database that we use as a running example. In this database, every tuple has an uncertain attribute (the **B** attributes) and these are indicated in Figure 4.4(a) by specifying the probabilities with which each attribute takes the assignments from its domain. In our proposed framework, we represent this uncertainty by associating a random variable with each of the uncertain attributes, and by using factors to capture the corresponding probability distributions and correlations if present.

For instance, s_2.**B** can be assigned the value 1 with probability 0.6 and the value 2 with probability 0.4 and we would represent this using the factor

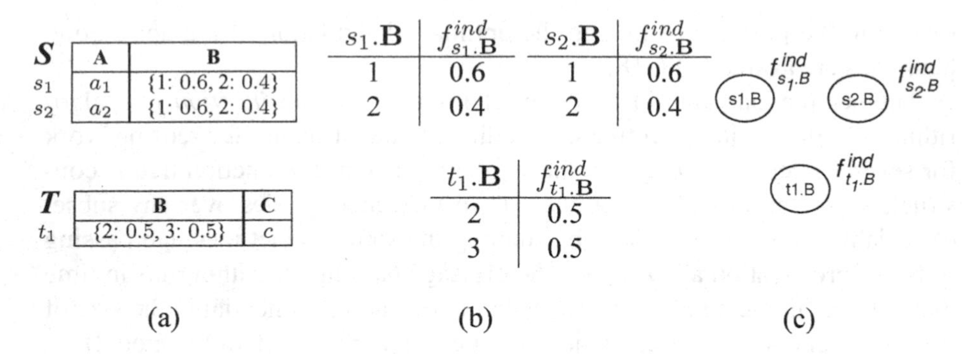

S

	A	B
s_1	a_1	{1: 0.6, 2: 0.4}
s_2	a_2	{1: 0.6, 2: 0.4}

T

	B	C
t_1	{2: 0.5, 3: 0.5}	c

(a)

$s_1.\mathbf{B}$	$f^{ind}_{s_1.\mathbf{B}}$
1	0.6
2	0.4

$s_2.\mathbf{B}$	$f^{ind}_{s_2.\mathbf{B}}$
1	0.6
2	0.4

$t_1.\mathbf{B}$	$f^{ind}_{t_1.\mathbf{B}}$
2	0.5
3	0.5

(b)

Figure 4.4. (a) A small database with uncertain attributes. For ease of exposition, we show the marginal pdfs over the attribute values in the table; this information can be derived from the factors. (b) Factors corresponding to the database assuming complete independence. (c) Graphical representation of the factors.

$f_{s_2.\mathbf{B}}$ shown in Figure 4.4(b). We show all three required factors $f_{s_1.\mathbf{B}}(s_1.\mathbf{B})$, $f_{s_2.\mathbf{B}}(s_2.\mathbf{B})$ and $f_{t_1.\mathbf{B}}(t_1.\mathbf{B})$ in Figure 4.4(b). Here we assume that the attributes are independent of each other. If, for instance, $s_2.\mathbf{B}$ and $t_1.\mathbf{B}$ were correlated, we would capture that using a factor $f_{t_1.\mathbf{B},s_2.\mathbf{B}}(t_1.\mathbf{B}, s_2.\mathbf{B})$ (detailed example below).

In addition to the random variables which denote uncertain attribute values, we can introduce tuple *existence* random variables $s_1.e$, $s_2.e$, and $t_1.e$, to capture tuple uncertainty. These are boolean-valued random variables and can have associated factors. In Figure 4.4, we assume the tuples are certain, so we don't show the existence random variables for the base tuples.

3.1 Possible World Semantics

We now define the semantics for our formulation of a probabilistic database. Let $\mathcal{X}$ denote the set of random variables associated with database $\mathcal{D} = \langle \mathcal{R}, \mathcal{P} \rangle$. Possible world semantics define a probabilistic database $\mathcal{D}$ as a probability distribution over deterministic databases (possible worlds) [14] each of which is obtained by assigning $\mathcal{X}$ a joint assignment $\mathbf{x} \in \times_{X \in \mathcal{X}} dom(X)$. The probability associated with the possible world obtained from the joint assignment $\mathbf{x}$ is given by the distribution defined by the PGM $\mathcal{P}$ (Definition 2.3).

For the example shown in Figure 4.4, each possible world is obtained by assigning all three random variables $s_1.\mathbf{B}$, $s_2.\mathbf{B}$ and $t_1.\mathbf{B}$ assignments from their respective domains. Since each of the attributes can take 2 values, there are $2^3 = 8$ possible worlds. Figure 4.5 shows all 8 possible worlds with the corresponding probabilities listed under the column "prob.(ind.)" (indicating the independence assumption). The probability associated with each possible world is obtained by multiplying the appropriate numbers returned by the factors and normalizing if necessary. For instance, for the possible world ob-

possible world	prob (ind.)	prob. (implies)	prob. (diff.)	prob. (pos.corr.)
$D_1 : S = \{(a_1, 1), (a_2, 1)\}$ $T = \{(2, c)\}$	0.18	0.50	0.30	0.06
$D_2 : S = \{(a_1, 1), (a_2, 1)\}$ $T = \{(3, c)\}$	0.18	0.02	0.06	0.30
$D_3 : S = \{(a_1, 1), (a_2, 2)\}$ $T = \{(2, c)\}$	0.12	0	0.20	0.04
$D_4 : S = \{(a_1, 1), (a_2, 1)\}$ $T = \{(3, c)\}$	0.12	0.08	0.04	0.20
$D_5 : S = \{(a_1, 2), (a_2, 1)\}$ $T = \{(2, c)\}$	0.12	0	0	0.24
$D_6 : S = \{(a_1, 2), (a_2, 1)\}$ $T = \{(3, c)\}$	0.12	0.08	0.24	0
$D_7 : S = \{(a_1, 2), (a_2, 2)\}$ $T = \{(2, c)\}$	0.08	0	0	0.16
$D_8 : S = \{(a_1, 2), (a_2, 2)\}$ $T = \{(3, c)\}$	0.08	0.32	0.16	0

Figure 4.5. Possible worlds for example in Figure 4.4(a) and three other different types of correlations.

$$\Pr^{implies}(s_1.\mathbf{B}, s_2.\mathbf{B}, t_1.\mathbf{B}) = f^{implies}_{t_1.\mathbf{B}}(t_1.\mathbf{B}) f^{implies}_{t_1.\mathbf{B}, s_1.\mathbf{B}}(t_1.\mathbf{B}, s_1.\mathbf{B}) f^{implies}_{t_1.\mathbf{B}, s_2.\mathbf{B}}(t_1.\mathbf{B}, s_2.\mathbf{B})$$

$t_1.\mathbf{B}$	$f^{implies}_{t_1.\mathbf{B}}$
2	0.5
3	0.5

$t_1.\mathbf{B}$	$s_1.\mathbf{B}$	$f^{implies}_{t_1.\mathbf{B}, s_1.\mathbf{B}}$
2	1	1
2	2	0
3	1	0.2
3	2	0.8

$t_1.\mathbf{B}$	$s_2.\mathbf{B}$	$f^{implies}_{t_1.\mathbf{B}, s_2.\mathbf{B}}$
2	1	1
2	2	0
3	1	0.2
3	2	0.8

Figure 4.6. Factors for the probabilistic databases with "implies" correlations (we have omitted the normalization constant $\mathcal{Z}$ because the numbers are such that distribution is already normalized)

tained by the assignment $s_1.\mathbf{B} = 1$, $s_2.\mathbf{B} = 2$, $t_1.\mathbf{B} = 2$ (D_3 in Figure 4.5) the probability is $0.6 \times 0.4 \times 0.5 = 0.12$.

Let us now try to modify our example to illustrate how to represent correlations in a probabilistic database. In particular, we will try to construct three different databases containing the following dependencies:

- *implies*: $t_1.\mathbf{B} = 2$ implies $s_1.\mathbf{B} \neq 2$ and $s_2.\mathbf{B} \neq 2$, in other words, $(t_1.\mathbf{B} = 2) \implies (s_1.\mathbf{B} = 1) \wedge (s_2.\mathbf{B} = 1)$.

- *different*: $t_1\mathbf{B}$ and $s_1.\mathbf{B}$ cannot have the same assignment, in other words, $(t_1.\mathbf{B} = 2) \Leftrightarrow (s_1.\mathbf{B} = 1)$ or $(s_1.\mathbf{B} = 2) \Leftrightarrow (t_1.\mathbf{B} = 3)$.

- *positive correlation*: High positive correlation between $t_1.\mathbf{B}$ and $s_1.\mathbf{B}$ – if one is assigned 2 then the other is also assigned the same value with high probability.

Figure 4.5 shows four distributions over the possible worlds that each satisfy one of the above correlations (the columns are labeled with abbreviations of the names of the correlations, e.g., the column for positive correlation is labeled "pos. corr.").

To represent the possible worlds of our example database with the new correlations, we simply redefine the factors in the database appropriately. For example, Figure 4.6 represents the factors for the first case (*implies*). In this case, we use a factor on $t_1.B$ and $s_1.B$ to encode the correlation that $(t_1.\mathbf{B} = 2) \implies (s_1.\mathbf{B} = 1)$. Similarly, a factor on $t_1.B$ and $s_2.B$ is used to encode the other correlation.

Note that in Definition 3.1, we make no restrictions as to which random variables appear as arguments in a factor. Thus, if the user wishes, she may define a factor containing random variables from the same tuple, different tuples, tuples from different relations or tuple existence and attribute value random variables; thus, in our formulation we can express any kind of correlation that one might think of representing in a probabilistic database.

id	A	B
s_1	a_1	$\perp$
s_2	a_2	$\perp$

id	B	C
t_1	$\perp$	c

(a) Base Tables

fid	args	probs
f_1	1	"2,0.5;3,0.5"
f_2	2	"2,1,1;2,2,0 ..."
f_3	2	"2,1,1;2,2,0 ..."

(b) *factors* table

RV	fid	pos
$t_1.B$	$f_1 = f^{implies}_{t_1.B}$	1
$t_1.B$	$f_2 = f^{implies}_{t_1.B,s_1.B}$	1
$s_1.B$	$f_2 = f^{implies}_{t_1.B,s_1.B}$	2
$t_1.B$	$f_3 = f^{implies}_{t_1.B,s_2.B}$	1
$s_2.B$	$f_3 = f^{implies}_{t_1.B,s_2.B}$	2

(c) *factor-rvs* table

Figure 4.7. Representing the factors from Figure 4.6 using a relational database; shared factors can be represented by using an additional level of indirection.

3.2 Shared Factors

In many cases, the uncertainty in the data is defined using general statistics that *do not* vary on a per-tuple basis, and this leads to significant duplication of factors in the probabilistic database. For instance, when combining data from different sources in a data integration scenario, the sources may be assigned data quality values, which may be translated into tuple existence probabilities [1]; all tuples from the same source are then expected to have the same factor associated with them. If the uncertainties are derived from an attribute-level joint probability distribution (as shown in our earlier example in Figure 4.1), then many of the factors are expected to be identical.

Another source of shared correlations in probabilistic databases is the query evaluation approach itself. As we will see in the next section, while evaluating queries we first build an augmented PGM on the fly by introducing small factors involving the base tuples and the intermediate tuples. For instance, if tuples t and t' join to produce intermediate tuple r, we introduce a factor that encodes the correlation that r exists iff both t and t exist (an $\wedge$-factor). More importantly, such a factor is introduced whenever any pair of tuples join, thus leading to repeated copies of the same $\wedge$-factor.

We call such factors *shared factors* and explicitly capture them in our framework; furthermore, our inference algorithms actively identify and exploit such commonalities to reduce the query processing time [53].

3.3 Representing Probabilistic Relations

Earlier approaches represented probabilistic relations by storing uncertainty with each tuple in isolation. This is inadequate for our purpose since the same tuple can be involved in multiple factors, and the same factor can be associated with different sets of random variables. This necessitates an approach where the data and the uncertainty parts are stored separately. Figure 4.7 shows how we store the factors and associate them with the tuples in our current prototype implementation. We use an internal *id* attribute for each relation that is automatically generated when tuples are inserted into the relation; this attribute is used to identify the random variables corresponding to a tuple uniquely. We use $\perp$ to indicate uncertain attribute values (Figure 4.7(a)). Two additional tables are used to store the factors and their associations with the tuple variables:

- *factors:* This table stores a serialized representation of the factor along with some auxiliary information such as the number of arguments.

- *factor-rvs:* This normalized relation stores the association between factors and random variables; the random variables can be of two types: (1) attribute value random variables (e.g. $t_1.B$), or (2) existence random variables (e.g. $t_1.e$). Each row in this table indicates the participation of a random variable in a factor. Since the table is normalized, we also need to store the "position" of the random variable in the factor.

Note that this schema does not exploit shared factors (factors f_2 and f_3 are identical in the above example); they can be easily handled by adding one additional table.

4. Query Evaluation over Uncertain Data

Having defined our representation scheme, we now move our discussion to query evaluation. The main advantage of associating possible world semantics with a probabilistic database is that it lends precise semantics to the query evaluation problem. Given a user-submitted query q (expressed in some standard query language such as relational algebra) and a database $\mathcal{D}$, the result of evaluating q against $\mathcal{D}$ is defined to be the set of results obtained by evaluating q against each possible world of $\mathcal{D}$, augmented with the probabilities of the possible worlds. Relating back to our earlier examples, suppose we want to run the query $q = \prod_{\mathrm{C}}(S \bowtie_{\mathrm{B}} T)$. Figure 4.8(a) shows the set of results obtained from each set of possible worlds, augmented by the corresponding probabilities depending on which database we ran the query against.

Now, even though query evaluation under possible world semantics is clear and intuitive, it is typically not feasible to evaluate a query directly using these semantics. First and foremost among these issues is the size of the result. Since the number of possible worlds is exponential in the number of random variables in the database (to be more precise, it is equal to the product of the domain sizes of all random variables), if every possible world returns a different result, the result size itself will be very large. To get around this issue, it is traditional to compress the result before returning it to the user. One way of doing this is to collect all tuples from the set of results returned by possible world semantics and return these along with the sum of probabilities of the possible worlds that return the tuple as a result [14]. In Figure 4.8(a), there is only one tuple that is returned as a result and this tuple is returned by possible worlds D_3, D_5 and

possible world	query result	prob. (ind.)	prob. (implies)	prob. (diff.)	prob. (pos.corr.)
D_1	$\emptyset$	0.18	0.50	0.30	0.06
D_2	$\emptyset$	0.18	0.02	0.06	0.30
D_3	$\{c\}$	0.12	0	0.20	0.04
D_4	$\emptyset$	0.12	0.08	0.04	0.20
D_5	$\{c\}$	0.12	0	0	0.24
D_6	$\emptyset$	0.12	0.08	0.24	0
D_7	$\{c\}$	0.08	0	0	0.16
D_8	$\emptyset$	0.08	0.32	0.16	0

(a)

query result	$\Pr(D_3)+\Pr(D_5)+\Pr(D_7)$			
	ind.	implies	diff.	pos.corr.
$\{c\}$	0.32	0	0.20	0.40

(b)

Figure 4.8. Results running the query $\prod_{\mathrm{C}}(S \bowtie_{\mathrm{B}} T)$ on example probabilistic databases (Figures 4.4 and 4.5). The query returns a non-empty (and identical) result in possible worlds D_3, D_5, and D_7, and the final result probability is obtained by adding up the probabilities of those worlds.

D_7. In Figure 4.8(b), we show the resulting probabilities obtained by summing across these three possible worlds for each example database.

The second issue is related to the complexity of computing the results of a query from these first principles. Since the number of possible worlds is very large for any non-trivial probabilistic database, evaluating results directly by enumerating all of its possible worlds is going to be infeasible.

To solve this problem, we first make the connection between computing query results for a probabilistic database and the marginal probability computation problem for probabilistic graphical models.

DEFINITION 4.1 *Given a PGM* $\mathcal{P} = \langle \mathcal{F}, \mathcal{X} \rangle$ *and a random variable* $X \in \mathcal{X}$, *the* marginal probability *associated with the assignment* $X = x$, *where* $x \in dom(X)$, *is defined as* $\mu(x) = \sum_{\mathbf{x} \sim x} \Pr(\mathbf{x})$, *where* $\Pr(\mathbf{x})$ *denotes the distribution defined by the PGM and* $\mathbf{x} \sim x$ *denotes a joint assignment to* $\mathcal{X}$ *where* X *is assigned* x.

Since each possible world is obtained by a joint assignment to all random variables in the probabilistic database, there is an intuitive connection between computing marginal probabilities and computing result tuple probabilities by summing over all possible worlds. In the rest of this section, we make this connection more precise. We first show how to augment the PGM underlying the database such that the augmented PGM contains random variables representing result tuples. We can then express the probability computation associated with evaluating the query as a standard marginal probability computation problem; this allows us to use standard probabilistic inference algorithms to evaluate queries over probabilistic databases.

We first present an example to illustrate the basic ideas underlying our approach to augmenting the PGM underlying the database given a query, after that we discuss how to augment the PGM in the general case given any relational algebra query.

4.1 Example

Consider running the query $\prod_{\mathrm{C}}(S \bowtie_{\mathrm{B}} T)$ on the database presented in Figure 4.4(a). Our query evaluation approach is very similar to query evaluation in traditional database systems and is depicted in Figure 4.9. Just as in traditional database query processing, in Figure 4.9, we introduce intermediates tuples produced by the join (i_1 and i_2) and produce a result tuple (r_1) from the projection operation. What makes query processing for probabilistic databases different from traditional database query processing is the fact that we need to preserve the correlations among the random variables representing the intermediate and result tuples and the random variables representing the tuples they were produced from. In our example, there are three such correlations that we need to take care of:

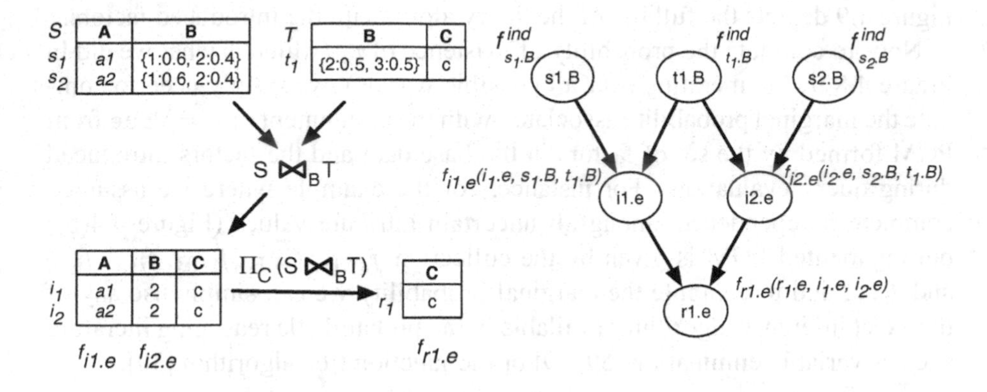

Figure 4.9. Evaluating $\prod_{\mathbf{C}}(S \bowtie_{\mathbf{B}} T)$ on database in Figure 4.4(a).

- i_1 (produced by the join between s_1 and t_1) exists or $i_1.e$ is `true` only in those possible worlds where both $s_1.\mathbf{B}$ and $t_1.\mathbf{B}$ are assigned the value 2.
- Similarly, $i_2.e$ is `true` only in those possible worlds where both $s_2.\mathbf{B}$ and $t_1.\mathbf{B}$ are assigned the value 2.
- Finally, r_1 (the result tuple produced by the projection) exists or $r_1.e$ is `true` only in those possible worlds that produce at least one of i_1 or i_2 or both.

To enforce these correlations, during query evaluation we introduce intermediate factors defined over appropriate random variables. For our example, we introduce the following three correlations:

- For the correlation among $i_1.e$, $s_1.\mathbf{B}$ and $t_1.\mathbf{B}$ we introduce the factor $f_{i_1.e}$ which is defined as:

$$f_{i_1.e}(i_1.e, s_1.\mathbf{B}, t_1.\mathbf{B}) = \begin{cases} 1 & \text{if } i_1.e \Leftrightarrow ((s_1.\mathbf{B} == 2) \wedge (t_1.\mathbf{B} == 2)) \\ 0 & \text{otherwise} \end{cases}$$

- Similarly, for the correlation among $i_2.e$, $s_2.\mathbf{B}$ and $t_1.\mathbf{B}$ we introduce the factor $f_{i_2.e}$ which is defined as:

$$f_{i_2.e}(i_2.e, s_2.\mathbf{B}, t_1.\mathbf{B}) = \begin{cases} 1 & \text{if } i_2.e \Leftrightarrow ((s_2.\mathbf{B} == 2) \wedge (t_1.\mathbf{B} == 2)) \\ 0 & \text{otherwise} \end{cases}$$

- For the correlation among $r_1.e$, $i_1.e$ and $i_2.e$, we introduce a factor $f_{r_1.e}$ capturing the `or` semantics:

$$f_{r_1.e}(r_1.e, i_1.e, i_2.e) = \begin{cases} 1 & \text{if } r_1.e \Leftrightarrow (i_1.e \vee i_2.e) \\ 0 & \text{otherwise} \end{cases}$$

Figure 4.9 depicts the full run of the query along with the introduced factors.

Now, to compute the probability of existence of r_1 (which is what we did in Figure 4.8 by enumerating over all possible worlds), we simply need to compute the marginal probability associated with the assignment $r_1.e = \texttt{true}$ from PGM formed by the set of factors in the base data and the factors introduced during query evaluation. For instance, for the example where we assumed complete independence among all uncertain attribute values (Figure 4.4(b)) our augmented PGM is given by the collection $f_{s1.\mathbf{B}}, f_{s2.\mathbf{B}}, f_{t1.\mathbf{B}}, f_{i1.e}, f_{i2.e}$ and $f_{r_1.e}$, and to compute the marginal probability we can simply use any of the exact inference algorithms available in the probabilistic reasoning literature such as variable elimination [59, 19] or the junction tree algorithm [30].

4.2 Generating Factors during Query Evaluation

Query evaluation for general relational algebra also follows the same basic ideas. In what follows, we modify the traditional relational algebra operators so that they not only generate intermediate tuples but also introduce intermediate factors which, combined with the factors on the base data, provide a PGM that can then be used to compute marginal probabilities of the random variables associated with result tuples of interest. We next describe the modified σ, $\times$, $\prod$, δ, $\cup$, $-$ and γ (aggregation) operators where we use $\emptyset$ to denote a special "null" symbol.

Select: Let $\sigma_c(R)$ denote the query we are interested in, where c denotes the predicate of the select operation. Every tuple $t \in R$ can be jointly instantiated with values from $\times_{a \in attr(R)} dom(t.a)$. If none of these instantiations satisfy c then t does not give rise to any result tuple. If even a single instantiation satisfies c, then we generate an intermediate tuple r that maps attributes from R to random variables, besides being associated with a tuple existence random variable $r.e$. We then introduce factors encoding the correlations among the random variables for r and the random variables for t. The first factor we introduce is $f^{\sigma}_{r.e}$, which encodes the correlations for $r.e$:

$$f^{\sigma}_{r.e}(r.e, t.e, \{t.a\}_{a \in attr(R)}) = \begin{cases} 1 & \text{if } t.e \wedge c(\{t.a\}_{a \in attr(R)}) \Leftrightarrow r.e \\ 0 & \text{otherwise} \end{cases}$$

where $c(\{t.a\}_{a \in attr R})$ is `true` if a joint assignment to the attribute value random variables of t satisfies the predicate c and `false` otherwise.

We also introduce a factor for $r.a$, $\forall a \in attr(R)$ (where $dom(r.A) = dom(t.A)$), denoted by $f^{\sigma}_{r.a}$. $f^{\sigma}_{r.a}$ takes $t.a, r.e$ and $r.a$ as arguments and can be defined as:

$$f^{\sigma}_{r.a}(r.a, r.e, t.a) = \begin{cases} 1 & \text{if } r.e \wedge (t.a == r.a) \\ 1 & \text{if } \overline{r.e} \wedge (r.a == \emptyset) \\ 0 & \text{otherwise} \end{cases}$$

Cartesian Product: Suppose R_1 and R_2 are the two relations involved in the Cartesian product operation. Let r denote the join result of two tuples $t_1 \in R_1$ and $t_2 \in R_2$. Thus r maps every attribute from $attr(R_1) \cup attr(R_2)$ to a random variable, besides being associated with a tuple existence random variable $r.e$. The factor for $r.e$, denoted by $f^{\times}_{r.e}$, takes $t_1.e$, $t_2.e$ and $r.e$ as arguments, and is defined as:

$$f^{\times}_{r.e}(r.e, t_1.e, t_2.e) = \begin{cases} 1 & \text{if } t_1.e \wedge t_2.e \Leftrightarrow r.e \\ 0 & \text{otherwise} \end{cases}$$

We also introduce a factor $f^{\times}_{r.a}$ for each $a \in attr(R_1) \cup attr(R_2)$, and this is defined exactly in the same fashion as $f^{\sigma}_{r.a}$. Basically, for $a \in attr(R_1)$ ($a \in attr(R_2)$), it returns 1 if $r.e \wedge (t_1.a == r.a)$ ($r.e \wedge (t_2.a == r.a)$) holds or if $\overline{r.e} \wedge (r.a == \emptyset)$ holds, and 0 otherwise.

Project (without duplicate elimination): Let $\prod_{\mathbf{a}}(R)$ denote the operation we are interested in where $\mathbf{a} \subseteq attr(R)$ denotes the set of attributes we want to project onto. Let r denote the result of projecting $t \in R$. Thus r maps each attribute $a \in \mathbf{a}$ to a random variable, besides being associated with $r.e$. The factor for $r.e$, denoted by $f^{\prod}_{r.e}$, takes $t.e$ and $r.e$ as arguments, and is defined as follows:

$$f^{\prod}_{r.e}(r.e, t.e) = \begin{cases} 1 & \text{if } t.e \Leftrightarrow r.e \\ 0 & \text{otherwise} \end{cases}$$

Each factor $f^{\prod}_{r.a}$, introduced for $r.a$, $\forall a \in \mathbf{a}$, is defined exactly as $f^{\sigma}_{r.a}$, in other words, $f^{\prod}_{r.a}(r.a, r.e, t.a) = f^{\sigma}_{r.a}(r.a, r.e, t.a)$.

Duplicate Elimination: Duplicate elimination is a slightly more complex operation because it can give rise to multiple intermediate tuples even if there was only one input tuple to begin with. Let R denote the relation from which we want to eliminate duplicates, then the resulting relation after duplicate elimination will contain tuples whose existence is uncertain, more precisely the resulting tuples' attribute values are known. Any element from $\bigcup_{t \in R} \times_{a \in attr(R)} dom(t.a)$ may correspond to the values of a possible result tuple. Let r denote any such result tuple whose attribute values are known, only $r.e$ is not `true` with certainty. Denote by r_a the value of attribute a in r. We only need to introduce the factor $f^{\delta}_{r.e}$ for $r.e$. To do this we compute the set of tuples from R that may give rise to r. Any tuple t that satisfies $\bigwedge_{a \in attr(R)}(r_a \in dom(t.a))$ may give rise to r. Let y_t^r be an intermediate random variable with $dom(y_t^r) = \{\texttt{true}, \texttt{false}\}$ such that y_t^r is `true` iff t gives rise to r and `false` otherwise. This is easily done by introducing a factor $f_{y_t^r}^{\delta}$ that takes $\{t.a\}_{a \in attr(R)}$, $t.e$ and y_t^r as arguments and is defined as:

$$f_{y_t^r}^{\delta}(y_t^r, \{t.a\}_{a \in attr(R)}, t.e) = \begin{cases} 1 & \text{if } t.e \wedge \bigwedge_a (t.a == r_a) \Leftrightarrow y_t^r \\ 0 & \text{otherwise} \end{cases}$$

where $\{t.a\}_{a \in attr(R)}$ denotes all attribute value random variables of t. We can then define $f_{r.e}^{\delta}$ in terms of y_t^r. $f_{r.e}^{\delta}$ takes as arguments $\{y_t^r\}_{t \in T_r}$, where T_r denotes the set of tuples that may give rise to r (contains the assignment $\{r_a\}_{a \in attr(R)}$ in its joint domain), and $r.e$, and is defined as:

$$f_{r.e}^{\delta}(r.e, \{y_t^r\}_{t \in T_r}) = \begin{cases} 1 & \text{if } \bigvee_{t \in T_r} y_t^r \Leftrightarrow r.e \\ 0 & \text{otherwise} \end{cases}$$

Union and set difference: These operators require set semantics. Let R_1 and R_2 denote the relations on which we want to apply one of these two operators, either $R_1 \cup R_2$ or $R_1 - R_2$. We will assume that both R_1 and R_2 are sets of tuples such that every tuple contained in them have their attribute values fixed and the only uncertainty associated with these tuples are with their existence (if not then we can apply a δ operation to convert them to this form). Now, consider result tuple r and sets of tuples T_r^1, containing all tuples from R_1 that match r's attribute values, and T_r^2, containing all tuples from R_2 that match r's attribute values. The required factors for $r.e$ can now be defined as follows:

$$f_{r.e}^{\cup}(r.e, \{t_1.e\}_{t_1 \in T_r^1}, \{t_2.e\}_{t_2 \in T_r^2}) = \begin{cases} 1 & \text{if } (\bigvee_{t \in T_r^1 \cup T_r^2} t.e) \Leftrightarrow r.e \\ 0 & \text{otherwise} \end{cases}$$

$$f_{r.e}^{-}(r.e, \{t_1.e\}_{t_1 \in T_r^1}, \{t_2.e\}_{t_2 \in T_r^2}) = \begin{cases} 1 & \text{if } ((\bigvee_{t \in T_r^1} t.e) \wedge \neg(\bigvee_{t \in T_r^2} t.e)) \Leftrightarrow r.e \\ 0 & \text{otherwise} \end{cases}$$

Aggregation operators: Aggregation operators are also easily handled using factors. Suppose we want to compute the `sum` aggregate on attribute a of relation R, then we simply define a random variable $r.a$ for the result and introduce a factor that takes as arguments $\{t.a\}_{t \in attr(R)}$ and $r.a$, and define the factor so that it returns 1 if $r.a == (\sum_{t \in R} t.a)$ and 0 otherwise. Thus for any aggregate operator γ and result tuple random variable $r.a$, we can

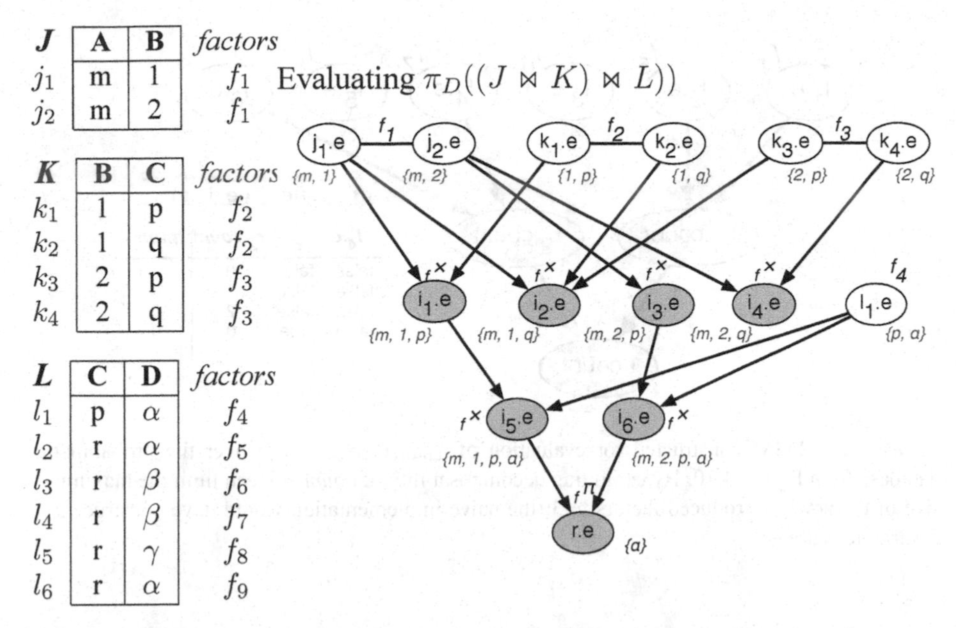

J	A	B	*factors*
j_1	m	1	f_1
j_2	m	2	f_1

K	B	C	*factors*
k_1	1	p	f_2
k_2	1	q	f_2
k_3	2	p	f_3
k_4	2	q	f_3

L	C	D	*factors*
l_1	p	α	f_4
l_2	r	α	f_5
l_3	r	β	f_6
l_4	r	β	f_7
l_5	r	γ	f_8
l_6	r	α	f_9

Figure 4.10. An example query evaluation over a 3-relation database with only tuple uncertainty but many correlations (tuples associated with the same factor are correlated with each other). The intermediate tuples are shown alongside the corresponding random variables. Tuples $l_2, \ldots, l_6$ do not participate in the query.

define the following factor:

$$f^{\gamma}_{r.a}(r.a, \{t.a\}_{t \in R}) = \begin{cases} 1 & \text{if } r.a == \gamma_{t \in R} t.a \\ 1 & \text{if } (r.a == \emptyset) \Leftrightarrow \bigwedge_{t \in R}(t.a == \emptyset) \\ 0 & \text{otherwise} \end{cases}$$

4.3 Query Evaluation as Inference

Given a query and a probabilistic database (and the corresponding PGM), we can use the procedures described in the previous section to construct an augmented PGM that contains random variables corresponding to the result tuples. Computing the result probabilities is simply a matter of evaluating marginal probability queries over this PGM. We can use any standard exact or approximate inference algorithm developed in the probabilistic reasoning community for this purpose, depending on our requirements of accuracy and speed. Note that the resulting PGM, and hence the complexity of inference, will depend on the query plan used for executing the query. We revisit this issue in Section 5.

Figure 4.10 shows the PGM generated when evaluating a multi-way join query over 3 relations; computing the result tuple probability is equivalent to

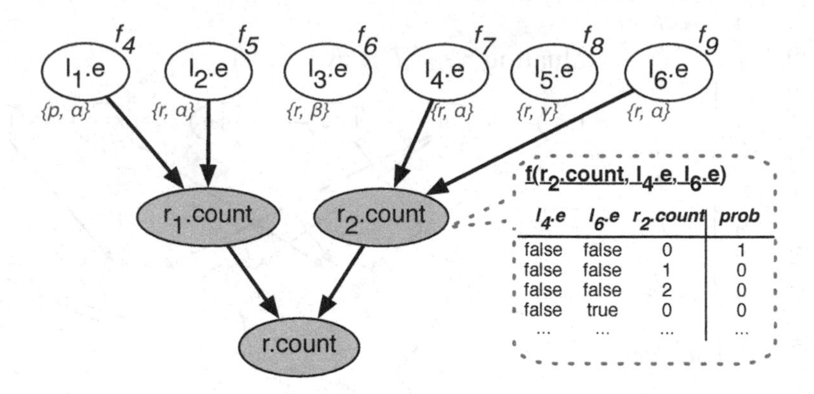

$l_4.e$	$l_6.e$	$r_2.count$	prob
false	false	0	1
false	false	1	0
false	false	2	0
false	true	0	0
...	...	...	...

Figure 4.11. PGM constructed for evaluation of ${}_{count}G(\sigma_{D=\alpha}(L))$ over the probabilistic database from Figure 4.10. By exploiting decomposability of *count*, we can limit the maximum size of the newly introduced factors to 3 (the naive implementation would have constructed a 5-variable factor).

computing the marginal probability distribution over the random variable $r.e$. Similarly, Figure 4.11 shows the PGM constructed in response to an aggregate query (details below).

4.4 Optimizations

For the above operator modifications, we have attempted to be completely general and hence the factors introduced may look slightly more complicated than need be. For example, it is not necessary that $f^{\sigma}_{r.E}$ take as arguments all random variables $\{t.a\}_{a \in attr(R)}$ (as defined above), it only needs to take those $t.a$ random variables as arguments which are involved in the predicate c of the σ operation. Also, given a theta-join we do not need to implement this as a Cartesian product followed by a select operation. It is straightforward to push the select operation into the Cartesian product factors and implement the theta-join directly by modifying $f^{\times}_{r.E}$ appropriately using c.

Another type of optimization that is extremely useful for aggregate computation, duplicate elimination and the set-theoretic operations ($\cup$ and $-$) is to exploit decomposable functions. A decomposable function is one whose result does not depend on the order in which the inputs are presented to it. For instance, $\vee$ is a decomposable function, and so are most of the aggregation operators including `sum`, `count`, `max` and `min`. The problem with some of the redefined relational algebra operators is that, if implemented naively, they may lead to large intermediate factors. For instance, while running a δ operation, if T_r contains n tuples for some r, then the factor $f^{\delta}_{r.e}$ will be of size 2^{n+1}. By exploiting decomposability of $\vee$ we can implement the same factor using a linear number of constant sized (3-argument) factors which may lead to significant speedups. We refer the interested reader to [50, 60] for more details. The only aggregation operator that is not decomposable is `avg`, but even in this case we can exploit the same ideas by implementing `avg` in terms of `sum` and `count` both of which are decomposable. Figure 4.11 shows the PGM constructed for an example aggregate query over the database from Figure 4.10.

Finally, one of the key ways we can reduce the complexity of query evaluation is by exploiting recurring (shared) factors. In recent work [53], we developed a general-purpose inference algorithm that can exploit such shared factors. Our algorithm identifies and exploits the symmetry present in the augmented PGM to significantly speed up query evaluation in most cases. We omit the details due to space constraints and refer the reader to [53] for further details.

5. Related Work and Discussion

Next we briefly discuss some of the closely related concepts in query evaluation over probabilistic databases, namely *safe plans* and *lineage*. We then briefly discuss the relationship of our approach to probabilistic relational models, lifted inference, and scalable inference using databases. We believe most of these represent rich opportunities for future research.

5.1 Safe Plans

One of the key results in query evaluation over probabilistic databases is the dichotomy of conjunctive query evaluation on tuple-independent probabilistic databases by Dalvi and Suciu [14, 15]. Briefly the result states that the complexity of evaluating a conjunctive query over tuple-independent probabilistic databases is either PTIME or #P-complete. For the former case, Dalvi and Suciu [14] also present an algorithm to find what are called *safe query plans*, that permit correct *extensional* evaluation of the query. We relate the notion of safe plans to our approach through the following theorem:

THEOREM 5.1 *When executing a query over a tuple-independent probabilistic database using a safe query plan, the resulting probabilistic graphical model is tree-structured (for which inference can be done in PTIME).*

Note that the dichotomy result presented in [15] reflects a worst-case scenario over all possible instances of a probabilistic database. In other words, even if a query does not have safe plan, for a specific probabilistic database instance, query evaluation may still be reasonably efficient. Our approach can easily capture this because in such cases the resulting PGM will either be tree-structured or have low tree-width, thus allowing us to execute the query efficiently. One of the important open problems in this area is developing algorithms for identifying query plans that result in PGMs with low tree-widths for a given probabilistic database and a given query.

5.2 Representing Uncertainty using Lineage

Several works [26, 58, 6, 48, 49] have proposed using explicit *boolean* formulas to capture the relationship between the base tuples and the intermediate tuples. In the Trio system [58, 6], such formulas are called *lineage*, and are computed during the query evaluation. The result tuple probabilities are then computed on demand by evaluating the lineage formulas. In recent work, Re et al. [49] presented techniques for approximate compression of such lineage formulas for more efficient storage and query evaluation.

The PGM constructed in our approach can be thought of as a generalization of such boolean formulas, since the PGM can represent more complex relationships than can be captured using a boolean formula. Further, the PGM naturally captures common subexpressions between the lineage formulas corresponding to different result tuples, and avoids re-computation during the inference process. Evaluation of boolean formulas can be seen as a special case of probabilistic inference, and thus techniques from exact or approximate inference literature can be directly applied to evaluating the lineage formulas as well. However lineage formula evaluation admits efficient approximation schemes (e.g. based on polynomial approximation [49]) that may not translate to general probabilistic graphical models.

5.3 Probabilistic Relational Models

Probabilistic relational models (PRMs) [25, 27] extend Bayesian networks with the concepts of objects, their properties and relations between them. In a way, they are to Bayesian networks as relational logic is to propositional logic. PRMs can also be thought of as a generalization of the probabilistic database framework that we presented in this chapter, and extending our approach to transparently and efficiently handle a PRM-based model is one of the important research directions that we plan to pursue in future. We begin with illustrating PRMs using a simple example, and then discuss the challenges in integrating them with our approach.

A PRM contains a relational component that describes the relational schema of the domain, and a probabilistic component that captures the probabilistic dependencies that hold in the domain. Figure 4.12 shows a simple example PRM over a relational schema containing three relations, *Author*, *Paper*, and *Review*. For simplicity the relationship *AuthorOf* is modeled as many-to-one (with a single author per paper), whereas the relationship *Reviewed* is many-to-many. Along with the relational schema, a PRM specifies a probabilistic model over the attributes of the relations. A key difference between Bayesian

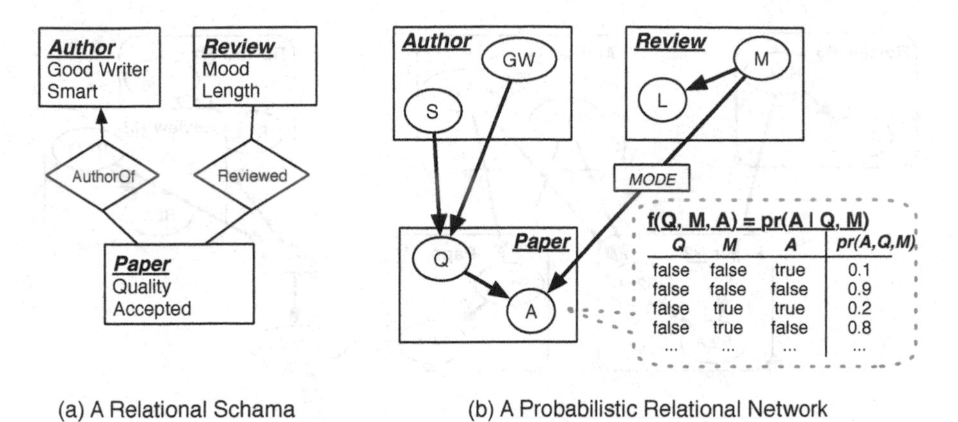

Figure 4.12. A probabilistic relational model defined over an example relational schema. Similar to Bayesian networks, the model parameters consist of conditional probability distributions for each node given its parents.

networks and PRMs is that an attribute in one relation may depend on an attribute in another relation. For example, the *quality* of a *paper* may depend on the properties of the *author* (as shown in the figure).

When defining a dependence across a many-to-one relationship, a mechanism to aggregate the attribute values must be specified as well. For instance, the *accepted* attribute for a paper is modeled as dependent on the *mood* attribute from the review relation. However a single paper may have multiple reviews, and we must somehow combine the values of *mood* attribute from those reviews; the example PRM uses the *MODE* of the attribute values for this purpose.

Now, given a relational skeleton that specifies the primary keys and foreign keys for the tuples, the PRM defines a probability distribution over the attributes of the tuples. Figure 4.13 shows an example of this, with two papers with keys *P1* and *P2*, both by the author *A1*. The PRM then specifies a joint probability distribution over the random variables as shown in the figure. If the skeleton also specifies the values of some of the attributes, those can be treated as evidence in a straightforward way.

PRMs can also capture uncertainty in the link structure (i.e., the key-foreign key dependencies). We refer the reader to [27] for more details.

Conceptually it is straightforward to extend our probabilistic model to allow the dependences to be defined using a PRM (shared factors is one step in that direction); the real challenge is doing inference over such models (see below). We are planning to explore closer integration between these two areas in the future.

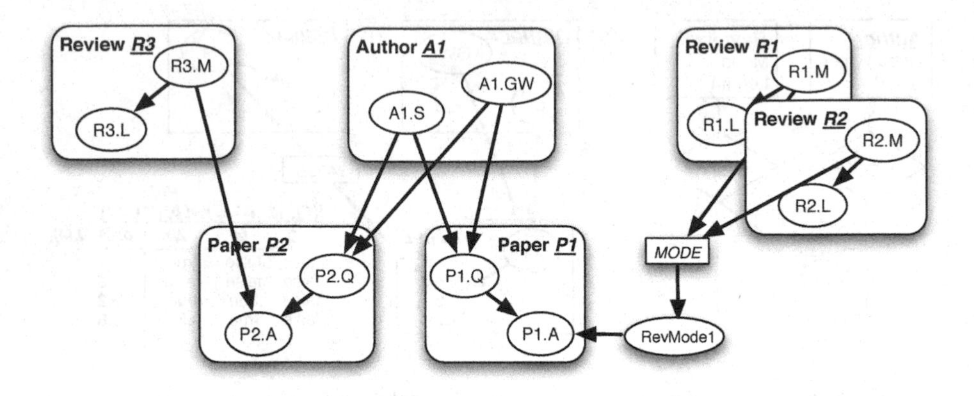

Figure 4.13. An instance of the example PRM with two papers: $P1, P2$, with the same author $A1$. For $P1$, we use an explicit random variable for representing the mode of $R1.M$ and $R2.M$. No such variable is needed for $P2$ since it only has one review.

5.4 Lifted Inference

Many first-order machine learning models such as PRMs allow defining rich, compact probability distributions over large collections of random variables. Inference over such models can be tricky, and the initial approaches to inference involved *grounding* out the graphical model by explicitly creating random variables (as shown in Figure 4.13) and then using standard inference algorithms. This can however result in very large graphical models, and can involve much redundant inference (since most of the factors are shared). Lifted inference techniques aim to address this situation by avoiding propositionalization (grounding) as much as possible [45, 46, 17, 55, 41, 53]. Most of this work assumes that the input is a first-order probabilistic model (such as a PRM). Poole [46] presents a modified version of the variable elimination algorithm [59] for this purpose. Braz et al. [17] and Milch et al. [41] present algorithms that look for specific types of structures in the first-order model, and exploit these for efficient inference. Singla et al. [55] develop a modified loopy belief propagation algorithm (for approximate inference) for lifted inference.

As discussed above, in our recent work [53], we developed a general-purpose lifted inference algorithm for probabilistic query evaluation. Our algorithm however does not operate on the first-order representation, and we are currently working on combining our approach with the techniques developed in the lifted inference literature.

5.5 Scalable Inference using a Relational Database

Finally a very related but at the same time fundamentally different problem is that of expressing inference tasks as database queries. Consider the Bayesian network shown in Figure 4.2, and consider the (inference) task of finding the marginal probability distribution over *income* (I). As seen before, this can be written as:

$$\Pr(I) = \Sigma_{L,A,D,E}\ f_1(L)f_2(A)f_3(D)f_4(E,A)f_5(I,E,D)$$

If the factors (CPDs) become very large, we might choose to store them as relations in a database (called *functional relations* by Bravo et al. [9]). For example, the relations corresponding to f_1 and f_5 may have schemas *F1(L, prob)*, and *F5(I, E, D, prob)* respectively. Then this inference task can be written as an SQL query as follows:

```
select I, sum(F1.prob * F2.prob * F3.prob * F4.prob * F5.prob)
from F1 join F2 join F3 join F4 join F5
group by I
```

This approach not only enables easy and persistent maintenance of Bayesian networks, but can also enable significant performance optimizations (we refer the reader to Bravo et al. [9] for a more detailed discussion).

However note that this approach is only suitable when the number of random variables is small (i.e. the size of the network is small), since each factor must be stored as a separate relation. The number of uncertain facts in a probabilistic database is likely to be very large and continuously changing, and storing each factor as a different relation would be infeasible in those cases. Second, the main "query/inference" tasks that need to be supported in the two scenarios are quite different. In probabilistic databases, the SQL queries operate on the values of the random variables, concatenating or aggregating them, whereas inference in Bayesian networks is typically concerned with marginalization and conditioning. Supporting both types of tasks in a unified manner remains one of the most important open problems in this area.

6. Conclusions

Graphical models are a versatile tool that have been applied to many database problems such as selectivity estimation [28, 21, 43, 31], sensor network data management [23], information extraction [12, 51], data integration [54, 29] to name a few. In this chapter, we presented a simple and intuitive framework for managing large-scale uncertain data using graphical models, that allows us to capture complex uncertainties and correlations in the data in a uniform manner. We showed how the problem of query evaluation in uncertain databases can be seen to be equivalent to probabilistic inference in an appropriately constructed graphical model. This equivalence enables us to employ the formidable machinery developed in the probabilistic reasoning literature over the years for answering queries over probabilistic databases. We believe it will also lead to a deeper understanding of how to devise more efficient inference algorithms for large-scale, structured probabilistic models.

Acknowledgments

This work was supported in part by the National Science Foundation under Grants No. 0438866 and 0546136. We thank Sunita Sarawagi who co-presented a tutorial on graphical models at VLDB 2007 with one of the authors, and Brian Milch for stimulating discussions regarding lifted inference.

References

[1] Periklis Andritsos, Ariel Fuxman, and Renee J. Miller. Clean answers over dirty databases. In *International Conference on Data Engineering (ICDE)*, 2006.

[2] Stefan Arnborg. Efficient algorithms for combinatorial problems on graphs with bounded decomposability - a survey. *BIT Numerical Mathematics*, 1985.

[3] Sanjeev Arulampalam, Simon Maskell, Neil Gordon, and Tim Clapp. A tutorial on particle filters for on-line non-linear/non-gaussian Bayesian tracking. *IEEE Transactions of Signal Processing*, 50(2), 2002.

[4] Daniel Barbara, Hector Garcia-Molina, and Daryl Porter. The management of probabilistic data. *IEEE Transactions on Knowledge and Data Engineering*, 1992.

[5] Catriel Beeri, Ronald Fagin, David Maier, and Mihalis Yannakakis. On the desirability of acyclic database schemes. *J. ACM*, 30(3):479–513, 1983.

[6] Omar Benjelloun, Anish Das Sarma, Alon Halevy, and Jennifer Widom. ULDBs: Databases with uncertainty and lineage. In *International Conference on Very Large Data Bases (VLDB)*, 2006.

[7] Umberto Bertele and Francesco Brioschi. *Nonserial Dynamic Programming.* Academic Press, New York, 1972.

[8] Jihad Boulos, Nilesh Dalvi, Bhushan Mandhani, Chris Re, Shobhit Mathur, and Dan Suciu. Mystiq: A system for finding more answers by using probabilities. In *ACM SIGMOD International conference on Management of Data*, 2005.

[9] Héctor Corrada Bravo and Raghu Ramakrishnan. Optimizing MPF queries: decision support and probabilistic inference. In *ACM SIGMOD International conference on Management of Data*, pages 701–712, 2007.

[10] Roger Cavallo and Michael Pittarelli. The theory of probabilistic databases. In *International Conference on Very Large Data Bases (VLDB)*, 1987.

[11] Reynold Cheng, Dmitri Kalashnikov, and Sunil Prabhakar. Evaluating probabilistic queries over imprecise data. In *ACM SIGMOD International conference on Management of Data*, 2003.

[12] William W. Cohen and Sunita Sarawagi. Exploiting dictionaries in named entity extraction: Combining semi-markov extraction processes and data integration methods. In *SIGKDD*, 2004.

[13] Robert G. Cowell, A. Philip Dawid, Steffen L. Lauritzen, and David J. Spiegelhater. *Probabilistic Networks and Expert Systems*. Springer, 1999.

[14] Nilesh Dalvi and Dan Suciu. Efficient query evaluation on probabilistic databases. In *International Conference on Very Large Data Bases (VLDB)*, 2004.

[15] Nilesh Dalvi and Dan Suciu. Management of probabilistic data: Foundations and challenges. In *PODS*, 2007.

[16] Anish Das Sarma, Omar Benjelloun, Alon Halevy, and Jennifer Widom. Working models for uncertain data. In *International Conference on Data Engineering (ICDE)*, 2006.

[17] Rodrigo de Salvo Braz, Eyal Amir, and Dan Roth. Lifted first-order probabilistic inference. In *International Joint Conferences on Artificial Intelligence (IJCAI)*, 2005.

[18] Rina Dechter. Constraint networks. *Encyclopedia of Artificial Intelligence*, 1992.

[19] Rina Dechter. Bucket elimination: A unifying framework for probabilistic inference. In *Uncertainty in Artificial Intelligence (UAI)*, 1996.

[20] Amol Deshpande, Minos Garofalakis, and Michael Jordan. Efficient stepwise selection in decomposable models. In *Proceedings of the 17th Annual Conference on Uncertainty in Artificial Intelligence (UAI)*, pages 128–135, 2001.

[21] Amol Deshpande, Minos Garofalakis, and Rajeev Rastogi. Independence is Good: Dependency-Based Histogram Synopses for High-Dimensional Data. In *ACM SIGMOD International conference on Management of Data*, 2001.

[22] Amol Deshpande, Carlos Guestrin, Sam Madden, Joseph M. Hellerstein, and Wei Hong. Model-driven data acquisition in sensor networks. In *International Conference on Very Large Data Bases (VLDB)*, 2004.

[23] Amol Deshpande, Carlos Guestrin, and Samuel Madden. Using probabilistic models for data management in acquisitional environments. In *Conference on Innovative Data Systems Research (CIDR)*, 2005.

[24] Debabrata Dey and Sumit Sarkar. A probabilistic relational model and algebra. *ACM Transactions on Database Systems (TODS)*, 1996.

[25] Nir Friedman, Lise Getoor, Daphne Koller, and Avi Pfeffer. Learning probabilistic relational models. In *International Joint Conferences on Artificial Intelligence (IJCAI)*, 1999.

[26] Norbert Fuhr and Thomas Rolleke. A probabilistic relational algebra for the integration of information retrieval and database systems. *ACM Transactions on Information Systems (TODS)*, 1997.

[27] Lise Getoor and Ben Taskar, editors. *Introduction to Statistical Relational Learning*. MIT Press, Cambridge, MA, USA, 2007.

[28] Lise Getoor, Ben Taskar, and Daphne Koller. Selectivity estimation using probabilistic models. In *ACM SIGMOD International conference on Management of Data*, 2001.

[29] Rahul Gupta and Sunita Sarawagi. Creating probabilistic databases from information extraction models. In *International Conference on Very Large Data Bases (VLDB)*, 2006.

[30] Cecil Huang and Adnan Darwiche. Inference in belief networks: A procedural guide. *International Journal of Approximate Reasoning*, 1994.

[31] Ihab F. Ilyas, Volker Markl, Peter Haas, Paul Brown, and Ashraf Aboulnaga. Cords: automatic discovery of correlations and soft functional dependencies. In *SIGMOD*, 2004.

[32] Tomasz Imielinski and Witold Lipski, Jr. Incomplete information in relational databases. *Journal of the ACM*, 1984.

[33] Tommi Jaakkola and Michael I. Jordan. Variational probabilistic inference and the QMR-DT network. *Journal of Artificial Intelligence Research*, 10:291–322, 1999.

[34] T. S. Jayram, Rajasekar Krishnamurthy, Sriram Raghavan, Shivakumar Vaithyanathan, and Huaiyu Zhu. Avatar information extraction system. In *IEEE Data Engineering Bulletin*, 2006.

[35] Michael I. Jordan, editor. *Learning in graphical models.* MIT Press, Cambridge, MA, USA, 1999.

[36] Michael I. Jordan, Zoubin Ghahramani, Tommi S. Jaakkola, and Lawrence K. Saul. An introduction to variational methods for graphical models. *Machine Learning*, 1999.

[37] Rudolph E. Kalman. A new approach to linear filtering and prediction problems. *Transactions of the ASME–Journal of Basic Engineering*, 82(Series D):35–45, 1960.

[38] Bhargav Kanagal and Amol Deshpande. Online filtering, smoothing and probabilistic modeling of streaming data. In *ICDE*, 2008.

[39] Laks V. S. Lakshmanan, Nicola Leone, Robert Ross, and V. S. Subrahmanian. Probview: a flexible probabilistic database system. *ACM Transactions on Database Systems (TODS)*, 1997.

[40] Blackford Middleton, Michael Shwe, David Heckerman, Max Henrion, Eric Horvitz, Harold Lehmann, and Gregory Cooper. Probabilistic diagnosis using a reformulation of the internist-1/qmr knowledge base. *Methods of Information in Medicine*, 30:241–255, 1991.

[41] Brian Milch, Luke Zettlemoyer, Kristian Kersting, Michael Haimes, and Leslie Kaelbling. Lifted probabilistic inference with counting formulas. In *Association for the Advancement of Artificial Intelligence (AAAI)*, 2008.

[42] Kevin P. Murphy, Yair Weiss, and Michael I. Jordan. Loopy belief propagation for approximate inference: An empirical study. In *Uncertainty in Artificial Intelligence (UAI)*, pages 467–475, 1999.

[43] Dmitry Pavlov, Heikki Mannila, and Padhraic Smyth. Beyond independence: Probabilistic models for query approximation on binary transaction data. *IEEE TKDE*, 2003.

[44] Judaea Pearl. *Probabilistic Reasoning in Intelligent Systems.* Morgan Kaufmann, 1988.

[45] Avi Pfeffer, Daphne Koller, Brian Milch, and Ken Takusagawa. SPOOK: A system for probabilistic object-oriented knowledge representation. In *Uncertainty in Artificial Intelligence (UAI)*, 1999.

[46] David Poole. First-order probabilistic inference. In *International Joint Conferences on Artificial Intelligence (IJCAI)*, 2003.

[47] Lawrence R. Rabiner. A tutorial on hidden Markov models and selected applications in speech recognition. In *Proceedings of the IEEE*, 77(2):257–286, 1989.

[48] Chris Re, Nilesh Dalvi, and Dan Suciu. Efficient top-k query evaluation on probabilistic data. In *International Conference on Data Engineering (ICDE)*, 2007.

[49] Chris Re and Dan Suciu. Approximate lineage for probabilistic databases. In *International Conference on Very Large Data Bases (VLDB)*, 2008.

[50] Irina Rish. *Efficient Reasoning in Graphical Models*. PhD thesis, University of California, Irvine, 1999.

[51] Sunita Sarawagi. Efficient inference on sequence segmentation models. In *ICML*, 2006.

[52] Prithviraj Sen and Amol Deshpande. Representing and querying correlated tuples in probabilistic databases. In *International Conference on Data Engineering (ICDE)*, 2007.

[53] Prithviraj Sen, Amol Deshpande, and Lise Getoor. Exploiting shared correlations in probabilistic databases. In *International Conference on Very Large Data Bases (VLDB)*, 2008.

[54] Parag Singla and Pedro Domingos. Multi-relational record linkage. In *Proceedings of 3rd Workshop on Multi-Relational Data Mining at ACM SIGKDD*, Seattle, WA, 2004.

[55] Parag Singla and Pedro Domingos. Lifted first-order belief propagation. In *Association for the Advancement of Artificial Intelligence (AAAI)*, 2008.

[56] Padhraic Smyth. Belief networks, hidden Markov models, and Markov random fields: a unifying view. *Pattern Recognition Letters*, 18(11-13), 1997.

[57] Greg Welch and Gary Bishop. An introduction to Kalman filter. `http://www.cs.unc.edu/~welch/kalman/kalmanIntro.html`, 2002.

[58] Jennifer Widom. Trio: A system for integrated management of data, accuracy, and lineage. In *Conference on Innovative Data Systems Research (CIDR)*, 2005.

[59] Nevin Lianwen Zhang and David Poole. A simple approach to Bayesian network computations. In *Canadian Conference on Artificial Intelligence*, 1994.

[60] Nevin Lianwen Zhang and David Poole. Exploiting causal independence in Bayesian network inference. *Journal of Artificial Intelligence Research*, 1996.

Chapter 5

TRIO: A SYSTEM FOR DATA, UNCERTAINTY, AND LINEAGE

Jennifer Widom
Dept. of Computer Science
Stanford University
widom@cs.stanford.edu

Abstract

This chapter covers the *Trio* database management system. Trio is a robust prototype that supports *uncertain data* and *data lineage*, along with the standard features of a relational DBMS. Trio's new *ULDB* data model is an extension to the relational model capturing various types of uncertainty along with data lineage, and its *TriQL* query language extends SQL with a new semantics for uncertain data and new constructs for querying uncertainty and lineage. Trio's data model and query language are implemented as a translation-based layer on top of a conventional relational DBMS, with some stored procedures for functionality and increased efficiency. Trio provides both an API and a full-featured graphical user interface.

Acknowledgments. Contributors to the Trio project over the years include (alphabetically) Parag Agrawal, Omar Benjelloun, Ashok Chandra, Julien Chaumond, Anish Das Sarma, Alon Halevy, Chris Hayworth, Ander de Keijzer, Raghotham Murthy, Michi Mutsuzaki, Tomoe Sugihara, Martin Theobald, and Jeffrey Ullman. Funding has been provided by the National Science Foundation and the Boeing and Hewlett-Packard Corporations.

Keywords: Uncertainty, Trio, ULDB, Lineage

Introduction

Trio is a new kind of database management system (DBMS): one in which *data*, *uncertainty* of the data, and data *lineage* are all first-class citizens. Combining data, uncertainty, and lineage yields a data management platform that is useful for data integration, data cleaning, information extraction systems, scientific and sensor data management, approximate and hypothetical query processing, and other modern applications.

The databases managed by Trio are called *ULDBs*, for *Uncertainty-Lineage Databases*. ULDBs extend the standard relational model. Queries are expressed using *TriQL* (pronounced "treacle"), a strict extension to SQL. We have built a robust prototype system that supports a substantial fraction of the TriQL language over arbitrary ULDBs. The remainder of this Introduction briefly motivates the ULDB data model, the TriQL language, and the prototype system. Details are then elaborated in the rest of the chapter.

Examples in this chapter are based on a highly simplified "crime-solver" application, starting with two *base tables*:

- `Saw(witness,color,car)` contains (possibly uncertain) crime vehicle sightings.
- `Drives(driver,color,car)` contains (possibly uncertain) information about cars driven.

We will derive additional tables by posing queries over these tables.

The ULDB Data Model. Uncertainty is captured by tuples that may include several *alternative* possible values for some (or all) of their attributes, with optional *confidence* values associated with each alternative. For example, if a witness saw a vehicle that was a blue Honda with confidence 0.5, a red Toyota with confidence 0.3, or a blue Mazda with confidence 0.2, the sighting yields one tuple in table `Saw` with three alternative values for attributes `color, car`. Furthermore, the presence of tuples may be uncertain, again with optionally specified confidence. For example, another witness may have 0.6 confidence that she saw a crime vehicle, but if she saw one it was definitely a red Mazda. Based on alternative values and confidences, each ULDB represents multiple *possible-instances* (sometimes called *possible-worlds*), where a possible-instance is a regular relational database.

Lineage, sometimes called *provenance*, associates with a data item information about its derivation. Broadly, lineage may be *internal*, referring to data within the ULDB, or *external*, referring to data outside the ULDB, or to other data-producing entities such as programs or devices. As a simple example of internal lineage, we may generate a table `Suspects` by joining tables `Saw` and `Drives` on attributes `color, car`. Lineage associated with a value in `Suspects` identifies the `Saw` and `Drives` values from which it was derived. A useful feature of internal lineage is that the confidence of a value in `Suspects` can be computed from the confidence of the data in its lineage (Section 4). If we generate further tables—`HighSuspects`, say—by issuing queries involving `Suspects` (perhaps together with other data), we get transitive lineage information: data in `HighSuspects` is derived from data in `Suspects`, which in turn is derived from data in `Saw` and `Drives`. Trio supports arbitrarily complex layers of internal lineage.

As an example of external lineage, table `Drives` may be populated from various car registration databases, and lineage can be used to connect the data to its original source. Although Trio supports some preliminary features for external lineage, this chapter describes internal lineage only.

The TriQL Query Language. Section 1.5 specifies a precise generic semantics for any relational query over a ULDB, and Section 2 provides an operational description of Trio's SQL-based query language that conforms to the generic semantics. Intuitively, the result of a relational query Q on a ULDB U is a result R whose possible-instances correspond to applying Q to each possible-instance of U. Internal lineage connects the data in result R to the data from which it was derived, as in the `Suspects` join query discussed above. Confidence values in query results are, by default, defined in a standard probabilistic fashion.

In addition to adapting SQL to Trio's possible-instances semantics in a straightforward and natural manner, TriQL includes a number of new features specific to uncertainty and lineage:

- Constructs for querying lineage, e.g., "find all witnesses contributing to `Jimmy` being a high suspect."
- Constructs for querying uncertainty, e.g., "find all high-confidence sightings," or "find all sightings with at least three different possible cars."
- Constructs for querying lineage and uncertainty together. e.g., "find all suspects whose lineage contains low-confidence sightings or drivers."
- Special types of aggregation suitable for uncertain databases, e.g., "find the expected number of distinct suspects."
- Query-defined result confidences, e.g., combine confidence values of joining tuples using *max* instead of multiplication.
- Extensions to SQL's data modification commands, e.g., to add new alternative values to an existing tuple, or to modify confidence values.
- Constructs for restructuring a ULDB relation, e.g, "flatten" or reorganize alternative values.

The Trio Prototype. The Trio prototype system is primarily layered on top of a conventional relational DBMS. From the user and application standpoint,

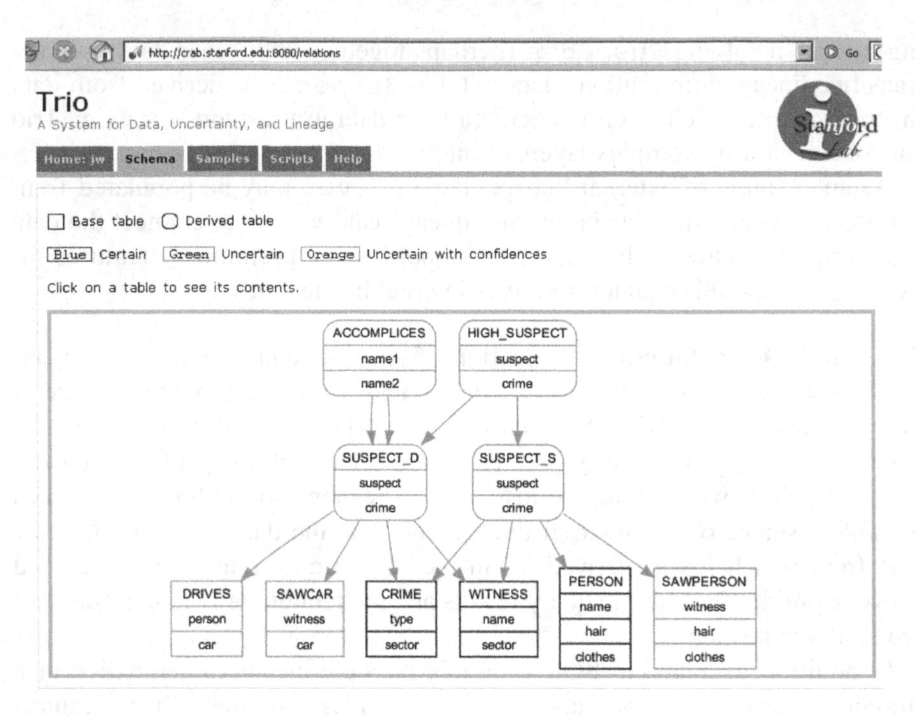

Figure 5.1. TrioExplorer Screenshot.

the Trio system appears to be a "native" implementation of the ULDB model, TriQL query language, and other features. However, Trio encodes the uncertainty and lineage in ULDB databases in conventional relational tables, and it uses a translation-based approach for most data management and query processing. A small number of stored procedures are used for specific functionality and increased efficiency.

The Trio system offers three interfaces: a typical DBMS-style API for applications, a command-line interface called *TrioPlus*, and a a full-featured graphical user interface called *TrioExplorer*. A small portion of the TrioExplorer interface is depicted in Figure 5.1. (The screenshot shows a *schema-level lineage graph*—discussed in Section 5—for a somewhat more elaborate crime-solver application than the running example in this chapter.) The Trio prototype is described in more detail in Section 6.

1. ULDBs: Uncertainty-Lineage Databases

The ULDB model is presented primarily through examples. A more formal treatment appears in [2]. ULDBs extend the standard SQL (multiset) relational model with:

1. *alternative values*, representing uncertainty about the contents of a tuple

2. *maybe* ('?') annotations, representing uncertainty about the presence of a tuple
3. numerical *confidence* values optionally attached to alternatives
4. *lineage*, connecting tuple-alternatives to other tuple-alternatives from which they were derived.

Each of these four constructs is specified next, followed by a specification of the semantics of relational queries on ULDBs.

1.1 Alternatives

ULDB relations have a set of *certain* attributes and a set of *uncertain* attributes, designated as part of the schema. Each tuple in a ULDB relation has one value for each certain attribute, and a set of possible values for the uncertain attributes. In table Saw, let witness be a certain attribute while color and car are uncertain. If witness Amy saw either a blue Honda, a red Toyota, or a blue Mazda, then in table Saw we have:

witness	**(color, car)**
Amy	(blue,Honda) ‖ (red,Toyota) ‖ (blue,Mazda)

This tuple logically yields three possible-instances for table Saw, one for each set of alternative values for the uncertain attributes. In general, the possible-instances of a ULDB relation R correspond to all combinations of alternative values for the tuples in R. For example, if a second tuple in Saw had four alternatives for (color,car), then there would be 12 possible-instances altogether.

Designating certain versus uncertain attributes in a ULDB relation is important for data modeling and efficient implementation. However, for presentation and formal specifications, sometimes it is useful to assume all attributes are uncertain (without loss of expressive power). For example, in terms of possible-instances, the Saw relation above is equivalent to:

(witness, color, car)
(Amy,blue,Honda) ‖ (Amy,red,Toyota) ‖ (Amy,blue,Mazda)

When treating all attributes as uncertain, we refer to the alternative values for each tuple as *tuple-alternatives*, or *alternatives* for short. In the remainder of the chapter we often use tuple-alternatives when the distinction between certain and uncertain attributes is unimportant.

1.2 '?' (Maybe) Annotations

Suppose a second witness, Betty, thinks she saw a car but is not sure. However, if she saw a car, it was definitely a red Mazda. In ULDBs, uncertainty about the existence of a tuple is denoted by a '?' annotation on the tuple. Betty's observation is thus added to table `Saw` as:

witness	(color, car)	
Amy	(blue,Honda) || (red,Toyota) || (blue,Mazda)	
Betty	(red,Mazda)	?

The '?' on the second tuple indicates that this entire tuple may or may not be present (so we call it a *maybe-tuple*). Now the possible-instances of a ULDB relation include not only all combinations of alternative values, but also all combinations of inclusion/exclusion for the maybe-tuples. This `Saw` table has six possible-instances: three choices for Amy's `(color,car)` times two choices for whether or not Betty saw anything. For example, one possible-instance of `Saw` is the tuples `(Amy,blue,Honda)`, `(Betty,red,Mazda)`, while another instance is just `(Amy,blue,Mazda)`.

1.3 Confidences

Numerical *confidence* values may be attached to the alternative values in a tuple. Suppose Amy's confidence in seeing the Honda, Toyota, or Mazda is 0.5, 0.3, and 0.2 respectively, and Betty's confidence in seeing a vehicle is 0.6. Then we have:

witness	(color, car)	
Amy	(blue,Honda):0.5 || (red,Toyota):0.3 || (blue,Mazda):0.2	
Betty	(red,Mazda):0.6	?

Reference [2] formalizes an interpretation of these confidence values in terms of probabilities. (Other interpretations may be imposed, but the probabilistic one is the default for Trio.) Thus, if Σ is the sum of confidences for the alternative values in a tuple, then we must have $\Sigma \leq 1$, and if $\Sigma < 1$ then the tuple must have a '?'. Implicitly, '?' is given confidence $(1 - \Sigma)$ and denotes the probability that the tuple is not present.

Now each possible-instance of a ULDB relation itself has a probability, defined as the product of the confidences of the tuple-alternatives and '?''s comprising the instance. It can be shown (see [2]) that for any ULDB relation:

1. The probabilities of all possible-instances sum to 1.
2. The confidence of a tuple-alternative (respectively a '?') equals the sum of probabilities of the possible-instances containing this alternative (respectively not containing any alternative from this tuple).

An important special case of ULDBs is when every tuple has only one alternative with a confidence value that may be < 1. This case corresponds to the traditional notion of *probabilistic databases*.

In Trio each ULDB relation R is specified at the schema level as either *with confidences*, in which case R must include confidence values on all of its data, or *without confidences*, in which case R has no confidence values. However, it is permitted to mix relations with and without confidence values, both in a database and in queries.

1.4 Lineage

Lineage in ULDBs is recorded at the granularity of alternatives: lineage connects a tuple-alternative to those tuple-alternatives from which it was derived. (Recall we are discussing only internal lineage in this chapter. External lineage also can be recorded at the tuple-alternative granularity, although for some lineage types coarser granularity is more appropriate; see [12] for a discussion.) Specifically, lineage is defined as a function λ over tuple-alternatives: $\lambda(t)$ is a boolean formula over the tuple-alternatives from which the alternative t was derived.

Consider again the join of `Saw` and `Drives` on attributes `color`, `car`, followed by a projection on `driver` to produce a table `Suspects(person)`. Assume all attributes in `Drives` are uncertain. (Although not shown in the tiny sample data below, we might be uncertain what car someone drives, or for a given car we might be uncertain who drives it.) Let column ID contain a unique identifier for each tuple, and let (i, j) denote the jth tuple-alternative of the tuple with identifier i. (That is, (i, j) denotes the tuple-alternative comprised of i's certain attributes together with the jth set of alternative values for its uncertain attributes.) Here is some sample data for all three tables, including lineage formulas for the derived data in `Suspects`. For example, the lineage of the Jimmy tuple-alternative in table `Suspects` is a conjunction of the second alternative of `Saw` tuple 11 with the second alternative of `Drives` tuple 21.

Saw

ID	witness	(color, car)
11	Cathy	(blue,Honda) ‖ (red,Mazda)

Drives

ID	Drives (driver, color, car)	
21	(Jimmy,red,Honda) ‖ (Jimmy,red,Mazda)	?
22	(Billy,blue,Honda)	
23	(Hank,red,Mazda)	

Suspects

ID	person		
31	Jimmy	?	$\lambda(31,1) = (11,2) \wedge (21,2)$
32	Billy	?	$\lambda(32,1) = (11,1) \wedge (22,1)$
33	Hank	?	$\lambda(33,1) = (11,2) \wedge (23,1)$

An interesting and important effect of lineage is that it imposes restrictions on the possible-instances of a ULDB: A tuple-alternative with lineage can be present in a possible-instance only if its lineage formula is satisfied by the presence (or, in the case of negation, absence) of other alternatives in the same possible-instance. Consider the derived table `Suspects`. Even though there is a '?' on each of its three tuples, not all combinations are possible. If Jimmy is present in `Suspects` then alternative 2 must be chosen for tuple 11, and therefore Hank must be present as well. Billy is present in `Suspects` only if alternative 1 is chosen for tuple 11, in which case neither Jimmy nor Hank can be present.

Thus, once a ULDB relation R has lineage to other relations, it is possible that not all combinations of alternatives and '?' choices in R correspond to valid possible-instances. The above ULDB has six possible-instances, determined by the two choices for tuple 11 times the three choices (including '?') for tuple 21.

Now suppose we have an additional base table, `Criminals`, containing a list of known criminals, shown below. Joining `Suspects` with `Criminals` yields the `HighSuspects` table on the right:

`Criminals`

ID	person
41	Jimmy
42	Frank
43	Hank

`HighSuspects`

ID	person		
51	Jimmy	?	$\lambda(51,1) = (31,1) \wedge (41,1)$
52	Hank	?	$\lambda(52,1) = (33,1) \wedge (43,1)$

Now we have multilevel (transitive) lineage relationships, e.g., $\lambda(51,1) = (31,1) \wedge (41,1)$ and $\lambda(31,1) = (11,2) \wedge (21,2)$. Lineage formulas specify direct derivations, but when the alternatives in a lineage formula are themselves derived from other alternatives, it is possible to recursively expand a lineage formula until it specifies base alternatives only. (Since we are not considering external lineage, base data has no lineage of its own.) As a very simple example, $\lambda(51,1)$'s expansion is $((11,2) \wedge (21,2)) \wedge (41,1)$.

Note that arbitrary lineage formulas may not "work" under our model—consider for example a tuple with one alternative and no '?' whose lineage (directly or transitively) includes the conjunction of two different alternatives of the same tuple. The tuple must exist because it doesn't have a '?', but it can't exist because its lineage formula can't be satisfied. Reference [2] formally defines *well-behaved* lineage (which does not permit, for example, the situation just described), and shows that internal lineage generated by relational queries is always well-behaved. Under well-behaved lineage, the possible-instances of an entire ULDB correspond to the possible-instances of the base data (data with no lineage of its own), as seen in the example above. With well-behaved lineage our interpretation of confidences carries over directly: combining confidences on the base data determines the probabilities of the possible-instances,

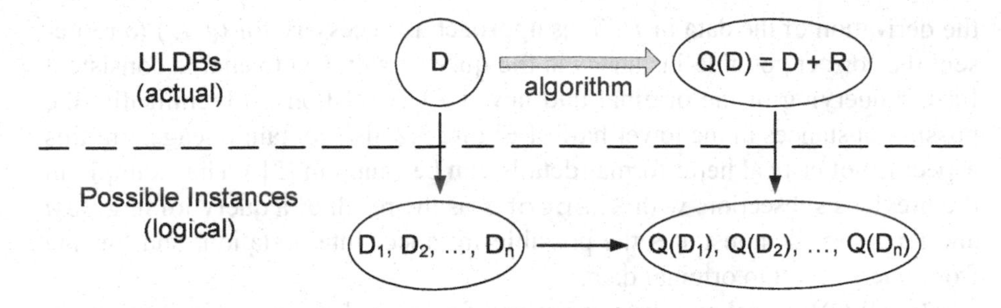

Figure 5.2. Relational Queries on ULDBs.

just as before. The confidence values associated with derived data items are discussed later in Section 4.

Finally, note that lineage formulas need not be conjunctive. As one example, suppose `Drives` tuple 23 contained `Billy` instead of `Hank`, and the `Suspects` query performed duplicate-eliminating projection. Then the query result is:

ID	person		
61	Jimmy	?	$\lambda(61,1) = (11,2) \wedge (21,2)$
62	Billy		$\lambda(62,1) = ((11,1) \wedge (22,1)) \vee ((11,2) \wedge (23,1))$

Note that the lineage formula for tuple 62 is always satisfied since one alternative of base tuple 11 must always be picked. Thus, there is no '?' on the tuple.

1.5 Relational Queries

In this section we formally define the semantics of any relational query over a ULDB. Trio's SQL-based query language will be presented in Section 2. The semantics for relational queries over ULDBs is quite straightforward but has two parts: (1) the possible-instances interpretation; and (2) lineage in query results.

Refer to Figure 5.2. Consider a ULDB D whose possible-instances are $D_1, D_2, \ldots, D_n$, as shown on the left side of the figure. If we evaluate a query Q on D, the possible-instances in the result of Q should be $Q(D_1)$, $Q(D_2), \ldots, Q(D_n)$, as shown in the lower-right corner. For example, if a query Q joins tables `Saw` and `Drives`, then logically it should join all of the possible-instances of these two ULDB relations. Of course we would never actually generate all possible-instances and operate on them, so a query processing algorithm follows the top arrow in Figure 5.2, producing a query result $Q(D)$ that represents the possible-instances.

A ULDB query result $Q(D)$ contains the original relations of D, together with a new *result relation* R. Lineage from R into the relations of D reflects the derivation of the data in R. This approach is necessary for $Q(D)$ to represent the correct possible-instances in the query result, and to enable consistent further querying of the original and new ULDB relations. (Technically, the possible-instances in the lower half of Figure 5.2 also contain lineage, but this aspect is not critical here; formal details can be found in [2].) The example in the previous subsection, with `Suspects` as the result of a query joining `Saw` and `Drives`, demonstrates the possible-instances interpretation, and lineage from query result to original data.

The ULDB model and the semantics of relational queries over it has been shown (see [2]) to exhibit two desirable and important properties:

- **Completeness:** Any finite set of possible-instances conforming to a single schema can be represented as a ULDB database.
- **Closure:** The result of any relational query over any ULDB database can be represented as a ULDB relation.

2. TriQL: The Trio Query Language

This section describes *TriQL*, Trio's SQL-based query language. Except for some additional features described later, TriQL uses the same syntax as SQL. However, the interpretation of SQL queries must be modified to reflect the semantics over ULDBs discussed in the previous section.

As an example, the join query producing `Suspects` is written in TriQL exactly as expected:

```
SELECT Drives.driver as person INTO Suspects
FROM Saw, Drives
WHERE Saw.color = Drives.color AND Saw.car = Drives.car
```

If this query were executed as regular SQL over each of the possible-instances of `Saw` and `Drives`, as in the lower portion of Figure 5.2, it would produce the expected set of possible-instances in its result. More importantly, following the operational semantics given next, this query produces a result table `Suspects`, including lineage to tables `Saw` and `Drives`, that correctly represents those possible-instances.

This section first specifies an operational semantics for basic SQL query blocks over arbitrary ULDB databases. It then introduces a number of additional TriQL constructs, with examples and explanation for each one.

2.1 Operational Semantics

We provide an operational description of TriQL by specifying direct evaluation of a generic TriQL query over a ULDB, corresponding to the upper arrow in Figure 5.2. We specify evaluation of single-block queries:

```
SELECT attr-list [ INTO new-table ]
FROM T1, T2, ..., Tn
WHERE predicate
```

The operational semantics of additional constructs are discussed later, when the constructs are introduced. Note that in TriQL, the result of a query has confidence values only if all of the tables in the query's `FROM` clause have confidence values. (Sections 2.8 and 2.9 introduce constructs that can be used in the `FROM` clause to logically add confidence values to tables that otherwise don't have them.)

Consider the generic TriQL query block above; call it Q. Let *schema*(Q) denote the composition *schema*(`T1`) $\uplus$ *schema*(`T2`) $\uplus \cdots \uplus$ *schema*(`Tn`) of the `FROM` relation schemas, just as in SQL query processing. The `predicate` is evaluated over tuples in *schema*(Q), and the `attr-list` is a subset of *schema*(Q) or the symbol "$\star$", again just as in SQL.

The steps below are an operational description of evaluating the above query block. As in SQL database systems, a query processor would rarely execute the simplest operational description since it could be woefully inefficient, but any query plan or execution technique (such as our translation-based approach described in Section 6) must produce the same result as this description.

1 Consider every combination $t_1, t_2, \ldots, t_n$ of tuples in `T1`, `T2`,..., `Tn`, one combination at a time, just as in SQL.

2 Form a "super-tuple" T whose tuple-alternatives have schema *schema*(Q). T has one alternative for each combination of tuple-alternatives in t_1, $t_2, \ldots, t_n$.

3 If any of $t_1, t_2, \ldots, t_n$ has a '?', add a '?' to T.

4 Set the lineage of each alternative in T to be the conjunction of the alternatives $t_1, t_2, \ldots, t_n$ from which it was constructed.

5 Retain from T only those alternatives satisfying the `predicate`. If no alternatives satisfy the predicate, we're finished with T. If any alternative does not satisfy the predicate, add a '?' to T if it is not there already.

6 If `T1`, `T2`,..., `Tn` are all tables with confidence values, then either compute the confidence values for T's remaining alternatives and store them (*immediate confidence computation*), or set the confidence values to `NULL` (*lazy confidence computation*). See Sections 2.8 and 4 for further discussion.

7 Project each alternative of T onto the attributes in `attr-list`, generating a tuple in the query result. If there is an `INTO` clause, insert T into table `new-table`.

It can be verified easily that this operational semantics produces the `Suspects` result table shown with example data in Section 1.4. More generally it conforms to the "square diagram" (Figure 5.2) formal semantics given in Section 1.5. Later we will introduce constructs that do not conform to the square diagram because they go beyond relational operations.

Note that this operational semantics generates result tables in which, by default, all attributes are uncertain—it constructs result tuples from full tuple-alternatives. In reality, it is fairly straightforward to deduce statically, based on a query and the schemas of its input tables (specifically which attributes are certain and which are uncertain), those result attributes that are guaranteed to be certain. For example, if we joined `Saw` and `Drives` without projection, attribute `witness` in the result would be certain.

2.2 Querying Confidences

TriQL provides a built-in function `Conf()` for accessing confidence values. Suppose we want our `Suspects` query to only use sightings having confidence > 0.5 and drivers having confidence > 0.8. We write:

```
SELECT Drives.driver as person INTO Suspects
FROM Saw, Drives
WHERE Saw.color = Drives.color AND Saw.car = Drives.car
  AND Conf(Saw) > 0.5 AND Conf(Drives) > 0.8
```

In the operational semantics, when we evaluate the `predicate` over the alternatives in T in step 6, `Conf(Ti)` refers to the confidence associated with the t_i component of the alternative being evaluated. Note that this function may trigger confidence computations if confidence values are being computed lazily (recall Section 2.1).

Function `Conf()` is more general than as shown by the previous example—it can take any number of the tables appearing in the `FROM` clause as arguments. For example, `Conf(T1,T3,T5)` would return the "joint" confidence of the t_1, t_3, and t_5 components of the alternative being evaluated. If t_1, t_3, and t_5 are independent, their joint confidence is the product of their individual confidences. If they are nonindependent—typically due to shared lineage—then the computation is more complicated, paralleling confidence computation for query results discussed in Section 4 below. As a special case, `Conf(*)` is shorthand for `Conf(T1,T2,...,Tn)`, which normally corresponds to the confidence of the result tuple-alternative being constructed.

2.3 Querying Lineage

For querying lineage, TriQL introduces a built-in predicate designed to be used as a join condition. If we include predicate `Lineage`(T_1, T_2) in the `WHERE` clause of a TriQL query with ULDB tables T_1 and T_2 in its `FROM` clause, then we are constraining the joined T_1 and T_2 tuple-alternatives to be connected, directly or transitively, by lineage. For example, suppose we want to find all witnesses contributing to Hank being a high suspect. We can write:

```
SELECT S.witness
FROM HighSuspects H, Saw S
WHERE Lineage(H,S) AND H.person = 'Hank'
```

In the `WHERE` clause, `Lineage(H,S)` evaluates to true for any pair of tuple-alternatives t_1 and t_2 from `HighSuspects` and `Saw` such that t_1's lineage directly or transitively includes t_2. Of course we could write this query directly on the base tables if we remembered how `HighSuspects` was computed, but the `Lineage()` predicate provides a more general construct that is insensitive to query history.

Note that the the `Lineage()` predicate does not take into account the structure of lineage formulas: $\texttt{lineage}(T_1, T_2)$ is true for tuple-alternatives t_2 and t_2 if and only if, when we expand t_1's lineage formula using the lineage formulas of its components, t_2 appears at some point in the expanded formula. Effectively, the predicate is testing whether t_2 had any effect on t_1.

Here is a query that incorporates both lineage and confidence; it also demonstrates the "==>" shorthand for the `Lineage()` predicate. The query finds persons who are suspected based on high-confidence driving of a Honda:

```
SELECT Drives.driver
FROM Suspects, Drives
WHERE Suspects ==> Drives
  AND Drives.car = 'Honda' AND Conf(Drives) > 0.8
```

2.4 Duplicate Elimination

In ULDBs, duplicates may appear "horizontally"—when multiple alternatives in a tuple have the same value—and "vertically"—when multiple tuples have the same value for one or more alternatives. As in SQL, `DISTINCT` is used to merge vertical duplicates. A new keyword `MERGED` is used to merge horizontal duplicates. In both cases, merging can be thought of as an additional final step in the operational evaluation of Section 2.1. (`DISTINCT` subsumes `MERGED`, so the two options never co-occur.)

As a very simple example of horizontal merging, consider the query:

```
SELECT MERGED Saw.witness, Saw.color FROM Saw
```

The query result on our sample data with confidences (recall Section 1.3) is:

witness	color	
Amy	blue:0.7 || red:0.3	
Betty	red: 0.6	?

Without merging, the first result tuple would have two `blue` alternatives with confidence values 0.5 and 0.2. Note that confidences are summed when horizontal duplicates are merged. In terms of the formal semantics in Section 1.5, specifically the square diagram of Figure 5.2, merging horizontal duplicates in the query answer on the top-right of the square corresponds cleanly to merging duplicate possible-instances on the bottom-right.

A query with vertical duplicate-elimination was discussed at the end of Section 1.4, where `DISTINCT` was used to motivate lineage with disjunction.

2.5 Aggregation

For starters, TriQL supports standard SQL grouping and aggregation following the relational possible-instances semantics of Section 1.5. Consider the following query over the `Drives` data in Section 1.4:

```
SELECT car, COUNT(*) FROM Drives GROUP BY car
```

The query result is:

ID	car	count
71	Honda	1 ‖ 2
72	Mazda	1 ‖ 2

$\lambda(71,1) = (22,1) \wedge \neg\,(21,1)$
$\lambda(71,2) = (21,1) \wedge (22,1)$
$\lambda(72,1) = (23,1) \wedge \neg\,(21,2)$
$\lambda(72,2) = (21,2) \wedge (23,1)$

Note that attribute `car` is a certain attribute, since we're grouping by it. Also observe that lineage formulas in this example include negation.

In general, aggregation can be an exponential operation in ULDBs (and in other data models for uncertainty): the aggregate result may be different in every possible-instance, and there may be exponentially many possible-instances. (Consider for example `SUM` over a table comprised of 10 maybe-tuples. The result has 2^{10} possible values.) Thus, TriQL includes three additional options for aggregate functions: a *low* bound, a *high* bound, and an *expected* value; the last takes confidences into account when present. Consider for example the following two queries over the `Saw` data with confidences from Section 1.3. Aggregate function `ECOUNT` asks for the expected value of the `COUNT` aggregate.

```
SELECT color, COUNT(*) FROM Saw GROUP BY car
SELECT color, ECOUNT(*) FROM Saw GROUP BY car
```

The answer to the first query (omitting lineage) considers all possible-instances:

color	count	
blue	1:0.7	?
red	1:0.54 ‖ 2:0.18	?

The '?' on each tuple intuitively corresponds to a possible count of 0. (Note that zero counts never appear in the result of a SQL `GROUP BY` query.) The second query returns just one expected value for each group:

color	ecount
blue	0.7
red	0.9

It has been shown (see [9]) that expected aggregates are equivalent to taking the weighted average of the alternatives in the full aggregate result (also taking zero values into account), as seen in this example. Similarly, low and high bounds for aggregates are equivalent to the lowest and highest values in the full aggregate result.

In total, Trio supports 20 different aggregate functions: four versions (*full*, *low*, *high*, and *expected*) for each of the five standard functions (*count*, *min*, *max*, *sum*, *avg*).

2.6 Reorganizing Alternatives

TriQL has two constructs for reorganizing the tuple-alternatives in a query result:

- *Flatten* turns each tuple-alternative into its own tuple.
- *GroupAlts* regroups tuple-alternatives into new tuples based on a set of attributes.

As simple examples, and omitting lineage (which in both cases is a straightforward one-to-one mapping from result alternatives to source alternatives), "`SELECT FLATTEN * FROM Saw`" over the simple one-tuple `Saw` table from Section 1.4 gives:

witness	color	car	
Cathy	blue	Honda	?
Cathy	red	Mazda	?

and "`SELECT GROUPALTS(color,car) * FROM Drives`" gives:

color	car	person	
red	Honda	Jimmy	?
red	Mazda	Jimmy || Hank	
blue	Honda	Billy	

With `GROUPALTS`, the specified grouping attributes are certain attributes in the answer relation. For each set of values for these attributes, the corresponding tuple in the result contains the possible values for the remaining (uncertain) attributes as alternatives. '?' is present whenever all of the tuple-alternatives contributing to the result tuple are uncertain.

FLATTEN is primarily a syntactic operation—if lineage is retained (i.e., if the query does not also specify NoLineage, discussed below), then there is no change to possible-instances as a result of including FLATTEN in a query. GROUPALTS, on the other hand, may drastically change the possible-instances; it does not fit cleanly into the formal semantics of Section 1.5.

2.7 Horizontal Subqueries

"Horizontal" subqueries in TriQL enable querying across the alternatives that comprise individual tuples. As a contrived example, we can select from table Saw all Honda sightings where it's also possible the sighting was a car other than a Honda (i.e., all Honda alternatives with a non-Honda alternative in the same tuple).

```
SELECT * FROM Saw
WHERE car = 'Honda' AND EXISTS [car <> 'Honda']
```

Over the simple one-tuple Saw table from Section 1.4, the query returns just the first tuple-alternative, (Cathy,blue,Honda), of tuple 11.

In general, enclosing a subquery in [] instead of () causes the subquery to be evaluated over the "current" tuple, treating its alternatives as if they are a relation. Syntactic shortcuts are provided for common cases, such as simple filtering predicates as in the example above. More complex uses of horizontal subqueries introduce a number of subtleties; full details and numerous examples can be found in [11]. By their nature, horizontal subqueries query "across" possible-instances, so they do not follow the square diagram of Figure 5.2; they are defined operationally only.

2.8 Query-Defined Result Confidences

A query result includes confidence values only if all of the tables in its FROM clause have confidence values. To assign confidences to a table T for the purpose of query processing, "UNIFORM T" can be specified in the FROM clause, in which case confidence values are logically assigned across the alternatives and '?' in each of T's tuples using a uniform distribution.

Result confidence values respect a probabilistic interpretation, and they are computed by the system on-demand. (A "COMPUTE CONFIDENCES" clause can be added to a query to force confidence computation as part of query execution.) Algorithms for confidence computation are discussed later in Section 4. A query can override the default result confidence values, or add confidence values to a result that otherwise would not have them, by assigning values in its SELECT clause to the reserved attribute name conf. Furthermore, a special "value" UNIFORM may be assigned, in which case confidence values are assigned uniformly across the alternatives and '?' (if present) of each result tuple.

As an example demonstrating query-defined result confidences as well as UNIFORM in the FROM clause, suppose we generate suspects by joining the Saw table with confidences from Section 1.3 with the Drives table from Section 1.4. We decide to add uniform confidences to table Drives, and we prefer result confidences to be the lesser of the two input confidences, instead of their (probabilistic) product. Assuming a built-in function lesser, we write:

```
SELECT person, lesser(Conf(Saw),Conf(Drives)) AS conf
FROM Saw, UNIFORM Drives
WHERE Saw.color = Drives.color AND Saw.car = Drives.car
```

Let the two tuples in table Saw from Section 1.3 have IDs 81 and 82. The query result, including lineage, is:

ID	person		
91	Billy:0.5	?	$\lambda(91,1) = (81,1) \wedge (22,1)$
92	Jimmy:0.333	?	$\lambda(92,1) = (82,1) \wedge (21,2)$
93	Hank:0.6	?	$\lambda(93,1) = (82,1) \wedge (23,1)$

With probabilistic confidences, Jimmy would instead have confidence 0.2. Had we used greater() instead of lesser(), the three confidence values would have been 1.0, 0.6, and 1.0 respectively.

With the "AS Conf" feature, it is possible to create confidence values in a tuple whose sum exceeds 1. ("1.1 AS Conf," assigning confidence value 1.1 to each result tuple-alternative, is a trivial example.) Although the Trio prototype does not forbid this occurrence, a warning is issued, and anomalous behavior with respect to confidence values—either the newly created values, or later ones that depend on them—may subsequently occur.

2.9 Other TriQL Query Constructs

TriQL contains a number of additional constructs not elaborated in detail in this chapter, as follows. For comprehensive coverage of the TriQL language, see [11].

- TriQL is a strict superset of SQL, meaning that (in theory at least) every SQL construct is available in TriQL: subqueries, set operators, like predicates, and so on. Since SQL queries are relational, the semantics of any SQL construct over ULDBs follows the semantics for relational queries given in Section 1.5.
- One SQL construct not strictly relational is Order By. TriQL includes Order By, but only permits ordering by certain attributes and/or the special "attribute" Confidences, which for ordering purposes corresponds to the total confidence value (excluding '?') in each result tuple.

- In addition to built-in function `Conf()` and predicate `Lineage()`, TriQL offers a built-in predicate `Maybe()`. In a query, `Maybe(T)` returns true if and only if the tuple-alternative from table `T` being evaluated is part of a maybe-tuple, i.e., its tuple has a '?'.

- Horizontal subqueries (Section 2.7) are most useful in the `FROM` clause, but they are permitted in the `SELECT` clause as well. For example, the query "`SELECT [COUNT(*)] FROM Saw`" returns the number of alternatives in each tuple of the `Saw` table.

- As discussed in Section 2.8, preceding a table `T` in the `FROM` clause with keyword `UNIFORM` logically assigns confidence values to the tuple-alternatives in `T` for the duration of the query, using a uniform distribution. Similarly, "`UNIFORM AS conf`" in the `SELECT` clause assigns confidence values to query results using a uniform distribution. Another option for both uses is keyword `SCALED`. In this case, table `T` (respectively result tuples) must already have confidence values, but they are scaled logically for the duration of the query (respectively in the query result) so each tuple's total confidence is 1 (i.e., ?'s are removed). For example, if a tuple has two alternatives with confidence values 0.3 and 0.2, the `SCALED` confidences would be 0.6 and 0.4.

- Finally, three query qualifiers, `NoLineage`, `NoConf`, and `NoMaybe` may be used to signal that the query result should not include lineage, confidence values, or ?'s, respectively.

3. Data Modifications in Trio

Data modifications in Trio are initiated using TriQL's `INSERT`, `DELETE`, and `UPDATE` commands, which are in large part analogous to those in SQL. Additional modifications specific to the ULDB model are supported by extensions to these commands. The three statement types are presented in the following three subsections, followed by a discussion of how Trio incorporates *versioning* to support data modifications in the presence of derived relations with lineage.

3.1 Inserts

Inserting entirely new tuples into a ULDB poses no unusual semantic issues. (Inserting new alternatives into existing tuples is achieved through the `UPDATE` command, discussed below.) Trio supports both types of SQL `INSERT` commands:

```
INSERT INTO table-name VALUES tuple-spec
INSERT INTO table-name subquery
```

The tuple-spec uses a designated syntax to specify a complete Trio tuple to be inserted, including certain attributes, alternative values for uncertain attributes, confidence values, and/or '?,' but no lineage. The subquery is any TriQL query whose result tuples are inserted, together with their lineage (unless NoLineage is specified in the subquery; Section 2.9).

3.2 Deletes

Deletion also follows standard SQL syntax:

```
DELETE FROM table-name WHERE predicate
```

This command deletes each tuple-alternative satisfying the predicate. (Deleting a tuple-alternative is equivalent to deleting one alternative for the uncertain attributes; Section 1.1.) If all alternatives of a tuple are deleted, the tuple itself is deleted. A special qualifier "AdjConf" can be used to redistribute confidence values on tuples after one or more alternatives are deleted; without AdjConf, deleted confidence values implicitly move to '?.'

3.3 Updates

In addition to conventional updates, the TriQL UPDATE command supports updating confidence values, adding and removing '?'s, and inserting new alternatives into existing tuples. Consider first the standard SQL UPDATE command:

```
UPDATE table-name SET attr-list = expr-list WHERE predicate
```

This command updates every tuple-alternative satisfying the predicate, setting each attribute in the attr-list to the result of the corresponding expression in the expr-list.

There is one important restriction regarding the combination of certain and uncertain attributes. Consider as an example the following command, intended to rename as "Doris" every witness who saw a blue Honda:

```
UPDATE Saw SET witness = 'Doris'
WHERE color = 'blue' AND car = 'Honda'
```

In the Saw table of Section 1.1, the WHERE predicate is satisfied by some but not all of the (color,car) alternatives for witness Amy. Thus, it isn't obvious whether Amy should be be modified. Perhaps the best solution would be to convert witness to an uncertain attribute:

(witness,color, car)		
(Doris,blue,Honda)	(Amy,red,Toyota)	(Amy,blue,Mazda)

However, Trio treats attribute types (certain versus uncertain) as part of the fixed schema, declared at CREATE TABLE time. A similar ambiguity can arise if the expression on the right-hand-side of the SET clause for a certain attribute produces different values for different alternatives. Hence, UPDATE commands are permitted to modify certain attributes only if all references to uncertain attributes, function Conf(), and predicate Lineage() in the WHERE predicate, and in every SET expression corresponding to a certain attribute, occur within horizontal subqueries. This restriction ensures that the predicate and the expression always evaluate to the same result for all alternatives of a tuple. For our example, the following similar-looking command updates every witness who *may* have seen a blue Honda to be named "Doris":

```
UPDATE Saw SET witness = 'Doris'
WHERE [color = 'blue' AND car = 'Honda']
```

To update confidence values, the special attribute conf may be specified in the attr-list of the UPDATE command. As with query-defined result confidences (Section 2.8), there is no guarantee after modifying conf that confidence values in a tuple sum to ≤ 1; a warning is issued when they don't, and anomalous behavior may subsequently occur. Finally, the special keywords UNIFORM or SCALED may be used as the expression corresponding to attribute conf in the SET clause, to modify confidence values across each tuple using uniform or rescaled distributions—analogous to the use of these keywords with "AS Conf" (Sections 2.8 and 2.9).

A variation on the UPDATE command is used to add alternatives to existing tuples:

```
UPDATE table-name ALTINSERT expression WHERE predicate
```

To ensure the predicate either holds or doesn't on entire tuples, once again all references to uncertain attributes, Conf(), and Lineage() must occur within horizontal subqueries. For each tuple satisfying the predicate, alternatives are added to the tuple, based on the result of evaluating the expression. Like the INSERT command (Section 3.1), the expression can be "VALUES tuple-spec" to specify a single alternative, or a subquery producing zero or more alternatives. Either way, the schema of the alternatives to add must match the schema of the table's uncertain attributes only. If adding alternatives to an existing tuple creates duplicates, by default horizontal duplicate-elimination does not occur, but it can be triggered by specifying UPDATE MERGED. As with other constructs that affect confidence values, creating tuples whose confidences sum to > 1 results in a warning.

Finally, the following self-explanatory UPDATE commands can be used to add and remove ?'s. These commands may only be applied to tables without confidences, and once again, in the predicate all references to uncertain attributes, Conf(), and Lineage() must be within horizontal subqueries.

```
UPDATE table-name ADDMAYBE WHERE predicate
UPDATE table-name DELMAYBE WHERE predicate
```

3.4 Data Modifications and Versioning

Trio query results include lineage identifying the input data from which the results were derived. Lineage is not only a user-level feature—it is needed for on-demand confidence computation, and it is critical for capturing the correct possible-instances in a query result (Section 1.4).

Suppose we run our `Suspects` query, store the result, then modifications occur to some alternatives in table `Saw` that are referenced by lineage in table `Suspects`. There are two basic options for handling such modifications:

(1) *Propagate* modifications to all derived tables, effectively turning query results into materialized views.

(2) *Don't propagate* modifications, allowing query results to become "stale" with respect to the data from which they were derived originally.

Option (1) introduces a variation on the well-known *materialized view maintenance problem*. It turns out Trio's lineage feature can be used here for broad applicability and easy implementation of the most efficient known techniques; see [6].

With option (2), after modifications occur, lineage formulas may contain incorrect or "dangling" pointers. Trio's solution to this problem is to introduce a lightweight *versioning* system: Modified data is never removed, instead it remains in the database as part of a previous version. The lineage formula for a derived tuple-alternative t may refer to alternatives in the current version and/or previous versions, thus accurately reflecting the data from which t was derived. Details of Trio's versioning system and how it interacts with data modifications and lineage can be found in [6].

4. Confidence Computation

Computing confidence values for query results is one of the most interesting and challenging aspects of Trio. In general, efficient computation of correct result confidence values in uncertain and probabilistic databases is known to be a difficult problem. Trio uses two interrelated techniques to address the problem:

1. By default, confidence values are not computed during query evaluation. Instead, they are computed on demand: when requested through one of Trio's interfaces, or as needed for further queries. This approach has two benefits: (a) Computing confidence values as part of query evaluation constrains how queries may be evaluated, while lazy computation frees the system to select any valid relational query execution plan. (See [7] for detailed discussion.) (b) If a confidence value is never needed, its potentially expensive computation is never performed.

2. On-demand confidence computation is enabled by Trio's lineage feature. Specifically, the confidence of an alternative in a query result can be computed through lineage, as described below. Furthermore, a number of optimizations are possible to speed up the computation, also discussed below.

Suppose a query Q is executed producing a result table T, and consider tuple-alternative t in T. Assume all tables in query Q have confidence values (perhaps not yet computed), so t should have a confidence value as well. Formally, the confidence value assigned to t should represent the total probability of the possible-instances of result table T that contain alternative t (recall Section 1.3). It has been shown (see [7]) that this probability can be computed as follows:

1. Expand t's lineage formula recursively until it refers to base alternatives only: If $\lambda(t)$ refers to base alternatives only, stop. Otherwise, pick one t_i in $\lambda(t)$ that is not a base alternative, replace t_i with $(\lambda(t_i))$, and continue expanding.

2. Let f be the expanded formula from step 1. If f contains any sets $t_1, \ldots, t_n$ of two or more alternatives from the same tuple (a possible but unlikely case), then $t_1, \ldots, t_n$'s confidence values are modified for the duration of the computation, and clauses are added to f to encode their mutual exclusion; details are given in [7].

3. The confidence value for alternative t is the probability of formula f computed using the confidence values for the base alternatives comprising f.

It is tempting to expand formula $\lambda(t)$ in step 1 only as far as needed to obtain confidence values for all of the alternatives mentioned in the formula. However, expanding to the base alternatives is required for correctness in the general case. Consider for example the following scenario, where t_3, t_4, and t_5 are base alternatives.

$$\lambda(t) = t_1 \wedge t_2 \quad \lambda(t_1) = t_3 \wedge t_4 \quad \lambda(t_2) = t_3 \wedge t_5$$
$$\texttt{Conf}(t_3) = \texttt{Conf}(t_4) = \texttt{Conf}(t_5) = 0.5$$

Based on the specified confidences, we have $\texttt{Conf}(t_1) = \texttt{Conf}(t_2) = 0.25$. If we computed $\texttt{Conf}(t)$ using $t_1 \wedge t_2$ we would get 0.0625, whereas the correct value expanding to the base alternatives is 0.125. As this example demonstrates, lineage formulas must be expanded all the way to base alternatives because derived alternatives may not be probabilistically independent.

Trio incorporates some optimizations to the basic confidence-computation algorithm just described:

- Whenever confidence values are computed, they are *memoized* for future use.

- There are cases when it is not necessary to expand a lineage formula all the way to its base alternatives. At any point in the expansion, if all of the alternatives in the formula are known to be independent, and their confidences have already been computed (and therefore memoized), there is no need to go further. Even when confidences have not been computed, independence allows the confidence values to be computed separately and then combined, typically reducing the overall complexity. Although one has to assume nonindependence in the general case, independence is common and often can be easy to deduce and check, frequently at the level of entire tables.

- We have developed algorithms for *batch* confidence computation that are implemented through SQL queries. These algorithms are appropriate and efficient when confidence values are desired for a significant portion of a result table.

Reference [7] provides detailed coverage of the confidence-computation problem, along with our algorithms, optimizations, implementation in the Trio prototype.

5. Additional Trio Features

TriQL queries and data modifications are the typical way of interacting with Trio data, just as SQL is used in a standard relational DBMS. However, uncertainty and lineage in ULDBs introduce some interesting features beyond just queries and modifications.

Lineage. As TriQL queries are executed and their results are stored, and additional queries are posed over previous results, complex lineage relationships can arise. Data-level lineage is used for confidence computation (Section 4) and `Lineage()` predicates; it is also used for *coexistence checks* and *extraneous data removal*, discussed later in this section. The *TrioExplorer* graphical user interface supports data-level lineage tracing through special buttons next to each displayed alternative; the textual and API interfaces provide corresponding functionality.

Trio also maintains a schema-level lineage graph (specifically a DAG), with tables as nodes and edges representing lineage relationships. This graph is used when translating queries with `Lineage()` predicates (Section 6.7), and for determining independence to optimize confidence computation (Section 4). This graph also is helpful for for users to understand the tables in a database and their interrelationships. A schema-level lineage graph was depicted in the Figure 5.1 screenshot showing the *TrioExplorer* interface.

Coexistence Checks. A user may wish to select a set of alternatives from one or more tables and ask whether those alternatives can all coexist. Two alternatives from the same tuple clearly cannot coexist, but the general case must take into account arbitrarily complex lineage relationships as well as tuple alternatives. For example, if we asked about alternatives (11,2) and (32,1) in our sample database of Section 1.4, the system would tell us these alternatives cannot coexist.

Checking coexistence is closely related to confidence computation. To check if alternatives t_1 and t_2 can coexist, we first expand their lineage formulas to reference base alternatives only, as in step 1 of confidence computation (Section 4). Call the expanded formulas f_1 and f_2. Let f_3 be an additional formula that encodes mutual exclusion of any alternatives from the same tuple appearing in f_1 and/or f_2, as in step 2 of confidence computation. Then t_1 and t_2 can coexist if and only if formula $f_1 \wedge f_2 \wedge f_3$ is satisfiable. Note that an equivalent formulation of this algorithm creates a "dummy" tuple t whose lineage is $t_1 \wedge t_2$. Then t_1 and t_2 can coexist if and only if $\texttt{Conf}(t) > 0$. This formulation shows clearly the relationship between coexistence and confidence computation, highlighting in particular that our optimizations for confidence computation in Section 4 can be used for coexistence checks as well.

Extraneous Data Removal. The natural execution of TriQL queries can generate *extraneous data*: an alternative is extraneous if it can never be chosen (i.e., its lineage requires presence of multiple alternatives that cannot coexist); a '?' annotation is extraneous if its tuple is always present. It is possible to check for extraneous alternatives and ?'s immediately after query execution (and, sometimes, as part of query execution), but checking can be expensive. Because we expect extraneous data and ?'s to be relatively uncommon, and users may not be concerned about them, by default Trio supports extraneous data removal as a separate operation, similar to garbage collection.

Like coexistence checking, extraneous data detection is closely related to confidence computation: An alternative t is extraneous if and only if $\texttt{Conf}(t) = 0$. A '?' on a tuple u is extraneous if and only if the confidence values for all of u's alternatives sum to 1.

6. The Trio System

Figure 5.3 shows the basic three-layer architecture of the Trio system. The core system is implemented in Python and mediates between the underlying relational DBMS and Trio interfaces and applications. The Python layer presents a simple Trio API that extends the standard Python DB 2.0 API for database

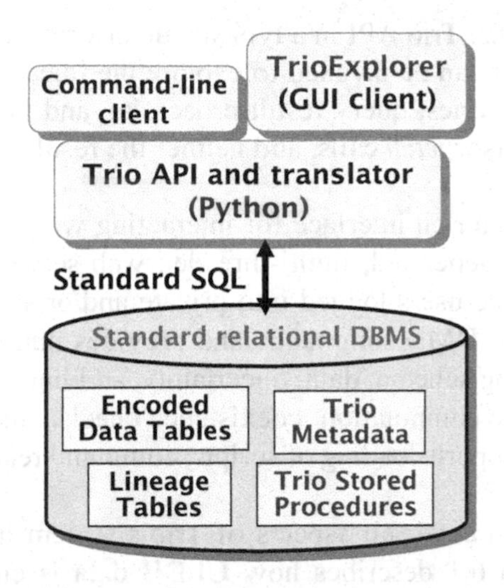

Figure 5.3. Trio Basic System Architecture.

access (Python's analog of JDBC). The Trio API accepts TriQL queries and modification commands in addition to regular SQL, and query results may be ULDB tuples as well as regular tuples. The API also exposes the other Trio-specific features described in Section 5. Using the Trio API, we built a generic command-line interactive client (*TrioPlus*) similar to that provided by most DBMS's, and the *TrioExplorer* graphical user interface shown earlier in Figure 5.1.

Trio DDL commands are translated via Python to SQL DDL commands based on the encoding to be described in Section 6.1. The translation is fairly straightforward, as is the corresponding translation of `INSERT` statements and bulk load.

Processing of TriQL queries proceeds in two phases. In the *translation* phase, a TriQL parse tree is created and progressively transformed into a tree representing one or more standard SQL statements, based on the data encoding scheme. In the *execution* phase, the SQL statements are executed against the relational database encoding. Depending on the original TriQL query, Trio stored procedures may be invoked and some post-processing may occur. For efficiency, most additional runtime processing executes within the DBMS server. Processing of TriQL data modification commands is similar, although a single TriQL command often results in a larger number of SQL statements, since several relational tables in the encoding (Section 6.1) may all need to be modified.

TriQL query results can either be *stored* or *transient*. Stored query results (indicated by an `INTO` clause in the query) are placed in a new persistent table, and lineage relationships from the query's result data to data in the query's input tables also is stored persistently. Transient query results (no `INTO` clause) are accessed through the Trio API in a typical cursor-oriented fashion, with an additional method that can be invoked to explore the lineage of each returned tuple. For transient queries, query result processing and lineage creation occurs in response to cursor *fetch* calls, and neither the result data nor its lineage are persistent.

TrioExplorer offers a rich interface for interacting with the Trio system. It implements a Python-generated, multi-threaded web server using *CherryPy*, and it supports multiple users logged into private and/or shared databases. It accepts Trio DDL and DML commands and provides numerous features for browsing and exploring schema, data, uncertainty, and lineage. It also enables on-demand confidence computation, coexistence checks, and extraneous data removal. Finally, it supports loading of scripts, command recall, and other user conveniences.

It is not possible to cover all aspects of Trio's system implementation in this chapter. Section 6.1 describes how ULDB data is encoded in regular relations. Section 6.2 demonstrates the basic query translation scheme for `SELECT-FROM-WHERE` statements, while Sections 6.3–6.9 describe translations and algorithms for most of TriQL's additional constructs.

6.1 Encoding ULDB Data

We now describe how ULDB databases are encoded in regular relational tables. For this discussion we use *u-tuple* to refer to a tuple in the ULDB model, i.e., a tuple that may include alternatives, '?', and confidence values, and *tuple* to denote a regular relational tuple.

Let $T(A_1, \ldots, A_n)$ be a ULDB table. We store the data portion of T in two relational tables, T_C and T_U. Table T_C contains one tuple for each u-tuple in T. T_C's schema consists of the certain attributes of T, along with two additional attributes:

- `xid` contains a unique identifier assigned to each u-tuple in T.
- `num` contains a number used to track efficiently whether or not a u-tuple has a '?', when T has no confidence values. (See Section 6.2 for further discussion.)

Table T_U contains one tuple for each tuple-alternative in T. Its schema consists of the uncertain attributes of T, along with three additional attributes:

- `aid` contains a unique identifier assigned to each alternative in T.
- `xid` identifies the u-tuple that this alternative belongs to.
- `conf` stores the confidence of the alternative, or `NULL` if this confidence value has not (yet) been computed, or if T has no confidences.

Clearly several optimizations are possible: Tables with confidence values can omit the `num` field, while tables without confidences can omit `conf`. If a table T with confidences has no certain attributes, then table T_C is not needed since it would contain only `xid`'s, which also appear in T_U. Conversely, if T contains no uncertain attributes, then table T_U is not needed: attribute `aid` is unnecessary, and attribute `conf` is added to table T_C. Even when both tables are present, the system automatically creates a virtual view that joins the two tables, as a convenience for query translation (Section 6.2).

The system always creates indexes on T_C.`xid`, T_U.`aid`, and T_U.`xid`. In addition, Trio users may create indexes on any of the original data attributes $A_1, \ldots, A_n$ using standard `CREATE INDEX` commands, which are translated by Trio to `CREATE INDEX` commands on the appropriate underlying tables.

The lineage information for each ULDB table T is stored in a separate relational table. Recall the lineage $\lambda(t)$ of a tuple-alternative t is a boolean formula. The system represents lineage formulas in *disjunctive normal form* (DNF), i.e., as a disjunction of conjunctive clauses, with all negations pushed to the "leaves." Doing so allows for a uniform representation: Lineage is stored in a single table T_L(`aid`, `src_aid`, `src_table`,`flag`), indexed on `aid` and `src_aid`. A tuple (t_1, t_2, T_2, f) in T_L denotes that T's alternative t_1 has alternative t_2 from table T_2 in its lineage. Multiple lineage relationships for a given alternative are conjunctive by default; special values for `flag` and (occasionally) "dummy" entries are used to encode negation and disjunction. By far the most common type of lineage is purely conjunctive, which is represented and manipulated very efficiently with this scheme.

Example. As one example that demonstrates many aspects of the encoding, consider the aggregation query result from Section 2.5. Call the result table R. Recall that attribute `car` is certain while attribute `count` is uncertain. The encoding as relational tables follows, omitting the lineage for result tuple 72 since it parallels that for 71.

R_C:

xid	num	car
71	2	Honda
72	2	Mazda

R_U:

aid	xid	count
711	71	1
712	71	2
721	72	1
722	72	2

R_L:

aid	src_aid	src_table	flag
711	221	Drives	NULL
711	211	Drives	neg
712	211	Drives	NULL
712	221	Drives	neg

For readability, unique `aid`'s are created by concatenating `xid` and alternative number. The values of 2 in attribute `R_C.num` indicate no '?'s (see Section 6.2), and R_U.conf is omitted since there are no confidence values. The remaining attributes should be self-explanatory given the discussion of the encoding above. In addition, the system automatically creates a virtual view joining tables R_C and R_U on xid.

6.2 Basic Query Translation Scheme

Consider the Suspects query from the beginning of Section 2, first in its transient form (i.e., without CREATE TABLE). The Trio Python layer translates the TriQL query into the following SQL query, sends it to the underlying DBMS, and opens a cursor on the result. The translated query refers to the virtual views joining Saw_C and Saw_U, and joining Drives_C, and Drives_U; call these views Saw_E and Drives_E ("E" for encoding) respectively.

```
SELECT Drives_E.driver,
       Saw_E.aid, Drives_E.aid, Saw_E.xid, Drives_E.xid,
       (Saw_E.num * Drives_E.num) AS num
FROM Saw_E, Drives_E
WHERE Saw_E.color = Drives_E.color AND Saw_E.car = Drives_E.car
ORDER BY Saw_E.xid, Drives_E.xid
```

Let *Tfetch* denote a cursor call to the Trio API for the original TriQL query, and let *Dfetch* denote a cursor call to the underlying DBMS for the translated SQL query. Each call to *Tfetch* must return a complete u-tuple, which may entail several calls to *Dfetch*: Each tuple returned from *Dfetch* on the SQL query corresponds to one alternative in the TriQL query result, and the set of alternatives with the same returned Saw_E.xid and Drives_E.xid pair comprise a single result u-tuple (as specified in the operational semantics of Section 2.1). Thus, on *Tfetch*, Trio collects all SQL result tuples for a single Saw_E.xid/Drives_E.xid pair (enabled by the ORDER BY clause in the SQL query), generates a new xid and new aid's, and constructs and returns the result u-tuple.

Note that the underlying SQL query also returns the aid's from Saw_E and Drives_E. These values (together with the table names) are used to construct the lineage for the alternatives in the result u-tuple. Recall that the num field is used to encode the presence or absence of '?': Our scheme maintains the invariant that an alternative's u-tuple has a '?' if and only if its num field exceeds the u-tuple's number of alternatives, which turns out to be efficient to maintain for most queries. This example does not have result confidence values, however even if it did, result confidence values by default are not computed until they are explicitly requested (recall Section 4). When a "COMPUTE CONFIDENCES" clause is present, *Tfetch* invokes confidence computation before returning its result tuple. Otherwise, *Tfetch* returns placeholder NULLs for all confidence values.

For the stored (`CREATE TABLE`) version of the query, Trio first issues DDL commands to create the new tables, indexes, and virtual view that will encode the query result. Trio then executes the same SQL query shown above, except instead of constructing and returning u-tuples one at a time, the system directly inserts the new alternatives and their lineage into the result and lineage tables, already in their encoded form. All processing occurs within a stored procedure on the database server, thus avoiding unnecessary round-trips between the Python module and the underlying DBMS.

The remaining subsections discuss how TriQL constructs beyond simple `SELECT-FROM-WHERE` statements are translated and executed. All translations are based on the data encoding scheme of Section 6.1; many are purely "add-ons" to the basic translation just presented.

6.3 Duplicate Elimination

Recall from Section 2.4 that TriQL supports "horizontal" duplicate-elimination with the `MERGED` option, as well as conventional `DISTINCT`. In general, either type of duplicate-elimination occurs as the final step in a query that may also include filtering, joins, and other operations. Thus, after duplicate-elimination, the lineage of each result alternative is a formula in DNF (recall Section 6.1): disjuncts are the result of merged duplicates, while conjunction within each disjunct represents a tuple-alternative's derivation prior to merging; a good example can be seen at the end of Section 1.4. How Trio encodes DNF formulas in lineage tables was discussed briefly in Section 6.1.

Merging horizontal duplicates and creating the corresponding disjunctive lineage can occur entirely within the *Tfetch* method: All alternatives for each result u-tuple, together with their lineage, already need to be collected within *Tfetch* before the u-tuple is returned. Thus, when `MERGED` is specified, *Tfetch* merges all duplicate alternatives and creates the disjunctive lineage for them, then returns the modified u-tuple.

`DISTINCT` is more complicated, requiring two phases. First, a translated SQL query is produced as if `DISTINCT` were not present, except the result is ordered by the data attributes instead of `xid`'s; this query produces a temporary result T. One scan through T is required to merge duplicates and create disjunctive lineage, then T is reordered by `xid`'s to construct the correct u-tuples in the final result.

6.4 Aggregation

Recall from Section 2.5 that TriQL supports 20 different aggregation functions: four versions (*full*, *low*, *high*, and *expected*) for each of the five standard functions (*count*, *min*, *max*, *sum*, *avg*). All of the *full* functions and some of the other options cannot be translated to SQL queries over the encoded data, and thus are implemented as stored procedures. (One of them, *expected average*, is implemented as an approximation, since finding the exact answer based on possible-instances can be extremely expensive [9].) Many of the options, however, can be translated very easily. Consider table `Saw` with confidence values. Then the TriQL query:

```
SELECT color, ECOUNT(*) FROM Saw GROUP BY car
```

is translated based on the encoding to:

```
SELECT color, SUM(conf) FROM Saw_E GROUP BY car
```

A full description of the implementation of Trio's 20 aggregate functions can be found in [9].

6.5 Reorganizing Alternatives

Recall *Flatten* and *GroupAlts* from Section 2.6. The translation scheme for queries with *Flatten* is a simple modification to the basic scheme in which each result alternative is assigned its own `xid`. *GroupAlts* is also a straightforward modification: Instead of the translated SQL query grouping by `xid`'s from the input tables to create result u-tuples, it groups by the attributes specified in `GROUPALTS` and generates new `xid`'s.

6.6 Horizontal Subqueries

Horizontal subqueries are very powerful yet surprisingly easy to implement based on our data encoding. Consider the example from Section 2.7:

```
SELECT * FROM Saw
WHERE car = 'Honda' AND EXISTS [car <> 'Honda']
```

First, syntactic shortcuts are expanded. In the example, `[car <> 'Honda']` is a shortcut for `[SELECT * FROM Saw WHERE car<>'Honda']`. Here, `Saw` within the horizontal subquery refers to the `Saw` alternatives in the current u-tuple being evaluated [11]. In the translation, the horizontal subquery is replaced with a standard SQL subquery that adds aliases for inner tables and a condition correlating `xid`'s with the outer query:

```
... AND EXISTS (SELECT * FROM Saw_E S
                WHERE car <> 'Honda' AND S.xid = Saw_E.xid)
```

`S.xid=Saw_E.xid` restricts the horizontal subquery to operate on the data in the current u-tuple. Translation for the general case involves a fair amount of context and bookkeeping to ensure proper aliasing and ambiguity checks, but all horizontal subqueries, regardless of their complexity, have a direct translation to regular SQL subqueries with additional `xid` equality conditions.

6.7 Built-In Predicates and Functions

Trio has three built-in predicates and functions: `Conf()` introduced in Section 2.2, `Maybe()` introduced in Section 2.9, and `Lineage()` introduced in Section 2.3.

Function `Conf()` is implemented as a stored procedure. If it has just one argument T, the procedure first examines the current T_E.conf field to see if a value is present. (Recall from Section 6.1 that T_E is the encoded data table, typically a virtual view over tables T_C and T_U.) If so, that value is returned. If T_E.conf is NULL, on-demand confidence computation is invoked (see Section 4); the resulting confidence value is stored permanently in T_E and returned.

The situation is more complicated when `Conf()` has multiple arguments, or the special argument "$\star$" as an abbreviation for all tables in the query's FROM list (recall Section 2.2). The algorithm for arguments $T_1, \ldots, T_k$ logically constructs a "dummy" tuple-alternative t whose lineage is the conjunction of the current tuple-alternatives from $T_1, \ldots, T_k$ being considered. It then computes t's confidence, which provides the correct result for the current invocation of Conf($T_1, \ldots, T_k$). In the case of Conf($\star$), the computed values usually also provide confidence values for the query result, without a need for on-demand computation.

The `Maybe()` and `Lineage()` predicates are incorporated into the query translation phase. Predicate `Maybe()` is straightforward: It translates to a simple comparison between the num attribute and the number of alternatives in the current u-tuple. (One subtlety is that `Maybe()` returns true even when a tuple's question mark is "extraneous"—that is, the tuple in fact always has an alternative present, due to its lineage. See Section 5 for a brief discussion.)

Predicate Lineage(T_1,T_2) is translated into one or more SQL EXISTS subqueries that check if the lineage relationship holds: Schema-level lineage information is used to determine the possible table-level "paths" from T_1 to T_2. Each path produces a subquery that joins lineage tables along that path, with T_1 and T_2 at the endpoints; these subqueries are then OR'd to replace predicate Lineage(T_1,T_2) in the translated query.

As an example, recall table HighSuspects in Section 1.4, derived from table Suspects, which in turn is derived from table Saw. Then predicate `Lineage(HighSuspects, Saw)` would be translated into one subquery as follows, recalling the lineage encoding described in Section 6.1.

```
EXISTS (SELECT *
  FROM HighSuspects_L L1, Suspects_L L2
  WHERE HighSuspects.aid = L1.aid
  AND L1.src_table = 'Suspects' AND L1.src_aid = L2.aid
  AND L2.src_table = 'Saw' AND L2.src_aid = Saw.aid )
```

6.8 Query-Defined Result Confidences

The default probabilistic interpretation of confidence values in query results can be overridden by including "*expression* AS conf" in the SELECT clause of a TriQL query (Section 2.8). Since Trio's data encoding scheme uses a column called conf to store confidence values, "AS conf" clauses simply pass through the query translation phase unmodified.

6.9 Remaining Constructs

We briefly describe implementation of the remaining TriQL constructs and features.

- **Rest of SQL.** As mentioned in Section 2.9, since TriQL is a superset of SQL, any complete TriQL implementation must handle all of SQL. In our translation-based scheme, some constructs (e.g., LIKE predicates) can be passed through directly to the underlying relational DBMS, while others (e.g., set operators, some subqueries) can involve substantial rewriting during query translation to preserve TriQL semantics. At the time of writing this chapter, the Trio prototype supports all of the constructs discussed or used by examples in this chapter, as well as set operators UNION, INTERSECT, and EXCEPT.

- **Order By.** Because ordering by xid's is an important part of the basic query translation (Section 6.2), ORDER BY clauses in TriQL require materializing the result first, then ordering by the specified attributes. When special "attribute" Confidences (Section 2.9) is part of the ORDER BY list, "COMPUTE CONFIDENCES" (Section 2.8) is logically added to the query, to ensure the conf field contains actual values, not placeholder NULLs, before sorting occurs.

- **UNIFORM and SCALED.** The keywords UNIFORM (Section 2.8) and SCALED (Section 2.9) can be used in a TriQL FROM clause to add or modify confidences on an input table, or with "AS conf" to specify confidences on the result. The "AS conf" usage is easy to implement within the *Tfetch* procedure (Section 6.2): *Tfetch* processes entire u-tuples one at a time and can easily add or modify confidence values before returning them.

  UNIFORM and SCALED in the FROM clause are somewhat more complex: Confidence computation for the query result must occur during query processing (as opposed to on-demand), to ensure result confidence values take into account the modifier(s) in the FROM clause. (Alternatively, special flags could be set, then checked during later confidence computation, but Trio does not use this approach.) Special processing again occurs in *Tfetch*, which logically adds or modifies confidence values on input alternatives when computing confidence values for the query result.

- **NoLineage, NoConf, and NoMaybe.** These TriQL options are all quite easy to implement: `NoLineage` computes confidence values for the query result as appropriate (since no lineage is maintained by which to compute confidences later), then essentially turns the query result into a Trio base table. `NoConf` can only be specified in queries that otherwise would include confidence values in the result; now the result is marked as a Trio table without confidences (and, of course, does not compute confidence values except as needed for query processing). Finally, `NoMaybe` can only be specified in queries that produce results without confidences; all ?'s that otherwise would be included in the result are removed by modifying the `num` field in the encoding (Section 6.1).

- **Data modifications and versioning.** Recall from Section 3.4 that Trio supports a lightweight versioning system, in order to allow data modifications to base tables that are not propagated to derived tables, while still maintaining "meaningful" lineage on the derived data. Implementation of the versioning system is quite straightforward: If a ULDB table T is versioned, *start-version* and *end-version* attributes are added to encoded table T_U (Section 6.1). A query over versioned tables can produce a versioned result with little overhead, thanks to the presence of lineage. Alternatively, queries can request *snapshot* results, as of the current or a past version. Data modifications often simply manipulate versions rather than modify the data, again with little overhead. For example, deleting an alternative t from a versioned table T translates to modifying t's *end-version* in T_U. Reference [6] provides details of how the Trio system implements versions, data modifications, and the propagation of modifications to derived query results when desired.

References

[1] P. Agrawal, O. Benjelloun, A. Das Sarma, C. Hayworth, S. Nabar, T. Sugihara, and J. Widom. Trio: A system for data, uncertainty, and lineage. In *Proc. of Intl. Conference on Very Large Databases (VLDB)*, pages 1151–1154, Seoul, Korea, September 2006. *Demonstration description.*

[2] O. Benjelloun, A. Das Sarma, A. Halevy, and J. Widom. ULDBs: Databases with uncertainty and lineage. In *Proc. of Intl. Conference on Very Large Databases (VLDB)*, pages 953–964, Seoul, Korea, September 2006.

[3] O. Benjelloun, A. Das Sarma, C. Hayworth, and J. Widom. An introduction to ULDBs and the Trio system. *IEEE Data Engineering Bulletin, Special Issue on Probabilistic Databases*, 29(1):5–16, March 2006.

[4] A. Das Sarma, P. Agrawal, S. Nabar, and J. Widom. Towards special-purpose indexes and statistics for uncertain data. In *Proc. of the Workshop on Management of Uncertain Data*, Auckland, New Zealand, August 2008.

[5] A. Das Sarma, O. Benjelloun, A. Halevy, and J. Widom. Working models for uncertain data. In *Proc. of Intl. Conference on Data Engineering (ICDE)*, Atlanta, Georgia, April 2006.

[6] A. Das Sarma, M. Theobald, and J. Widom. Data modifications and versioning in Trio. Technical report, Stanford University InfoLab, March 2008. Available at: `http://dbpubs.stanford.edu/pub/2008-5`.

[7] A. Das Sarma, M. Theobald, and J. Widom. Exploiting lineage for confidence computation in uncertain and probabilistic databases. In *Proc. of Intl. Conference on Data Engineering (ICDE)*, Cancun, Mexico, April 2008.

[8] A. Das Sarma, J.D. Ullman, and J. Widom. Schema design for uncertain databases. Technical report, Stanford University InfoLab, November 2007. Available at: `http://dbpubs.stanford.edu/pub/2007-36`.

[9] R. Murthy and J. Widom. Making aggregation work in uncertain and probabilistic databases. In *Proc. of the Workshop on Management of Uncertain Data*, pages 76–90, Vienna, Austria, September 2007.

[10] M. Mutsuzaki, M. Theobald, A. de Keijzer, J. Widom, P. Agrawal, O. Benjelloun, A. Das Sarma, R. Murthy, , and T. Sugihara. Trio-One: Layering uncertainty and lineage on a conventional DBMS. In *Proc. of Conference on Innovative Data Systems Research (CIDR)*, Pacific Grove, California, 2007.

[11] TriQL: The Trio Query Language. Available from: `http://i.stanford.edu/trio`.

[12] J. Widom. Trio: A system for integrated management of data, accuracy, and lineage. In *Proc. of Conference on Innovative Data Systems Research (CIDR)*, Pacific Grove, California, 2005.

Chapter 6

MAYBMS: A SYSTEM FOR MANAGING LARGE UNCERTAIN AND PROBABILISTIC DATABASES

Christoph Koch
Cornell University, Ithaca, NY
koch@cs.cornell.edu

Abstract MayBMS is a state-of-the-art probabilistic database management system that has been built as an extension of Postgres, an open-source relational database management system. MayBMS follows a principled approach to leveraging the strengths of previous database research for achieving scalability. This chapter describes the main goals of this project, the design of query and update language, efficient exact and approximate query processing, and algorithmic and systems aspects.

Acknowledgments. My collaborators on the MayBMS project are Dan Olteanu (Oxford University), Lyublena Antova (Cornell), Jiewen Huang (Oxford), and Michaela Goetz (Cornell). Thomas Jansen and Ali Baran Sari are alumni of the MayBMS team. I thank Dan Suciu for the inspirational talk he gave at a Dagstuhl seminar in February of 2005, which triggered my interest in probabilistic databases and the start of the project. I am also indebted to Joseph Halpern for insightful discussions. The project was previously supported by German Science Foundation (DFG) grant KO 3491/1-1 and by funding provided by the Center for Bioinformatics (ZBI) at Saarland University. It is currently supported by NSF grant IIS-0812272, a KDD grant, and a gift from Intel.

Keywords: Probabilistic database, system, query language, updates

1. Introduction

Database systems for uncertain and probabilistic data promise to have many applications. Query processing on uncertain data occurs in the contexts of data warehousing, data integration, and of processing data extracted from the Web. Data cleaning can be fruitfully approached as a problem of reducing uncertainty in data and requires the management and processing of large amounts of uncertain data. Decision support and diagnosis systems employ hypothetical (what-if) queries. Scientific databases, which store outcomes of scientific experiments, frequently contain uncertain data such as incomplete observations or imprecise measurements. Sensor and RFID data is inherently uncertain. Applications in the contexts of fighting crime or terrorism, tracking moving objects, surveillance, and plagiarism detection essentially rely on techniques for processing and managing large uncertain datasets. Beyond that, many further potential applications of probabilistic databases exist and will manifest themselves once such systems become available.

Inference in uncertain data is a field in which the Artificial Intelligence research community has made much progress in the past years. Some of the most exciting AI applications, such as using graphical models in biology, belong to this area. While a number of papers on uncertain data and probabilistic databases have been written within the data management research community over the past decades, this area has moved into the focus of research interest only very recently, and work on scalable systems has only just started.

The *MayBMS* project* aims at creating a probabilistic database management system by leveraging techniques developed by the data management research community. The MayBMS project is founded on the thesis that a principled effort to use previous insights from databases will allow for substantial progress towards more robust and scalable systems for managing and querying large uncertain datasets. This will have a positive impact on current applications such as in computational science and will allow for entirely new data management applications.

Central themes in our research include the creation of foundations of query languages for probabilistic databases by developing analogs of relational algebra [22, 21] and SQL [6, 8] and the development of efficient query processing techniques [5, 25, 3, 23, 24, 17]. In practice, the efficient evaluation of queries on probabilistic data requires approximation techniques, and another important goal was to understand which approximation guarantees can be made for complex, realistic query languages [22, 15].

We have worked on developing a complete database management system for uncertain and probabilistic data. Apart from data representation and storage mechanisms, a query language, and query processing techniques, our work covers query optimization, an update language, concurrency control and recovery, and APIs for uncertain data.

MayBMS stands alone as a complete probabilistic database management system that supports a very powerful, compositional query language for which nevertheless worst-case efficiency and result quality guarantees can be made. Central to this is our choice of essentially using probabilistic versions of con-

*MayBMS is read as "maybe-MS", like DBMS.

ditional tables [18] as the representation system, but in a form engineered for admitting the efficient evaluation and automatic optimization of most operations of our language using robust and mature relational database technology [3].

The structure of this chapter is as follows. Section 2 sketches our model of probabilistic databases. Section 3 outlines desiderata that have guided the design of our query languages. Section 4 introduces our query algebra and illustrates it by examples. The section also gives an overview over theoretical results, in particular on expressiveness, that have been achieved for this algebra. Section 5 introduces U-relations, the representation system of MayBMS. Section 6 shows how most of the operations of our algebra can be evaluated efficiently using mature relational database techniques. Moreover, the problem of efficiently processing the remaining operations is discussed and an overview of the known results on the (worst-case) complexity of the query algebra is given. Section 7 presents the query and update language of MayBMS, which is based on our algebra but uses an extension of SQL as syntax. Section 8 discusses further systems issues. Section 9 concludes.

This chapter is meant to provide an overview over the MayBMS project and some topics are covered in a sketchy fashion. For details on the various techniques, experiments, and the theoretical contributions, the reader is referred to the original technical papers on MayBMS that can be found in the references.

2. Probabilistic Databases

Informally, our model of probabilistic databases is the following. The schema of a probabilistic database is simply a relational database schema. Given such a schema, a probabilistic database is a finite set of database instances of that schema (called possible worlds), where each world has a weight (called probability) between 0 and 1 and the weights of all worlds sum up to 1. In a subjectivist Bayesian interpretation, one of the possible worlds is "true", but we do not know which one, and the probabilities represent degrees of belief in the various possible worlds. Note that this is only the conceptual model. The physical representation of the set of possible worlds in the MayBMS system is quite different (see Section 5).

Given a schema with relation names $R_1, \ldots, R_k$. We use $sch(R_l)$ to denote the attributes of relation schema R_l. Formally, a *probabilistic database* is a *finite* set of structures

$$\mathbf{W} = \{\langle R_1^1, \ldots, R_k^1, p^{[1]}\rangle, \ldots, \langle R_1^n, \ldots, R_k^n, p^{[n]}\rangle\}$$

of relations $R_1^i, \ldots, R_k^i$ and numbers $0 < p^{[i]} \leq 1$ such that

$$\sum_{1 \leq i \leq n} p^{[i]} = 1.$$

We call an element $\langle R_1^i, \ldots, R_k^i, p^{[i]} \rangle \in \mathbf{W}$ a *possible world*, and $p^{[i]}$ its probability. We use superscripts for indexing possible worlds. To avoid confusion with exponentiation, we sometimes use bracketed superscripts $\cdot^{[i]}$. We call a relation R *complete* or *certain* if its instantiations are the same in all possible worlds of $\mathbf{W}$, i.e., if $R^1 = \cdots = R^n$.

Tuple *confidence* refers to the probability of the event $\vec{t} \in R$, where R is one of the relation names of the schema, with

$$\Pr[\vec{t} \in R] = \sum_{1 \leq i \leq n:\ \vec{t} \in R^i} p^{[i]}.$$

3. Query Language Desiderata

At the time of writing this, there is no accepted standard query language for probabilistic databases. In fact, we do not even agree today what use cases and functionality such systems should support. It seems to be proper to start the query language discussion with the definition of design *desiderata*. The following are those used in the design of MayBMS.

1 Efficient query evaluation.

2 The right degree of expressive power. The language should be powerful enough to support important queries. On the other hand, it should not be too strong, because expressiveness generally comes at a price: high evaluation complexity and infeasibility of query optimization. Can a case be made that some language is in a natural way a probabilistic databases analog of the relationally complete languages (such as relational algebra) – an expressiveness yardstick?

3 Genericity. The semantics of a query language should be independent from details of how the data is represented. Queries should behave in the same way no matter how the probabilistic data is stored. This is a basic requirement that is even part of the traditional definition of what constitutes a query (cf. e.g. [1]), but it is nontrivial to achieve for probabilistic databases [6, 4].

4 The ability to transform data. Queries on probabilistic databases are often interpreted quite narrowly in the literature. It is the author's view that queries in general should be compositional mappings between databases, in this case probabilistic databases. This is a property taken for granted in relational databases. It allows for the definition of clean database update languages.

5 The ability to introduce additional uncertainty. This may appear to be a controversial goal, since uncertainty is commonly considered undesirable, and probabilistic databases are there to deal with it by providing

useful functionality *despite* uncertainty. However, it can be argued that an uncertainty-introduction operation is important for at least three reasons: (1) for compositionality, and to allow construction of an uncertain database from scratch (as part of the update language); (2) to support what-if queries; and (3) to extend the hypothesis space modeled by the probabilistic database. The latter is needed to accommodate the results of experiments or new evidence, and to define queries that map from prior to posterior probabilistic databases. This is a nontrivial issue, and will be discussed in more detail later.

The next section introduces a query algebra and argues that it satisfies each of these desiderata.

4. The Algebra

This section covers the core query algebra of MayBMS: *probabilistic world-set algebra* (probabilistic WSA) [6, 22, 21]. Informally, probabilistic world-set algebra consists of the operations of relational algebra, an operation for computing tuple confidence conf, and the repair-key operation for *introducing* uncertainty. The operations of relational algebra are evaluated individually, in "parallel", in each possible world. The operation conf(R) computes, for each tuple that occurs in relation R in at least one world, the sum of the probabilities of the worlds in which the tuple occurs. The result is a certain relation, or viewed differently, a relation that is the same in all possible worlds. Finally, repair-key$_{\vec{A}@P}(R)$, where $\vec{A}, P$ are attributes of R, conceptually nondeterministically chooses a maximal repair of key $\vec{A}$. This operation turns a possible world R^i into the set of worlds consisting of all possible *maximal repairs* of key $\vec{A}$. A repair of key $\vec{A}$ in relation R^i is a subset of R^i for which $\vec{A}$ is a key. It uses the numerically-valued column P for weighting the newly created alternative repairs.

Formally, probabilistic world-set algebra consists of the following operations:

- The operations of relational algebra (selection σ, projection π, product $\times$, union $\cup$, difference $-$, and attribute renaming ρ), which are applied in each possible world independently.

  The semantics of operations Θ on probabilistic database $\mathbf{W}$ is

  $$[\![\Theta(R_l)]\!](\mathbf{W}) := \{\langle R_1, \ldots, R_k, \Theta(R_l), p\rangle \mid \langle R_1, \ldots, R_k, p\rangle \in \mathbf{W}\}$$

  for unary operations ($1 \leq l \leq k$). For binary operations, the semantics is

  $$[\![\Theta(R_l, R_m)]\!](\mathbf{W}) := \{\langle R_1, \ldots, R_k, \Theta(R_l, R_m), p\rangle \mid \langle R_1, \ldots, R_k, p\rangle \in \mathbf{W}\}.$$

Selection conditions are Boolean combinations of atomic conditions (i.e., negation is permitted even in the positive fragment of the algebra). Arithmetic expressions may occur in atomic conditions and in the arguments of π and ρ. For example, $\rho_{A+B\to C}(R)$ in each world adds up the A and B values of each tuple of R and keeps them in a new C attribute.

- An operation for computing tuple confidence,

$$[\![\text{conf}(R_l)]\!](\mathbf{W}) := \{\langle R_1, \ldots, R_k, S, p\rangle \mid \langle R_1, \ldots, R_k, p\rangle \in \mathbf{W}\}$$

where, w.l.o.g., $P \notin sch(R_l)$, and

$$S = \{\langle \vec{t}, P : \Pr[\vec{t} \in R_l]\rangle \mid \vec{t} \in \bigcup_i R_l^i\},$$

with schema $sch(S) = sch(R_l) \cup \{P\}$. The result of conf$(R_l)$, the relation S, is the same in all possible worlds, i.e., it is a certain relation. By our definition of probabilistic databases, each possible world has nonzero probability. As a consequence, conf does not return tuples with probability 0.

For example, on probabilistic database

R^1	A	B
	a	b
	b	c

$p^{[1]} = .3$

R^2	A	B
	a	b
	c	d

$p^{[2]} = .2$

R^3	A	B
	a	c
	c	d

$p^{[3]} = .5$

conf(R) computes, for each possible tuple, the sum of the weights of the possible worlds in which it occurs, here

conf(R)	A	B	P
	a	b	.5
	a	c	.5
	b	c	.3
	c	d	.7

- An uncertainty-introducing operation, *repair-key*, which can be thought of as sampling a maximum repair of a key for a relation. Repairing a key of a complete relation R means to compute, as possible worlds, all subset-maximal relations obtainable from R by removing tuples such that a key constraint is satisfied. We will use this as a method for constructing probabilistic databases, with probabilities derived from relative weights attached to the tuples of R.

  We say that relation R' is a *maximal repair* of a functional dependency (fd, cf. [1]) for relation R if R' is a maximal subset of R which satisfies that functional dependency, i.e., a subset $R' \subseteq R$ that satisfies the fd such that there is no relation R'' with $R' \subset R'' \subseteq R$ that satisfies the fd. Let $\vec{A}, B \in sch(R_l)$. For each possible world $\langle R_1, \ldots, R_k, p\rangle \in \mathbf{W}$, let column B of R contain only numerical values greater than 0 and let R_l satisfy the fd $(sch(R_l) - B) \rightarrow sch(R_l)$. Then,

$$[\![\text{repair-key}_{\vec{A}@B}(R_l)]\!](\mathbf{W}) := \Big\{\langle R_1, \ldots, R_k, \pi_{sch(R_l)-B}(\hat{R}_l), \hat{p}\rangle \mid \langle R_1, \ldots, R_k, p\rangle \in \mathbf{W}, \hat{R}_l \text{ is a maximal repair of fd } \vec{A} \rightarrow sch(R_l), \hat{p} = p \cdot \prod_{\vec{t} \in \hat{R}_l} \frac{\vec{t}.B}{\sum_{\vec{s} \in R_l : \vec{s}.\vec{A} = \vec{t}.\vec{A}} \vec{s}.B} \Big\}$$

Such a repair operation, apart from its usefulness for the purpose implicit in its name, is a powerful way of constructing probabilistic databases from complete relations.

EXAMPLE 6.1 Consider the example of tossing a biased coin twice. We start with a certain database

R	Toss	Face	FProb
	1	H	.4
	1	T	.6
	2	H	.4
	2	T	.6

$p = 1$

that represents the possible outcomes of tossing the coin twice. We turn this into a probabilistic database that represents this information using alternative possible worlds for the four outcomes using the query $S :=$ repair-key$_{\text{Toss@FProb}}(R)$. The resulting possible worlds are

S^1	Toss	Face
	1	H
	2	H

S^2	Toss	Face
	1	H
	2	T

S^3	Toss	Face
	1	T
	2	H

S^4	Toss	Face
	1	T
	2	T

with probabilities $p^{[1]} = p \cdot \frac{.4}{.4+.6} \cdot \frac{.4}{.4+.6} = .16$, $p^{[2]} = p^{[3]} = .24$, and $p^{[4]} = .36$. □

Coins	Type	Count
	fair	2
	2headed	1

Faces	Type	Face	FProb
	fair	H	.5
	fair	T	.5
	2headed	H	1

Tosses	Toss
	1
	2

R^f	Type
	fair

R^{dh}	Type
	2headed

$S^{f.HH}$	Type	Toss	Face
	fair	1	H
	fair	2	H

$p^{f.HH} = 1/6$

$S^{f.HT}$	Type	Toss	Face
	fair	1	H
	fair	2	T

$p^{f.HT} = 1/6$

S^{dh}	Type	Toss	Face
	2headed	1	H
	2headed	2	H

$p^{dh} = 1/3$

$S^{f.TH}$	Type	Toss	Face
	fair	1	T
	fair	2	H

$p^{f.TH} = 1/6$

$S^{f.TT}$	Type	Toss	Face
	fair	1	T
	fair	2	T

$p^{f.TT} = 1/6$

Ev	Toss	Face
	1	H
	2	H

Q	Type	P
	fair	$(1/6)/(1/2) = 1/3$
	2headed	$(1/3)/(1/2) = 2/3$

Figure 6.1. Tables of Example 6.2.

The fragment of probabilistic WSA which excludes the difference operation is called *positive* probabilistic WSA.

Computing possible and certain tuples is redundant with conf:

$$\begin{aligned} \text{poss}(R) &:= \pi_{sch(R)}(\text{conf}(R)) \\ \text{cert}(R) &:= \pi_{sch(R)}(\sigma_{P=1}(\text{conf}(R))) \end{aligned}$$

EXAMPLE 6.2 A bag of coins contains two fair coins and one double-headed coin. We take one coin out of the bag but do not look at its two faces to determine its type (fair or double-headed) for certain. Instead we toss the coin twice to collect evidence about its type.

We start out with a complete database (i.e., a relational database, or a probabilistic database with one possible world of probability 1) consisting of three relations, Coins, Faces, and Tosses (see Figure 6.1 for all tables used in this example). We first pick a coin from the bag and model that the coin be either fair or double-headed. In probabilistic WSA this is expressed as

$$R := \text{repair-key}_{\emptyset @ \text{Count}}(\text{Coins}).$$

This results in a probabilistic database of two possible worlds,

$$\{\langle \text{Coins, Faces}, R^f, p^f = 2/3\rangle, \langle \text{Coins, Faces}, R^{dh}, p^{dh} = 1/3\rangle\}.$$

The possible outcomes of tossing the coin twice can be modeled as

$$S := \text{repair-key}_{\text{Toss@FProb}}(R \bowtie \text{Faces} \times \text{Tosses}).$$

This turns the two possible worlds into five, since there are four possible outcomes of tossing the fair coin twice, and only one for the double-headed coin.

Let $T := \pi_{\text{Toss,Face}}(S)$. The posterior probability that a coin of type x was picked, given the *evidence* Ev (see Figure 6.1) that both tosses result in H, is

$$\Pr[x \in R \mid T = Ev] = \frac{\Pr[x \in R \wedge T = Ev]}{\Pr[T = Ev]}.$$

Let A be a relational algebra expression for the Boolean query $T = Ev$. Then we can compute a table of pairs $\langle x, \Pr[x \in R \mid T = Ev]\rangle$ as

$$Q := \pi_{\text{Type},P_1/P_2 \to P}(\rho_{P \to P_1}(\text{conf}(R \times A)) \times \rho_{P \to P_2}(\text{conf}(A))).$$

The prior probability that the chosen coin was fair was 2/3; after taking the evidence from two coin tosses into account, the posterior probability Pr[the coin is fair | both tosses result in H] is only 1/3. Given the evidence from the coin tosses, the coin is now more likely to be double-headed. □

EXAMPLE 6.3 We redefine the query of Example 6.2 such that repair-key is only applied to certain relations. Starting from the database obtained by computing R, with its two possible worlds, we perform the query $S_0 :=$ repair-key$_{\text{Type,Toss@FProb}}$(Faces × Tosses) to model the possible outcomes of tossing the chosen coin twice. The probabilistic database representing these repairs consists of eight possible worlds, with the two possible R relations of Example 6.2 and, independently, four possible S_0 relations. Let $S := R \bowtie S_0$. While we now have eight possible worlds rather than five, the four worlds in which the double-headed coin was picked all agree on S with the one world in which the double-headed coin was picked in Example 6.2, and the sum of their probabilities is the same as the probability of that world. It follows that the new definition of S is equivalent to the one of Example 6.2 and the rest of the query is the same. □

Discussion. The repair-key operation admits an interesting class of queries: Like in Example 6.2, we can start with a probabilistic database of prior probabilities, add further evidence (in Example 6.2, the result of the coin tosses) and then compute interesting posterior probabilities. The adding of further evidence may require extending the hypothesis space first. For this, the repair-key operation is essential. Even though our goal is not to update the database, we have to be able to introduce uncertainty just to be able to model new evidence – say, experimental data. Many natural and important probabilistic database queries cannot be expressed without the repair-key operation. The coin tossing example was admittedly a toy example (though hopefully easy to understand). Real applications such as diagnosis or processing scientific data involve technically similar questions.

Regarding our desiderata, it is quite straightforward to see that probabilistic WSA is generic (3): see also the proof for the non-probabilistic language in [6]. It is clearly a data transformation query language (4) that supports powerful queries for defining databases. The repair-key operation is our construct for uncertainty introduction (5). The evaluation efficiency (1) of probabilistic WSA is studied in Section 6. The expressiveness desideratum (2) is discussed next.

An expressiveness yardstick. In [6] a non-probabilistic version of world-set algebra is introduced. It replaces the confidence operation with an operation poss for computing possible tuples. Using poss, repair-key, and the operations of relational algebra, powerful queries are expressible. For instance, the certain answers of a query on an uncertain database can be computed using poss and difference. Compared to the poss operation described above, the operation of [6] is more powerful. The syntax is $\text{poss}_{\vec{A}}(Q)$, where $\vec{A}$ is a set of column names of Q. The operation partitions the set of possible worlds into the groups of those worlds that agree on $\pi_{\vec{A}}(Q)$. The result in each world is the set of tuples possible in Q within the world's group. Thus, this operation supports the grouping of possible worlds just like the group-by construct in SQL supports the grouping of tuples.

The main focus of [6] is to study the fragment of (non-probabilistic) WSA in which repair-key is replaced by the choice-of operation, defined as:

$$\text{choice-of}_{\vec{A}@P}(R) := R \bowtie \text{repair-key}_{\emptyset @P}(\pi_{\vec{A},P}(R)).$$

The choice-of operation introduces uncertainty like the repair-key operation, but can only cause a polynomial, rather than exponential, increase of the number of possible worlds. This language has the property that query evaluation on enumerative representations of possible worlds is in PTIME (see Section 6 for more on this). Moreover, it is *conservative* over relational algebra in the sense that any query that starts with a certain database (a classical relational database) and produces a certain database is equivalent to a relational algebra query and can be efficiently rewritten into relational algebra. This is a non-trivial result, because in this language we can produce uncertain intermediate results consisting of many possible worlds using the choice-of operator. This allows us to express and efficiently answer hypothetical (what-if) queries.

(Full non-probabilistic) WSA consists of the relational algebra operations, repair-key, and $\text{poss}_{\vec{A}}$. In [21], it is shown that WSA precisely captures second-order logic. Leaving aside inessential details about interpreting second-order logic over uncertain databases – it can be done in a clean way – this result shows that a query is expressible in WSA if and only if it is expressible in second-order logic. WSA seems to be the first algebraic (i.e., variable and quantifier-free) language known to have exactly the same expressive power as second-order logic.

More importantly for us, it can be argued that this establishes WSA as the natural analog of relational algebra for uncertain databases. Indeed, while it is well known that useful queries (such as transitive closure or counting queries, cf. [1]) cannot be expressed in it, relational algebra is a very popular expressiveness yardstick for relational query languages (and query languages that are as expressive as relational algebra are called *relationally complete*). Relational algebra is also exactly as expressive as the *domain-independent* first-order queries [1], also known as the *relational calculus*. Second-order logic is just first-order logic extended by (existential) quantification over relations ("Does there exist a relation R such that ϕ holds?", where ϕ is a formula). This is the essence of (what-if) reasoning over uncertain data. For example, the query of Example 6.2 employed what-if reasoning over relations twice via the repair-key operation, first considering alternative choices of coin and then alternative outcomes to coin tossing experiments.

It is unknown whether probabilistic WSA as defined in this chapter can express all the queries of WSA (with $\text{poss}_{\vec{A}}$). Given the known data complexity bounds for the two languages (see Section 6) alone, there is no reason to assume that this is not the case. On the other hand, it seems unlikely, and a mapping from WSA to probabilistic WSA, if it exists, must be nontrivial.

It would be easy to define a sufficiently strong extension of probabilistic WSA by just generalizing conf to a world-grouping $\text{conf}_{\vec{A}}$ operation. In this chapter, this is not done because we do not know how to obtain any even just moderately efficient implementation of this operation (or of $\text{poss}_{\vec{A}}$) on succinct data representations.

5. Representing Probabilistic Data

This section discusses the method used for representing and storing probabilistic data and correlations in MayBMS. We start by motivating the problem of finding a practical representation system.

EXAMPLE 6.4 Consider a census scenario, in which a large number of individuals manually fill in forms. The data in these forms subsequently has to be put into a database, but no matter whether this is done automatically using OCR or by hand, some uncertainty may remain about the correct values for some of the answers. Figure 6.2 shows two simple filled in forms. Each one contains the social security number, name, and marital status of one person.

The first person, Smith, seems to have checked marital status "single" after first mistakenly checking "married", but it could also be the opposite. The second person, Brown, did not answer the marital status question. The social security numbers also have several possible readings. Smith's could be 185 or

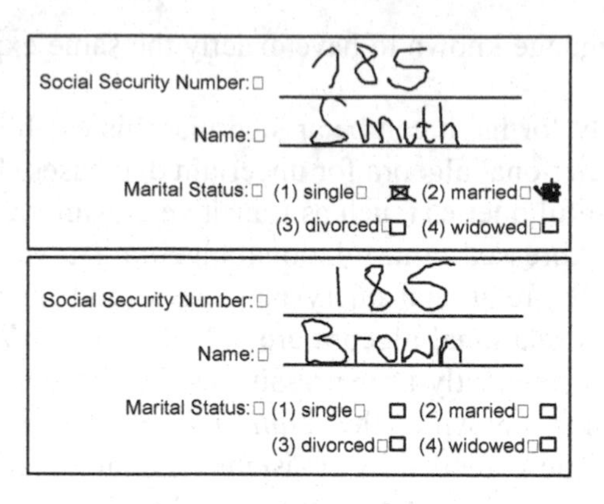
Social Security Number: 185
Name: Smith
Marital Status: (1) single (2) married
(3) divorced (4) widowed

Social Security Number: 185
Name: Brown
Marital Status: (1) single (2) married
(3) divorced (4) widowed

Figure 6.2. Two census forms.

785 (depending on whether Smith originally is from the US or from Europe) and Brown's may either be 185 or 186.

In an SQL database, uncertainty can be managed using null values, using a table

(TID)	SSN	N	M
t_1	null	Smith	null
t_2	null	Brown	null

Using nulls, information is lost about the values considered possible for the various fields. Moreover, it is not possible to express correlations such as that, while social security numbers may be uncertain, no two distinct individuals can have the same. In this example, we can exclude the case that both Smith and Brown have social security number 185. Finally, we cannot store probabilities for the various alternative possible worlds. □

This leads to three natural desiderata for a representation system: (*) Expressiveness, that is, the power to represent all (relevant) probabilistic databases, (*) succinctness, that is, space-efficient storage of the uncertain data, and (*) efficient real-world query processing.

Often there are many rather (but not quite) independent local alternatives in probabilistic data, which multiply up to a very large number of possible worlds. For example, the US census consists of many dozens of questions for about 300 million individuals. Suppose forms are digitized using OCR and the resulting data contains just two possible readings for 0.1% of the answers before cleaning. Then, there are on the order of $2^{10,000,000}$ possible worlds, and each one will take close to one Terabyte of data to store. Clearly, we need a way of representing this data that is much better than a naive enumeration of possible worlds.

Also, the repair-key operator of probabilistic world-set algebra in general causes an exponential increase in the number of possible worlds.

There is a trade-off between succinctness on one hand and efficient processing on the other. Computing confidence conf(Q) of conjunctive queries Q on tuple-independent databases is #P-hard – one such hard query [13] (in datalog notation [1]) is $Q \leftarrow R(x), S(x,y), T(y)$. At the same time, much more expressive queries can be evaluated efficiently on nonsuccinct representations (enumerations of possible worlds) [6]. Query evaluation in probabilistic databases is not hard because of the presence of probabilities, but because of the succinct storage of alternative possible worlds! We can still have the goal of doing well in practice.

Conditional tables. MayBMS uses a purely relational representation system for probabilistic databases called *U-relational databases*, which is based on probabilistic versions of the classical *conditional tables* (c-tables) of the database literature [18]. Conditional tables are a relational representation system based on the notion of *labeled null values* or *variables*, that is, null values that have a name. The name makes it possible to use the same variable x in several fields of a database, indicating that the value of x is unknown but must be the same in all those fields in which x occurs. Tables with variables are also known as *v-tables*.

Formally, c-tables are v-tables extended by a column for holding a local condition. That is, each tuple of a c-table has a Boolean condition constructed using "and", "or", and "not" from atomic conditions of the form $x = c$ or $x = y$, where c are constants and x and y are *variables*. Possible worlds are determined by functions θ that map each variable that occurs in at least one of the local conditions in the c-tables of the database to a constant. The database in that possible world is obtained by (1) selecting those tuples whose local condition ϕ satisfies the variable assignment θ, i.e., that becomes true if each variable x in ϕ is replaced by $\theta(x)$, (2) replacing all variables y in the value fields of these tuples by $\theta(y)$, and (3) projecting away the local condition column.

Conditional tables are sometimes defined to include a notion of *global condition*, which we do not use: We want each probabilistic database to have at least one possible world.

Conditional tables are a so-called *strong representation system*: They are closed under the application of relational algebra queries. The set of worlds obtained by evaluating a relational algebra query in each possible world represented by a conditional table can again be straightforwardly represented by a conditional table. Moreover, the local conditions are in a sense the most natural and simple formalism possible to represent the result of queries on data with labeled nulls. The local conditions just represent the information necessary to preserve correctness and can also be understood to be just data provenance information [10].

U-Relational Databases. In our model, probabilistic databases are finite sets of possible worlds with probability weights. It follows that each variable naturally has a finite domain, the set of values it can take across all possible worlds. This has several consequences. First, variables can be considered *finite random variables*. Second, only allowing for variables to occur in local conditions, but not in attribute fields of the tuples, means no restriction of expressiveness. Moreover, we may assume without loss of generality that each atomic condition is of the form $x = c$ (i.e., we never have to compare variables).

If we start with a c-table in which each local condition is a conjunction of no more than k atomic conditions, then a positive relational algebra query on this uncertain database will result in a c-table in which each local condition is a conjunction of no more than k' atoms, where k' only depends on k and the query, but not on the data. If k is small, it is reasonable to actually hard-wire it in the schema, and represent local conditions by k pairs of columns to store atoms of the form $x = c$.

These are the main ideas of our representation system, U-relations. Random variables are assumed independent in the *current* MayBMS system, but as we will see, this means no restriction of generality. Nevertheless, it is one goal of future work to support graphical models for representing more correlated joint probability distributions below our U-relations. This would allow us to represent *learned* distributions in the form of e.g. Bayesian networks directly in the system (without the need to map them to local conditions) and run queries on top, representing the inferred correlations using local conditions. The latter seem to be better suited for representing the incremental correlations constructed by queries.

One further idea employed in U-relational databases is to use vertical partitioning [9, 26] for representing *attribute-level uncertainty*, i.e., to allow to decompose tuples in case several fields of a tuple are independently uncertain.

EXAMPLE 6.5 The set of tables shown in Figure 6.3 is a U-relational database representation for the census data scenario of Example 6.4, extended by suitable probabilities for the various alternative values the fields can take (represented by table W). □

Formally, a U-relational database consists of a set of independent random variables with finite domains (here, x, y, v, w), a set of U-relations, and a ternary table W (the *world-table*) for representing distributions. The W table stores, for each variable, which values it can take and with what probability. The schema of each U-relation consists of a *set* of pairs (V_i, D_i) of *condition*

$U_{R[SSN]}$	V	D	TID	SSN
	x	1	t_1	185
	x	2	t_1	785
	y	1	t_2	185
	y	2	t_2	186

$U_{R[M]}$	V	D	TID	M
	v	1	t_1	1
	v	2	t_1	2
	w	1	t_2	1
	w	2	t_2	2
	w	3	t_2	3
	w	4	t_2	4

$U_{R[N]}$	TID	N
	t_1	Smith
	t_2	Brown

W	V	D	P
	x	1	.4
	x	2	.6
	y	1	.7
	y	2	.3
	v	1	.8
	v	2	.2
	w	1	.25
	w	2	.25
	w	3	.25
	w	4	.25

Figure 6.3. A U-relational database.

columns representing variable assignments and a set of *value columns* for representing the data values of tuples.

The semantics of U-relational databases is as follows. Each possible world is identified by a valuation θ that assigns one of the possible values to each variable. The probability of the possible world is the product of weights of the values of the variables. A tuple of a U-relation, stripped of its condition columns, is in a given possible world if its variable assignments are consistent with θ. Attribute-level uncertainty is achieved through vertical decompositioning, so one of the value columns is used for storing tuple ids and undoing the vertical decomposition on demand.

EXAMPLE 6.6 Consider the U-relational database of Example 6.5 and the possible world

$$\theta = \{x \mapsto 1, y \mapsto 2, v \mapsto 1, w \mapsto 1\}.$$

The probability weight of this world is $.4 \cdot .3 \cdot .8 \cdot .25 = .024$. By removing all the tuples whose condition columns are inconsistent with θ and projecting away the condition columns, we obtain the relations

$R[SSN]$	TID	SSN
	t_1	185
	t_2	186

$R[M]$	TID	M
	t_1	1
	t_2	1

$R[N]$	TID	N
	t_1	Smith
	t_2	Brown

which are just a vertically decomposed version of R in the chosen possible world. That is, R is $R[SSN] \bowtie R[M] \bowtie R[N]$ in that possible world. □

Properties of U-relations. U-relational databases are a *complete* representation system for (finite) probabilistic databases [3]. This means that any probabilistic database can be represented in this formalism. In particular, it follows that U-relations are closed under query evaluation using any generic query language, i.e., starting from a represented database, the query result can again be represented as a U-relational database. Completeness also implies that any (finite) correlation structure among tuples can be represented, despite the fact that we currently assume that the random variables that our correlations are constructed from (using tuple conditions) are independent: The intuition that some form of graphical model for finite distributions may be more powerful (i.e., able to represent distributions that cannot be represented by U-relations) is *false*.

Historical Note. The first prototype of MayBMS [5, 7, 25] did not use U-relations for representations, but a different representation system called *world-set decompositions* [5]. These representations are based on factorizations of the space of possible worlds. They can also be thought of as shallow Bayesian networks. The problem with this approach is that some selection operations can cause an exponential blowup of the representations. This problem is not shared by U-relations, even though they are strictly more succinct than world-set decompositions. This was the reason for introducing U-relations in [3] and developing a new prototype of MayBMS based on U-relations.

6. Conceptual Query Evaluation, Rewritings, and Asymptotic Efficiency

This section gives a complete solution for efficiently evaluating a large fragment of probabilistic world-set algebra using relational database technology. Then we discuss the evaluation of the remaining operations of probabilistic WSA, namely difference and tuple confidence. Finally, an overview of known worst-case computational complexity results is given.

Translating queries down to the representation relations. Let *rep* be the *representation function*, which maps a U-relational database to the set of possible worlds it represents. Our goal is to give a reduction that maps any positive relational algebra query Q over probabilistic databases represented as U-relational databases T to an equivalent positive relational algebra query $\overline{Q}$ of polynomial size such that

$$rep(\overline{Q}(T)) = \{Q(\mathcal{A}^i) \mid \mathcal{A}^i \in rep(T)\}$$

where the $\mathcal{A}^i$ are relational database instances (possible worlds).

The following is such a reduction, which maps the operations of positive relational algebra, poss, and repair-key to relational algebra over U-relational representations:

$$\begin{aligned}
[\![R \times S]\!] &:= \pi_{(U_R.\overline{VD} \cup U_S.\overline{VD}) \to \overline{VD}, sch(R), sch(S)}(\\
&\qquad U_R \bowtie_{U_R.\overline{VD} \text{ consistent with } U_S.\overline{VD}} U_S) \\
[\![\sigma_\phi R]\!] &:= \sigma_\phi(U_R) \\
[\![\pi_{\vec{B}} R]\!] &:= \pi_{\overline{VD},\vec{B}}(R) \\
[\![R \cup S]\!] &:= U_R \cup U_S \\
[\![\text{poss}(R)]\!] &:= \pi_{sch(R)}(U_R).
\end{aligned}$$

The consistency test for conditions can be expressed simply using Boolean conditions (see Example 6.8, and [3]). Note that the product operation, applied to two U-relations of k and l (V_i, D_i) column pairs, respectively, returns a U-relation with $k + l$ (V_i, D_i) column pairs.

For simplicity, let us assume that the elements of $\pi_{\langle \vec{A} \rangle}(U_R)$ are not yet used as variable names. Moreover, let us assume that the B value column of U_R, which is to provide weights for the alternative values of the columns $sch(R) - (\vec{A} \cup B)$ for each tuple $\vec{a}$ in $\pi_{\langle \vec{A} \rangle}(U_R)$, are probabilities, i.e., sum up to one for each $\vec{a}$ and do not first have to be normalized as described in the definition of the semantics of repair-key in Section 4. The operation $S :=$ repair-key$_{\vec{A}@B}(R)$ for complete relation R is translated as

$$U_S := \pi_{\langle \vec{A} \rangle \to V, \langle (sch(R) - \vec{A}) - \{B\} \rangle \to D, sch(R)} U_R$$

with

$$W := W \cup \pi_{\langle \vec{A} \rangle \to V, \langle (sch(R) - \vec{A}) - \{B\} \rangle \to D, B \to P} U_R.$$

Here, $\langle \cdot \rangle$ turns tuples of values into atomic values that can be stored in single fields.

That is, repair-key starting from a complete relation is just a projection/copying of columns, even though we may create an exponential number of possible worlds.

EXAMPLE 6.7 Consider again the relation R of Example 6.1, which represents information about tossing a biased coin twice, and the query $S :=$ repair-key$_{\text{Toss@FProb}}(R)$. The result is

U_S	V	D	Toss	Face	FProb
	1	H	1	H	.4
	1	T	1	T	.6
	2	H	2	H	.4
	2	T	2	T	.6

W	V	D	P
	1	H	.4
	1	T	.6
	2	H	.4
	2	T	.6

as a U-relational database. □

The projection technique only works if the relation that repair-key is applied to is certain. However, for practical purposes, this is not a restriction of expressive power (cf. [21], and see also Example 6.3).

The next example demonstrates the application of the rewrite rules to compile a query down to relational algebra on the U-relations.

EXAMPLE 6.8 We revisit our census example with U-relations $U_{R[SSN]}$ and $U_{R[N]}$. We ask for possible names of persons who have SSN 185, $\text{poss}(\pi_N(\sigma_{SSN=185}(R)))$. To undo the vertical partitioning, the query is evaluated as $\text{poss}(\pi_N(\sigma_{SSN=185}(R[SSN] \bowtie R[N])))$. We rewrite the query using our rewrite rules into $\pi_N(\sigma_{SSN=185}(U_{R[SSN]} \bowtie_{\psi \wedge \phi} U_{R[N]}))$, where ψ ensures that we only generate tuples that occur in some worlds,

$$\psi := (U_{R[SSN]}.V = U_{R[N]}.V \Rightarrow U_{R[SSN]}.D = U_{R[N]}.D),$$

and ϕ ensures that the vertical partitioning is correctly undone,

$$\phi := (U_{R[SSN]}.TID = U_{R[N]}.TID).$$

□

Properties of the relational-algebra reduction. The relational algebra rewriting down to positive relational algebra on U-relations has a number of nice properties. First, since relational algebra has PTIME (even AC_0) data complexity, the query language of positive relational algebra, repair-key, and poss on probabilistic databases represented by U-relations has the same. The rewriting is in fact a *parsimonious translation*: The number of algebra operations does not increase and each of the operations selection, projection, join, and union remains of the same kind. Query plans are hardly more complicated than the input queries. As a consequence, we were able to observe that off-the-shelf relational database query optimizers do well in practice [3].

Thus, for all but two operations of probabilistic world-set algebra, it seems that there is a very efficient solution that builds on relational database technology. These remaining operations are confidence computation and relational algebra difference.

Approximate confidence computation. To compute the confidence in a tuple of data values occurring possibly in several tuples of a U-relation, we have to compute the probability of the disjunction of the local conditions of all these tuples. We have to eliminate duplicate tuples because we are interested in the probability of the data tuples rather than some abstract notion of tuple identity that is really an artifact of our representation. That is, we have to compute the probability of a DNF, i.e., the sum of the weights of the worlds identified with valuations θ of the random variables such that the DNF becomes true under θ. This problem is #P-complete [16, 13]. The result is not the sum of the probabilities of the individual conjunctive local conditions, because they may, intuitively, "overlap".

EXAMPLE 6.9 Consider a U-relation with schema $\{V, D\}$ (representing a nullary relation) and two tuples $\langle x, 1\rangle$, and $\langle y, 1\rangle$, with the W relation from Example 6.5. Then the confidence in the nullary tuple $\langle\rangle$ is $\Pr[x \mapsto 1 \vee y \mapsto 1] = \Pr[x \mapsto 1] + \Pr[y \mapsto 1] - \Pr[x \mapsto 1 \wedge y \mapsto 1] = .82$. □

Confidence computation can be efficiently approximated by Monte Carlo simulation [16, 13, 22]. The technique is based on the Karp-Luby fully polynomial-time randomized approximation scheme (FPRAS) for counting the number of solutions to a DNF formula [19, 20, 12]. There is an efficiently computable unbiased estimator that in expectation returns the probability p of a DNF of n clauses (i.e., the local condition tuples of a Boolean U-relation) such that computing the average of a polynomial number of such Monte Carlo steps (= calls to the Karp-Luby unbiased estimator) is an (ϵ, δ)-approximation for the probability: If the average $\hat{p}$ is taken over at least $\lceil 3 \cdot n \cdot \log(2/\delta)/\epsilon^2 \rceil$ Monte Carlo steps, then $\Pr\big[|p - \hat{p}| \geq \epsilon \cdot p\big] \leq \delta$. The paper [12] improves upon this by determining smaller numbers (within a constant factor from optimal) of necessary iterations to achieve an (ϵ, δ)-approximation.

Avoiding the difference operation. Difference $R - S$ is conceptually simple on c-tables. Without loss of generality, assume that S does not contain tuples $\langle\vec{a}, \psi_1\rangle, \ldots, \langle\vec{a}, \psi_n\rangle$ that are duplicates if the local conditions are disregarded. (Otherwise, we replace them by $\langle\vec{a}, \psi_1 \vee \cdots \vee \psi_n\rangle$.) For each tuple $\langle\vec{a}, \phi\rangle$ of R, if $\langle\vec{a}, \psi\rangle$ is in S then output $\langle\vec{a}, \phi \wedge \neg\psi\rangle$; otherwise, output $\langle\vec{a}, \phi\rangle$. Testing whether a tuple is possible in the result of a query involving difference is already NP-hard [2]. For U-relations, we in addition have to turn $\phi \wedge \neg\psi$ into a DNF to represent the result as a U-relation. This may lead to an exponentially large output and a very large number of $\vec{V}\vec{D}$ columns may be required to represent the conditions. For these reasons, MayBMS currently does not implement the difference operation.

In many practical applications, the difference operation can be avoided. Difference is only hard on uncertain relations. On such relations, it can only lead to displayable query results in queries that close the possible worlds semantics using conf, computing a single certain relation. Probably the most important application of the difference operation is for encoding universal constraints, for example in data cleaning. But if the confidence operation is applied on top of a universal query, there is a trick that will often allow to rewrite the query into an existential one (which can be expressed in positive relational algebra plus conf, without difference) [22].

EXAMPLE 6.10 The example uses the census scenario and the uncertain relation R discussed earlier. Consider the query of finding, for each TID t_i and

SSN s, the confidence in the statement that s is the correct SSN for the individual associated with the tuple identified by t_i, assuming that social security numbers uniquely identify individuals, that is, assuming that the functional dependency $SSN \rightarrow TID$ (subsequently called ψ) holds. In other words, the query asks, for each TID t_i and SSN s, to find the probability $\Pr[\phi \mid \psi]$, where $\phi(t_i, s) = \exists t \in R\ t.TID = t_i \wedge t.SSN = s$. Constraint ψ can be thought of as a data cleaning constraint that ensures that the SSN fields in no two distinct census forms (belonging to two different individuals) are interpreted as the same number.

We compute the desired conditional probabilities, for each possible pair of a TID and an SSN, as $\Pr[\phi \mid \psi] = \Pr[\phi \wedge \psi]/\Pr[\psi]$. Here ϕ is existential (expressible in positive relational algebra) and ψ is an equality-generating dependency (i.e., a special universal query) [1]. The trick is to turn relational difference into the subtraction of probabilities, $\Pr[\phi \wedge \psi] = \Pr[\phi] - \Pr[\phi \wedge \neg\psi]$ and $\Pr[\psi] = 1 - \Pr[\neg\psi]$, where $\neg\psi = \exists t, t' \in R\ t.SSN = t'.SSN \wedge t.TID \neq t'.TID$ is existential (with inequalities). Thus $\neg\psi$ and $\phi \wedge \neg\psi$ are expressible in positive relational algebra. This works for a considerable superset of the equality-generating dependencies [22], which in turn subsume useful data cleaning constraints, such as *conditional functional dependencies* [11].

Let $R_{\neg\psi}$ be the relational algebra expression for $\neg\psi$,

$$\pi_{\emptyset}(R \bowtie_{TID=TID' \wedge SSN \neq SSN'} \rho_{TID \rightarrow TID'; SSN \rightarrow SSN'}(R)),$$

and let S be

$$\rho_{P \rightarrow P_\phi}(\mathrm{conf}(R)) \bowtie \rho_{P \rightarrow P_{\phi \wedge \neg\psi}}(\mathrm{conf}(R \times R_{\neg\psi})) \times \rho_{P \rightarrow P_{\neg\psi}}(\mathrm{conf}(R_{\neg\psi})).$$

The overall example query can be expressed as

$$T := \pi_{TID, SSN, (P_\phi - P_{\phi \wedge \neg\psi})/(1 - P_{\neg\psi}) \rightarrow P}(S).$$

For the example table R given above, S and T are

S	TID	SSN	P_ϕ	$P_{\phi \wedge \neg\psi}$	$P_{\neg\psi}$
	t_1	185	.4	.28	.28
	t_1	785	.6	0	.28
	t_2	185	.7	.28	.28
	t_2	186	.3	0	.28

T	TID	SSN	P
	t_1	185	1/6
	t_1	785	5/6
	t_2	185	7/12
	t_2	186	5/12

Complexity Overview. Figure 6.4 gives an overview over the known complexity results for the various fragments of probabilistic WSA. Two different representations are considered, non-succinct representations that basically consist of enumerations of the possible worlds [6] and succinct representations: U-relational databases. In the non-succinct case, only the repair-key operation, which may cause an exponential explosion in the number of possible worlds,

Language Fragment	Complexity	Reference
On non-succinct representations:		
RA + conf + possible + choice-of	**in PTIME** (SQL)	[22]
RA + possible + repair-key	**NP-&coNP-hard**, **in** $\mathbf{P}^{NP}$	[6] [21]
RA + possible$_Q$ + repair-key	**PHIER-compl.**	[21]
On U-relations:		
Pos.RA + repair-key + possible	**in AC0**	[3]
RA + possible	**co-NP-hard**	Abiteboul et al. [2]
Conjunctive queries + conf	**#P-hard**	Dalvi, Suciu [13]
Probabilistic WSA	**in** $\mathbf{P}^{\#P}$	[22]
Pos.RA + repair-key + possible + approx.conf + egds	**in PTIME**	[22]

Figure 6.4. Complexity results for (probabilistic) world-set algebra. RA denotes relational algebra.

makes queries hard. All other operations, including confidence computation, are easy. In fact, we may add much of SQL – for instance, aggregations – to the language and it still can be processed efficiently, even by a reduction of the query to an SQL query on a suitable non-succinct relational representation.

When U-relations are used as representation system, the succinctness causes both difference [2] and confidence computation [13] independently to make queries NP-hard. Full probabilistic world-set algebra is essentially not harder than the language of [13], even though it is substantially more expressive.

It is worth noting that repair-key by itself, despite the blowup of possible worlds, does not make queries hard. For the language consisting of positive relational algebra, repair-key, and poss, we have shown by construction that it has PTIME complexity: We have given a positive relational algebra rewriting to queries on the representations earlier in this section. Thus queries are even in the highly parallelizable complexity class AC_0.

The final result in Figure 6.4 concerns the language consisting of the positive relational algebra operations, repair-key, (ϵ, δ)-approximation of confidence computation, and the generalized equality generating dependencies of [22] for which we can rewrite difference of uncertain relations to difference of confidence values (see Example 6.10). The result is that queries of that language that close the possible worlds semantics – i.e., that use conf to compute a certain relation – are in PTIME overall. In [22], a stronger result than just the claim that each of the operations of such a query is individually in PTIME is proven. It is shown that, leaving aside a few pitfalls, global approximation guarantees can be achieved in polynomial time, i.e., results of entire queries in this language can be approximated arbitrarily closely in polynomial time.

This is a non-obvious result because the query language is compositional and selections can be made based on approximated confidence values. Clearly, in a query $\sigma_{P=0.5}(\text{approx.conf}(R))$, an approximated P value will almost always be slightly off, even if the exact P value is indeed 0.5, and the selection of tuples made based on whether P is 0.5 is nearly completely arbitrary. In [22, 15], it is shown that this is essentially an unsurmountable problem. All we can tell is that if P is very different from 0.5, then the probability that the tuple should be in the answer is very small. If atomic selection conditions on (approximated) probabilities usually admit ranges such as $P < 0.5$ or $0.4 < P < 0.6$, then query approximation will nevertheless be meaningful: we are able to approximate query results unless probability values are very close or equal to the constants used as interval bounds. (These special points are called *singularities* in [22].)

The results of [22] have been obtained for powerful conditions that may use arithmetics over several approximated attributes, which is important if conditional probabilities have to be checked in selection conditions or if several probabilities have to be compared. The algorithm that gives overall (ϵ, δ)-approximation guarantees in polynomial time is not strikingly practical. Further progress on this has been made in [15], but more work is needed.

7. The MayBMS Query and Update Language

This section describes the query and update language of MayBMS, which is based on SQL. In fact, our language is a generalization of SQL on classical relational databases. To simplify the presentation, a fragment of the full language supported in MayBMS is presented here.

The representation system used in MayBMS, U-relations, has classical relational tables as a special case, which we will call *typed-certain (t-certain) tables* in this section. Tables that are not t-certain are called uncertain. This notion of certainty is purely syntactic, and $\text{cert}(R) = \pi_{sch(R)}(\sigma_{P=1}(\text{conf}(R)))$ may well be equal to the projection of a U-relation U_R to its attribute (non-condition) columns despite R not being t-certain according to this definition.

Aggregates. In MayBMS, full SQL is supported on t-certain tables. Beyond t-certain tables, some restrictions are in place to assure that query evaluation is feasible. In particular, we do not support the standard SQL aggregates such as `sum` or `count` on uncertain relations. This can be easily justified: In general, these aggregates will produce exponentially many different numerical results in the various possible worlds, and there is no way of representing these results efficiently. However, MayBMS supports a different set of aggregate operations on uncertain relations. These include the computations of *expected* sums and counts (using aggregates `esum` and `ecount`).

Moreover, the confidence computation operation is an aggregate in the MayBMS query language. This is a deviation from the language flavor of our algebra, but there is a justification for this. The algebra presented earlier assumed a set-based semantics for relations, where operations such as projections automatically remove duplicates. In the MayBMS query language, just like in SQL, duplicates have to be eliminated explicitly, and confidence is naturally an aggregate that computes a single confidence value for each group of tuples that agree on (a subset of) the non-condition columns. By using aggregation syntax for `conf` and not supporting `select distinct` on uncertain relations, we avoid a need for conditions beyond the special conjunctions that can be stored with each tuple in U-relations.

All the aggregates on uncertain tables produce t-certain tables.

Duplicate tuples. SQL databases in general support multiset tables, i.e., tables in which there may be duplicate tuples. There is no conceptual difficulty at all in supporting multiset U-relations. In fact, since U-relations are just relations in which some columns are interpreted to have a special meaning (conditions), just storing them in a standard relational database management system (which supports duplicates in tables) yields support for multiset U-relations.

Syntax. The MayBMS query language is compositional and built from uncertain and t-certain queries. The uncertain queries are those that produce a possibly uncertain relation (represented by a U-relation with more than zero V and D columns). Uncertain queries can be constructed, inductively, from t-certain queries, `select-from-where` queries over uncertain tables, the multiset union of uncertain queries (using the SQL `union` construct), and statements of the form:

```
repair key <attributes> in <t-certain-query>
weight by <attribute>
```

Note that repair-key is a query, rather than an update statement. The `select-from-where` queries may use any t-certain subqueries in the conditions, plus uncertain subqueries in atomic conditions of the form `<tuple> in <uncertain-query>` that occur positively in the condition. (That is, if the condition is turned into DNF, these literals are not negated.)

The t-certain queries (i.e., queries that produce a t-certain table) are given by

- all constructs of SQL on t-certain tables and t-certain subqueries, extended by a new aggregate

  ```
  argmax(<argument-attribute>, <value-attribute>)
  ```

  which outputs all the `argument-attribute` values in the current group (determined by the group-by clause) whose tuples have a maximum `value-attribute` value within the group. Thus, this is the typical argmax construct from mathematics added as an SQL extension.

- `select-from-where-group-by` on uncertain queries using aggregates `conf`, `esum`, and `ecount`, but none of the standard SQL aggregates. There is an exact and an approximate version of the `conf` aggregate. The latter takes two parameters ϵ and δ (see the earlier discussion of the Karp-Luby FPRAS).

The aggregates `esum` and `ecount` compute expected sums and counts across groups of tuples. While it may seem that these aggregates are at least as hard as confidence computation (which is #P-hard), this is in fact not so. These aggregates can be efficiently computed exploiting linearity of expectation. A query

```
select A, esum(B) from R group by A;
```

is equivalent to a query

```
select A, sum(B * P) from R' group by A;
```

where `R'` is obtained from the U-relation of `R` by replacing each local condition $V_1, D_1, \ldots, V_k, D_k$ by the probability $\Pr[V_1 = D_1 \wedge \cdots \wedge V_k = D_k]$, not eliminating duplicates. That is, expected sums can be computed efficiently tuple by tuple, and only require to determine the probability of a conjunction, which is easy, rather than a DNF of variable assignments as in the case of the `conf` aggregate. The `ecount` aggregate is a special case of `esum` applied to a column of ones.

EXAMPLE 6.11 The query of Example 6.2 can be expressed in the query language of MayBMS as follows. Let `R` be `repair key in Coins weight by Count` and let `S` be

```
select R.Type, Toss, Face from
       (repair key Type, Toss
        in (select * from Faces, Tosses)
        weight by FProb) S0, R
where R.Type = S0.Type;
```

It is not hard to see that $\pi_{\text{Toss,Face}}(S) \neq Ev$ exactly if there exist tuples $\vec{s} \in S, \vec{t} \in Ev$ such that $\vec{s}$.Toss $= \vec{t}$.Toss and $\vec{s}$.Face $\neq \vec{t}$.Face. Let `C` be

```
select S.Type from S, Ev where
S.Toss = Ev.Toss and S.Face <> Ev.Face;
```

Then we can compute `Q` using the trick of Example 6.10 as

```
select Type, (P1-P2)/(1-P3) as P
from (select Type, conf() as P1 from S group by Type) Q1,
```

```
    (select Type, conf() as P2 from C group by Type) Q2,
    (select conf() as P3 from C) Q3
where Q1.Type = Q2.Type;
```

The argmax aggregate can be used to compute maximum-a-posteriori (MAP) and maximum-likelihood estimates. For example, the MAP coin type $\text{argmax}_{\text{Type}}$ Pr[evidence is twice heads $\wedge$ coin type is Type] can be computed as `select argmax(Type, P) from Q` because the normalizing factor `(1-P3)` has no impact on argmax. Thus, the answer in this example is the double-headed coin. (See table Q of Figure 6.1: The fair coin has $P = 1/3$, while the double-headed coin has $P = 2/3$.)

The maximum likelihood estimate

$$\text{argmax}_{\text{Type}} \text{ Pr[evidence is twice heads} \mid \text{coin type is Type]}$$

can be computed as

```
select argmax(Q.Type, Q.P/R'.P) from Q,
     (select Type, conf() as P from R) R'
where Q.Type = R'.Type;
```

Here, again, the result is 2headed, but this time with likelihood 1. (The fair coin has likelihood 1/4). □

Updates. MayBMS supports the usual schema modification and update statements of SQL. In fact, our use of U-relations makes this quite easy. An insertion of the form

```
insert into <uncertain-table> (<uncertain-query>);
```

is just the standard SQL insertion for tables we interpret as U-relations. Thus, the table inserted into must have the right number (that is, a sufficient number) of condition columns. Schema-modifying operations such as

```
create table <uncertain-table> as  (<uncertain-query>);
```

are similarly straightforward. A deletion

```
delete from <uncertain-table> where <condition>;
```

admits conditions that refer to the attributes of the current tuple and may use t-certain subqueries. Updates can be thought of as combinations of deletions and insertions, but in practice there are of course ways of implementing updates much more efficiently.

Conditioning. Apart from the basic update operations of SQL, MayBMS also supports an update operation `assert` for conditioning, or knowledge compilation. The assert operation takes a Boolean positive relational algebra query ϕ in SQL syntax as an argument, i.e., a select-from-where-union query without aggregation. It conditions the database using this *constraint* ϕ, i.e., conceptually it removes all the possible worlds in which ϕ evaluates to false and renormalizes the probabilities so that they sum up to one again.

Formally, the semantics is thus

$$[\![\text{assert}(\phi)]\!](\mathbf{W}) := \{(R_1, \ldots, R_k, p/p_0) \mid (R_1, \ldots, R_k, p) \in \mathbf{W}, \\ (R_1, \ldots, R_k) \vDash \phi, \; p_0 = \sum_{(R'_1,\ldots,R'_k,p)\in\mathbf{W},(R'_1,\ldots,R'_k)\vDash\phi} p\}.$$

If the condition is inconsistent with the database, i.e., would delete all possible worlds when executed, the assert operation fails with an error (and does not modify the database).

EXAMPLE 6.12 Consider the four possible worlds for the $R[SSN]$ relation of the census example.

$R[SSN]^1$	TID	SSN	$R[SSN]^2$	TID	SSN
	t_1	185		t_1	185
	t_2	185		t_2	186

$R[SSN]^3$	TID	SSN	$R[SSN]^4$	TID	SSN
	t_1	785		t_1	785
	t_2	185		t_2	186

To assert the functional dependency $R : SSN \to TID$, which states that no two individuals can have the same SSN, we can express the functional dependency as a Boolean query Q and execute assert(Q). This deletes the first of the four worlds and renormalizes the probabilities to sum up to one. □

Knowledge compilation using assert has obvious applications in areas such as data cleaning, where we may start with an uncertain database and then chase [1] a set of integrity constraints to reduce uncertainty. The assert operation can apply a set of constraints to a probabilistic database and materialize the cleaned, less uncertain database.

The assert operation is at least as hard as exact confidence operation (it is also practically no harder [23], and essentially the same algorithms can be used for both problems), but differently from confidence computation, the result has to be computed exactly and currently there is no clear notion of useful approximation to a cleaned database.

8. The MayBMS System

The MayBMS system has been under development since 2005 and has undergone several transformations. From the beginning, our choice was to develop MayBMS as an extension of the Postgres server backend. Two prototypes have been demonstrated at ICDE 2007 [7] and VLDB 2007 [8]. Currently, MayBMS is approaching its first release. MayBMS is open source and the source code is available through

http://maybms.sourceforge.net

The academic homepage of the MayBMS project is at

http://www.cs.cornell.edu/database/maybms/

Test data generators and further resources such as main-memory implementations of some of our algorithms have been made available on these Web pages as well.

We are aware of several research prototype probabilistic database management systems that are built as front-end applications of Postgres, but of no other system that aims to develop a fully integrated system. Our backend is accessible through several APIs, with efficient internal operators for computing and managing probabilistic data.

Representations, relational encoding, and query optimization. Our representation system, U-relations, is basically implemented as described earlier, with one small exception. With each pair of columns V_i, D_i in the condition, we also store a column P_i for the probability weight of alternative D_i for variable V_i, straight from the W relation. While the operations of relational algebra, as observed earlier, do not use probability values, confidence computation does. This denormalization (the extension by P_i columns) removes the need to look up any probabilities in the W table in our exact confidence computation algorithms.

Our experiments show that the relational encoding of positive relational algebra which is possible for U-relations is so simple – it is a parsimonious transformation, i.e., the number of relational algebra operations is not increased – that the standard Postgres query optimizer actually does well at finding good query plans (see [3]).

Approximate confidence computation. MayBMS implements both an approximation algorithm and several exact algorithms for confidence computation. The approximation algorithm is a combination of the Karp-Luby unbiased estimator for DNF counting [19, 20] in a modified version adapted for confidence computation in probabilistic databases (cf. e.g. [22]) and the

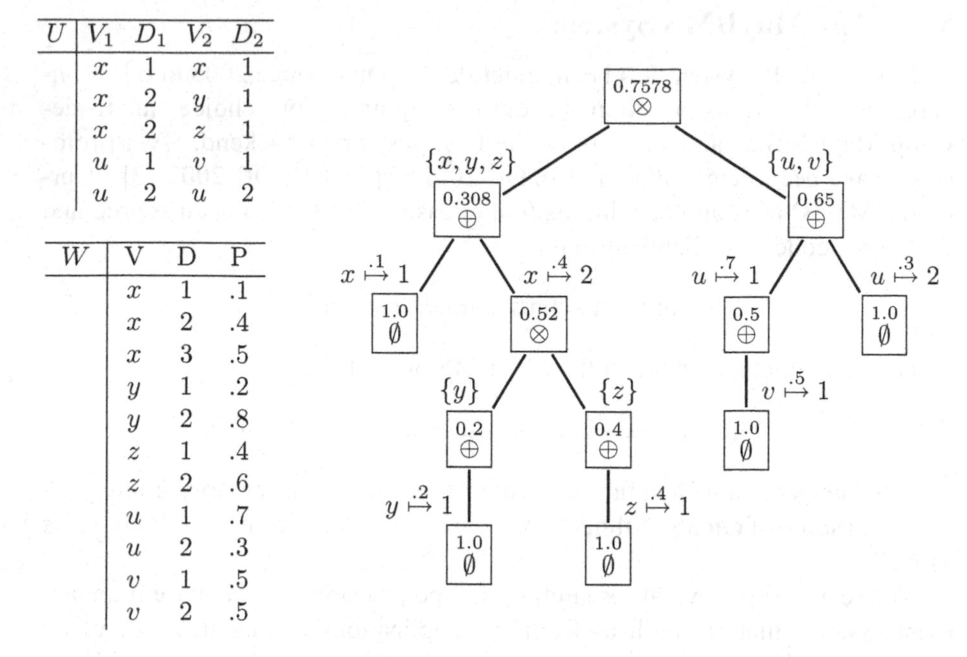

U	V_1	D_1	V_2	D_2
	x	1	x	1
	x	2	y	1
	x	2	z	1
	u	1	v	1
	u	2	u	2

W	V	D	P
	x	1	.1
	x	2	.4
	x	3	.5
	y	1	.2
	y	2	.8
	z	1	.4
	z	2	.6
	u	1	.7
	u	2	.3
	v	1	.5
	v	2	.5

Figure 6.5. Exact confidence computation.

Dagum-Karp-Luby-Ross optimal algorithm for Monte Carlo estimation [12]. The latter is based on sequential analysis and determines the number of invocations of the Karp-Luby estimator needed to achieve the required bound by running the estimator a small number of times to estimate its mean and variance. We actually use the probabilistic variant of a version of the Karp-Luby estimator described in the book [27] which computes fractional estimates that have smaller variance than the zero-one estimates of the classical Karp-Luby estimator.

Exact confidence computation. Our exact algorithm for confidence computation is described in [23]. It is based on an extended version of the Davis-Putnam procedure [14] that is the basis of the best exact Satisfiability solvers in AI. Given a DNF (of which each clause is a conjunctive local condition), the algorithm employs a combination of variable elimination (as in Davis-Putnam) and decomposition of the DNF into independent subsets of clauses (i.e., subsets that do not share variables), with cost-estimation heuristics for choosing whether to use the former (and for which variable) or the latter.

EXAMPLE 6.13 Consider the U-relation U representing a nullary table and the W table of Figure 6.5. The local conditions of U are $\Phi = \{\{x \mapsto 1\}, \{x \mapsto 2, y \mapsto 1\}, \{x \mapsto 2, z \mapsto 1\}, \{u \mapsto 1, v \mapsto 1\}, \{u \mapsto 2\}\}$.

The algorithm proceeds recursively. We first choose to exploit the fact that the Φ can be split into two independent sets, the first using only the variables $\{x, y, z\}$ and the second only using $\{u, v\}$. We recurse into the first set and eliminate the variable x. This requires us to consider two cases, the alternative values 1 and 2 for x (alternative 3 does not have to be considered because in each of the clauses to be considered, x is mapped to either 1 or 2. In the case that x maps to 2, we eliminate x from the set of clauses that are compatible with the variable assignment $x \mapsto 2$, i.e., the set $\{\{y \mapsto 1\}, \{z \mapsto 1\}\}$, and can decompose exploiting the independence of the two clauses. Once y and z are eliminated, respectively, the conditions have been reduced to "true". The alternative paths of the computation tree, shown in Figure 6.5, are processed analogously.

On returning from the recursion, we compute the probabilities of the subtrees in the obvious way. For two independent sets S_1, S_2 of clauses with probabilities p_1 and p_2, the probability of $S_1 \cup S_2$ is $1-(1-p_1)\cdot(1-p_2)$. For variable elimination branches, the probability is the sum of the products of the probabilities of the subtrees and the probabilities of the variable assignments used for elimination.

It is not hard to verify that the probability of Φ, i.e., the confidence in tuple $\langle\rangle$, is 0.7578. □

Our exact algorithm solves a #P-hard problem and exhibits exponential running time in the worst case. However, like some other algorithms for combinatorial problems, this algorithm shows a clear easy-hard-easy pattern. Outside a narrow range of variable-to-clause count ratios, it very pronouncedly outperforms the (polynomial-time) approximation techniques [23]. It is straightforward to extend this algorithm to condition a probabilistic database (i.e., to compute "assert") [23].

Hierarchical queries. The tuple-independent databases are those probabilistic databases in which, for each tuple, a probability can be given such that the tuple occurs in the database with that probability and the tuples are uncorrelated. It is known since the work of Dalvi and Suciu [13] that there is a class of conjunctive queries, the hierarchical queries Q, for which computing conf(Q) exactly on tuple-independent probabilistic databases is feasible in polynomial time.

In fact, these queries can essentially be computed using SQL queries that involve several nested aggregate-group-by queries. On the other hand, it was also shown in [13] that for any conjunctive query Q that is not hierarchical, computing conf(Q) is #P-hard with respect to data complexity. Dalvi and Suciu introduce the notion of *safe plans* that are at once certificates that a query is hierarchical and query plans with aggregation operators that can be used for evaluating the queries.

Dan Olteanu's group at Oxford has recently extended this work in three ways, and implemented it in MayBMS [17]. First, the observation is used that in the case that a query has a safe plan, it is not necessary to use that safe plan for query evaluation. Instead we can choose our plan from a large set of possible plans, some of which will be much better and use fewer levels of aggregation than the canonical safe plans of [13]. Second, a special low-level operator for processing these aggregations has been implemented, which reduces the number of data scans needed [17]. Finally, the fact is exploited that the #P-hardness result for any single nonhierarchical query of [13] only applies as long as the problem is that of evaluating the query on an arbitrary probabilistic database of suitable schema. If further information about permissible databases is available in the form of functional dependencies that the databases must satisfy, then a larger class of queries can be processed by our approach.

Olteanu and Huang [24] have also obtained results on polynomial-time confidence computation on fragments of conjunctive queries with inequalities, using a powerful framework based on Ordered Binary Decision Diagrams.

Updates, concurrency control and recovery. As a consequence of our choice of a purely relational representation system, these issues cause surprisingly little difficulty. U-relations are just relational tables and updates are just modifications of these tables that can be expressed using the standard SQL update operations. However, finding a suitable programming model and API for efficiently supporting programming access without exposing the user applications to internals of the representation system (which will differ among the various probabilistic DBMS) is a difficult problem. A full statement of this problem and some first results can be found in [4].

9. Conclusions and Outlook

The aim of the MayBMS system is to be the first robust and scalable probabilistic database system that can be used in real applications. By our choice of running the entire project as an open-source project with the goal of creating mature code and serious documentation for developers, we hope to be able to accelerate progress in the field by making a testbed for new algorithms available to the research community.

Our possibly most important goal is to extend MayBMS to support continuous distributions. The path towards this goal is clearly sketched by our use of, essentially, a class of conditional tables for data representation. Our representations will not be hard to generalize, but some of the advantages of U-relations will be lost. There will be a need for a special column type "condition" for storing the more general local conditions needed, which has implications on operator implementations and will require us to study query optimization closely:

We will not be able to rely as much on standard query optimizers to produce good plans as we currently do.

Another major goal is an extensive and careful experimental comparison of ours versus the graphical models approach, and to understand where the sweet spots of the two directions lie. More generally, it will be important to start working on a fair benchmark for probabilistic databases and, ideally, AI systems, even though it may still be too early to see the full set of dimensions that the space of systems will have, which is necessary to be able to define a benchmark that will remain fair and useful for some time.

A final grand goal is a query and update language specification that is a widely acceptable candidate for a future standard. This will be essential for wide acceptance of probabilistic databases. We expect our past work on the foundations of query algebras [6, 22, 21] to be useful in such an effort.

References

[1] S. Abiteboul, R. Hull, and V. Vianu. *Foundations of Databases*. Addison-Wesley, 1995.

[2] S. Abiteboul, P. Kanellakis, and G. Grahne. "On the Representation and Querying of Sets of Possible Worlds". *Theor. Comput. Sci.*, **78**(1):158–187, 1991.

[3] L. Antova, T. Jansen, C. Koch, and D. Olteanu. "Fast and Simple Relational Processing of Uncertain Data". In *Proc. ICDE*, 2008.

[4] L. Antova and C. Koch. "On APIs for Probabilistic Databases". In *Proc. 2nd International Workshop on Management of Uncertain Data*, Auckland, New Zealand, 2008.

[5] L. Antova, C. Koch, and D. Olteanu. "10^{10^6} Worlds and Beyond: Efficient Representation and Processing of Incomplete Information". In *Proc. ICDE*, 2007.

[6] L. Antova, C. Koch, and D. Olteanu. "From Complete to Incomplete Information and Back". In *Proc. SIGMOD*, 2007.

[7] L. Antova, C. Koch, and D. Olteanu. "MayBMS: Managing Incomplete Information with Probabilistic World-Set Decompositions". In *Proc. ICDE*, 2007.

[8] L. Antova, C. Koch, and D. Olteanu. "Query Language Support for Incomplete Information in the MayBMS System". In *Proc. VLDB*, 2007.

[9] D. S. Batory. "On Searching Transposed Files". *ACM Trans. Database Syst.*, **4**(4):531–544, 1979.

[10] O. Benjelloun, A. D. Sarma, C. Hayworth, and J. Widom. "An Introduction to ULDBs and the Trio System". *IEEE Data Engineering Bulletin*, 2006.

[11] P. Bohannon, W. Fan, F. Geerts, X. Jia, and A. Kementsietsidis. "Conditional Functional Dependencies for Data Cleaning". In *Proc. ICDE*, 2007.

[12] P. Dagum, R. M. Karp, M. Luby, and S. M. Ross. "An Optimal Algorithm for Monte Carlo Estimation". *SIAM J. Comput.*, **29**(5):1484–1496, 2000.

[13] N. Dalvi and D. Suciu. "Efficient query evaluation on probabilistic databases". *VLDB Journal*, **16**(4):523–544, 2007.

[14] M. Davis and H. Putnam. "A Computing Procedure for Quantification Theory". *Journal of ACM*, **7**(3):201–215, 1960.

[15] M. Goetz and C. Koch. "A Compositional Framework for Complex Queries over Uncertain Data", 2008. Under submission.

[16] E. Grädel, Y. Gurevich, and C. Hirsch. "The Complexity of Query Reliability". In *Proc. PODS*, pages 227–234, 1998.

[17] J. Huang, D. Olteanu, and C. Koch. "Lazy versus Eager Query Plans for Tuple-Independent Probabilistic Databases". In *Proc. ICDE*, 2009. To appear.

[18] T. Imielinski and W. Lipski. "Incomplete information in relational databases". *Journal of ACM*, **31**(4):761–791, 1984.

[19] R. M. Karp and M. Luby. "Monte-Carlo Algorithms for Enumeration and Reliability Problems". In *Proc. FOCS*, pages 56–64, 1983.

[20] R. M. Karp, M. Luby, and N. Madras. "Monte-Carlo Approximation Algorithms for Enumeration Problems". *J. Algorithms*, **10**(3):429–448, 1989.

[21] C. Koch. "A Compositional Query Algebra for Second-Order Logic and Uncertain Databases". Technical Report arXiv:0807.4620, 2008.

[22] C. Koch. "Approximating Predicates and Expressive Queries on Probabilistic Databases". In *Proc. PODS*, 2008.

[23] C. Koch and D. Olteanu. "Conditioning Probabilistic Databases". In *Proc. VLDB*, 2008.

[24] D. Olteanu and J. Huang. Conjunctive queries with inequalities on probabilistic databases. In *Proc. SUM*, 2008.

[25] D. Olteanu, C. Koch, and L. Antova. "World-set Decompositions: Expressiveness and Efficient Algorithms". *Theoretical Computer Science*, **403**(23):265–284, 2008.

[26] M. Stonebraker, D. J. Abadi, A. Batkin, X. Chen, M. Cherniack, M. Ferreira, E. Lau, A. Lin, S. Madden, E. J. O'Neil, P. E. O'Neil, A. Rasin, N. Tran, and S. B. Zdonik. "C-Store: A Column-oriented DBMS". In *Proc. VLDB*, pages 553–564, 2005.

[27] V. V. Vazirani. *Approximation Algorithms*. Springer, 2001.

Chapter 7

UNCERTAINTY IN DATA INTEGRATION

Anish Das Sarma
Stanford University, CA, USA
anish@cs.stanford.edu

Xin Dong
AT&T Labs-Research, NJ, USA
lunadong@research.att.com

Alon Halevy
Google Inc., CA, USA
halevy@google.com

Abstract Data integration has been an important area of research for several years. In this chapter, we argue that supporting modern data integration applications requires systems to handle uncertainty at every step of integration. We provide a formal framework for data integration systems with uncertainty. We define probabilistic schema mappings and probabilistic mediated schemas, show how they can be constructed automatically for a set of data sources, and provide techniques for query answering. The foundations laid out in this chapter enable bootstrapping a *pay-as-you-go* integration system completely automatically.

Keywords: data integration, uncertainty, pay-as-you-go, mediated schema, schema mapping

1. Introduction

Data integration and exchange systems offer a uniform interface to a multitude of data sources and the ability to share data across multiple systems. These systems have recently enjoyed significant research and commercial success [18, 19]. Current data integration systems are essentially a natural extension of traditional database systems in that queries are specified in a structured form and data are modeled in one of the traditional data models (relational, XML). In addition, the data integration system has exact knowledge of how the data in the sources map to the schema used by the data integration system.

In this chapter we argue that as the scope of data integration applications broadens, such systems need to be able to model uncertainty at their core. Uncertainty can arise for multiple reasons in data integration. First, the semantic mappings between the data sources and the mediated schema may be approximate. For example, in an application like Google Base [17] that enables anyone to upload structured data, or when mapping millions of sources on the deep web [28], we cannot imagine specifying exact mappings. In some domains (e.g., bioinformatics), we do not necessarily know what the exact mapping is. Second, data are often extracted from unstructured sources using information extraction techniques. Since these techniques are approximate, the data obtained from the sources may be uncertain. Third, if the intended users of the application are not necessarily familiar with schemata, or if the domain of the system is too broad to offer form-based query interfaces (such as web forms), we need to support keyword queries. Hence, another source of uncertainty is the transformation between keyword queries and a set of candidate structured queries. Finally, if the scope of the domain is very broad, there can even be uncertainty about the concepts in the mediated schema.

Another reason for data integration systems to model uncertainty is to support *pay-as-you-go* integration. Dataspace Support Platforms [20] envision data integration systems where sources are added with no effort and the system is constantly evolving in a pay-as-you-go fashion to improve the quality of semantic mappings and query answering. This means that as the system evolves, there will be uncertainty about the semanantic mappings to its sources, its mediated schema and even the semantics of the queries posed to it.

This chapter describes some of the formal foundations for data integration with uncertainty. We define probabilistic schema mappings and probabilistic mediated schemas, and show how to answer queries in their presence. With these foundations, we show that it is possible to completely automatically bootstrap a pay-as-you-go integration system.

This chapter is largely based on previous papers [10, 6]. The proofs of the theorems we state and the experimental results validating some of our claims can be found in there. We also place several other works on uncertainty in data integration in the context of the system we envision. In the next section, we begin by describing an architecture for data integration system that incorporates uncertainty.

2. Overview of the System

This section describes the requirements from a data integration system that supports uncertainty and the overall architecture of the system.

2.1 Uncertainty in data integration

A data integration system needs to handle uncertainty at three levels.

Uncertain mediated schema: The mediated schema is the set of schema terms in which queries are posed. They do not necessarily cover all the attributes appearing in any of the sources, but rather the aspects of the domain that the application builder wishes to expose to the users. Uncertainty in schema mappings can arise for several reasons. First, as we describe in Section 4, if the mediated schema is automatically inferred from the data sources in a pay-as-you-go integration system, there will be some uncertainty about the results. Second, when domains get broad, there will be some uncertainty about how to model the domain. For example, if we model all the topics in Computer Science there will be some uncertainty about the degree of overlap between different topics.

Uncertain schema mappings: Data integration systems rely on schema mappings for specifying the semantic relationships between the data in the sources and the terms used in the mediated schema. However, schema mappings can be inaccurate. In many applications it is impossible to create and maintain precise mappings between data sources. This can be because the users are not skilled enough to provide precise mappings, such as in personal information management [11], because people do not understand the domain well and thus do not even know what correct mappings are, such as in bioinformatics, or because the scale of the data prevents generating and maintaining precise mappings, such as in integrating data of the web scale [27]. Hence, in practice, schema mappings are often generated by semi-automatic tools and not necessarily verified by domain experts.

Uncertain data: By nature, data integration systems need to handle uncertain data. One reason for uncertainty is that data are often extracted from unstructured or semi-structured sources by automatic methods (e.g., HTML pages, emails, blogs). A second reason is that data may come from sources that are unreliable or not up to date. For example, in enterprise settings, it is common for informational data such as gender, racial, and income level to be dirty or missing, even when the transactional data is precise.

Uncertain queries: In some data integration applications, especially on the web, queries will be posed as keywords rather than as structured queries against a well defined schema. The system needs to translate these queries into some structured form so they can be reformulated with respect to the data sources.

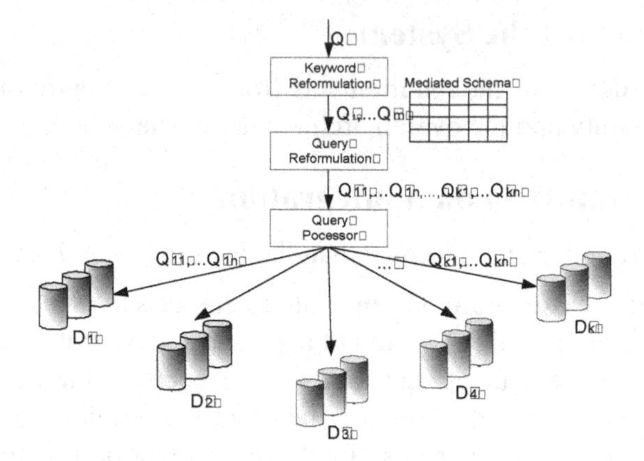

Figure 7.1. Architecture of a data-integration system that handles uncertainty.

At this step, the system may generate multiple candidate structured queries and have some uncertainty about which is the real intent of the user.

2.2 System architecture

Given the previously discussed requirements, we describe the architecture of a data integration system we envision that manages uncertainty at its core. We describe the system by contrasting it to a traditional data integration system.

The first and most fundamental characteristic of this system is that it is based on a probabilistic data model. This means that we attach probabilities to:

- tuples that we process in the system,
- schema mappings,
- mediated schemas, and
- possible interpretations of keyword queries posed to the system.

In contrast, a traditional data integration system includes a single mediated schema and we assume we have as single (and correct) schema mapping between the mediated schema and each source. The data in the sources is also assumed to be correct.

Traditional data integration systems assume that the query is posed in a structured fashion (i.e., can be translated to some subset of SQL). Here, we assume that queries can be posed as keywords (to accommodate a much broader class of users and applications). Hence, whereas traditional data integration systems begin by reformulating a query onto the schemas of the data sources, a data integration system with uncertainty needs to first reformulate a keyword query into a set of candidate structured queries. We refer to this step as *keyword reformulation*. Note that keyword reformulation is different from techniques for keyword search on structured data (e.g., [22, 1]) in that (a) it does not assume access to all the data in the sources or that the sources support keyword search, and (b) it tries to distinguish different structural elements in the query in order to pose more precise queries to the sources (e.g., realizing that in the keyword query "Chicago weather", "weather" is an attribute label and "Chicago" is an instance name). That being said, keyword reformulation should benefit from techniques that support answering keyword search on structured data.

The query answering model is different. Instead of necessarily finding *all* answers to a given query, our goal is typically to find the top-k answers, and rank these answers most effectively.

The final difference from traditional data integration systems is that our query processing will need to be more adaptive than usual. Instead of generating a query answering plan and executing it, the steps we take in query processing will depend on results of previous steps. We note that adaptive query processing has been discussed quite a bit in data integration [12], where the need for adaptivity arises from the fact that data sources did not answer as quickly as expected or that we did not have accurate statistics about their contents to properly order our operations. In our work, however, the goal for adaptivity is to get the answers with high probabilities faster.

The architecture of the system is shown in Figure 7.1. The system contains a number of data sources and a mediated schema (we omit probabilistic mediated schemas from this figure). When the user poses a query Q, which can be either a structured query on the mediated schema or a keyword query, the system returns a set of answer tuples, each with a probability. If Q is a keyword query, the system first performs keyword reformulation to translate it into a set of candidate structured queries on the mediated schema. Otherwise, the candidate query is Q itself.

2.3 Source of probabilities

A critical issue in any system that manages uncertainty is whether we have a reliable source of probabilities. Whereas obtaining reliable probabilities for such a system is one of the most interesting areas for future research, there is quite a bit to build on. For keyword reformulation, it is possible to train and test reformulators on large numbers of queries such that each reformulation result is given a probability based on its performance statistics. For information extraction, current techniques are often based on statistical machine learning methods and can be extended to compute probabilities of each extraction result. Finally, in the case of schema matching, it is standard practice for schema matchers to also associate numbers with the candidates they propose (e.g., [3, 7–9, 21, 26, 34, 35]). The issue here is that the numbers are meant only as a ranking mechanism rather than true probabilities. However, as schema matching techniques start looking at a larger number of schemas, one can imagine ascribing probabilities (or estimations thereof) to their measures.

2.4 Outline of the chapter

We begin by discussing probabilistic schema mappings in Section 3. We also discuss how to answer queries in their presence and how to answer top-k queries. In Section 4 we discuss probabilistic mediated schemas. We begin by motivating them and showing that in some cases they add expressive power to the resulting system. Then we describe an algorithm for generating probabilistic mediated schemas from a collection of data sources.

3. Uncertainty in Mappings

The key to resolving heterogeneity at the schema level is to specify schema mappings between data sources. These mappings describe the relationship between the contents of the different sources and are used to reformulate a query posed over one source (or a mediated schema) into queries over the sources that are deemed relevant. However, in many applications we are not able to provide all the schema mappings upfront. In this section we introduce probabilistic schema mappings (p-mappings) to capture uncertainty on mappings between schemas.

We start by presenting a running example for this section that also motivates p-mappings (Section 3.1). Then we present a formal definition of probabilistic schema mapping and its semantics (Section 3.2). Then, Section 3.3 describes algorithms for query answering with respect to probabilistic mappings and discusses the complexity. Next, Section 3.4 shows how to leverage previous work on schema matching to automatically create probabilistic mappings. In the end, Section 3.5 briefly describes various extensions to the basic definition and Section 3.6 describes other types of approximate schema mappings that have been proposed in the literature.

3.1 Motivating probabilistic mappings

EXAMPLE 7.1 *Consider a data source S, which describes a person by her email address, current address, and permanent address, and the mediated schema T, which describes a person by her name, email, mailing address, home address and office address:*

```
S=(pname, email-addr, current-addr, permanent-addr)
T=(name, email, mailing-addr, home-addr, office-addr)
```

A semi-automatic schema-mapping tool may generate three possible mappings between S and T, assigning each a probability. Whereas the three mappings all map pname *to* name*, they map other attributes in the source and the target differently. Figure 7.2(a) describes the three mappings using sets of*

	Possible Mapping	Prob
m_1 =	{(pname, name), (email-addr, email), (current-addr, mailing-addr), (permanent-addr, home-addr)}	0.5
m_2 =	{(pname, name), (email-addr, email), (permanent-addr, mailing-addr), (current-addr, home-addr)}	0.4
m_3 =	{(pname, name), (email-addr, mailing-addr), (current-addr, home-addr)}	0.1

(a)

pname	*email-addr*	*current-addr*	*permanent-addr*
Alice	alice@	Mountain View	Sunnyvale
Bob	bob@	Sunnyvale	Sunnyvale

(b)

Tuple (mailing-addr)	Prob
('Sunnyvale')	0.9
('Mountain View')	0.5
('alice@')	0.1
('bob@')	0.1

(c)

Figure 7.2. The running example: (a) a probabilistic schema mapping between S and T; (b) a source instance D_S; (c) the answers of Q over D_S with respect to the probabilistic mapping.

attribute correspondences. For example, mapping m_1 maps pname *to* name, email-addr *to* email, current-addr *to* mailing-addr, *and* permanent-addr *to* home-addr. *Because of the uncertainty about which mapping is correct, we consider all of these mappings in query answering.*

Suppose the system receives a query Q composed on the mediated schema and asking for people's mailing addresses:

```
Q: SELECT mailing-addr FROM T
```

Using the possible mappings, we can reformulate Q into different queries:

```
Q1: SELECT current-addr FROM S
Q2: SELECT permanent-addr FROM S
Q3: SELECT email-addr FROM S
```

If the user requires all possible answers, the system generates a single aggregation query based on Q_1, Q_2 and Q_3 to compute the probability of each returned tuple, and sends the query to the data source. Suppose the data source contains a table D_S as shown in Figure 7.2(b), the system will retrieve four answer tuples, each with a probability, as shown in Figure 7.2(c).

If the user requires only the top-1 answer (i.e., the answer tuple with the highest probability), the system decides at runtime which reformulated queries to execute. For example, after executing Q_1 and Q_2 at the source, the system *can already conclude that ('Sunnyvale') is the top-1 answer and can skip query* Q_3. □

3.2 Definition and Semantics

Schema mappings. We begin by reviewing non-probabilistic schema mappings. The goal of a schema mapping is to specify the semantic relationships between a *source schema* and a *target schema*. We refer to the source schema as $\bar{S}$, and a relation in $\bar{S}$ as $S = \langle s_1, \ldots, s_m \rangle$. Similarly, we refer to the target schema as $\bar{T}$, and a relation in $\bar{T}$ as $T = \langle t_1, \ldots, t_n \rangle$.

We consider a limited form of schema mappings that are also referred to as *schema matching* in the literature. Specifically, a schema matching contains a set of *attribute correspondences*. An attribute correspondence is of the form $c_{ij} = (s_i, t_j)$, where s_i is a *source attribute* in the schema S and t_j is a *target attribute* in the schema T. Intuitively, c_{ij} specifies that there is a relationship between s_i and t_j. In practice, a correspondence also involves a function that transforms the value of s_i to the value of t_j. For example, the correspondence (c-degree, temperature) can be specified as temperature=c-degree $*1.8+32$, describing a transformation from Celsius to Fahrenheit. These functions are irrelevant to our discussion, and therefore we omit them. This class of mappings are quite common in practice and already expose many of the novel issues involved in probabilistic mappings and In Section 3.5 we will briefly discuss extensions to a broader class of mappings.

Formally, relation mappings and schema mappings are defined as follows.

DEFINITION 7.2 (SCHEMA MAPPING) *Let $\bar{S}$ and $\bar{T}$ be relational schemas. A* relation mapping *M is a triple (S, T, m), where S is a relation in $\bar{S}$, T is a relation in $\bar{T}$, and m is a set of attribute correspondences between S and T.*

When each source and target attribute occurs in at most one correspondence in m, we call M a one-to-one relation mapping.

A schema mapping *$\overline{M}$ is a set of one-to-one relation mappings between relations in $\bar{S}$ and in $\bar{T}$, where every relation in either $\bar{S}$ or $\bar{T}$ appears at most once.* □

A pair of instances D_S and D_T *satisfies* a relation mapping m if for every source tuple $t_s \in D_S$, there exists a target tuple $t_t \in D_t$, such that for every attribute correspondence $(s, t) \in m$, the value of attribute s in t_s is the same as the value of attribute t in t_t.

EXAMPLE 7.3 *Consider the mappings in Example 7.1. The source database in Figure 7.2(b) (repeated in Figure 7.3(a)) and the target database in Figure 7.3(b) satisfy m_1.* □

pname	email-addr	current-addr	*permanent-addr*
Alice	alice@	Mountain View	Sunnyvale
Bob	bob@	Sunnyvale	Sunnyvale

(a)

name	email	*mailing-addr*	*home-addr*	*office-addr*
Alice	alice@	Mountain View	Sunnyvale	office
Bob	bob@	Sunnyvale	Sunnyvale	office

(b)

name	email	*mailing-addr*	*home-addr*	*office-addr*
Alice	alice@	Sunnyvale	Mountain View	office
Bob	email	bob@	Sunnyvale	office

(c)

Tuple (mailing-addr)	Prob
('Sunnyvale')	0.9
('Mountain View')	0.5
('alice@')	0.1
('bob@')	0.1

(d)

Tuple (mailing-addr)	Prob
('Sunnyvale')	0.94
('Mountain View')	0.5
('alice@')	0.1
('bob@')	0.1

(e)

Figure 7.3. Example 7.11: (a) a source instance D_S; (b) a target instance that is by-table consistent with D_S and m_1; (c) a target instance that is by-tuple consistent with D_S and $< m_2, m_3 >$; (d) $Q^{table}(D_S)$; (e) $Q^{tuple}(D_S)$.

Probabilistic schema mappings. Intuitively, a probabilistic schema mapping describes a probability distribution of a set of *possible* schema mappings between a source schema and a target schema.

Definition 7.4 (Probabilistic Mapping) *Let $\bar{S}$ and $\bar{T}$ be relational schemas. A* probabilistic mapping (p-mapping), *pM, is a triple $(S, T, \mathbf{m})$, where $S \in \bar{S}$, $T \in \bar{T}$, and $\mathbf{m}$ is a set $\{(m_1, Pr(m_1)), \ldots, (m_l, Pr(m_l))\}$, such that*

- *for $i \in [1, l]$, m_i is a one-to-one mapping between S and T, and for every $i, j \in [1, l]$, $i \neq j \Rightarrow m_i \neq m_j$.*
- *$Pr(m_i) \in [0, 1]$ and $\sum_{i=1}^{l} Pr(m_i) = 1$.*

A schema p-mapping, $\overline{pM}$, *is a set of p-mappings between relations in $\bar{S}$ and in $\bar{T}$, where every relation in either $\bar{S}$ or $\bar{T}$ appears in at most one p-mapping.* □

We refer to a non-probabilistic mapping as an *ordinary mapping*. A schema p-mapping may contain both p-mappings and ordinary mappings. Example 7.1 shows a p-mapping (see Figure 7.2(a)) that contains three possible mappings.

Semantics of probabilistic mappings. Intuitively, a probabilistic schema mapping models the uncertainty about which of the mappings in pM is the correct one. When a schema matching system produces a set of candidate matches, there are two ways to interpret the uncertainty: (1) a single mapping in pM is the correct one and it applies to all the data in S, or (2) several mappings are partially correct and each is suitable for a subset of tuples in S, though it is not known which mapping is the right one for a specific tuple. Figure 7.3(b) illustrates the first interpretation and applies mapping m_1. For the same example, the second interpretation is equally valid: some people may choose to use their current address as mailing address while others use their permanent address as mailing address; thus, for different tuples we may apply different mappings, so the correct mapping depends on the particular tuple.

We define query answering under both interpretations. The first interpretation is referred to as the *by-table* semantics and the second one is referred to as the *by-tuple* semantics of probabilistic mappings. Note that one cannot argue for one interpretation over the other; the needs of the application should dictate the appropriate semantics. Furthermore, the complexity results for query answering, which will show advantages to by-table semantics, should not be taken as an argument in the favor of by-table semantics.

We next define query answering with respect to p-mappings in detail and the definitions for schema p-mappings are the obvious extensions. Recall that given a query and an ordinary mapping, we can compute *certain answers* to the query with respect to the mapping. Query answering with respect to p-mappings is defined as a natural extension of certain answers, which we next review.

A mapping defines a relationship between instances of S and instances of T that are *consistent* with the mapping.

DEFINITION 7.5 (CONSISTENT TARGET INSTANCE) *Let* $M = (S, T, m)$ *be a relation mapping and* D_S *be an instance of* S.

An instance D_T *of* T *is said to be* consistent with D_S and M, *if for each tuple* $t_s \in D_S$, *there exists a tuple* $t_t \in D_T$, *such that for every attribute correspondence* $(a_s, a_t) \in m$, *the value of* a_s *in* t_s *is the same as the value of* a_t *in* t_t. □

For a relation mapping M and a source instance D_S, there can be an infinite number of target instances that are consistent with D_S and M. We denote by $Tar_M(D_S)$ the set of all such target instances. The set of answers to a query Q is the intersection of the answers on all instances in $Tar_M(D_S)$.

DEFINITION 7.6 (CERTAIN ANSWER) *Let* $M = (S, T, m)$ *be a relation mapping. Let* Q *be a query over* T *and let* D_S *be an instance of* S.

A tuple t *is said to be a* certain answer of Q with respect to D_S and M, *if for every instance* $D_T \in Tar_M(D_S)$, $t \in Q(D_T)$. □

By-table semantics: We now generalize these notions to the probabilistic setting, beginning with the by-table semantics. Intuitively, a p-mapping pM describes a set of possible worlds, each with a possible mapping $m \in pM$. In by-table semantics, a source table can fall in one of the possible worlds; that is, the possible mapping associated with that possible world applies to the whole source table. Following this intuition, we define target instances that are *consistent with* the source instance.

DEFINITION 7.7 (BY-TABLE CONSISTENT INST.) *Let* $pM = (S, T, \mathbf{m})$ *be a p-mapping and* D_S *be an instance of* S.

An instance D_T *of* T *is said to be* by-table consistent with D_S and pM, *if there exists a mapping* $m \in \mathbf{m}$ *such that* D_S *and* D_T *satisfy* m. □

Given a source instance D_S and a possible mapping $m \in \mathbf{m}$, there can be an infinite number of target instances that are consistent with D_S and m. We denote by $Tar_m(D_S)$ the set of all such instances.

In the probabilistic context, we assign a probability to every answer. Intuitively, we consider the certain answers with respect to each possible mapping in isolation. The probability of an answer t is the sum of the probabilities of the mappings for which t is deemed to be a certain answer. We define by-table answers as follows:

DEFINITION 7.8 (BY-TABLE ANS.) *Let* $pM = (S, T, \mathbf{m})$ *be a p-mapping. Let* Q *be a query over* T *and let* D_S *be an instance of* S.

Let t *be a tuple. Let* $\bar{m}(t)$ *be the subset of* $\mathbf{m}$, *such that for each* $m \in \bar{m}(t)$ *and for each* $D_T \in Tar_m(D_S)$, $t \in Q(D_T)$.

Let $p = \sum_{m \in \bar{m}(t)} Pr(m)$. *If* $p > 0$, *then we say* (t, p) *is a* by-table answer of Q with respect to D_S and pM. □

By-tuple semantics: If we follow the possible-world notions, in by-tuple semantics, different tuples in a source table can fall in different possible worlds; that is, different possible mappings associated with those possible worlds can apply to the different source tuples.

Formally, the key difference in the definition of by-tuple semantics from that of by-table semantics is that a consistent target instance is defined by a mapping *sequence* that assigns a (possibly different) mapping in $\mathbf{m}$ to each source tuple in D_S. (Without losing generality, in order to compare between such sequences, we assign some order to the tuples in the instance).

DEFINITION 7.9 (BY-TUPLE CONSISTENT INST.) *Let* $pM = (S, T, \mathbf{m})$ *be a p-mapping and let* D_S *be an instance of* S *with* d *tuples.*

An instance D_T *of* T *is said to be* by-tuple consistent with D_S and pM, *if there is a sequence* $\langle m^1, \ldots, m^d \rangle$ *such that* d *is the number of tuples in* D_S *and for every* $1 \leq i \leq d$,

- $m^i \in \mathbf{m}$, *and*
- *for the* i^{th} *tuple of* D_S, t_i, *there exists a target tuple* $t'_i \in D_T$ *such that for each attribute correspondence* $(a_s, a_t) \in m^i$, *the value of* a_s *in* t_i *is the same as the value of* a_t *in* t'_i. □

Given a mapping sequence $seq = \langle m^1, \ldots, m^d \rangle$, we denote by $Tar_{seq}(D_S)$ the set of all target instances that are consistent with D_S and seq. Note that if D_T is by-table consistent with D_S and m, then D_T is also by-tuple consistent with D_S and a mapping sequence in which each mapping is m.

We can think of every sequence of mappings $seq = \langle m^1, \ldots, m^d \rangle$ as a separate event whose probability is $Pr(seq) = \Pi_{i=1}^{d} Pr(m^i)$. (Section 3.5 relaxes this independence assumption and introduces *conditional mappings*.) If there are l mappings in pM, then there are l^d sequences of length d, and their probabilities add up to 1. We denote by $\mathbf{seq}_d(pM)$ the set of mapping sequences of length d generated from pM.

DEFINITION 7.10 (BY-TUPLE ANSWER) *Let* $pM = (S, T, \mathbf{m})$ *be a p-mapping. Let* Q *be a query over* T *and* D_S *be an instance of* S *with* d *tuples.*

Let t *be a tuple. Let* $\overline{seq}(t)$ *be the subset of* $\mathbf{seq}_d(pM)$, *such that for each* $seq \in \overline{seq}(t)$ *and for each* $D_T \in Tar_{seq}(D_S)$, $t \in Q(D_T)$.

Let $p = \sum_{seq \in \overline{seq}(t)} Pr(seq)$. *If* $p > 0$, *we call* (t, p) *a* by-tuple answer of Q with respect to D_S and pM. □

The set of by-table answers for Q with respect to D_S is denoted by $Q^{table}(D_S)$ and the set of by-tuple answers for Q with respect to D_S is denoted by $Q^{tuple}(D_S)$.

EXAMPLE 7.11 *Consider the p-mapping pM, the source instance* D_S, *and the query* Q *in the motivating example.*

In by-table semantics, Figure 7.3(b) shows a target instance that is consistent with D_S *(repeated in Figure 7.3(a)) and possible mapping* m_1. *Figure 7.3(d) shows the by-table answers of* Q *with respect to* D_S *and* pM. *As an example, for tuple* t =*('Sunnyvale'), we have* $\bar{m}(t) = \{m_1, m_2\}$, *so the possible tuple ('Sunnyvale', 0.9) is an answer.*

In by-tuple semantics, Figure 7.3(c) shows a target instance that is by-tuple consistent with D_S *and the mapping sequence* $< m_2, m_3 >$. *Figure 7.3(e) shows the by-tuple answers of* Q *with respect to* D_S *and* pM. *Note that the probability of tuple t=('Sunnyvale') in the by-table answers is different from that in the by-tuple answers. We describe how to compute the probabilities in detail in the next section.* □

3.3 Query Answering

This section studies query answering in the presence of probabilistic mappings. We start with describing algorithms for returning all answer tuples with probabilities, and discussing the complexity of query answering in terms of the size of the data (*data complexity*) and the size of the p-mapping (*mapping complexity*). We then consider returning the top-k query answers, which are the k answer tuples with the top probabilities.

By-table query answering. In the case of by-table semantics, answering queries is conceptually simple. Given a p-mapping $pM = (S, T, \mathbf{m})$ and an SPJ query Q, we can compute the certain answers of Q under each of the mappings $m \in \mathbf{m}$. We attach the probability $Pr(m)$ to every certain answer under m. If a tuple is an answer to Q under multiple mappings in $\mathbf{m}$, then we add up the probabilities of the different mappings.

Algorithm BYTABLE takes as input an SPJ query Q that mentions the relations $T_1, \ldots, T_l$ in the FROM clause. Assume that we have the p-mapping pM_i associated with the table T_i. The algorithm proceeds as follows.

Step 1: We generate the possible reformulations of Q (a reformulation query computes all certain answers when executed on the source data) by considering every combination of the form $(m^1, \ldots, m^l)$, where m^i is one of the possible mappings in pM_i. Denote the set of reformulations by $Q'_1, \ldots, Q'_k$. The probability of a reformulation $Q' = (m^1, \ldots, m^l)$ is $\Pi_{i=1}^{l} Pr(m^i)$.

Step 2: For each reformulation Q', retrieve each of the unique answers from the sources. For each answer obtained by $Q'_1 \cup \ldots \cup Q'_k$, its probability is computed by summing the probabilities of the Q''s in which it is returned.

Importantly, note that it is possible to express both steps as an SQL query with grouping and aggregation. Therefore, if the underlying sources support SQL, we can leverage their optimizations to compute the answers.

With our restricted form of schema mapping, the algorithm takes time polynomial in the size of the data and the mappings. We thus have the following complexity result.

THEOREM 7.12 *Let $\overline{pM}$ be a schema p-mapping and let Q be an SPJ query. Answering Q with respect to $\overline{pM}$ in by-table semantics is in PTIME in the size of the data and the mapping.* □

By-tuple query answering. To extend the by-table query-answering strategy to by-tuple semantics, we would need to compute the certain answers for every *mapping sequence* generated by pM. However, the number of such mapping sequences is exponential in the size of the input data. The following example shows that for certain queries this exponential time complexity is inevitable.

Tuple (mailing-addr)	Pr
('Sunnyvale')	0.94
('Mountain View')	0.5
('alice@')	0.1
('bob@')	0.1

(a)

Tuple (mailing-addr)	Pr
('Sunnyvale')	0.8
('Mountain View')	0.8

(b)

Figure 7.4. Example 7.13: (a) $Q_1^{tuple}(D)$ and (b) $Q_2^{tuple}(D)$.

EXAMPLE 7.13 *Suppose that in addition to the tables in Example 7.1, we also have* U(city) *in the source and* V(hightech) *in the target. The p-mapping for V contains two possible mappings: ({(city, hightech)}, .8) and (∅, .2).*

Consider the following query Q, which decides if there are any people living in a high-tech city.

```
Q: SELECT 'true'
   FROM T, V
   WHERE T.mailing-addr = V.hightech
```

An incorrect way of answering the query is to first execute the following two sub-queries Q_1 and Q_2, then join the answers of Q_1 and Q_2 and summing up the probabilities.

```
Q1: SELECT mailing-addr FROM T
Q2: SELECT hightech FROM V
```

Now consider the source instance D, where D_S is shown in Figure 7.2(a), and D_U has two tuples ('Mountain View') and ('Sunnyvale'). Figure 7.4(a) and (b) show $Q_1^{tuple}(D)$ and $Q_2^{tuple}(D)$. If we join the results of Q_1 and Q_2, we obtain for the true *tuple the following probability: $0.94 * 0.8 + 0.5 * 0.8 = 1.152$. However, this is incorrect. By enumerating all consistent target tables, we in fact compute 0.864 as the probability. The reason for this error is that on some target instance that is by-tuple consistent with the source instance, the answers to both Q_1 and Q_2 contain tuple ('Sunnyvale') and tuple ('Mountain View'). Thus, generating the tuple ('Sunnyvale') as an answer for both Q_1 and Q_2 and generating the tuple ('Mountain View') for both queries are not independent events, and so simply adding up their probabilities leads to incorrect results.*

Indeed, it is not clear if there exists a better algorithm to answer Q than by enumerating all by-tuple consistent target instances and then answering Q on each of them. □

In fact, it is proved that in general, answering SPJ queries in by-tuple semantics with respect to schema p-mappings is hard.

THEOREM 7.14 *Let Q be an SPJ query and let $\overline{pM}$ be a schema p-mapping. The problem of finding the probability for a by-tuple answer to Q with respect to $\overline{pM}$ is #P-complete with respect to data complexity and is in PTIME with respect to mapping complexity.* □

Recall that #P is the complexity class of some hard counting problems (*e.g.* , counting the number of variable assignments that satisfy a Boolean formula). It is believed that a #P-complete problem cannot be solved in polynomial time, unless $P = NP$.

Although by-tuple query answering in general is hard, there are two restricted but common classes of queries for which by-tuple query answering takes polynomial time. The first class of queries are those that include only a single subgoal being the target of a p-mapping; here, we refer to an occurrence of a table in the `FROM` clause of a query as a *subgoal* of the query. Relations in the other subgoals are either involved in ordinary mappings or do not require a mapping. Hence, if we only have uncertainty with respect to one part of the domain, our queries will typically fall in this class. The second class of queries can include multiple subgoals involved in p-mappings, but return the join attributes for such subgoals. We next illustrate these two classes of queries and query answering for them using two examples.

EXAMPLE 7.15 *Consider rewriting Q in the motivating example, repeated as follows:*

```
Q: SELECT mailing-addr FROM T
```

To answer the query, we first rewrite Q into query Q′ by adding the id *column:*

```
Q': SELECT id, mailing-addr FROM T
```

We then invoke BYTABLE *and generate the following SQL query to compute by-table answers for Q′:*

```
Qa: SELECT id, mailing-addr, SUM(pr)
    FROM (
      SELECT DISTINCT id, current-addr
             AS mailing-addr, 0.5 AS pr
      FROM S
      UNION ALL
      SELECT DISTINCT id, permanent-addr
             AS mailing-addr, 0.4 AS pr
      FROM S
      UNION ALL
      SELECT DISTINCT id, email-addr
```

```
                AS mailing-addr, 0.1 AS pr
      FROM S)
    GROUP BY id, mailing-addr
```

Finally, we generate the results using the following query.

```
Qu:  SELECT mailing-addr, NOR(pr) AS pr
     FROM Qa
     GROUP BY mailing-addr
```

where for a set of probabilities $pr_1, \ldots, pr_n$, *NOR computes* $1-\Pi_{i=1}^{n}(1-pr_i)$.
□

EXAMPLE 7.16 *Consider the schema p-mapping in Example 7.13. If we revise Q slightly be returning the join attribute, shown as follows, we can answer the query in polynomial time.*

```
Q':  SELECT V.hightech
     FROM T, V
     WHERE T.mailing-addr = V.hightech
```

We answer the query by dividing it into two sub-queries, Q_1 *and* Q_2, *as shown in Example 7.13. We can compute* Q_1 *with query* Q_u *(shown in Example 7.15) and compute* Q_2 *similarly with a query* Q'_u. *We compute by-tuple answers of* Q' *as follows:*

```
SELECT Qu'.hightech, Qu.pr*Qu'.pr
FROM Qu, Qu'
WHERE Qu.mailing-addr = Qu'.hightect
```

□

Top-K Query Answering. The main challenge in designing the algorithm for returning top-k query answers is to only perform the necessary reformulations at every step and halt when the top-k answers are found. We focus on top-k query answering for by-table semantics and the algorithm can be modified for by-tuple semantics.

Recall that in by-table query answering, the probability of an answer is the sum of the probabilities of the reformulated queries that generate the answer. Our goal is to reduce the number of reformulated queries we execute. The algorithm we describe next proceeds in a greedy fashion: it executes queries in descending order of probabilities. For each tuple t, it maintains the upper bound $p_{max}(t)$ and lower bound $p_{min}(t)$ of its probability. This process halts when it finds k tuples whose p_{min} values are higher than p_{max} of the rest of the tuples.

TOPKBYTABLE takes as input an SPJ query Q, a schema p-mapping $\overline{pM}$, an instance D_S of the source schema, and an integer k, and outputs the top-k answers in $Q^{table}(D_S)$. The algorithm proceeds in three steps.

Step 1: Rewrite Q according to $\overline{pM}$ into a set of queries $Q_1, \ldots, Q_n$, each with a probability assigned in a similar way as stated in Algorithm BYTABLE.

Step 2: Execute $Q_1, \ldots, Q_n$ in descending order of their probabilities. Maintain the following measures:

- The highest probability, $PMax$, for the tuples that have not been generated yet. We initialize $PMax$ to 1; after executing query Q_i and updating the list of answers (see third bullet), we decrease $PMax$ by $Pr(Q_i)$;
- The threshold th determining which answers are potentially in the top-k. We initialize th to 0; after executing Q_i and updating the answer list, we set th to the k-th largest p_{min} for tuples in the answer list;
- A list L of answers whose p_{max} is no less than th, and bounds p_{min} and p_{max} for each answer in L. After executing query Q_i, we update the list as follows: (1) for each $t \in L$ and $t \in Q_i(D_S)$, we increase $p_{min}(t)$ by $Pr(Q_i)$; (2) for each $t \in L$ but $t \notin Q_i(D_S)$, we decrease $p_{max}(t)$ by $Pr(Q_i)$; (3) if $PMax \geq th$, for each $t \notin L$ but $t \in Q_i(D_S)$, insert t to L, set p_{min} to $Pr(Q_i)$ and $p_{max}(t)$ to $PMax$.
- A list T of k tuples with top p_{min} values.

Step 3: When $th > PMax$ and for each $t \notin T$, $th > p_{max}(t)$, halt and return T.

EXAMPLE 7.17 *Consider Example 7.1 where we seek for top-1 answer. We answer the reformulated queries in order of* Q_1, Q_2, Q_3. *After answering* Q_1, *for tuple ("Sunnyvale") we have* $p_{min} = .5$ *and* $p_{max} = 1$, *and for tuple ("Mountain View") we have the same bounds. In addition,* $PMax = .5$ *and* $th = .5$.

In the second round, we answer Q_2. *Then, for tuple ("Sunnyvale") we have* $p_{min} = .9$ *and* $p_{max} = 1$, *and for tuple ("Mountain View") we have* $p_{min} = .5$ *and* $p_{max} = .6$. *Now* $PMax = .1$ *and* $th = .9$.

Because $th > PMax$ *and* th *is above the* p_{max} *for the ("Mountain View") tuple, we can halt and return ("Sunnyvale") as the top-1 answer.* □

3.4 Creating P-mappings

We now address the problem of generating a p-mapping between a source schema and a target schema. We begin by assuming we have a set of weighted correspondences between the source attributes and the target attributes. These weighted correspondences are created by a set of schema matching modules. However, as we explain shortly, there can be *multiple* p-mappings that are consistent with a given set of weighted correspondences, and the question is which of them to choose. We describe an approach to creating p-mappings that is based on choosing the mapping that maximizes the *entropy* of the probability assignment.

Computing weighted correspondences. A *weighted correspondence* between a pair of attributes specifies the degree of semantic similarity between them. Let $S(s_1, \ldots, s_m)$ be a source schema and $T(t_1, \ldots, t_n)$ be a target schema. We denote by $C_{i,j}, i \in [1, m], j \in [1, n]$, the weighted correspondence between s_i and t_j and by $w_{i,j}$ the weight of $C_{i,j}$. The first step is to compute a weighted correspondence between every pair of attributes, which can be done by applying existing schema matching techniques.

Although weighted correspondences tell us the degree of similarity between pairs of attributes, they do not tell us *which* target attribute a source attribute should map to. For example, a target attribute mailing-address can be both similar to the source attribute current-addr and to permanent-addr, so it makes sense to map either of them to mailing-address in a schema mapping. In fact, given a set of weighted correspondences, there could be a *set* of p-mappings that are consistent with it. We can define the one-to-many relationship between sets of weighted correspondences and p-mappings by specifying when a p-mapping is *consistent with* a set of weighted correspondences.

DEFINITION 7.18 (CONSISTENT P-MAPPING) *A p-mapping pM is* consistent with *a weighted correspondence $C_{i,j}$ between a pair of source and target attributes if the sum of the probabilities of all mappings $m \in pM$ containing correspondence (i, j) equals $w_{i,j}$; that is,*

$$w_{i,j} = \sum_{m \in pM, (i,j) \in m} Pr(m).$$

A p-mapping is consistent with *a set of weighted correspondences* **C** *if it is consistent with each weighted correspondence $C \in$* **C**. □

However, not every set of weighted correspondences admits a consistent p-mapping. The following theorem shows under which conditions a consistent p-mapping exists, and establishes a normalization factor for weighted correspondences that will guarantee the existence of a consistent p-mapping.

THEOREM 7.19 *Let* **C** *be a set of weighted correspondences between a source schema $S(s_1, \ldots, s_m)$ and a target schema $T(t_1, \ldots, t_n)$.*

- *There exists a consistent p-mapping with respect to* $\mathbf{C}$ *if and only if (1) for every* $i \in [1, m]$, $\sum_{j=1}^{n} w_{i,j} \leq 1$ *and (2) for every* $j \in [1, n]$, $\sum_{i=1}^{m} w_{i,j} \leq 1$.
- *Let*

$$M' = max\{max_i\{\sum_{j=1}^{n} w_{i,j}\}, max_j\{\sum_{i=1}^{m} w_{i,j}\}\}.$$

Then, for each $i \in [1, m]$, $\sum_{j=1}^{n} \frac{w_{i,j}}{M'} \leq 1$ *and for each* $j \in [1, n]$, $\sum_{i=1}^{m} \frac{w_{i,j}}{M'} \leq 1$. □

Based on Theorem 7.19, we normalize the weighted correspondences we generated as described previously by dividing them by M'; that is,

$$w'_{i,j} = \frac{w_{i,j}}{M'}.$$

Generating p-mappings. To motivate our approach to generating p-mappings, consider the following example. Consider a source schema (A, B) and a target schema (A', B'). Assume we have computed the following weighted correspondences between source and target attributes: $w_{A,A'} = 0.6$ and $w_{B,B'} = 0.5$ (the rest are 0).

As we explained above, there are an infinite number of p-mappings that are consistent with this set of weighted correspondences and below we list two:
pM_1:

```
m1: (A,A'), (B,B'): 0.3 m2: (A,A'): 0.3 m3:
   (B,B'): 0.2 m4: empty: 0.2
```

pM_2:

```
m1: (A,A'), (B,B'): 0.5
m2: (A,A'): 0.1
m3: empty: 0.4
```

In a sense, pM_1 seems better than pM_2 because it assumes that the similarity between A and A' is independent of the similarity between B and B'.

In the general case, among the many p-mappings that are consistent with a set of weighted correspondences $\mathbf{C}$, we choose the one with the *maximum entropy*; that is, the p-mappings whose probability distribution obtains the maximum value of $\sum_{i=1}^{l} -p_i * logp_i$. In the above example, pM_1 obtains the maximum entropy.

The intuition behind maximum entropy is that when we need to select among multiple possible distributions on a set of exclusive events, we choose the one that does not favor any of the events over the others. Hence, we choose the distribution that does not *introduce new information* that we didn't have apriori. The principle of maximum entropy is widely used in other areas such as natural language processing.

To create the p-mapping, we proceed in two steps. First, we enumerate all possible one-to-one schema mappings between S and M that contain a subset of correspondences in **C**. Second, we assign probabilities on each of the mappings in a way that maximizes the entropy of our result p-mapping.

Enumerating all possible schema mappings given **C** is trivial: for each subset of correspondences, if it corresponds to a one-to-one mapping, we consider the mapping as a possible mapping.

Given the possible mappings $m_1, \ldots, m_l$, we assign probabilities $p_1, \ldots, p_l$ to $m_1, \ldots, m_l$ by solving the following constraint optimization problem (OPT):

`maximize` $\sum_{k=1}^{l} -p_k * \log p_k$ `subject to:`

1 $\forall k \in [1, l], 0 \leq p_k \leq 1,$

2 $\sum_{k=1}^{l} p_k = 1$, and

3 $\forall i, j : \sum_{k \in [1,l],(i,j) \in m_k} p_k = w_{i,j}.$

We can apply existing technology in solving the OPT optimization problem. Although finding maximum-entropy solutions in general is costly, the experiments described in [6] show that the execution time is reasonable for a one-time process.

3.5 Broader Classes of Mappings

In this section we describe several practical extensions to the basic mapping language. The query answering techniques and complexity results we have described carry over to these techniques.

GLAV mappings: The common formalism for schema mappings, GLAV (a.k.a. tuple-generating dependencies), is based on expressions of the form

$$m : \forall \mathbf{x}(\varphi(\mathbf{x}) \rightarrow \exists \mathbf{y} \psi(\mathbf{x}, \mathbf{y})).$$

In the expression, φ is the body of a conjunctive query over $\bar{S}$ and ψ is the body of a conjunctive query over $\bar{T}$. A pair of instances D_S and D_T *satisfies* a GLAV mapping m if for every assignment of $\mathbf{x}$ in D_S that satisfies φ there exists an assignment of $\mathbf{y}$ in D_T that satisfies ψ.

We define *general p-mappings* to be triples of the form $pGM = (\bar{S}, \bar{T}, \mathbf{gm})$, where $\mathbf{gm}$ is a set $\{(gm_i, Pr(gm_i)) \mid i \in [1, n]\}$, such that for each $i \in [1, n]$, gm_i is a general GLAV mapping. The definition of by-table semantics for such mappings is a simple generalization of Definition 7.8 and query answering can be conducted in PTIME. Extending by-tuple semantics to arbitrary

GLAV mappings is much trickier than by-table semantics and would involve considering mapping sequences whose length is the product of the number of tuples in each source table, and the results are much less intuitive.

THEOREM 7.20 *Let pGM be a general p-mapping between a source schema $\bar{S}$ and a target schema $\bar{T}$. Let D_S be an instance of $\bar{S}$. Let Q be an SPJ query with only equality conditions over $\bar{T}$. The problem of computing $Q^{table}(D_S)$ with respect to pGM is in PTIME in the size of the data and the mapping.* □

Complex mappings: Complex mappings map a set of attributes in the source to a set of attributes in the target. For example, we can map the attribute address to the concatenation of street, city, and state.

Formally, a *set correspondence* between S and T is a relationship between a subset of attributes in S and a subset of attributes in T. Here, the function associated with the relationship specifies a single value for each of the target attributes given a value for each of the source attributes. Again, the actual functions are irrelevant to our discussion. A *complex mapping* is a triple (S, T, cm), where cm is a set of set correspondences, such that each attribute in S or T is involved in at most one set correspondence. A *complex p-mapping* is of the form $pCM = \{(cm_i, Pr(cm_i)) \mid i \in [1, n]\}$, where $\sum_{i=1}^{n} Pr(cm_i) = 1$.

THEOREM 7.21 *Let $\overline{pCM}$ be a complex schema p-mapping between schemas $\bar{S}$ and $\bar{T}$. Let D_S be an instance of $\bar{S}$. Let Q be an SPJ query over $\bar{T}$. The data complexity and mapping complexity of computing $Q^{table}(D_S)$ with respect to $\overline{pCM}$ are PTIME. The data complexity of computing $Q^{tuple}(D_S)$ with respect to $\overline{pCM}$ is #P-complete. The mapping complexity of computing $Q^{tuple}(D_S)$ with respect to $\overline{pCM}$ is in PTIME.* □

Union mapping: *Union mappings* specify relationships such as both attribute home-address and attribute office-address can be mapped to address. Formally, a *union mapping* is a triple $(S, T, \bar{m})$, where $\bar{m}$ is a set of mappings between S and T. Given a source relation D_S and a target relation D_T, we say D_S and D_T are consistent with respect to the union mapping if for each source tuple t and $m \in \bar{m}$, there exists a target tuple t', such that t and t' satisfy m. A *union p-mapping* is of the form $pUM = \{(\bar{m}_i, Pr(\bar{m}_i)) \mid i \in [1, n]\}$, where $\sum_{i=1}^{n} Pr(\bar{m}_i) = 1$.

Both by-table and by-tuple semantics apply to probabilistic union mappings.

THEOREM 7.22 *Let $\overline{pUM}$ be a union schema p-mapping between a source schema $\bar{S}$ and a target schema $\bar{T}$. Let D_S be an instance of $\bar{S}$. Let Q be a conjunctive query over $\bar{T}$. The problem of computing $Q^{table}(D_S)$ with respect to $\overline{pUM}$ is in PTIME in the size of the data and the mapping; the problem of computing $Q^{tuple}(D_S)$ with respect to $\overline{pUM}$ is in PTIME in the size of the mapping and #P-complete in the size of the data.* □

Conditional mappings: In practice, our uncertainty is often conditioned. For example, we may want to state that daytime-phone maps to work-phone with probability 60% if age ≤ 65, and maps to home-phone with probability 90% if age > 65.

We define a *conditional p-mapping* as the set $cpM = \{(pM_1, C_1), \ldots, \ldots(pM_n, C_n)\}$, where $pM_1, \ldots, pM_n$ are p-mappings, and $C_1, \ldots, C_n$ are pairwise disjoint conditions. Intuitively, for each $i \in [1, n]$, pM_i describes the probability distribution of possible mappings when condition C_i holds. Conditional mappings make more sense for by-tuple semantics. The following theorem shows that the complexity results carry over to such mappings.

THEOREM 7.23 *Let $\overline{cpM}$ be a conditional schema p-mapping between $\bar{S}$ and $\bar{T}$. Let D_S be an instance of $\bar{S}$. Let Q be an SPJ query over $\bar{T}$. The problem of computing $Q^{tuple}(D_S)$ with respect to $\overline{cpM}$ is in PTIME in the size of the mapping and #P-complete in the size of the data.* □

3.6 Other Types of Approximate Schema Mappings

There have been various models proposed to capture uncertainty on mappings between attributes. [15] proposes keeping the top-K mappings between two schemas, each with a probability (between 0 and 1) of being true. [16] proposes assigning a probability for matching of every pair of source and target attributes. This notion corresponds to weighted correspondences described in Section 3.4.

Magnani and Montesi [29] have empirically shown that top-k schema mappings can be used to increase the recall of a data integration process and Gal [14] described how to generate top-k schema matchings by combining the matching results generated by various matchers. The probabilistic schema mappings we described above are different as they contain all possible schema mappings that conform to the schema matching results and assigns probabilities to these mappings to reflect the likelihood that each mapping is correct. Nottelmann and Straccia [32] proposed generating probabilistic schema matchings that capture the uncertainty on each matching step. The probabilistic schema mappings we create not only capture our uncertainty on results of the matching step, but also take into consideration various combinations of attribute correspondences and describe a *distribution* of possible schema mappings where the probabilities of all mappings sum up to 1.

There have also been work studying how to use probabilistic models to capture uncertainty on mappings of schema object classes, such as DatabasePapers and AIPapers. Query answering can take such uncertainty into consideration in computing the coverage percentage of the returned answers and in ordering information sources to maximize the likelihood of obtaining answers early. In the relational model, an object class is often represented using a relational table; thus, these probabilistic models focus on mapping between tables rather than attributes in the tables.

Specifically, consider two object classes A and B. The goal of the probabilistic models is to capture the uncertainty on whether A maps to B. One method [13] uses probability $P(B|A)$, which is the probability that an instance of A is also an instance of B. Another method [29] uses a tuple $< A, B, R, P >$, where R is a set of mutually exclusive relationships between A and B, and P is a probability distribution over R. The possible relationships considered in this model include *equivalent* $=$, *subset-subsumption* $\subset$, *superset-subsumption* $\supset$, *overlapping* $\cap$, *disjointness* $\not\cap$, and *incompatibility* $\not\sim$.

4. Uncertainty in Mediated Schema

The mediated schema is the set of schema terms (e.g., relations, attribute names) in which queries are posed. They do not necessarily cover all the attributes appearing in any of the sources, but rather the aspects of the domain that are important for the integration application. When domains are broad, and there are multiple perspectives on them (e.g., a domain in science that is constantly under evolution), then there will be uncertainty about which is the correct mediated schema and about the meaning of its terms. When the mediated schema is created automatically by inspecting the sources in a pay-as-you-go system, there will also be uncertainty about the mediated schema.

In this section we first motivate the need for probabilistic mediated schemas (p-med-schemas) with an example (Section 4.1). In Section 4.2 we formally define p-med-schemas and relate them with p-mappings in terms of expressive power and semantics of query answering. Then in Section 4.3 we describe an algorithm for creating a p-med-schema from a set of data sources. Finally, Section 4.4 gives an algorithm for consolidating a p-med-schema into a single schema that is visible to the user in a pay-as-you-go system.

4.1 P-Med-Schema Motivating Example

Let us begin with an example motivating p-med-schemas. Consider a setting in which we are trying to automatically infer a mediated schema from a set of data sources, where each of the sources is a single relational table. In this context, the mediated schema can be thought of as a "clustering" source attributes, with similar attributes being grouped into the same cluster. The quality of query answers critically depends on the quality of this clustering. Because of the heterogeneity of the data sources being integrated, one is typically unsure of the semantics of the source attributes and in turn of the clustering.

EXAMPLE 7.24 *Consider two source schemas both describing people:*

```
S1(name, hPhone, hAddr, oPhone, oAddr)
S2(name, phone, address)
```

In S2, the attribute phone *can either be a home phone number or be an office phone number. Similarly,* address *can either be a home address or be an office address.*

Suppose we cluster the attributes of S1 and S2. There are multiple ways to cluster the attributes and they correspond to different mediated schemas. Below we list a few (in the mediated schemas we abbreviate hPhone *as* hP*,* oPhone *as* oP*,* hAddr *as* hA*, and* oAddr *as* oA*):*

M1({name}, {phone, hP, oP}, {address, hA, oA})
M2({name}, {phone, hP}, {oP}, {address, oA}, {hA})
M3({name}, {phone, hP}, {oP}, {address, hA}, {oA})
M4({name}, {phone, oP}, {hP}, {address, oA}, {hA})
M5({name}, {phone}, {hP}, {oP}, {address}, {hA}, {oA})

None of the listed mediated schemas is perfect. Schema M_1 groups multiple attributes from S1. M_2 seems inconsistent because phone *is grouped with* hPhone *while* address *is grouped with* oAddress*. Schemas M_3, M_4 and M_5 are partially correct but none of them captures the fact that* phone *and* address *can be either home phone and home address, or office phone and office address.*

Even if we introduce probabilistic schema mappings, none of the listed mediated schemas will return ideal answers. For example, using M_1 prohibits returning correct answers for queries that contain both hPhone *and* oPhone *because they are taken to be the same attribute. As another example, consider a query that contains* phone *and* address*. Using M_3 or M_4 as the mediated schema will unnecessarily favor home address and phone over office address and phone or vice versa. A system with M_2 will incorrectly favor answers that return a person's home address together with office phone number. A system with M_5 will also return a person's home address together with office phone, and does not distinguish such answers from answers with correct correlations.*

A probabilistic mediated schema will avoid this problem. Consider a probabilistic mediated schema $\mathbf{M}$ *that includes M_3 and M_4, each with probability 0.5. For each of them and each source schema, we generate a probabilistic mapping (Section 3). For example, the set of probabilistic mappings* $\mathbf{pM}$ *for S_1 is shown in Figure 7.5(a) and (b).*

Now consider an instance of S_1 with a tuple

```
('Alice', '123-4567', '123, A Ave.',
          '765-4321', '456, B Ave.')
```

and a query

Possible Mapping	Probability
{(name, name), (hP, hPP), (oP, oP), (hA, hAA), (oA, oA)}	0.64
{(name, name), (hP, hPP), (oP, oP), (oA, hAA), (hA, oA)}	0.16
{(name, name), (oP, hPP), (hP, oP), (hA, hAA), (oA, oA)}	0.16
{(name, name), (oP, hPP), (hP, oP), (oA, hAA), (hA, oA)}	0.04

(a)

Possible Mapping	Probability
{(name, name), (oP, oPP), (hP, hP), (oA, oAA), (hA, hA)}	0.64
{(name, name), (oP, oPP), (hP, hP), (hA, oAA), (oA, hA)}	0.16
{(name, name), (hP, oPP), (oP, hP), (oA, oAA), (hA, hA)}	0.16
{(name, name), (hP, oPP), (oP, hP), (hA, oAA), (oA, hA)}	0.04

(b)

Answer	Probability
('Alice', '123-4567', '123, A Ave.')	0.34
('Alice', '765-4321', '456, B Ave.')	0.34
('Alice', '765-4321', '123, A Ave.')	0.16
('Alice', '123-4567', '456, B Ave.')	0.16

(c)

Figure 7.5. The motivating example: (a) p-mapping for S_1 and M_3, (b) p-mapping for S_1 and M_4, and (c) query answers w.r.t. $\mathbf{M}$ and $\mathbf{pM}$. Here we denote {phone, hP} by hPP, {phone, oP} by oPP, {address, hA} by hAA, and {address, oA} by oAA.

```
SELECT name, phone, address
FROM People
```

The answer generated by our system with respect to $\mathbf{M}$ *and* $\mathbf{pM}$ *is shown in Figure 7.5(c). (As we describe in detail in the following sections, we allow users to compose queries using any attribute in the source.) Compared with using one of* M_2 *to* M_5 *as a mediated schema, our method generates better query results in that (1) it treats answers with home address and home phone and answers with office address and office phone equally, and (2) it favors answers with the correct correlation between address and phone number.* □

4.2 Probabilistic Mediated Schema

Consider a set of source schemas $\{S_1, \ldots, S_n\}$. We denote the attributes in schema $S_i, i \in [1, n]$, by $attr(S_i)$, and the set of all source attributes as $\mathcal{A}$. That is, $\mathcal{A} = attr(S_1) \cup \cdots \cup attr(S_n)$. We denote a mediated schema for the set of sources $\{S_1, \ldots, S_n\}$ by $M = \{A_1, \ldots, A_m\}$, where each of the A_i's is called a *mediated attribute*. The mediated attributes are *sets* of attributes from the sources, i.e., $A_i \subseteq \mathcal{A}$; for each $i, j \in [1, m], i \neq j \Rightarrow A_i \cap A_j = \emptyset$.

Note that whereas in a traditional mediated schema an attribute has a name, we do not deal with naming of an attribute in our mediated schema and allow users to use any source attribute in their queries. (In practice, we can use the most frequent source attribute to represent a mediated attribute when exposing the mediated schema to users.) If a query contains an attribute $a \in A_i, i \in [1, m]$, then when answering the query we replace a everywhere with A_i.

A *probabilistic mediated schema* consists of a set of mediated schemas, each with a probability indicating the likelihood that the schema correctly describes the domain of the sources. We formally define probabilistic mediated schemas as follows.

DEFINITION 7.25 (PROBABILISTIC MEDIATED SCHEMA) *Let* $\{S_1, \ldots, S_n\}$ *be a set of schemas. A* probabilistic mediated schema (p-med-schema) for $\{S_1, \ldots, S_n\}$ *is a set*

$$\mathbf{M} = \{(M_1, Pr(M_1)), \ldots, (M_l, Pr(M_l))\}$$

where

- *for each* $i \in [1, l]$, M_i *is a mediated schema for* $S_1, \ldots, S_n$, *and for each* $i, j \in [1, l], i \neq j$, M_i *and* M_j *correspond to different clusterings of the source attributes;*
- $Pr(M_i) \in (0, 1]$, *and* $\Sigma_{i=1}^{l} Pr(M_i) = 1$. □

Semantics of queries: Next we define the semantics of query answering with respect to a p-med-schema and a set of p-mappings for each mediated schema in the p-med-schema. Answering queries with respect to p-mappings returns a set of answer tuples, each with a probability indicating the likelihood that the tuple occurs as an answer. We consider by-table semantics here. Given a query Q, we compute answers by first answering Q with respect to each possible mapping, and then for each answer tuple t summing up the probabilities of the mappings with respect to which t is generated.

We now extend this notion for query answering that takes p-med-schema into consideration. Intuitively, we compute query answers by first answering the query with respect to each possible mediated schema, and then for each answer tuple taking the sum of its probabilities weighted by the probabilities of the mediated schemas.

DEFINITION 7.26 (QUERY ANSWER) *Let S be a source schema and* $\mathbf{M} = \{(M_1, Pr(M_1)), \ldots, (M_l, Pr(M_l))\}$ *be a p-med-schema. Let* $\mathbf{pM} = \{pM(M_1), \ldots, pM(M_l)\}$ *be a set of p-mappings where $pM(M_i)$ is the p-mapping between S and M_i. Let D be an instance of S and Q be a query.*

Let t be a tuple. Let $Pr(t|M_i), i \in [1, l]$, be the probability of t in the answer of Q with respect to M_i and $pM(M_i)$. Let $p = \Sigma_{i=1}^{l} Pr(t|M_i) * Pr(M_i)$. If $p > 0$, then we say (t, p) is a by-table answer with respect to $\mathbf{M}$ *and* $\mathbf{pM}$.

We denote all by-table answers by $Q_{\mathbf{M},\mathbf{pM}}(D)$. □

We say that query answers A_1 and A_2 are *equal* (denoted $A_1 = A_2$) if A_1 and A_2 contain exactly the same set of tuples with the same probability assignments.

Expressive power: A natural question to ask at this point is whether probabilistic mediated schemas provide any added expressive power compared to deterministic ones. Theorem 7.27 shows that if we consider *one-to-many* schema mappings, where one source attribute can be mapped to multiple mediated attributes, then any combination of a p-med-schema and p-mappings can be equivalently represented using a deterministic mediated schema with p-mappings, but may not be represented using a p-med-schema with deterministic schema mappings. Note that we can easily extend the definition of query answers to one-to-many mappings as one mediated attribute can correspond to no more than one source attribute.

THEOREM 7.27 (SUBSUMPTION) *The following two claims hold.*

1. *Given a source schema S, a p-med-schema* $\mathbf{M}$, *and a set of p-mappings* $\mathbf{pM}$ *between S and possible mediated schemas in* $\mathbf{M}$, *there exists a deterministic mediated schema T and a p-mapping pM between S and T, such that $\forall D, Q : Q_{\mathbf{M},\mathbf{pM}}(D) = Q_{T,pM}(D)$.*
2. *There exists a source schema S, a mediated schema T, a p-mapping pM between S and T, and an instance D of S, such that for any p-med-schema* $\mathbf{M}$ *and any set* $\mathbf{m}$ *of deterministic mappings between S and possible mediated schemas in* $\mathbf{M}$, *there exists a query Q such that $Q_{\mathbf{M},\mathbf{m}}(D) \neq Q_{T,pM}(D)$.* □

In contrast, Theorem 7.28 shows that if we restrict our attention to one-to-one mappings, then a probabilistic mediated schema *does* add expressive power.

THEOREM 7.28 *There exists a source schema S, a p-med-schema* $\mathbf{M}$, *a set of one-to-one p-mappings* $\mathbf{pM}$ *between S and possible mediated schemas in*

M, *and an instance D of S, such that for any deterministic mediated schema T and any one-to-one p-mapping pM between S and T, there exists a query Q such that,* $Q_{\mathbf{M},\mathbf{pM}}(D) \neq Q_{T,pM}(D)$. □

Constructing one-to-many p-mappings in practice is much harder than constructing one-to-one p-mappings. And, when we are restricted to one-to-one p-mappings, p-med-schemas grant us more expressive power while keeping the process of mapping generation feasible.

4.3 P-med-schema Creation

We now show how to create a probabilistic mediated schema **M**. Given source tables $S_1, \ldots, S_n$, we first construct the multiple schemas $M_1, \ldots, M_p$ in **M**, and then assign each of them a probability.

We exploit two pieces of information available in the source tables: (1) pairwise similarity of source attributes; and (2) statistical co-occurrence properties of source attributes. The former will be used for creating multiple mediated schemas, and the latter for assigning probabilities on each of the mediated schemas.

The first piece of information tells us when two attributes are likely to be similar, and is generated by a collection of schema matching modules. This information is typically given by some pairwise attribute similarity measure, say s. The similarity $s(a_i, a_j)$ between two source attributes a_i and a_j depicts how closely the two attributes represent the same real-world concept.

The second piece of information tells us when two attributes are likely to be different. Consider for example, source table schemas

```
S1: (name,address,email-address)
S2: (name,home-address)
```

Pairwise string similarity would indicate that attribute address can be similar to both email-address and home-address. However, since the first source table contains address and email-address together, they cannot refer to the same concept. Hence, the first table suggests address is different from email-address, making it more likely that address refers to home-address.

Creating Multiple Mediated Schemas: The creation of the multiple mediated schemas constituting the p-med-schema can be divided conceptually into three steps. First, we remove infrequent attributes from the set of all source attributes; that is, attribute names that do not appear in a large fraction of source tables. This step ensures that our mediated schema contains only information that is relevant and central to the domain. In the second step we construct a weighted graph whose nodes are the attributes that survived the filter of the first step. An edge in the graph is labeled with the pairwise similarity between

Algorithm 1 Generate all possible mediated schemas.

0: **Input**: Source schemas $S_1, \ldots, S_n$.
   Output: A set of possible mediated schemas.
1: Compute $\mathcal{A} = \{a_1, \ldots, a_m\}$, the set of all source attributes;
2: **for each** $(j \in [1, m])$
   Compute frequency $f(a_j) = \frac{|\{i \in [1,n] | a_j \in S_i\}|}{n}$;
3: Set $\mathcal{A} = \{a_j | j \in [1, m], f(a_j) \geq \theta\}$; *//$\theta$ is a threshold*
4: Construct a weighted graph $G(V, E)$, where (1) $V = \mathcal{A}$, and (2) for each $a_j, a_k \in \mathcal{A}, s(a_j, a_k) \geq \tau - \epsilon$, there is an edge (a_j, a_k) with weight $s(a_j, a_k)$;
5: Mark all edges with weight less than $\tau + \epsilon$ as *uncertain*;
6: **for each** (uncertain edge $e = (a_1, a_2) \in E$)
   Remove e from E if (1) a_1 and a_2 are connected by a path with only certain edges, or (2) there exists $a_3 \in V$, such that a_2 and a_3 are connected by a path with only certain edges and there is an uncertain edge (a_1, a_3);
7: **for each** (subset of uncertain edges)
   Omit the edges in the subset and compute a mediated schema where each connected component in the graph corresponds to an attribute in the schema;
8: **return** distinct mediated schemas.

the two nodes it connects. Finally, several possible clusterings of nodes in the resulting weighted graph give the various mediated schemas.

Algorithm 1 describes the various steps in detail. The input is the set of source schemas creating $S_1, \ldots, S_n$ and a pairwise similarity function s, and the output is the multiple mediated schemas in **M**. Steps 1–3 of the algorithm find the attributes that occur frequently in the sources. Steps 4 and 5 construct the graph of these high-frequency attributes. We allow an error ϵ on the threshold τ for edge weights. We thus have two kinds of edges: *certain edges*, having weight at least $\tau + \epsilon$, and *uncertain edges*, having weight between $\tau - \epsilon$ and $\tau + \epsilon$.

Steps 6-8 describe the process of obtaining multiple mediated schemas. Specifically, a mediated schema in **M** is created for every subset of the uncertain edges. For every subset, we consider the graph resulting from omitting that subset from the graph. The mediated schema includes a mediated attribute for each connected component in the resulting graph. Since, in the worst case, the number of resulting graphs is exponential in the number of uncertain edges, the parameter ϵ needs to be chosen carefully. In addition, Step 6 removes uncertain edges that when omitted will not lead to different mediated schemas. Specifically, we remove edges that connect two nodes already connected by certain edges. Also, we consider only one among a set of uncertain edges that

Algorithm 2 Assign probabilities to possible mediated schemas.

0: **Input:** Possible mediated schemas $M_1, \ldots, M_l$ and source schemas $S_1, \ldots, S_n$.
Output: $Pr(M_1), \ldots, Pr(M_l)$.
1: **for each** $(i \in [1, l])$
Count the number of source schemas that are consistent with M_i, denoted as c_i;
2: **for each** $(i \in [1, l])$ Set $Pr(M_i) = \frac{c_i}{\sum_{i=1}^{l} c_i}$.

connect a particular node with a set of nodes that are connected by certain edges.

Probability Assignment: The next step is to compute probabilities for possible mediated schemas that we have generated. As a basis for the probability assignment, we first define when a mediated schema is *consistent with* a source schema. The probability of a mediated schema in **M** will be the proportion of the number of sources with which it is consistent.

DEFINITION 7.29 (CONSISTENCY) *Let M be a mediated schema for sources $S_1, \ldots, S_n$. We say M is* consistent with *a source schema $S_i, i \in [1, n]$, if there is no pair of attributes in S_i that appear in the same cluster in M.*

Intuitively, a mediated schema is consistent with a source only if it does not group distinct attributes in the source (and hence distinct real-world concepts) into a single cluster. Algorithm 2 shows how to use the notion of consistency to assign probabilities on the p-med-schema.

4.4 Consolidation

To complete the fully automatic setup of the data integration system, we consider the problem of consolidating a probabilistic mediated schema into a single mediated schema and creating p-mappings to the consolidated schema. We require that the answers to queries over the consolidated schema be equivalent to the ones over the probabilistic mediated schema.

The main reason to consolidate the probabilistic mediated schema into a single one is that the user expects to see a single schema. In addition, consolidating to a single schema has the advantage of more efficient query answering: queries now need to be rewritten and answered based on only one mediated schema. We note that in some contexts, it may be more appropriate to show the application builder a set of mediated schemas and let her select one of them (possibly improving on it later on).

Algorithm 3 Consolidate a p-med-schema.

0: **Input:** Mediated schemas $M_1, \ldots, M_l$.
   Output: A consolidated single mediated schema T.
1: Set $T = M_1$.
2: **for** $(i = 2, \ldots, l)$ modify T as follows:
3:     **for each** (attribute A' in M_i)
4:         **for each** (attribute A in T)
5:             Divide A into $A \cap A'$ and $A - A'$;
6: **return** T.

Consolidating a p-med-schema: Consider a p-med-schema $\mathbf{M} = \{(M_1, Pr(M_1)), \ldots, (M_l, Pr(M_l))\}$. We consolidate $\mathbf{M}$ into a single mediated schema T. Intuitively, our algorithm (see Algorithm 3) generates the "coarsest refinement" of the possible mediated schemas in $\mathbf{M}$ such that every cluster in any of the M_i's is equal to the union of a set of clusters in T. Hence, any two attributes a_i and a_j will be together in a cluster in T if and only if they are together in every mediated schema of $\mathbf{M}$. The algorithm initializes T to M_1 and then modifies each cluster of T based on clusters from M_2 to M_l.

EXAMPLE 7.30 *Consider a p-med-schema $M = \{M_1, M_2\}$, where M_1 contains three attributes $\{a_1, a_2, a_3\}$, $\{a_4\}$, and $\{a_5, a_6\}$, and M_2 contains two attributes $\{a_2, a_3, a_4\}$ and $\{a_1, a_5, a_6\}$. The target schema T would then contain four attributes: $\{a_1\}$, $\{a_2, a_3\}$, $\{a_4\}$, and $\{a_5, a_6\}$.* □

Note that in practice the consolidated mediated schema is the same as the mediated schema that corresponds to the weighted graph with only certain edges. Here we show the general algorithm for consolidation, which can be applied even if we do not know the specific pairwise similarities between attributes.

Consolidating p-mappings: Next, we consider consolidating p-mappings specified w.r.t. $M_1, \ldots, M_l$ to a p-mapping w.r.t. the consolidated mediated schema T. Consider a source S with p-mappings $pM_1, \ldots, pM_l$ for $M_1, \ldots, M_l$ respectively. We generate a single p-mapping pM between S and T in three steps. First, we modify each p-mapping $pM_i, i \in [1, l]$, between S and M_i to a p-mapping pM_i' between S and T. Second, we modify the probabilities in each pM_i'. Third, we consolidate all possible mappings in pM_i''s to obtain pM. The details are as follows.

1. **For each** $i \in [1, l]$, **modify p-mapping** pM_i**:** Do the following for every possible mapping m in pM_i:

    - For every correspondence $(a, A) \in m$ between source attribute a and mediated attribute A in M_i, proceed as follows. (1) Find the set of all mediated attributes B in T such that $B \subset A$. Call this set $\overline{B}$. (2) Replace (a, A) in m with the set of all (a, B)'s, where $B \in \overline{B}$.

Call the resulting p-mapping pM_i'.

2. **For each** $i \in [1, l]$, **modify probabilities in** pM_i'**:** Multiply the probability of every schema mapping in pM_i' by $Pr(M_i)$, which is the probability of M_i in the p-med-schema. (Note that after this step the sum of probabilities of all mappings in pM_i' is not 1.)
3. **Consolidate** pM_i'**'s:** Initialize pM to be an empty p-mapping (i.e., with no mappings). For each $i \in [1, l]$, *add* pM_i' to pM as follows:

   - For each schema mapping m in pM_i' with probability p: *if* m *is in* pM, *with probability* p', modify the probability of m in pM to $(p + p')$; *if* m *is not in* pM, then add m to pM with probability p.

   The resulting p-mapping, pM, is the final consolidated p-mapping. The probabilities of all mappings in pM add to 1.

Note that Step 2 can map one source attribute to multiple mediated attributes; thus, the mappings in the result pM are one-to-many mappings, and so typically different from the p-mapping generated directly on the consolidated schema. The following theorem shows that the consolidated mediated schema and the consolidated p-mapping are equivalent to the original p-med-schema and p-mappings.

THEOREM 7.31 (MERGE EQUIVALENCE) *For all queries* Q*, the answers obtained by posing* Q *over a p-med-schema* $\mathbf{M} = \{M_1, \ldots, M_l\}$ *with p-mappings* $pM_1, \ldots, pM_l$ *is equal to the answers obtained by posing* Q *over the consolidated mediated schema* T *with consolidated p-mapping* pM. □

4.5 Other approaches

He and Chang [21] considered the problem of generating a mediated schema for a set of web sources. Their approach was to create a mediated schema that is statistically maximally *consistent* with the source schemas. To do so, they assume that the source schemas are created by a *generative model* applied to some mediated schema, which can be thought of as a probabilistic mediated schema. The probabilistic mediated schema we described in this chapter has several advantages in capturing heterogeneity and uncertainty in the domain. We can express a wider class of attribute clusterings, and in particular clusterings that capture attribute correlations. Moreover, we are able to combine attribute matching and co-occurrence properties for the creation of the probabilistic mediated schema, allowing for instance two attributes from one source to have a nonzero probability of being grouped together in the mediated schema. Also, the approach for p-med-schema creation described in this chapter is independent of a specific schema-matching technique, whereas the approach in [21] is tuned for constructing generative models and hence must rely on statistical properties of source schemas.

Magnani et al. [30] proposed generating a set of alternative mediated schemas based on probabilistic relationships between *relations* (such as an Instructor relation intersects with a Teacher relation but is disjoint with a Student relation) obtained by sampling the overlapping of data instances. Here we focus on matching attributes within relations. In addition, our approach allows exploring various types of evidence to improve matching and we assign probabilities to the mediated schemas we generate.

Chiticariu et. al. [5] studied the generation of multiple mediated schemas for an existing set of data sources. They consider multi-table data sources, not considered in this chapter, but explore interactive techniques that aid humans in arriving at the mediated schemas.

There has been quite a bit of work on automatically creating mediated schemas that focused on the theoretical analysis of the semantics of merging schemas and the choices that need to be made in the process [2, 4, 23, 25, 31, 33]. The goal of these work was to make as many decisions automatically as possible, but where some ambiguity arises, refer to input from a designer.

5. Future Directions

The investigation of data integration with uncertainty is only beginning. This chapter described some of the fundamental concepts on which such systems will be built, but there is a lot more to do.

The main challenge is to build actual data integration systems that incorporate uncertainty and thereby uncover a new set of challenges, such as efficiency and understanding what are the common types of uncertainty that arise in data integration applications, so techniques can be tailored for these cases.

The work we described showed how to create p-mediated schemas and schema mappings automatically. This is only a way to bootstrap a pay-as-you-go integration system. The next challenge is to find methods to improve it over time (see [24] for a first work on doing so). We would also like to incorporate multi-table sources, rather than only single-table ones as we described so far.

Finally, when we have many data sources, the sources tend to be redundant and contain dependencies (and therefore not offer independent sources of evidence). An important line of work is to discover these dependencies and use them to provide more precise answers to queries. We are currently exploring how the formalism and techniques from this chapter can be extended to consider uncertain and interdependent data sources, and how query answering can be performed efficiently even in the presence of dependencies.

References

[1] S. Agrawal, S. Chaudhuri, and G. Das. DBXplorer: A system for keyword-based search over relational databases. In *Proc. of ICDE*, pages 5–16, 2002.

[2] C. Batini, M. Lenzerini, and S. B. Navathe. A comparative analysis of methodologies for database schema integration. In *ACM Computing Surveys*, pages 323–364, 1986.

[3] J. Berlin and A. Motro. Database schema matching using machine learning with feature selection. In *Proc. of the 14th Int. Conf. on Advanced Information Systems Eng. (CAiSE02)*, 2002.

[4] P. Buneman, S. Davidson, and A. Kosky. Theoretical aspects of schema merging. In *Proc. of EDBT*, 1992.

[5] L. Chiticariu, P. G. Kolaitis, and L. Popa. Interactive generation of integrated schemas. In *Proc. of ACM SIGMOD*, 2008.

[6] A. Das Sarma, X. Dong, and A. Halevy. Bootstrapping pay-as-you-go data integration systems. Technical report, 2008. Available at http://dbpubs.stanford.edu/pub/2008-8.

[7] R. Dhamankar, Y. Lee, A. Doan, A. Y. Halevy, and P. Domingos. iMAP: Discovering complex semantic matches between database schemas. In *Proc. of ACM SIGMOD*, 2004.

[8] H. Do and E. Rahm. COMA - a system for flexible combination of schema matching approaches. In *Proc. of VLDB*, 2002.

[9] A. Doan, J. Madhavan, P. Domingos, and A. Y. Halevy. Learning to map between ontologies on the Semantic Web. In *Proc. of the Int. WWW Conf.*, 2002.

[10] X. Dong, A. Y. Halevy, and C. Yu. Data integration with uncertainty. In *Proc. of VLDB*, 2007.

[11] Xin Dong and Alon Y. Halevy. A platform for personal information management and integration. In *Proc. of CIDR*, 2005.

[12] Alon Levy (Ed.). Data engineering special issue on adaptive query processing, june 2000. *IEEE Data Eng. Bull.*, 23(2), 2000.

[13] D. Florescu, D. Koller, and Alon Y. Levy. Using probabilistic information in data integration. In *Proc. of VLDB*, 1997.

[14] A. Gal. Why is schema matching tough and what can we do about it? *SIGMOD Record*, 35(4):2–5, 2007.

[15] A. Gal, G. Modica, H. Jamil, and A. Eyal. Automatic ontology matching using application semantics. *AI Magazine*, 26(1):21–31, 2005.

[16] Avigdor Gal, Ateret Anaby-Tavor, Alberto Trombetta, and Danilo Montesi. A framework for modeling and evaluating automatic semantic reconciliation. 2003.

[17] GoogleBase. http://base.google.com/, 2005.

[18] A. Y. Halevy, N. Ashish, D. Bitton, M. J. Carey, D. Draper, J. Pollock, A. Rosenthal, and V. Sikka. Enterprise information integration: successes, challenges and controversies. In *SIGMOD*, 2005.

[19] A. Y. Halevy, A. Rajaraman, and J. J. Ordille. Data integration: The teenage years. In *VLDB*, 2006.

[20] Alon Y. Halevy, Michael J. Franklin, and David Maier. Principles of dataspace systems. In *PODS*, 2006.

[21] B. He and K. C. Chang. Statistical schema matching across web query interfaces. In *Proc. of ACM SIGMOD*, 2003.

[22] V. Hristidis and Y. Papakonstantinou. DISCOVER: Keyword search in relational databases. In *Proc. of VLDB*, pages 670–681, 2002.

[23] R. Hull. Relative information capacity of simple relational database schemata. In *Proc. of ACM PODS*, 1984.

[24] S. Jeffery, M. Franklin, and A. Halevy. Pay-as-you-go user feedback for dataspace systems. In *Proc. of ACM SIGMOD*, 2008.

[25] L. A. Kalinichenko. Methods and tools for equivalent data model mapping construction. In *Proc. of EDBT*, 1990.

[26] J. Kang and J. Naughton. On schema matching with opaque column names and data values. In *Proc. of ACM SIGMOD*, 2003.

[27] J. Madhavan, S. Cohen, X. Dong, A. Halevy, S. Jeffery, D. Ko, and C. Yu. Web-scale data integration: You can afford to pay as you go. In *Proc. of CIDR*, 2007.

[28] J. Madhavan, D. Ko, L. Kot, V. Ganapathy, A. Rasmussen, and A. Halevy. Google's deep-web crawl. In *Proc. of VLDB*, 2008.

[29] M. Magnani and D. Montesi. Uncertainty in data integration: current approaches and open problems. In *VLDB workshop on Management of Uncertain Data*, pages 18–32, 2007.

[30] M. Magnani, N. Rizopoulos, P. Brien, and D. Montesi. Schema integration based on uncertain semantic mappings. *Lecture Notes in Computer Science*, pages 31–46, 2005.

[31] R. J. Miller, Y. Ioannidis, and R. Ramakrishnan. The use of information capacity in schema integration and translation. In *Proc. of VLDB*, 1993.

[32] H. Nottelmann and U. Straccia. Information retrieval and machine learning for probabilistic schema matching. *Information Processing and Management*, 43(3):552–576, 2007.

[33] R. Pottinger and P. Bernstein. Creating a mediated schema based on initial correspondences. In *IEEE Data Eng. Bulletin*, pages 26–31, Sept 2002.

[34] E. Rahm and P. A. Bernstein. A survey of approaches to automatic schema matching. *VLDB Journal*, 10(4):334–350, 2001.

[35] J. Wang, J. Wen, F. H. Lochovsky, and W. Ma. Instance-based schema matching for Web databases by domain-specific query probing. In *Proc. of VLDB*, 2004.

Chapter 8

SKETCHING AGGREGATES OVER PROBABILISTIC STREAMS

Erik Vee
Yahoo! Research
701 First Avenue, Sunnyvale, CA 94089
erikvee@yahoo-inc.com

Abstract The datastream model of computation has proven a valuable tool in developing algorithms for processing large amounts of data in small space. This survey examines an extension of this model that deals with uncertain data, called the *probabilistic stream model.* As in the standard setting, we are presented with a stream of items, with no random access to the data. However, each item is represented by a probability distribution function, allowing us to model the uncertainty associated with each element. We examine the computation of several aggregates in the probabilistic stream setting, including the frequency moments of the stream, average, minimum, and quantiles. The key difficulty in these computations is the fact that the stream represents an exponential number of possible worlds, and even simple numbers like the length of the stream can be different in different possible worlds. Obtaining accurate, reliable estimates can be very non-trivial.

Keywords: probabilistic streams, uncertain data, aggregates

1. Introduction

The amount of data that computers store and process is growing ever larger. Despite the dramatic increase in the speed of processors, there is still a need for algorithms that can deal with data extremely efficiently, both in time and space. To address these needs, the data stream model of computation was proposed. In this model, we are presented a sequence of items that arrive one at a time. The algorithm is restricted in its memory use, typically bounded to be polylogarithmic in the number of items, so any processing of the items must be done at the time of arrival; there is no random access. We generally lift the restriction of computing exact answers in this context, so the goal of such data stream is to produce a good approximation while being extremely time and space efficient. Since such algorithms take a single pass over their inputs, they clearly are very "database friendly," and so are quite useful in processing large data sets. And in some applications, such as sensor networks or packet routing, there is so much data arriving so quickly that data stream algorithms such as this are also a practical necessity.

The model of data streaming algorithms has proven to be enormously successful in dealing with large amounts of data. But increasingly, systems are being built that must deal with inherently uncertain information: data that is approximate, noisy, and incomplete. Thus, researchers have proposed a variant of the data streaming model that focuses on uncertain data, called *probabilistic streams*. In the probabilistic stream model, the items presented are actually represented by a distribution of possible values, together with a probability that the item is not actually in the stream. Although it is fairly clear why noisy, uncertain data would need to be represented as a distribution over possible values, it may be less obvious why there should be a chance that the item does not actually exist. However, this problem is ubiquitous. In sensor nets, we may have spurious shadow readings. In automated data collection, data can be mis-categorized. In data cleansing, fuzzy tuples may be assigned multiple values, each with an associated probability. As a simple example, consider the following.

EXAMPLE 8.1 *We have a database of books, together with their category and their price. Recently, the type "Fantasy/Sci-Fi" was split into two categories, and some of the older books have not been relabeled. To cleanse the data, the database system labels each of these books with both categories, each associated with some probability. So for instance, Book1 has cost $4.50, and is labeled "Fantasy" with associated probability 0.3 and "Sci-Fi" with associated probability 0.7; Book2 has been properly labeled "Sci-Fi" (with associated probability 1) at a cost of $6.00. Now, consider a query asking for the average price of all Sci-Fi books. In this simple example, the algorithm estimating the average price is presented two items. The first exists (with respect to the query) only with probability 0.7, and the second exists with probability 1.*

Very recently, researchers have been considering problems such as the one in the above example in a database context. Implicit in some of this work is a data stream-like model [2, 5, 18]. Here, we focus on the formally defined probabilistic stream model proposed in [15]. The precise details of this formal model are given in Section 2.

One of the most salient assumptions of this model is that the stream of items seen take on their probabilistic values as *independent events*. That is, the value of the i-th item seen is independent of the value of the j-th item. This has the advantage of a simple representation, understandable semantics, and much more manageable mathematics. In addition, many intuitive situations can be well-modeled in this framework.

Of course, we are limiting ourselves somewhat by constraining items to be independent. There are simple extensions to this model that work when the data items have limited independence. For example, if the probability distribution of each data item depends on the same k primitive random variables (together with an independent random variable that is private to each data item), then we can rewrite the stream as 2^k streams, each with an associated probability, and each satisfying our original independence assumption. We then use a data stream algorithm for each stream, combining the results. Of course, such solutions are not very satisfying, and have very limited applicability. Currently, however, no work in the probabilistic stream context has improvements beyond this basic idea.

1.1 Aggregates over probabilistic streams

Throughout this survey, we focus on computing aggregates— such as average, median, and minimum— over probabilistic streams. We note that there has also been interesting work on clustering uncertain streams [1], as well as on processing more complex event queries over streams of uncertain data [19]. However, the focus of this chapter will be simply on aggregate estimation, primarily the work found in [15, 8, 16].

Although the research in those papers deals with aggregates over probabilistic streams, the focus is somewhat different in each. The work of [15, 16] is motivated by the problem of answering queries over uncertain data in the OLAP model.* They propose the probabilistic stream model of computation that we use here, and they study a number of aggregates that are useful for a typical database user. The work of [8] is motivated by continuous data streams, giving algorithms that capture essential features of the stream, such as quantiles, heavy-hitters, and frequency moments. Of course, there is significant overlap in the two approaches. Together, these lines of research yield streaming algorithms for a wide range of aggregates.

One of the issues that these papers address is that it is not immediately obvious what an aggregate like average (AVG) means with respect to uncertain data. For instance, a probabilistic stream describes many possible worlds, and the value of AVG in each possible world is different. One vein of research uses the expected value of the aggregate as the principle value. This decision

*The OLAP model treats database items as points in a high-dimensional space, in which each dimension is a hierarchy. So for example, one dimension may refer to Location, which is divided by state, then by county, then by city. Users may ask queries about specific points (e.g. city = 'Sunnyvale') or about regions, like state, which map to multiple points.

follows in part due to the work of Burdick, et. al [6, 7], who argue that any value reported by a database system operating over uncertain data must satisfy some minimum requirements, which the expected value meets. Of course, the expectation is also established and well-understood mathematically, which makes it a more intuitive quantity. The work of [8] additionally studies the variance of the aggregates (in this case, the frequency moments of the stream). We formally define these notions, as well as each of the aggregates we study, in Section 2.

1.2 Organization

This survey is organized as follows. Section 2 defines the model of probabilistic streams, as well as the problems we study A general overview of the techniques used, some discussion of the results, and a summary of the running time and space requirements of the algorithms is found in Section 3. We then turn to some of the details of each algorithm. As a warm-up, Section 3 briefly describes the algorithms for SUM and COUNT. Section 4 discusses the universal sampling method of [8]. Section 5 deals with the frequency moments DISTINCT and REPEAT-RATE. Next, we consider quantiles and heavy-hitters in Section 6. The algorithm for MIN (and MAX) is in Section 7. Finally, we discuss AVG in Section 8. Section 9 wraps up the survey.

2. The Probabilistic Stream Model

An algorithm in the probabilistic stream model is presented with a sequence of items, each represented by a probability distribution function (pdf) which tells what value the item takes. The algorithm is restricted in its memory use, so as each item arrives, it must process it immediately. There is no random access of the items.

Formally, a probabilistic stream is a sequence of tuples $\langle\theta_1, \theta_2, ..., \theta_n\rangle$, where each tuple θ_i defines the distribution function of a random variable X_i, where the domain of X_i is $[m] \cup \{\perp\}$. (Here $\perp$ represents the event that the corresponding item does not exist in the datastream, and we use the notation $[k]$ to denote the set $\{1, 2, ..., k\}$.) For each $i \in [n]$, θ_i is written as at most ℓ pairs $(j, p_i(j))$ for $j \in [m]$; the value of $p_i(j)$ is defined to be 0 for any j not appearing in a pair for θ_i. The probability of X_i taking on a value j is then defined as $\Pr(X_i = j) = p_i(j)$. Further, since X_i takes on the value $\perp$ if it does not take on a value $j \in [m]$, we have that $\Pr(X_i = \perp) = 1 - \sum_j p_i(j)$. We assume that each X_i takes on its value independent of the other random variables. This corresponds to the popular block-model in probabilistic databases.

For convenience, we will often say that the i-th item *appears* if $X_i \neq \perp$ (and does not appear, otherwise), and define the *expected length* to be the number of i such that $X_i \neq \perp$, in expectation. Likewise, we say the i-th item takes on value t if $X_i = t$. Notice that the i-th item (or i-th tuple) always refers to the i-th tuple of the probabilistic stream, regardless of whether previous items have taken on the value $\perp$ or not.

Throughout, we assume that every $p_i(j)$ is representable using $O(\log m)$ space. We will sometimes report the space usage in terms of a number of registers, each with the ability to hold a single number (which we may assume is representable in $O(\log mn)$ space).

The *update time* is the time spent processing each pair of the pdf. So, for example, an update time of $O(1)$ means that the algorithm will spend total time of $O(\ell n)$ to process a stream of length n with block-size ℓ. Note that we actually spend $O(\ell)$ time per item in the stream. Since ℓ is generally a small constant, this distinction is not crucial.

One might imagine other encodings of the probability distribution functions (pdfs) for the random variables X_i. Indeed, any compactly representable pdf would be appropriate for data streaming applications, and much of the following work would apply to such variations. However, the above representation θ_i works for a wide range of aggregates. In fact, for some aggregates (e.g., SUM, COUNT, and AVG), the only value actually needed from the pdf is its expected value (together with the probability of $\perp$). Hence, using just $\ell = 1$ tuple is sufficient to compute these aggregate values.

Possible Worlds. It will occasionally be helpful to imagine different realizations of the random variables in the probabilistic stream as describing different possible worlds. Note, however, that when we consider the stream arising in a possible world, we ignore all items that evaluated to $\perp$, since, in essence, they do not exist in that world. Hence, it is possible to obtain the same stream under different realizations of the random variables. Consider the following simple example.

EXAMPLE 8.2 *Let* $\theta_1 = \{(1, \frac{1}{3})\}$ *and* $\theta_2 = \{(1, \frac{1}{7}), (2, \frac{2}{7})\}$. *Then there are* five *possible worlds for the stream* $\langle\theta_1, \theta_2\rangle$*:* $\langle 1, 1\rangle$ *occurring with probability* $\frac{1}{21}$, $\langle 1, 2\rangle$ *occurring with probability* $\frac{2}{21}$, $\langle 1\rangle$ *occurring with probability* $\frac{4}{21} + \frac{2}{21}$, $\langle 2\rangle$ *occurring with probability* $\frac{2}{21}$, *and the empty stream* $\langle\rangle$ *occurring with probability* $\frac{8}{21}$. *Note that the possible world* $\langle 1\rangle$ *can occur under two different realizations of the random variables in the probabilistic stream.*

Given probabilistic stream $\mathcal{S}$, we call each realizable stream of $\mathcal{S}$ a *grounded stream*, following the notation of [8]. We denote the set of grounded streams for $\mathcal{S}$ by $\mathrm{grnd}(\mathcal{S})$. Note that it is a simple matter to calculate the probability of a given realization of the random variables in $\mathcal{S}$: Let A denote a sequence $a_1, a_2, \ldots, a_n$, with $a_i \in [m] \cup \{\perp\}$ for each i, and let $\mathcal{S}(A)$ denote the grounded stream obtained by setting each $X_i = a_i$. Then we see

$$\Pr\left(X_i = a_i \text{ for all } i \in [n]\right) = \prod_{i \in [n]} p_i(a_i)$$

However, the probability that $\mathcal{S}$ actually realizes the grounded stream $\mathcal{S}(A)$ is somewhat different, in general, since there may be more than one setting of the random variables of $\mathcal{S}$ that produce the same grounded stream. Abusing notation somewhat, let $\mathcal{S}$ denote the random variable that describes the realization of the probabilistic stream. Then

$$\Pr\left(\mathcal{S} = \mathcal{S}(A)\right) = \sum_{B:\mathcal{S}(B)=\mathcal{S}(A)} \prod_{i \in [n]} p_i(b_i)$$

where B is a sequence of values in $[m] \cup \{\perp\}$ and b_i is the ith element in sequence B.

2.1 Problem definitions

We first define the general idea of the expectation and variance of an aggregation function. Let f be the aggregation function we wish to compute. Given a probabilistic stream $\mathcal{S}$, which we again treat as a random variable itself, we wish to report the *expected value* of f over the stream, denoted $\mathrm{E}\left(f(\mathcal{S})\right)$. Likewise, for certain applications we may be interested in reporting the variance of f, taken over the stream, denoted $\mathrm{Var}\left(f(\mathcal{S})\right)$. Following our earlier-defined notation, we have

$$\begin{aligned}
\mathrm{E}\left(f(\mathcal{S})\right) &= \sum_{B} f(\mathcal{S}(B)) \cdot \prod_{i \in [n]} p_i(b_i) \\
\mathrm{Var}\left(f(\mathcal{S})\right) &= \mathrm{E}\left(f^2(\mathcal{S})\right) - \mathrm{E}^2\left(f(\mathcal{S})\right) \\
&= \sum_{B} f^2(\mathcal{S}(B)) \cdot \prod_{i \in [n]} p_i(b_i) - \mathrm{E}^2\left(f(\mathcal{S})\right)
\end{aligned}$$

where again, B is a sequence of values in $[m] \cup \{\perp\}$ and b_i is the ith element in sequence B. Clearly, although the above expressions are correct, they are evaluated over an exponential number of sequences and are not practical to compute explicitly. The main goal of aggregate estimation is to approximate the corresponding expressions as efficiently as possible.

The main aggregates of interest to us are SUM, COUNT, MIN(and MAX), AVG, the frequency moments F_k for $k = 0, 1, 2$ (which includes DISTINCTand REPEAT-RATE), and quantiles including ϕ-HeavyHitters, ϕ-Quantiles, and MEDIAN. The value of SUM, COUNT, MAX, MIN, and

AVG are each the expected value of their respective deterministic counterpart. Specifically, SUM is the expected sum of the items in the data stream, COUNT is the expected number of items (recalling that each item has a probability of not appearing in the stream), MIN and MAX refer to the expected value of the smallest item and largest item in the stream, respectively, and AVG is the expected average over all items appearing in the stream. In symbols,

$$\mathsf{SUM} = \mathrm{E}\left(\sum_{i\in[n]:X_i\neq\perp} X_i\right) \qquad \mathsf{COUNT} = \mathrm{E}\left(\sum_{i\in[n]:X_i\neq\perp} 1\right)$$

$$\mathsf{MAX} = \mathrm{E}\left(\max_{i\in[n]:X_i\neq\perp}\{X_i\}\right) \qquad \mathsf{MIN} = \mathrm{E}\left(\min_{i\in[n]:X_i\neq\perp}\{X_i\}\right)$$

$$\mathsf{AVG} = \mathrm{E}\left(\frac{\sum_{i\in[n]:X_i\neq\perp} X_i}{\sum_{i\in[n]:X_i\neq\perp} 1}\right)$$

Note that AVG, MIN, and MAX are not well-defined in the case that the realized stream of $\mathcal{S}$ is the empty stream; in the case of AVG this causes a division by 0, and the minimum/maximum of an empty set is not well-defined. For convenience, we will simply assume that the probability of the empty stream is 0. In [16], the definition is modified to be the expectation *given* that the stream is non-empty. All of the work we summarize in this chapter goes through under either definition, with only minor modifications. Specifically, set $\rho = 1/\Pr(\mathcal{S}\text{ is non-empty})$. Then in each case, the estimate differs only by the factor ρ.

We give approximations to each of these quantities. For $\delta \geq 0, \varepsilon > 0$, we say a value $\widetilde{V}$ is an (ε,δ)-approximation to a value V if $V(1-\varepsilon) \leq \widetilde{V} \leq V(1+\varepsilon)$ with probability at least $1-\delta$, taken over its random coin tosses. In each of the algorithms for SUM, COUNT, MIN, MAX, and AVG, we give deterministic algorithms. That is, we discuss algorithms yielding $(\varepsilon, 0)$-approximations.

2.2 Frequency Moments and Quantiles

Recall the definition of the frequency moment F_k over a sequence $\mathcal{A}$ of data (with domain $[m]$): For each t in the domain $[m]$, let f_t denote frequency of t, i.e., the number of times t appears in the sequence. Then

$$F_k(\mathcal{A}) = \sum_{t\in[m]} f_t^k$$

where 0^0 is defined to be 0, for convenience. The frequency moments for $k = 0, 1, 2$ are of special interest, and often go by other names. The 0th frequency moment, F_0, is the number of distinct values in the sequence, and we denote its expected value by DISTINCT. The first frequency moment, F_1, is simply the number of items in the sequence. And F_2 is often referred to as the "repeat-rate;" we denote its expected value by REPEAT-RATE. Again letting $\mathcal{S}$ denote the random variable describing the realization of the probabilistic stream, we see f_t and $F_k(\mathcal{S})$ are themselves random variables as well. Thus, we have the following definitions.

$$\text{DISTINCT} = \mathrm{E}\left(F_0(\mathcal{S})\right)$$
$$\text{REPEAT-RATE} = \mathrm{E}\left(F_2(\mathcal{S})\right)$$

and COUNT $= \mathrm{E}\left(F_1(\mathcal{S})\right)$, which was also defined earlier. We will also be interested in the variance of the frequency moments, which we denote simply as $\mathrm{Var}\left(F_k(\mathcal{S})\right)$. As above, we will give algorithms yielding (ε, δ)-approximations to these quantities. However, for these estimates, δ will be non-zero, meaning that with some probability (according to the random coin tosses of the algorithm), the algorithm will fail to yield a good estimate. Note that there is not randomness due to the probabilistic stream, which is actually specified as a (non-random) set of tuples.

Finally, we will be interested in the quantiles and heavy-hitters of the stream, which are not so crisply defined in terms of expectation. Rather than finding the expected value of these aggregates, we instead find a value that is good, in expectation. Specifically, for any probabilistic stream $\mathcal{S}$ and value $t \in [m]$, let $\widetilde{f}_t = \mathrm{E}_{\mathcal{S}}\left(f_t\right)$. That is, $\widetilde{f}_t$ is the expected number of times that element t appears, taken over the possible ground streams of $\mathcal{S}$. Notice that $\sum_{t \in [m]} \widetilde{f}_t =$ COUNT. Then, given $\varepsilon > 0, \phi > 0$, we say an element s is an *ε-approximation to the ϕ-Quantiles problem* if

$$(\phi - \varepsilon)\text{COUNT} \leq \sum_{t=1}^{s} \widetilde{f}_t \leq (\phi + \varepsilon)\text{COUNT}$$

In the case that $\phi = \frac{1}{2}$, we call the ϕ-Quantiles problem the MEDIAN problem.

For the ε-approximate ϕ-HeavyHitters problem, we wish to return *all* elements s for which

$$\widetilde{f}_s \geq (\phi + \varepsilon)\text{COUNT}$$

and no items s for which $\widetilde{f}_s \leq (\phi - \varepsilon)$COUNT. The ϕ-HeavyHitters problem is the only problem we consider for which the answer is not a single number.

Abusing notation somewhat, we say that an algorithm for ϕ-Quantiles [resp., ϕ-HeavyHitters] is an (ε, δ)-approximation if, with probability at least $1 - \delta$, it is an ε-approximation to ϕ-Quantiles [resp., solves the ε-approximate ϕ-HeavyHitters problem], as defined in the previous paragraphs.

3. Overview of techniques and summary of results

Before going into the technical details of the stream algorithms for each of the aggregates, we give an overview of the general class of techniques used. Very broadly speaking, researchers studying aggregates over probabilistic streams have used three principle techniques: (1) reducing the probabilistic stream problem to a deterministic datastream problem, (2) adapting known datastream algorithms to the probabilistic stream world, and (3) approximating the mathematical expressions describing the aggregate over probabilistic streams by simpler expressions that are calculable over datastreams.

We might wonder whether there is a *universal* method of reducing a probabilistic stream problem to a deterministic one. Currently, no such method is known that works with high probability and small error. However, there is a general method based on sampling. Dubbed *universal sampling* [8], the idea is to instantiate a sample set of possible worlds described by the probabilistic stream. We then run, in parallel, standard datastreaming algorithms over each of these ground streams, which are themselves deterministic streams, and report the median value returned by each of these parallel instances. If we have an algorithm that estimates a given aggregate over deterministic datastreams, then this method does yield an algorithm that estimates that same aggregate over probabilistic streams. In fact, the expected value produced is as good as the original algorithm's answer. Unfortunately, the general guarantee on the variance of this procedure is not necessarily bounded. So, although universal sampling works for certain problems, it does not give a reliable answer in small space for every problem. The technical details of this are given in Section 4.

Universal sampling instantiates multiple ground streams based on the distribution of the probabilistic stream. For specific problems, we can produce deterministic streams based on the distribution of the probabilistic streams that capture exactly the needed information for the given problem, but in a much simpler way. As an example, the probabilistic stream for MEDIAN can be reduced to a deterministic stream by repeating each item a number of times equal to its expected number of appearances, multiplied by some value k [16]. The median of this new deterministic datastream is then MEDIAN for the probabilistic stream, affected only by the round-off error due to fractional numbers of appearances. (The larger k, the smaller this round-off error.) The details of this are given in Section 6. A similar technique is also used for DISTINCT [16]. Here, rather than creating a number of parallel instances of the stream, we intertwine all of the instances into a single stream in such a way that elements from separate instances are necessarily distinct. Thus, an algorithm estimating the number of distinct elements over deterministic datastreams will yield an estimate over probabilistic streams. Note here that rather than running, say ℓ, instances of the problem in parallel, we instead run a single algorithm on a single stream that is ℓ times as long. The reduction is described in Section 5.

We discuss another reduction, which additionally relies on the linearity of expectation. Estimating both quantiles and heavy-hitters can be done in the deterministic stream world using count-min sketches, a method first described by [10]. The count-min sketch allows us to maintain approximate information on how many times each element has been seen (even allowing fractional arrivals). So, by a simple reduction of the stream, we may instead track the number of arrivals, in expectation; in the above notation, this is simply approximating $\widetilde{f}_t$ for each t. This information is enough to approximate the values for ϕ-Quantiles and ϕ-HeavyHitters [8]. We go through this technique in more detail in Section 6.

In addition to direct reductions, it sometimes proves useful to adapt techniques from the deterministic stream world to probabilistic streams. In the case of REPEAT-RATE, also known as the second frequency moment, F_2, we may utilize the celebrated algorithm of Alon, Matias, and Szegedy [3]. Here, the reduction is not immediate, but the same technique may be modified to solve the probabilistic stream problem. Rather than maintaining the sum of a number of hashed values (as in [3]), we instead maintain the *expected* sum of a number of hashed values, together with a small correction term. This provides an unbiased estimator of the REPEAT-RATE, which we can further show has low variance, thus giving a good solution with high probability. We describe the solution, based on the work of [16, 8], in Section 5.

For several aggregates, standard reductions do not seem to work. To estimate the value of MIN (or MAX), we must first analyze the expression describing its value. We then devise datastream algorithms to compute the expression, based on the work of [15]. In order to produce an algorithm operating in small space, we further approximate these expressions using a standard binning technique. This work is described in Section 7.

Another aggregate requiring more in-depth analysis is AVG. Here, we begin by rewriting the expression for AVG as an *integral*, using generating function techniques described by [15]. We then approximate the integrand as a polynomial following [16]; this approximation is highly non-trivial, since the standard Taylor-series type expansion fails to be a good estimate. The necessary coefficients of the polynomial are maintainable in a datastream fashion, and it is simple to integrate any polynomial once its coefficients are known. Thus, we obtain an algorithm for estimating AVG over probabilistic streams.

Techniques for short and long streams. One common feature of many algorithms over probabilistic streams is that they work differently, depending on the expected length of the stream. For example, if we knew that the expected length of a probabilistic stream was very large, the the simple estimate

SUM/COUNT is a good approximation of AVG. On the other hand, for short streams, this is a very poor estimate. Note that "short" in this case means the expected number of items appearing is small. Thus, we may have a stream with many probabilistic items, each appearing with very small probability. So simply remembering all the items and constructing an exact solution does not work in general.

One intuition, which turns out to be true quite often, is that for very long (in expectation) streams, aggregates taken over the stream tend to look like the aggregates taken over the "expected value" of the stream. One incarnation of this is the universal sampling result of [8] shown in Section 4. It shows that for many aggregation functions, simply sampling a ground stream from the set of possible streams will produce an aggregate value that is very close to the expected aggregate value, with high probability.

On the other hand, we sometimes need specific techniques to handle streams with short expected length. In practice, most items will appear with some reasonably large probability (e.g. at least 1%). Hence, many streams with short expected length will also have a small number of items, and exact algorithms will often work quickly. So many of the techniques we present here that are tailored to the short-stream case will be of more theoretical interest. Still, in certain applications where many items appear with very low probability, such techniques will be necessary to guarantee good approximations.

Approximation and randomization. As is standard in datastreaming problems, many of the algorithms we describe here will produce approximate results and utilize randomization. In fact, it is provably impossible to calculate some of the aggregates we study exactly in small space. Exact computation of frequency moments is known to take $\Omega(n)$ space, even over deterministic streams, by a simple reduction to the well-known communication complexity problem of set-disjointness. Indeed, even approximating within $1+\varepsilon$ over deterministic streams takes $\Omega(n^{1-5/k})$ for any constant ε [3]; even for non-constant $\varepsilon = \Omega(n^{-1/2})$, approximating any frequency moments takes $\Omega(1/\varepsilon^2)$ space [20]. Likewise, approximating DISTINCT within $1+\varepsilon$, even over deterministic streams, must take at least $\Omega(\varepsilon^{-1})$ space [4]. Although it is trivial to compute the average value of a deterministic stream in $O(1)$ space, [15] show that calculating AVG exactly over a probabilistic stream takes $\Omega(n)$ space.

Thus, almost all studied datastream algorithms (in the deterministic stream world) use random bits and return approximate answers. What is surprising is that the algorithms for estimating MIN, MAX, and AVG (as well as SUM and COUNT) are deterministic. Perhaps this can best be explained by the fact that the corresponding algorithm in the deterministic datastream world for each of these aggregates is also deterministic. Indeed, each of MIN, MAX, AVG, SUM, and COUNT can be computed exactly using $O(1)$ space and no randomness when the stream has no uncertainty.

A second somewhat surprising aspect is that the error in each of the algorithms is lower than might first be guessed. Indeed, most Chernoff-based sampling techniques yield something on the order of $\Omega(1/\varepsilon^2)$ space needed for approximation factor of $(1+\varepsilon)$. The space needed for MIN and MAX scales with $1/\varepsilon$, while the space for AVG (as well as SUM and COUNT) is even less. Again, although these are very attractive features in the algorithm, they can be traced to the nature of the solutions. Rather than sampling or using randomness, they depend on deterministic approximations of the basic expressions describing the aggregates we compute. In the case of quantiles (including MEDIAN) and heavy-hitters, we again see space of $O(1/\varepsilon)$. This can be traced to the count-min sketch structure that is used.

This reduction in space from $O(\varepsilon^{-2})$ to $O(\varepsilon^{-1})$ is more than academic. For ε values around 1%, this is a savings of two orders of magnitude, bringing the datastream algorithms into the realm of the truly practical.

Summary of results. Table 8.1 summarizes the results we discuss. Not shown in the table are SUM and COUNT, which were first discussed by [6, 7], but are trivial to compute in $O(\log mn)$ space and with no chance of error. We also do not list the method of universal sampling, which uses $O(\frac{1}{\varepsilon^2}\frac{\mathrm{Var}_G(F(G))}{\mathrm{E}_G(F(G))^2}\log(1/\delta))$ grounded stream samples. The entries for DISTINCT ignore factors in $\log\varepsilon^{-1}$ and $\log\log mn$. In addition, the update time is actually the amortized time. The authors of [4] present another algorithm using space $O(\varepsilon^{-2}\log mn\log\delta^{-1})$ and the same non-amortized update time (up to $\log\varepsilon^{-1}$ and $\log\log mn$ factors). Also notice that there are two lines for AVG. We discuss one algorithm here. The other algorithm, which has faster update time but larger memory consumption, is a simple variant of this.

As we have remarked earlier, most of these algorithms work in time proportional to ε^{-1}, making them useable in practice, even for small ε. The algorithms for the frequency moments need larger ε in order to run in small space and time over real data.

Warm-up: SUM and COUNT. Before going through each of the algorithms in detail, we begin with a trivial example. As observed many times before [6, 7, 15, 16, 8], both SUM and COUNT can be computed easily, due to linearity of expectation. Specifically, SUM $= \sum_{i\in[n]}\mathrm{E}\left(X_i|X_i\neq\perp\right)\Pr\left(X_i\neq\perp\right)$ and COUNT $= \sum_{i\in[n]}\Pr\left(X_i\neq\perp\right)$, essentially treating the value of $\perp$ as 0. Clearly, both of these values are computable exactly in datastreaming fashion.

It may be helpful to think of these problems as reducing to the problem of finding the sum of elements in a deterministic datastream. In the case of SUM,

Problem	Space	Update Time	Authors
DISTINCT	$\widetilde{O}((\varepsilon^{-2} + \log mn) \log \delta^{-1})$	$\widetilde{O}(\varepsilon^{-3} \log mn \log \delta^{-1})$	[16]
REPEAT-RATE	$O(\varepsilon^{-2} \log mn \log \delta^{-1})$	$O(\varepsilon^{-2} \log \delta^{-1})$	[16, 8]
ϕ-Quantiles	$O(\varepsilon^{-1} \log mn \log \delta^{-1})$	$O(\log \delta^{-1})$	[8]
ϕ-HeavyHitters	$O(\varepsilon^{-1} \log mn \log \delta^{-1})$	$O(\log \delta^{-1})$	[8]
MIN(MAX)	$O(\varepsilon^{-1} \log mn)$	$O(\log \ell)$	[15]
AVG	$O(\log \varepsilon^{-1} \log mn)$	$O(\log \varepsilon^{-1})$	[16]
AVG	$O(\varepsilon^{-1} \log mn)$	$O(1)$	[16]

Table 8.1. Summary of results. Each algorithm gives an ε-approximation with probability of failure δ, except for MIN, MAX, and AVG, which all have no chance of failure. Note that the algorithms for DISTINCT and REPEAT-RATE assume that $\ell = 1$ for each block.

the i-th element has value $\mathrm{E}\left(X_i | X_i \neq \perp\right) \cdot \Pr\left(X_i \neq \perp\right)$, where for COUNT, the i-th element has value $\Pr\left(X_i \neq \perp\right)$.

4. Universal Sampling

One of the most fundamental questions we might ask about probabilistic streams is whether we can simply convert them to deterministic streams in some general fashion, then use a known standard datastream algorithm to solve the problem. To tackle this, we examine one of the simplest methods we might try: Since the probabilistic stream describes a distribution over possible worlds, just pick a possible world from this distribution, and report the value of the aggregate in this world. We will see that this *universal sampling* approach is actually quite reasonable, although the variance is too large for it to be useful in every situation.

Let F be the aggregate we are computing, and suppose that there is a datastream algorithm $\mathcal{A}$ that estimates F over deterministic streams. Sampling from the space of possible worlds is a simple matter. As each pdf of the probabilistic stream $\mathcal{S}$ arrives, simply choose a value for the corresponding item (or omit it, if that value is $\perp$) according to the pdf. In this way, we choose a grounded stream G according to the distribution given by $\mathcal{S}$.

Clearly, we obtain an unbiased estimate of $\mathrm{E}_{G \in \mathcal{S}}\left(F(G)\right)$ in this way, and our variance is $\mathrm{Var}_{G \in \mathcal{S}}\left(F(G)\right)$. Thus, by a standard Chernoff-bound argument, if choose grounded streams $G_1, G_2, ..., G_k$ (with $k = O(\frac{1}{\varepsilon^2} \frac{\mathrm{Var}_G(F(G))}{\mathrm{E}_G(F(G))^2}))$ in this way and take the average value of $F(G_i)$, then we estimate $\mathrm{E}_{G \in \mathcal{S}}\left(F(G)\right)$ within $(1 + \varepsilon)$ with constant probability.

Again, by a standard argument, we can reduce the probability of failure to δ by repeating the above experiment. Let $t = O(\log(1/\delta))$, and let G_{ij} be a randomly chosen grounded stream for $i \in [k]$ and $j \in [t]$. Let $\widetilde{F}_j = \frac{1}{k}\sum_{i\in[k]} F(G_{ij})$, the average value of F taken over the j-th experiment. We output the median of $\{\widetilde{F}_1, ..., \widetilde{F}_t\}$. Note that each value of $F(G_{ij})$ can be calculated in parallel, taking a total of $O(\frac{1}{\varepsilon^2}\frac{\mathrm{Var}_G(F(G))}{\mathrm{E}_G(F(G))^2}\log(1/\delta))$ grounded streams. Thus, we have the following theorem.

THEOREM 8.3 ([8]) *Let F be an aggregate function over deterministic datastreams, and let $\mathcal{S}$ be a probabilistic stream. Then for any $\varepsilon > 0, \delta > 0$, we can approximate* $\mathrm{E}\left(F(\mathcal{S})\right)$ *using* $k = O(\frac{1}{\varepsilon^2}\frac{\mathrm{Var}_G(F(G))}{\mathrm{E}_G(F(G))^2}\log(1/\delta))$ *grounded stream samples according to $\mathcal{S}$. If we have a streaming algorithm that (ε', δ')-approximates F over deterministic datastreams using space S and update time T, then this yields a $(O(\varepsilon + \varepsilon'), \delta + \delta')$-approximation algorithm for F over probabilistic streams, taking space $O(kS)$ and update time $O(kT)$.*

For a variety of aggregates, it will be the case that $\mathrm{Var}_{G\in\mathcal{S}}\left(F(G)\right) << \mathrm{E}_{G\in\mathcal{S}}\left(F(G)\right)$, and universal sampling will give a good estimate. However, this is not true in general. Further, we will see that for many of the aggregates we study, we can actually use space much less that $O(1/\varepsilon^2)$.

A similar technique to the one above works for estimating $\mathrm{Var}\left(F(\mathcal{S})\right)$.

5. Frequency moments: DISTINCT and REPEAT-RATE

We now examine algorithms for estimating frequency moments F_0 and F_2. The technique for F_0, or DISTINCT, reduces the probabilistic stream to a single deterministic stream (which is $O(\varepsilon^{-3}\log\delta^{-1})$ times as long as the original stream). For F_2, or REPEAT-RATE, the technique is a modification of the one first given in [3]. Note that in the following, we assume that each block has size $\ell = 1$.

5.1 DISTINCT

We first turn to the aggregate DISTINCT, or F_0. We describe the reduction technique from [16]. As with many probabilistic-stream algorithms, we must consider two cases. In the first case, the value of DISTINCT is at least $\varepsilon/2$. In the second, DISTINCT is less than $\varepsilon/2$.

In the first case, assume that DISTINCT $\geq \varepsilon/2$. Given the probabilistic stream $\mathcal{S}$, we will produce a deterministic stream, as follows:

- Set a constant c to be sufficiently large. (We will choose the value of c momentarily.)

- For each $k \in [c]$ and each tuple $(j, p_i(j))$ in the stream, put $jc + k$ in the constructed deterministic stream with probability $p_i(j)$.

Notice that we have expanded the domain of the elements in the constructed stream to be $[c(m+1)]$. Essentially, we have made c separate streams— which we can see by focusing our our attention to those elements of the constructed stream that are congruent to k modulo c for each $k \in [c]$. Each of these streams is one realization of the probabilistic stream, hence the expected value of F_0 for the probabilistic stream is also expected to be the F_0 value for each of these streams.

So, we simply estimate the value of F_0 for the constructed stream and divide by c; this is tantamount to taking c separate streams, then taking their average value of F_0. So we see, using standard Chernoff arguments, that for c large enough, we obtain a good estimate of DISTINCT for the probabilistic stream. In our case, $c = 54\varepsilon^{-3} \ln(4/\delta)$ is large enough.

In the second case, we use COUNT to approximate the value of DISTINCT. In particular, it is clear that COUNT $\geq$ DISTINCT. Furthermore, it is not hard to show that if COUNT $\leq \ln(1+\varepsilon)$, then $(1+\varepsilon)$COUNT $\leq$ DISTINCT, hence COUNT is a $(1+\varepsilon)$ approximation to DISTINCT. Specifically, by direct calculation we have

$$\text{DISTINCT} = \sum_{t \in [m]} \left(1 - \prod_{i \in [n]} (1 - p_i(t)) \right)$$
$$\geq 1 - \prod_{i \in [n]} p_i(\perp) \geq 1 - e^{-\text{COUNT}}$$
$$\geq (1+\varepsilon)\text{COUNT} \quad \text{for COUNT} \leq \ln(1+\varepsilon)$$

On the other hand, we see that if COUNT $> \ln(1+\varepsilon)$, then DISTINCT $\geq \varepsilon/2$. Thus, we have the following. Since there is an algorithm working over deterministic streams taking space $O(\varepsilon^{-2} \log mn \log \delta^{-1})$ and update time $O(\log \varepsilon^{-1} \log mn \log \delta^{-1})$ [4], we have the following.

THEOREM 8.4 ([16]) *For any $\varepsilon > 0, \delta > 0$, there is a probabilistic stream algorithm for block-size $\ell = 1$ that (ε, δ)-approximates DISTINCT, using space $O(\varepsilon^{-2} \log mn \log \delta^{-1})$ for the sketch and with update time of $O(\varepsilon^{-3} \log \varepsilon^{-1} \log mn \log^2 \delta^{-1})$.*

We note that the paper of [4] also provides an algorithm taking space of $O((\varepsilon^{-2} + \log mn) \log \delta^{-1})$ (ignoring $\log \log mn$ and $\log \varepsilon^{-1}$ factors) and approximately the same update time, in an amortized sense.

5.2 REPEAT-RATE

In this section, we describe an algorithm to estimate the second-frequency moment, or REPEAT-RATE, over probabilistic streams. The algorithm, essentially a generalization of the one estimating F_2 over deterministic streams given in [3], was discovering independently by both [8] and [16].

We first sketch the original algorithm of [3], working over deterministic streams. Let $\mathcal{H}$ denote a uniform family of 4-wise independent hash functions, with each $h \in \mathcal{H}$ mapping elements of $[m]$ to $\{-1, 1\}$. That is, for $x_1, x_2, x_3, x_4 \in [m]$, the values $h(x_1), h(x_2), h(x_3)$, and $h(x_4)$ are independent over the random choice of $h \in \mathcal{H}$. For a deterministic stream of values $\mathcal{B} = b_1, b_2, ..., b_n$, the estimator for F_2 (with respect to h) is defined to be $Z_h(\mathcal{B}) = (\sum_{i \in n} h(b_i))^2$. It can be shown that $\mathrm{E}_h\left(Z_h(\mathcal{B})\right) = F_2(\mathcal{B})$, and that the variance, $\mathrm{Var}_h\left(Z_h(\mathcal{B})\right) \leq 2F_2(\mathcal{B})^2$. We choose $O(\frac{1}{\varepsilon^2})$ values for h, and maintain each value Z_h in a datastreaming fashion, in parallel. By standard Chernoff bound arguments, this gives an estimate of $F_2(\mathcal{B})$ within $(1+\varepsilon)$ of the correct value, with constant probability. Again, a standard argument shows that we can reduce our chance of failure to at most δ by repeating this experiment $O(\log(1/\delta))$ times and taking the median value of each experiment, similarly to the procedure discussed in Section 4. Note that each of these experiments can be done in parallel. Thus, using $O(\varepsilon^{-2}\log(1/\delta))$ registers, we can estimate F_2 over a deterministic stream within $(1+\varepsilon)$, with probability at least $(1-\delta)$.

The approach of [8, 16] for probabilistic streams first notes that the *expectation* (taken over the possible ground streams that probabilistic stream $\mathcal{S}$ may take on) of the above estimator is itself an unbiased estimator of REPEAT-RATE over the probabilistic stream. That is, $\mathrm{E}_{G \in \mathcal{S}}\left(\mathrm{E}_h\left(Z_h(G)\right)\right)$ is an unbiased estimator for REPEAT-RATE. In fact, [16] goes further, arguing that many unbiased estimators for deterministic streams can be similarly augmented to become unbiased estimators for probabilistic streams. Unfortunately, it is not immediate that such estimators are easily calculable, nor that the variance is reasonably bounded. We now show that for REPEAT-RATE, that is the case.

Given a probabilistic stream $\mathcal{S}$ where the i-th item is either a_i with probability p_i or $\perp$ otherwise, we again let f_j be the random variable describing the number of occurrences of item j is the realization of $\mathcal{S}$. Likewise, we set $\widetilde{f}_j = \mathrm{E}_{\mathcal{S}}\left(f_j\right)$. Letting $h \in \mathcal{H}$ be as above, define the following variables:

$$U_h = \sum_{i \in [n]} h(a_i) p_i = \sum_{j \in [m]} h(j) \widetilde{f}_j$$

$$V_h = \sum_{i \in [n]} p_i(1-p_i)$$

Clearly, U_h and V_h are easy to maintain in a datastreaming fashion. We define

$$\widetilde{Z}_h = (U_h)^2 + V_h$$

and claim that $\widetilde{Z}_h = \mathrm{E}_{G \in \mathcal{S}}(Z_h)$.

To this end, we have

$$\begin{aligned}
\mathrm{E}_h\left(\widetilde{Z}_h\right) &= \mathrm{E}\left((\sum_{j\in[m]} h(j)\widetilde{f}_j)^2 \right) + \sum_{i\in[n]} p_i(1-p_i) \\
&= \sum_{j\in[m]} (\widetilde{f}_j)^2 + \mathrm{E}_h\left(\sum_{k\neq \ell} h(k)h(\ell)\widetilde{f}_k\widetilde{f}_\ell \right) + \sum_{i\in[n]} p_i(1-p_i) \\
&= \sum_{j\in[m]} (\widetilde{f}_j)^2 + \sum_{i\in[n]} p_i(1-p_i)
\end{aligned}$$

where the last line follows from the fact that $h(x)^2 = 1$, and $\mathrm{E}_h(h(x)) = 0$ for any $x \in [m]$ (as well as the 2-wise independence of h). Now, by the summation of variances, we know that $\mathrm{Var}_{\mathcal{S}}(f_j) = \sum_{i:a_i=j} p_i(1-p_i)$. But $\mathrm{Var}_{\mathcal{S}}(f_j) = \mathrm{E}_{\mathcal{S}}((f_j)^2) - (\widetilde{f}_j)^2$. Hence,

$$\begin{aligned}
\mathrm{E}_h\left(\widetilde{Z}_h\right) &= \sum_{j\in[m]} (\widetilde{f}_j)^2 + \sum_{i\in[n]} p_i(1-p_i) = \mathrm{E}_{\mathcal{S}}\left(\sum_{j\in[m]} f_j^2 \right) \\
&= \textsf{REPEAT-RATE}\,.
\end{aligned}$$

We next need to argue that the variance of $\widetilde{Z}_h$ is small. We now use the 4-wise independence of h, together with the fact that $h(x)^2 = 1$ and $\mathrm{E}_h(h(x)) = 0$ to find

$$\begin{aligned}
\mathrm{Var}_h\left(\widetilde{Z}_h\right) &= \mathrm{Var}_h(U_h^2) = \mathrm{E}_h(U_h^4) - \mathrm{E}_h(U_h^2)^2 \\
&= \sum_{j\in[m]} \widetilde{f}_j^4 + 3\sum_{k\neq\ell} \widetilde{f}_k^2\widetilde{f}_\ell^2 - (\sum_{j\in[m]} \widetilde{f}_j^2)^2 \\
&= 2\sum_{k\neq\ell} \widetilde{f}_k^2\widetilde{f}_\ell^2 \leq 2(\sum_j \widetilde{f}_j^2)^2 \leq 2\mathrm{E}_h\left(\widetilde{Z}_h\right)^2
\end{aligned}$$

Thus, using the precise argument as with deterministic streams, we can approximate REPEAT-RATE within $(1+\varepsilon)$ and probability at least $(1-\delta)$ by calculating $\widetilde{Z}_h$ for $O(\varepsilon^{-2}\log(1/\delta))$ different instances of h. We have the following.

THEOREM 8.5 ([8, 16]) *For any $\varepsilon > 0, \delta > 0$, there is a probabilistic stream algorithm for block-size $\ell = 1$ that (ε, δ)-approximates REPEAT-RATE, using space $O(\varepsilon^{-2} \log \delta^{-1} \log mn)$ and taking update time $O(\varepsilon^{-2} \log \delta^{-1})$.*

6. Heavy-Hitters, Quantiles, and MEDIAN

We now examine an algorithm of Cormode and Garofalakis [8], which reduces the problem of quantile estimation for probabilistic streams to one over deterministic streams.

In the deterministic setting, Cormode and Muthukrishan [10] provide what they call *count-min sketches* (CM-sketches). The CM-sketch algorithm processes a deterministic stream of tuples, where the i-th tuple has the form $(j_i, c_i), j_i \in [m]$. Conceptually, this tells us that the j_i-th item from $[m]$ has just appeared c_i more times (although we allow c_i to be any positive number, not just an integer). It is clear that using m registers, we can track the precise count for each of the m items. Thus, we can answer questions such as which items appear most frequently, or a ϕ-quantile, easily. Amazingly, the CM-sketch allows us to answer these same questions approximately, using space just $O(\frac{1}{\varepsilon} \log \delta^{-1} \log mn)$ and update time $O(\log \delta^{-1})$, where ε is an approximation error and δ is the probability that the algorithm fails. Specifically, let $C_j = \sum_{i:j_i=j} c_i$. Then for any ϕ and $\varepsilon > 0$, we can (with probability $(1-\delta)$) report all items j for which $C_j \geq (\phi + \varepsilon) \sum_k C_k$ while simultaneously reporting no item j for which $C_j \leq (\phi - \varepsilon) \sum_k C_k$ using space $O(\frac{1}{\varepsilon} \log \delta^{-1} \log mn)$ [10] (i.e. the ϕ-Heavy hitters problem for deterministic streams). Likewise, for any ϕ and $\varepsilon > 0$, we can (with probability $(1-\delta)$) report an item j for which $(\phi - \varepsilon) \sum_{k:C_k<C_j} C_k \leq C_j \leq (\phi + \varepsilon) \sum_{k:C_k<C_j} C_k$ [10] (i.e. the ϕ-quantile problem for deterministic streams). In both cases, the datastream algorithm works without knowing ϕ. It is a parameter that may be chosen at query time.

Now, consider the heavy hitter problem over a probabilistic stream $\mathcal{S}$ of length n. Recall that for $j \in [m]$, we define $\widetilde{f}_j = \mathrm{E}\left(|\{i \in [n] | X_i = j\}\right)$, and our goal in the ϕ-HeavyHitters problem is to return all $j \in [m]$ such that

$$\widetilde{f}_j \geq (\phi + \varepsilon)\text{COUNT}$$

and no j such that $\widetilde{f}_j \leq (\phi - \varepsilon)\text{COUNT}$. But notice that as each tuple of the probabilistic stream arrives, say $\langle (j_i^{(1)}, p_i^{(1)}), \ldots, (j_i^{(\ell)}, p_i^{(\ell)}) \rangle$, we can treat it as a set of ℓ pairs appearing in a deterministic stream, where $p_i^{(k)}$ plays the role of c_i. Thus we can use the CM-sketch algorithm on the corresponding deterministic stream, which produces the answer to the ϕ-HeavyHitters problem over probabilistic streams.

Notice that an analogous argument applies to the ϕ-Quantiles problem over probabilistic streams— each pdf can be treated as a set of pairs in a deterministic stream. Thus, we have the following result, first shown in [8].

THEOREM 8.6 ([8]) *Let $\varepsilon > 0, \delta > 0$. Given a probabilistic stream $\mathcal{S}$ of length n in which each element takes on a value from $[m] \cup \{\bot\}$, there are datastream algorithms that (ε, δ)-approximate ϕ-HeavyHitters and ϕ-Quantiles for any $\phi > \varepsilon$. Both use the same sketch structure, which takes space $O(\frac{1}{\varepsilon}\log(1/\delta)\log mn)$ and requires update time $O(\log \delta^{-1})$. The value of ϕ may be chosen after the sketch is created.*

MEDIAN. Although MEDIAN is a special case of ϕ-Quantiles with $\phi = \frac{1}{2}$, we briefly describe an algorithm proposed in [16], since it illustrates another reduction of a probabilistic stream to a deterministic stream.

In this case, we produce a deterministic stream in which every item appears essentially k times the expected number that it appears in the probabilistic stream. We choose k large enough so that the round-off is sufficiently small. Specifically, following the technique of [16], let $k = \lceil 2n\varepsilon^{-1} \rceil$, and do the following:

- For each tuple $(j, p_i(j))$ in the probabilistic stream, produce $\lfloor kp_i(j) \rfloor$ items with value j in the deterministic stream.

Note that if items could appear a fractional number of times, then the median of the resulting deterministic stream would be precisely the median of the probabilistic stream. However, since we need to round to the nearest integer value for each tuple in the probabilistic stream, we need to ensure that k is large enough so this error is small. Note that $kp_i(j) \geq \lfloor kp_i(j) \rfloor \geq kp_i(j) - 1$, so the accumulated error after n items in the probabilistic stream is still at most $n/k = \varepsilon/2$. This is enough to guarantee that the algorithm for median over deterministic streams, such as the one of [14], approximates the probabilistic stream version of MEDIAN.

7. A Binning Technique for MIN and MAX

We now examine the problem of estimating MIN and MAX over a probabilistic stream. Since the solution to the two problems is entirely analogous, we focus solely on MIN throughout this section. Recall that the definition of the problem asks for the expected value of the minimum element of the stream:

$$\mathsf{MIN} = \mathrm{E}\left(\min_{i \in [n]: X_i \neq \bot} \{X_i\}\right)$$

We will compute an estimate for this value by first analyzing an exact formulation, then using a binning technique to provide an approximation. To this end, we now give an exact algorithm using $O(m)$ space, where again, each item in the probabilistic stream takes on a value from $[m] \cup \{\perp\}$. Although the original paper of [15] handles the case where the support of each pdf is size up to ℓ, here we focus on the case where each item is either $\perp$ or a single value from $[m]$, simply for readability and ease of understanding. That is, the i-th tuple is represented by $\langle a_i, p_i \rangle$, meaning that $X_i = a_i$ or $\perp$, and the probability that $X_i = a_i$ is p_i. As with the other sections, we will assume that the probability that the probabilistic stream is empty is 0.

Let $X_{\min} = \min_{i \in [n]} X_i$, where we treat the value of $\perp$ as being $m + 1$. Note that $X_{\min}$ is itself a random variable, and that it takes on a value from $[m]$, since we assumed that the stream is non-empty. Our goal then is to find $\mathrm{E}(X_{\min})$. We will maintain several values incrementally. For each $a \in [m]$, and for $r \in [n]$, define the following:

$$X_{\min}^{(r)} = \min_{i \in [r]} X_i \text{ where } X_i = \perp \text{ is treated as } m + 1.$$

$$P_a^{(r)} = \Pr\left(X_{\min}^{(r)} = a\right)$$

$$Q_a^{(r)} = \Pr\left(X_{\min}^{(r)} > a\right)$$

Clearly, $\mathrm{E}\left(X_{\min}^{(r)}\right) = \sum_{a \in [m]} P_a^{(r)} \cdot a + (m+1) Q_m^{(r)}$. The value of $Q_a^{(r)}$ will be used as a helper variable in computing $P_a^{(r)}$. In particular, define $P_a^{(0)} = 0$ and $Q_a^{(0)} = 0$ for all $a \in [m]$. Then it is straightforward to compute the value of $P_a^{(r)}$, given the values of $P_a^{(r-1)}$ and $Q_a^{(r-1)}$ for $r \geq 1$. We have

$$P_a^{(r)} = P_a^{(r-1)} \cdot \Pr\left(X_r \geq a \text{ or } X_r = \perp\right) + Q_a^{(r-1)} \cdot Pr X_r = a$$

$$Q_a^{(r)} = Q_a^{(r-1)} \cdot \Pr\left(X_r > a \text{ or } X_r = \perp\right)$$

Note that these values are easy to maintain in a datastreaming fashion. Hence, using $O(m)$ space, and with update time $O(m)$ per item, we can calculate the value of $\mathrm{E}(X_{\min})$ exactly.

However, using the ideas found in [15], we can maintain different values that allow us to calculate $\mathrm{E}(X_{\min})$, and which take just $O(1)$ update time per item. For each $a \in [m]$, define the sequences $U_a^{(\cdot)}$ and $V_a^{(\cdot)}$ as follows. Let $U_a^{(0)} = 0$, $V_a^{(0)} = 1$, and for $r > 0$, let

$$U_a^{(r)} = U_a^{(r-1)} + \frac{\Pr\left(X_r = a\right)}{1 - \Pr\left(X_r < a\right)} V_a^{(r-1)}$$

$$V_a^{(r)} = \frac{1 - \Pr\left(X_r \leq a\right)}{1 - \Pr\left(X_r < a\right)} V_a^{(r-1)}$$

Notice that these values are simple to maintain in a datastreaming fashion. Furthermore, suppose that the r-th tuple is $\langle a_r, p_r \rangle$, so that $\Pr\left(X_r = a\right) = 0$

for all $a \neq a_r$. Hence, $U_a^{(r)} = U_a^{(r-1)}$ and $V_a^{(r)} = V_a^{(r-1)}$ for all $a \neq a_r$. That is, only $U_{a_r}^{(r)}$ and $V_{a_r}^{(r)}$ need to be updated on the r-th item. Thus, we can maintain each of the values in $O(1)$ update time.

The important point, however, is that we can use these maintained values to reconstruct the value of $\mathrm{E}(X_{\min})$. It can be shown via a telescoping product that

$$Q_a^{(r)} = \prod_{a' \leq a} V_{a'}^{(r)}$$

We can show by induction that

$$P_a^{(r)} = U_a^{(r)} \prod_{a' < a} V_{a'}^{(r)}$$

The algebraic details are omitted here. Hence, we have provided a datastreaming algorithm working in $O(m)$ memory and taking just $O(1)$ time to update per item. Note that in order to reconstruct the answer after the stream has passed, we need $O(m)$ time to calculate the value of $\mathrm{E}(X_{\min})$. In general, [15] show that when the pdf has support of size ℓ, there is an algorithm taking time $O(\ell \log \ell)$ per pdf, i.e. update time $O(\log \ell)$.

Of course, taking $O(m)$ memory is too space-intensive. We now present the binning technique described in [15], which provides a $(1+\varepsilon)$ approximation, using just $O(\frac{1}{\varepsilon} \log m)$ space. This has the additional advantage that the technique works even when the support of the pdf is not a finite, discrete set. The technique itself is the standard geometric binning idea: Let $\varepsilon > 0$, and let the i-th bin refer to the interval $[(1+\varepsilon)^i, (1+\varepsilon)^{i+1})$ for $i = 0, 1, ..., \lfloor \log m / \log(1+\varepsilon) \rfloor$. Clearly, these bins encompass the set $[m]$, and there are $1 + \lfloor \log m / \log(1+\varepsilon) \rfloor = O(\frac{1}{\varepsilon} \log m)$ of them.

So, we proceed as before, with the following small modification: We treat every item as though it took on a value $(1+\varepsilon)^i$ for some i. Specifically, if an item takes on value a with $a \in [(1+\varepsilon)^i, (1+\varepsilon)^{1+i})$, round a down to $(1+\varepsilon)^i$. In this way, we have decreased the domain to size $O(\frac{1}{\varepsilon} \log m)$, and only produced at error of $(1+\varepsilon)$. Thus, the memory requirements, and the time to reconstruct the estimate of $\mathrm{E}(X_{\min})$ are both $O(\frac{1}{\varepsilon} \log m)$. We have the following theorem.

THEOREM 8.7 ([15]) *Let $\varepsilon > 0$. There is a probabilistic stream algorithm working in memory $O(\frac{1}{\varepsilon} \log mn)$, and having update time $O(1)$, that produces a $(1+\varepsilon)$-estimate of* MIN. *If the tuples of the probabilistic stream have size ℓ, then there is an algorithm with the same memory requirements, taking update time $O(\log \ell)$. In both cases, reconstructing the answer from the sketch takes $O(\frac{1}{\varepsilon} \log m)$ time.*

8. Estimating AVG using generating functions

As the final aggregate we consider, we now explore a probabilistic stream algorithm to estimate the value of AVG. The analysis utilizes the method of generating functions to produce a mathematical expression that can be approximated in a datastream fashion. Using this expression we show first that estimating AVG by the simple expression SUM/COUNT is actually quite good when the probabilistic stream is long in expectation. We then turn to a probabilistic stream algorithm that works even when the stream is short. But first, we discuss the generating function technique.

8.1 Generating functions

The expressions denoting the actual value of various aggregates are general sums over an exponential number of terms. This is to be expected: the number of possible worlds for a given stream is generally exponential in its length. When this sum takes certain forms, we may use generating functions to simply it. In the technique we describe here, taken from the work of [16], the expression becomes the integral of a function that is easy to compute in a datastreaming fashion. Thus, we have traded one problem for another. However, as we see in the next subsection, estimating the value of the integral generated in the case of AVG can be done quite efficiently.

Let U_i, V_i be random variables for $i \in [n]$ such that U_i is independent of U_j and of V_j for all $i \neq j$. (It may be the case that U_i and V_i are not independent of each other.) Define the value RATIO by

$$\mathsf{RATIO} = \mathrm{E}\left(\frac{\sum_i U_i}{\sum_i V_i} \;\middle|\; \sum_i V_i \neq 0\right)$$

The difficulty in estimating RATIO stems from the fact that the denominator is itself a random variable. In order to address this difficulty, we use generating functions to remove the denominator. Let

$$G(x) = \mathrm{E}\left(\frac{\sum_i U_i}{\sum_i V_i} \cdot x^{\sum_i V_i} \;\middle|\; \sum_i V_i \neq 0\right)$$

Note that $G(1) = \mathsf{RATIO}$ and $G(0) = 0$. Using the fact that differentiation is a linear operator, we see that

$$\begin{aligned}
xG'(x) &= \mathrm{E}\left(\frac{\sum_i U_i}{\sum_i V_i} \cdot (\sum_j V_j) x^{\sum_j V_j} \,\middle|\, \sum_i V_i \neq 0\right) \\
&= \mathrm{E}\left((\sum_i U_i) x^{\sum_j V_j} \,\middle|\, \sum_i V_i \neq 0\right) \\
&= \mathrm{E}\left((\sum_i U_i) x^{\sum_j V_j}\right) \cdot \frac{1}{\Pr\left(\sum_i V_i \neq 0\right)} \\
&= \sum_i \mathrm{E}\left(U_i x^{V_i}\right) \prod_{j \neq i} \mathrm{E}\left(x^{V_j}\right) \cdot \frac{1}{\Pr\left(\sum_i V_i \neq 0\right)}
\end{aligned}$$

where the last line follows from the fact that U_i, V_i are independent of the V_j for $j \neq i$. Using the fact that $\mathsf{RATIO} = \int_0^1 G'(x)\mathrm{dx}$, we have the following theorem, proven in [15].

THEOREM 8.8 ([15]) *Let U_1, V_i, and RATIO be defined as above. Then*

$$\mathit{RATIO} = \frac{1}{\Pr\left(\sum_i V_i \neq 0\right)} \int_0^1 \frac{1}{x} \sum_i \mathrm{E}\left(U_i x^{V_i}\right) \prod_{j \neq i} \mathrm{E}\left(x^{V_j}\right) dx \,.$$

Although we do not provide a general method for evaluating the integral, we note that the integrand is calculable in a datastreaming fashion, so long as $\mathrm{E}\left(U_i x^{V_i}\right)$ and $\mathrm{E}\left(x^{V_i}\right)$ are efficiently calculable. In [15], the authors propose a datastreaming algorithm that estimates the integral using multiple passes. In the next subsection, we will see a method for estimating the integral in the special case of AVG. However, this relies on a more technical analysis of the integral we obtain.

8.2 Estimating AVG

We now apply the techniques of the previous section to derive an expression for AVG in terms of an integral. In this case, we are given a stream $X_1, ..., X_n$, and we wish to compute $\mathrm{E}\left(\sum_{i \in [n]: X_i \neq \perp} X_i / N \mid N > 0\right)$, where N is the number of $X_i \neq \perp$. (Recall that $N > 0$ with probability 1 for us.) Hence, we appeal to Theorem 8.8 by defining $U_i = X_i$ for $X_i \neq \perp$ and $U_i = 0$ otherwise, and defining $V_i = 1$ if $X_i \neq \perp$ and $V_i = 0$ otherwise. For convenience, let $a_i = \mathrm{E}\left(X_i \mid X_i \neq \perp\right)$. That is, let a_i be the expected value of the

i-th item, given that it is not bottom. Also, let $p_i = \Pr(X_i \neq \perp)$. Note that $\mathrm{E}\left(U_i x^{V_i}\right) = a_i p_i x$, while $\mathrm{E}\left(x^{V_i}\right) = p_i x + 1 - p_i$. Thus, we see

$$\begin{aligned}\mathsf{AVG} &= \int_0^1 \sum_i \frac{\mathrm{dx}}{x} \mathrm{E}\left(U_i x^{V_i}\right) \prod_{j \neq i} \mathrm{E}\left(x^{V_j}\right) \\ &= \int_0^1 \sum_{i: X_i \neq \perp} a_i p_i \frac{1}{p_i x + 1 - p_i} \prod_j (p_j x + 1 - p_j) \mathrm{dx}\end{aligned}$$

We have the following theorem, first shown in [15].

THEOREM 8.9 ([15]) *Let $\mathcal{S}$ be a probabilistic stream in which the i-th item is not $\perp$ with probability p_i, and its expected value, given that it is not $\perp$, is a_i. Further, assume that the probability of $\mathcal{S}$ being the empty stream is 0. Then*

$$\mathsf{AVG}(\mathcal{S}) = \int_0^1 \sum_{i \in [n]} \frac{a_i p_i}{1 - p_i + p_i x} \cdot \prod_{j \in [n]} (1 - p_j + p_j x) dx \,.$$

A key property of AVG that the above theorem shows is that the only information we need about the pdf describing each item is its expected value given that it is not $\perp$ (i.e. a_i) and the probability that it is not $\perp$ (i.e. p_i).

Using Theorem 8.9, Jayram, et. al [15] provide a multipass streaming algorithm to estimate AVG. Denoting the integrand by $f(x)$, their method finds a set of points $0 = x_0, x_1, ..., x_k = 1$ such that $f(x_i) \approx (1+\varepsilon) f(x_{i-1})$ for each $i \in [k]$. Since f is increasing, it is then a simple matter to estimate the integral:

$$\sum_{i \in [k]} (x_i - x_{i-1}) f(x_{i-1}) \leq \int_0^1 f(x) \mathrm{dx} \leq \sum_{i \in [k]} (x_i - x_{i-1}) f(x_i)$$

Since the left-hand side and the right-hand side are within (approximately) $(1+\varepsilon)$ of each other, this shows that they are both within $(1+\varepsilon)$ of the integral. Note, too, that once the values of x_i are known, it is a simple matter to evaluate $f(x_i)$ for each i in a datastreaming fashion. The difficulty, then, is to find the proper x_i values. But note that the x_i values satisfy (approximately) the equality $f(x_i) = f(0)(1+\varepsilon)^i$. In one pass, we can calculate $f(0)$, and hence, we can use binary search to find the x_i's that approximately satisfy the equality in a logarithmic (in the accuracy of the approximation) number of steps. Thus, [15] provides a $O(\log \varepsilon^{-1})$-pass algorithm to estimate the value of the integral from AVG. It is not hard to see that this method generalizes somewhat, working for any f that is increasing and calculable in one pass. The total memory requirement is then $O(k)$ registers, and the update time is also $O(k)$ per item, where k is the number of points x_i used in the calculation. Further, we see that k is the smallest integer such that $f(0)(1+\varepsilon)^k > f(1)$, hence $k = O(\frac{1}{\varepsilon} \log(f(1)/f(0))) = O(\frac{1}{\varepsilon} \log n)$.

Although the method of [15] is fairly general, it requires multiple passes over the data. In work building on this, [16] produce a single-pass algorithm working in smaller space and with smaller update time. Their method relies on a careful analysis of the integral for the specific case of AVG. We now examine their technique. The first step is rewriting the integral in a more useable form. Given this new form, we analyze, in the next subsection, the algorithm that approximates AVG by using SUM/COUNT, showing that if the stream has sufficiently many items in expectation, then this is a good estimate. Following that subsection, we do a more in-depth analysis to yield an algorithm for estimating AVG, regardless of the length of the stream.

But first, we rewrite the integral somewhat. Let $z = 1 - x$, and perform a change of variable with a little algebra to find the following.

$$\mathsf{AVG} = \int_0^1 \prod_{j\in[n]} (1 - p_j z) \sum_{i\in[n]} \frac{a_i p_i}{1 - p_i z} \mathrm{dz} \tag{8.1}$$

$$= \int_0^1 g(z)h(z)\mathrm{dz} \tag{8.2}$$

where we define

$$g(z) = \prod_{i\in[n]} (1 - p_i z)\ , \ h(z) = \sum_{i\in[n]} \frac{a_i p_i}{1 - p_i z}\ , \qquad \text{and } f(z) = g(z)h(z)\ .$$

Notice that both $g(z)$ and $h(z)$ are well-defined and easy to compute in a single pass for $z > 0$.

One might wonder whether simply estimating the integral by approximating $f(z)$ would be sufficient. Unfortunately, this approach fails. In fact, it is not hard to check that the coefficients of the Taylor series expansion about $z = 0$ grow exponentially. On the other hand, $f(z)$ is most interesting around $z = 0$. As we will see shortly, $f(z)$ drops to 0 exponentially as z moves away from 0.

Instead, we will approximate $h(z)$ with a Taylor series expansion, and approximate the *logarithm* of $g(z)$ with a Taylor series expansion. We will see in the coming subsections that the Taylor series, with a little work, will estimate these functions well. Even so, we will not be quite done. Recall that we need to evaluate the integral of $g(z)h(z)$, and we have approximated the logarithm of $g(z)$, hence written $g(z)$ as exp(low-degree polynomial). To overcome this obstacle, we then approximate our approximation of $g(z)$ as a polynomial, which allows us to integrate.

But first, we turn to the analysis of SUM/COUNT, which will serve as a warm-up to the more detailed analysis that appears later.

8.3 Approximating AVG by SUM/COUNT

Following the notation of [16], define $P = \mathsf{COUNT}$. We will show in this section that if P is sufficiently large, then SUM/COUNT is a good approximation of AVG. This estimate of AVG was first proposed by [6] in a non-datastream context, although they did not provide any bounds on its accuracy. Despite its guarantees for long streams, the simple estimate SUM/COUNT fails to be a good estimate in many simple cases. For example, consider a stream with two items: the first appearing with probability 1 and having value 1 if it appears, the second item appearing with probability 10^{-6} and having value 10^6 if it appears. The true value of AVG equals $1 \cdot (1 - 10^{-6}) + \frac{1}{2}(10^6 + 1) \cdot (10^{-6}) \approx \frac{3}{2}$, while the value of SUM/COUNT equals $(1 + 10^6 \cdot 10^{-6})/(1 + 10^{-6}) \approx 2$.

We now prove that for large P, we have a good approximation. The key idea throughout will be to estimate the product of terms using exponential expressions. For example, we have

$$\int_0^1 \prod_{j \neq i} (1 - p_j z) \mathrm{dz} \leq \int_0^1 \exp(\sum_{j \neq i} -p_j z) \mathrm{dz} \leq \int_0^1 \exp(-z(P-1)) \mathrm{dz}$$
$$= \left(\frac{\exp(-z(P-1))}{-(P-1)} \bigg|_0^1 \right. = \frac{1}{P-1}$$

Hence, using Equation 8.2, we see

$$\mathsf{AVG} = \int_0^1 \sum_{i \in [n]} a_i p_i \prod_{j \neq i} (1 - p_j z) \mathrm{dz}$$
$$\leq \mathsf{SUM} \frac{1}{P-1} = \frac{\mathsf{SUM}}{\mathsf{COUNT}} \left(1 + \frac{1}{P-1} \right)$$

For the other direction, we will utilize the inequality $1 - x \geq (1 - v)^{x/v}$ for any $x \in [0, v]$. We again estimate the product of terms using exponential expressions:

$$g(z) = \prod_{i \in [n]} (1 - p_i z) \geq \prod_{i \in [n]} (1 - v)^{p_i z / v} = (1 - v)^{Pz/v}$$

We additionally lower-bound $h(z) \geq \sum_{i \in [n]} a_i p_i = \mathsf{SUM}$. Hence, we see that

$$\mathsf{AVG} = \int_0^1 g(z) h(z) \mathrm{dz} \geq \int_0^v \mathsf{SUM} \cdot (1 - v)^{Pz/v} \mathrm{dz}$$
$$= \mathsf{SUM} \left(\frac{(1 - v)^{Pz/v}}{\frac{P}{v} \ln(1 - v)} \bigg|_0^v \right. = \frac{\mathsf{SUM}}{\mathsf{COUNT}} \cdot \frac{v}{\ln(\frac{1}{1-v})} (1 - (1 - v)^P)$$

This last expression is somewhat difficult to work with. Using a Taylor series expansion, we see that $\ln(\frac{1}{1-v}) \leq v + v^2$ for $v \in (0, 1/2]$. We also note that $(1-v)^P \leq e^{vP}$. Since we are interested in streams for which P is not small, we may assume that $P \geq e$. Set $v = \ln P/P$. Then continuing, we see

$$\begin{aligned} \mathsf{AVG} &\geq \frac{\mathsf{SUM}}{\mathsf{COUNT}} \cdot \frac{v}{v+v^2}(1-e^{vP}) \geq \frac{\mathsf{SUM}}{\mathsf{COUNT}} \cdot (1 - \frac{\ln P}{P})(1-P) \\ &\geq \frac{\mathsf{SUM}}{\mathsf{COUNT}} \cdot (1 - 2\frac{\ln P}{P}) \end{aligned}$$

So we see that AVG is within $(1 + O(\ln P/P))$ of SUM/COUNT (for both upper and lower bounds). With a little algebra, we arrive at the following theorem.

THEOREM 8.10 ([16]) *Let $\varepsilon < 1$, and let $\mathcal{S}$ be a probabilistic stream for which $P = \mathsf{COUNT}(\mathcal{S}) \geq \frac{4}{\varepsilon}\ln(2/\varepsilon)$, and such that the probability that $\mathcal{S}$ is the empty stream is 0. Then*

$$(1-\varepsilon)\frac{\mathsf{SUM}}{\mathsf{COUNT}} \leq \mathsf{AVG} \leq (1+\varepsilon)\frac{\mathsf{SUM}}{\mathsf{COUNT}}$$

Thus, if the stream is very long, or we are not concerned with the guaranteed accuracy of the algorithm, the simplistic SUM/COUNT method is quite good. However, we will see that using a somewhat more complicated method, we can get much better estimates while still using small space and update time.

As we mentioned earlier, the generating-function method used here allows us to obtain stronger bounds than one might first guess. In particular, using a Chernoff-based analysis, one would expect error bounds on the order of $O(1/\varepsilon^2)$. The detailed analysis here yields error bounds scaling with $1/\varepsilon$. For a reasonable value of ε, say around 1% error, this is two orders of magnitude improvement.

Estimating AVG for any stream. We now analyze a method of approximating AVG that works even for probabilistic streams whose expected number of items is very small. Although in this section, we will assume that the probability of the stream being empty is 0, it is worthwhile to point out that in the original work of [16], this was not necessarily the case. They allowed the stream to be empty with some non-zero probability. As mentioned early, this results in the minor modification of the estimates by multiplying by the value $\rho = 1/\Pr(\mathcal{S} \text{ is non-empty})$. However, for simplicity, we will restrict ourselves to the case where $\mathcal{S}$ is always non-empty.

The first step in the analysis is to restrict the interval we need to approximate somewhat. Let $z_0 = \min\{1, \frac{1}{P}\ln(2P/\varepsilon)\}$. (Note that since $\mathcal{S}$ is never empty, it must be the case that $P \geq 1$. In [16], the authors also consider the case

$P < 1$.) We will now argue that $\int_0^{z_0} f(z)\mathrm{d}z$ is a good approximation to AVG. To this end, note that f is decreasing, which can be checked by seeing that its derivative is negative on $(0, 1]$. Thus, we see for $z_0 = \frac{1}{P}\ln(2P/\varepsilon)$,

$$\int_{z_0}^{1} f(z)\mathrm{d}z \le f(z_0)(1 - z_0) = \prod_i (1 - p_i z_0) \sum_i \frac{a_i p_i}{1 - p_i z}(1 - z_0)$$

$$\le e^{-Pz_0}\mathsf{SUM} = \frac{\varepsilon}{2}\frac{\mathsf{SUM}}{\mathsf{COUNT}} \le \varepsilon\mathsf{AVG}$$

where the last inequality follows from the fact that $\mathsf{AVG} \ge \frac{1}{2}\mathsf{SUM}/\mathsf{COUNT}$ when $P \ge 4\ln 2$, as Theorem 8.10 shows.†

Thus, we see that $\int_0^{z_0} f(z)\mathrm{d}z$ is within $(1 - \varepsilon)$ of the actual value of AVG. So we only need to obtain approximations of $g(z)$ and $h(z)$ that are good on the interval $[0, z_0]$. As we mentioned earlier, we use the Taylor series expansions for $h(z)$ and $\ln g(z)$. We will show that for $z_0 < 1$, these Taylor series are very good approximations for their respective functions. However, when P is extremely small, (e.g. $P = O(\ln(1/\varepsilon))$), then $z_0 = 1$. In this case, the Taylor series expansions fail to be good approximations near $z = 1$. To remedy this, we use the following trick: We remember every (a_i, p_i) pair for which $p_i > \theta$, for some θ we choose. We then consider the Taylor series expansion only for those items with $p_i \le \theta$, ignoring those with large p_i. In this case, the coefficients (which are functions of the p_i) will decrease exponentially fast for higher-degree terms. Thus, this modified Taylor series is again a good approximation to the stream of low-probability items. To obtain an approximation for the entire stream, we simply combine the Taylor series with the items that we explicitly remembered. Note that the number of items we remember is relatively small, at most P/θ; once P grows large enough, we know that $z_0 < 1$, and we no longer have to remember the items. The authors of [16] consider other θ values, which allow the algorithm to trade off faster update times for somewhat higher cost in memory. For $\theta = \varepsilon^{-1/2}$, the algorithm has just $O(1)$ update time, while using $O(\varepsilon^{-1})$ registers.

Rather than going through the technical details of this idea, we will assume throughout the rest of this section that $z_0 = \frac{1}{P}\ln(2P/\varepsilon) < 1/e$ and set $\theta = 1/e$. We refer the interested reader to [16] for the full proof when z_0 is larger, and for other values of θ.

The algorithm we use maintains several simple values. Define

$$P_k = \sum_{i\in[n]} p_i^k \quad \text{and} \quad A_k = \sum_{i\in[n]} a_i p_i^k$$

†To be somewhat more precise, the proof from the previous section actually shows that $\mathsf{SUM}/\mathsf{COUNT} \le 2\mathsf{AVG}$ whenever $z_0 = \frac{1}{P}\ln(2P/\varepsilon) < 1$, which is true for us.

We will maintain P_k and A_k for $k = 1, 2, ..., O(\ln(1/\varepsilon))$. Note that maintaining each of these values is straightforward to do in a datastreaming fashion, in $O(1)$ time. Hence, maintaining all of these values leads to an algorithm with $O(\ln(1/\varepsilon))$ update time per item, with the number of registers equal to $O(\ln(1/\varepsilon))$ as well.

Our goal, again, is to estimate $\int_0^{z_0} g(z)h(z)\mathrm{d}z$, where $z_0 = \frac{1}{P}\ln(2P/\varepsilon) < 1/e$. We now write the Taylor series expansion for $\ln g(z)$ and $h(z)$:

$$\ln g(z) = \sum_{i\in[n]} \ln(1 - p_i z) = -\sum_{i\in[n]}\sum_{j\geq 1} \frac{(p_i z)^j}{j} = -\sum_{j\geq 0} P_j z^j/j$$

$$h(z) = \sum_{i\in[n]} \frac{a_i p_i}{1 - p_i z} = \sum_{i\in[n]} a_i p_i \sum_{j\geq 0}(p_i z)^j = \sum_{j\geq 0} A_{j+1} z^j$$

Thus, we define

$$\widetilde{g}_{k_0}(z) = \exp\left(-\sum_{j=1}^{k_0} P_j z^j/j\right) \qquad \text{with } k_0 = 2\ln(2/\varepsilon)$$

$$\widetilde{h}_{k_1}(z) = \sum_{j=0}^{k_1} A_{j+1} z^j \qquad \text{with } k_1 = \ln(2/\varepsilon)$$

Although we omit the technical details here, it is not hard to show that $\widetilde{g}_{k_0}(z)$ and $\widetilde{h}_{k_1}(z)$ are $(1+\varepsilon)$ approximations of their respective functions. The key observation in the argument is that z^j decreases exponentially in j, since $z \leq z_0 < 1/e$. (In the case that $z_0 > 1/e$, we instead examine P_j, showing that it decreases exponentially as well when we consider only items with value less than $\theta = 1/e$.)

Unfortunately, integrating $\widetilde{g}_{k_0}(z) \cdot \widetilde{h}_{k_1}(z)$ has no closed-form solution. So we now approximate $\widetilde{g}_{k_0}(z)$ by expanding each exponential term.

$$\widetilde{g}_{k_0}(z) = \exp\left(-\sum_{j=1}^{k_0} P_j z^j/j\right) = \prod_{j=0}^{k_0} \exp(-P_j z^j/j)$$
$$= \prod_{j=1}^{k_0}\sum_{\ell\geq 0} \frac{1}{\ell!}\left(\frac{-P_j z^j}{j}\right)^{\ell}$$

We again omit the details here, but the authors of [16] show that the above expression can be truncated to obtain a $(1+\varepsilon)$ approximation of $\widetilde{g}_{k_0}(z)$. In particular, they show the following.

LEMMA 8.11 ([16]) *Let $0 < \varepsilon < 1/2$, and define P, P_i, z_0, and $\widetilde{g}_{k_0}(z)$ as above. Further, let k_2 be the smallest integer greater than $5\ln(2P/\varepsilon)$. Then* *for $P \geq 1$, and $z \leq z_0$, we have*

$$\widetilde{g}_{k_0}(z) \leq \prod_{j=1}^{k_0} \sum_{\ell=0}^{k_2} \frac{1}{\ell!} \left(\frac{-P_j z^j}{j} \right)^{\ell} \leq (1+\varepsilon)\widetilde{g}_{k_0}(z)\,.$$

Putting this all together, we see that the algorithm simply needs to maintain the values A_k, P_k for $k = 1, ..., O(\log(1/\varepsilon))$, which can easily be maintained in a datastreaming fashion. To estimate that value of AVG, we use the expression

$$\int_0^{z_0} \mathrm{d}z \sum_{j=0}^{k_1} A_{j+1} z^j \prod_{j=1}^{k_0} \sum_{\ell=0}^{k_2} \frac{1}{\ell!} \left(\frac{-P_j z^j}{j} \right)^{\ell}$$

with k_0, k_1, k_2 as defined above. Notice that the integrand is actually a polynomial, so evaluating the integral is simple once the polynomial is reconstructed. As we stated earlier, the algorithm uses just $O(\log \varepsilon^{-1})$ registers for its sketch, the values P_k and A_k for $k = 1, ..., O(\log \varepsilon^{-1})$. Each register holds a number representable in $O(\log \varepsilon^{-1} \log mn)$ bits. Thus, we have the following.

THEOREM 8.12 *Let $\varepsilon > 0$. There is a probabilistic stream algorithm working in memory $O(\log^2 \varepsilon^{-1} \log(mn))$, and having update time $O(\log \varepsilon^{-1})$, that produces an $(\varepsilon, 0)$-approximation of* AVG. *Reconstructing the answer from the sketch takes $O(\log^3 \varepsilon^{-1} \log\log \varepsilon^{-1})$ time.*

As previously mentioned, the paper of [16] extends this theorem, additionally showing that there is a variant of this algorithm working in $O(1)$ update time, but taking space $O(\varepsilon^{-1} \log mn)$. The time to reconstruct the answer from the sketch is $O(\varepsilon^{-1} \log^2 \varepsilon^{-1})$. Remarkably, this datastreaming algorithm is deterministic, so there is no chance of failure.

9. Discussion

This survey examined several techniques for estimating aggregates over probabilistic streams. For some aggregates, it is possible to directly reduce the probabilistic stream to a deterministic one and to estimate the aggregate using known datastreaming algorithms. In the case of REPEAT-RATE, rather than using a direct reduction, we instead used the ideas of the algorithm for deterministic streams. But for aggregates such as MIN and AVG, it was necessary to analyze the mathematical expressions describing the expectation directly.

The algorithms themselves are surprisingly efficient. For all the aggregates other than the frequency-moments, both the space of update time of the corresponding algorithms are proportional to ε^{-1} (or better). This is a contrast to many datastreaming algorithms, which sometimes have $\Omega(\varepsilon^{-2})$ or more update time and space requirements. Although in some cases, these lower bounds are provable, it still means that the algorithms themselves will be orders of magnitude slower than ones working with update time $O(\varepsilon^{-1})$. And while algorithms working with update time $O(\varepsilon^{-1})$ are truly practical in general, algorithms that are significantly slower than this are not always as useful. The memory requirements are not always stringent in a database context, but they are sometimes crucial in streaming applications over large networks.

On the theoretical side of things, we would expect a $O(\varepsilon^{-2})$ bound for algorithms using sampling; indeed, these are typically of Chernoff-bound based analysis. It is remarkable that for all but the frequency moments (which themselves have provable lower bounds of $\Omega(\varepsilon^{-2})$ for REPEAT-RATE [20] and $\Omega(\varepsilon^{-1})$ for DISTINCT [4]), the algorithms presented for probabilistic streams beat these bounds.

A second surprise is that several of the algorithms we describe are deterministic. Indeed, this is somewhat of a rarity in the datastream literature. These deterministic algorithms, for MIN and AVG (and trivially, for SUM and COUNT) are all based on analysis of the expression describing the expectation. Further, they are all trivial to compute (deterministically) in the deterministic streaming world. Nevertheless, it is somewhat counterintuitive that we have deterministic algorithms for probabilistic streams.

There are several directions for future work. Of course, as we process more uncertain data, there will be a greater need to understand and produce algorithms for probabilistic streams; any function that is computable over deterministic streams is a potential research problem when mapped to probabilistic streams. Beyond this, however, there is a deeper question: what can we solve in the deterministic world that cannot be solved in the uncertain world? The universal sampling approach of [8] is a move towards answering that question, showing that there is a procedure for estimating any aggregate over probabilistic streams that can be estimated over deterministic streams— albeit with potentially large variance. On the negative side, it is known that answering even conjunctive queries over probabilistic databases is $\#P$-hard [11]. There is still a great gap in our understanding of efficient algorithms over uncertain data.

References

[1] C. C. Aggarwal and P. S. Yu. A framework for clustering uncertain data streams. In *ICDE*, pages 150–159, 2008.

[2] P. Agrawal, O. Benjelloun, A. D. Sarma, C. Hayworth, S. U. Nabar, T. Sugihara, and J. Widom. Trio: A system for data, uncertainty, and lineage. In *VLDB*, pages 1151–1154, 2006.

[3] N. Alon, Y. Matias, and M. Szegedy. The space complexity of approximating the frequency moments. In *STOC*, pages 20–29, 1996.

[4] Z. Bar-Yossef, T. S. Jayram, R. Kumar, D. Sivakumar, and L. Trevisan. Counting distinct elements in a data stream. In *RANDOM*, pages 1–10, 2002.

[5] O. Benjelloun, A. D. Sarma, A. Y. Halevy, and J. Widom. Uldbs: Databases with uncertainty and lineage. In *VLDB*, pages 953–964, 2006.

[6] D. Burdick, P. Deshpande, T. S. Jayram, R. Ramakrishnan, and S. Vaithyanathan. Olap over uncertain and imprecise data. In *VLDB*, pages 970–981, 2005.

[7] D. Burdick, P. M. Deshpande, T. S. Jayram, R. Ramakrishnan, and S. Vaithyanathan. Olap over uncertain and imprecise data. *VLDB J.*, 16(1):123–144, 2007.

[8] G. Cormode and M. N. Garofalakis. Sketching probabilistic data streams. In *SIGMOD Conference*, pages 281–292, 2007.

[9] G. Cormode and A. McGregor. Approximation algorithms for clustering uncertain data. In *PODS*, pages 191–200, 2008.

[10] G. Cormode and S. Muthukrishnan. An improved data stream summary: The count-min sketch and its applications. In *LATIN*, pages 29–38, 2004.

[11] N. N. Dalvi and D. Suciu. Efficient query evaluation on probabilistic databases. In *VLDB*, pages 864–875, 2004.

[12] N. N. Dalvi and D. Suciu. Efficient query evaluation on probabilistic databases. *VLDB J.*, 16(4):523–544, 2007.

[13] P. Flajolet and G. N. Martin. Probabilistic counting algorithms for data base applications. *J. Comput. Syst. Sci.*, 31(2):182–209, 1985.

[14] M. Greenwald and S. Khanna. Space-efficient online computation of quantile summaries. In *SIGMOD Conference*, pages 58–66, 2001.

[15] T. S. Jayram, S. Kale, and E. Vee. Efficient aggregation algorithms for probabilistic data. In *SODA*, pages 346–355, 2007.

[16] T. S. Jayram, A. McGregor, S. Muthukrishnan, and E. Vee. Estimating statistical aggregates on probabilistic data streams. In *PODS*, pages 243–252, 2007.

[17] B. Kanagal and A. Deshpande. Online filtering, smoothing and probabilistic modeling of streaming data. In *ICDE*, pages 1160–1169, 2008.

[18] C. Re, N. N. Dalvi, and D. Suciu. Efficient top-k query evaluation on probabilistic data. In *ICDE*, pages 886–895, 2007.

[19] C. Ré, J. Letchner, M. Balazinska, and D. Suciu. Event queries on correlated probabilistic streams. In *SIGMOD Conference*, pages 715–728, 2008.

[20] D. P. Woodruff. Optimal space lower bounds for all frequency moments. In *SODA*, pages 167–175, 2004.

[21] Q. Zhang, F. Li, and K. Yi. Finding frequent items in probabilistic data. In *SIGMOD Conference*, pages 819–832, 2008.

Chapter 9

PROBABILISTIC JOIN QUERIES IN UNCERTAIN DATABASES

A Survey of Join Methods for Uncertain Data

Hans-Peter Kriegel, Thomas Bernecker, Matthias Renz and Andreas Zuefle
Ludwig-Maximilians University Munich
Oettingenstr. 67, 80538 Munich, Germany
{ kriegel,bernecker,renz,zuefle } @dbs.ifi.lmu.de

Abstract The join query is a very important database primitive. It combines two datasets $\mathcal{R}$ and $\mathcal{S}$ ($\mathcal{R} = \mathcal{S}$ in case of a self-join) based on some query predicate into one set such that the new set contains pairs of objects of the two original sets. In various application areas, e.g. sensor databases, location-based services or face recognition systems, joins have to be computed based on vague and uncertain data. As a consequence, in the recent decade a lot of approaches that address the management and efficient query processing of uncertain data have been published. They mainly differ in the representation of the uncertain data, the distance measures or other types of object comparisons, the types of queries, the query predicates and the representation of the result. Only a few approaches directly address join queries on uncertain data. This chapter gives an overview of probabilistic join approaches. First, it surveys the categories that occur in general queries on uncertain data and secondly, it exemplarily sketches some join approaches on uncertain data from different categories.

Keywords: probabilistic query processing, uncertainty models, similarity join, spatial join

1. Introduction

In many modern application ranges, e.g. spatio-temporal query processing of moving objects [20], sensor databases [19] or personal identification systems [57], usually only uncertain data is available. For instance, in the area of mobile and location-based services, the objects continuously change their positions such that information about the exact location is almost impossible to obtain. An example of a location-based service and in particular of a spatial join is to notify moving people on their cell-phone if one of their friends enters their vicinity. In the area of multimedia databases, e.g. image or music databases, or in the area of personal identification systems based on face recognition and fingerprint analysis, there often occurs the problem that a feature value cannot exactly be determined. This uncertain data can be handled by assigning confidence intervals to the feature values, or by specifying probability density functions indicating the likelihoods of certain feature values.

A join query combines two datasets $\mathcal{R}$ and $\mathcal{S}$ ($\mathcal{R} = \mathcal{S}$ in case of a self-join) based on some query predicate into one set such that the new set contains pairs of objects of the two original sets. Formally,

DEFINITION 9.1 (JOIN QUERY) *Given two relations $\mathcal{R}$ and $\mathcal{S}$ and a predicate $\theta : \mathcal{R} \times \mathcal{S} \rightarrow \{true, false\}$. A join query on $\mathcal{R}$ and $\mathcal{S}$ is defined as follows:*

$$\mathcal{R} \bowtie_\theta \mathcal{S} = \{(r, s) \in \mathcal{R} \times \mathcal{S} | \theta(r, s) = true\}.$$

In order to join uncertain objects by traditional join methods, a non-uncertain result of the join predicate is required. However, if a query predicate is applied to uncertain attributes of uncertain objects, usually no unique answer whether the query predicate is fulfilled can be given. In this case, the vague information has to be aggregated in order to make the join predicate evaluable. Obviously, aggregation goes hand in hand with information loss. For instance, we have no information about how uncertain the similarity between two uncertain objects is. Even if we had one, it would be of no use because traditional join algorithms cannot handle this additional information.

This chapter gives an overview of probabilistic join approaches. They mainly differ in the representation of the uncertain data, the distance measure or other types of object comparisons, the types of queries, the query predicates and the representation of the result. First, the following section (Section 2) gives a rough overview of traditional join methods originally defined for non-uncertain data, but which form an important foundation for several join approaches defined for uncertain data. Section 3 surveys different uncertainty models and shows how existing join queries (and queries in general) on uncertain data can be categorized. Section 4 exemplarily sketches existing probabilistic join approaches on uncertain data which are representatives of different probabilistic join categories.

2. Traditional Join Approaches

A variety of different algorithms have been proposed for joining relations in the traditional case where the data is not uncertain. The following section gives a brief overview over existing approaches and serves as a base for the generalization to the case of uncertain data.

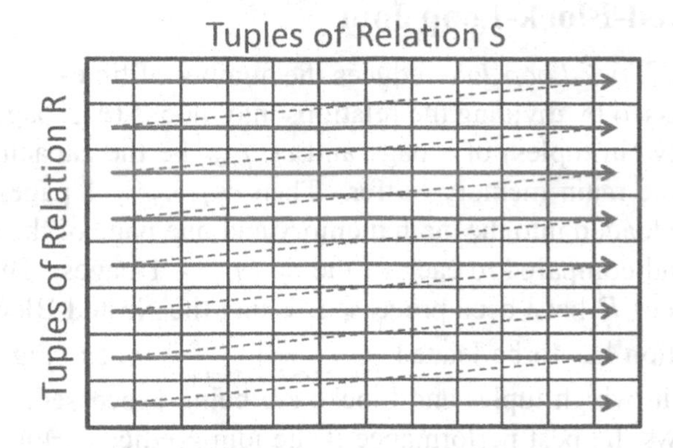

Figure 9.1. Order of accessed tuple pairs using the simple Nested-Loop Join

2.1 Simple Nested-Loop Join

The brute-force approach to perform a join on two relations $\mathcal{R}$ and $\mathcal{S}$ is called the *Simple Nested-Loop Join* and works as follows:
The first tuple r of Relation $\mathcal{R}$ ($\mathcal{R}$ is called the outer relation) is compared to each tuple s of $\mathcal{S}$ ($\mathcal{S}$ is the inner relation). Whenever a pair (r, s) satisfies the join predicate, both tuples are joined and added to the result relation. This is repeated for each tuple in the outer relation. Figure 9.1 shows, in matrix notation, the order in which the tuples of both relations are joined. Note that each tuple in the inner relation has to be loaded $|\mathcal{R}|$ times, which is unacceptable from a computational point of view, especially when both relations are too large to fit into the main memory.

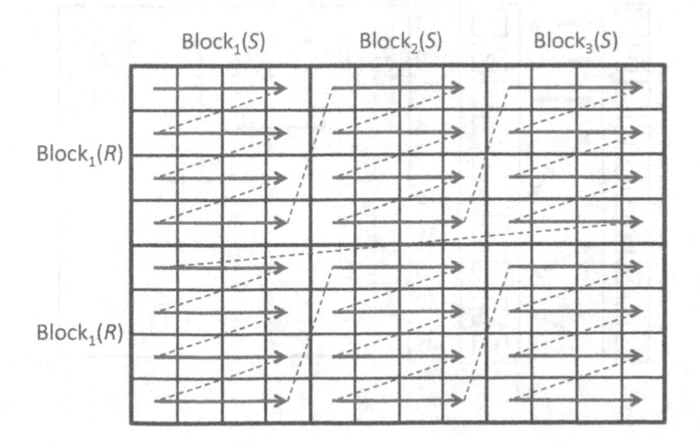

Figure 9.2. Order of accessed blocks and tuples using the Nested-Block-Loop Join. Here, Relation $\mathcal{R}$ is split into two blocks and Relation $\mathcal{S}$ is split into three blocks.

2.2 Nested-Block-Loop Join

The *Nested-Block-Loop Join* reduces the number of times the inner relation has to be accessed by dividing the relations into blocks (e.g. pages). Let c_{block} be the capacity (in tuples) of a page and c_{buffer} be the capacity (in number of pages) of the main memory buffer. Then $c_{buffer} - 1$ pages of the outer relation $\mathcal{R}$ are loaded into the main memory and one page of the inner relation $\mathcal{S}$ is loaded and compared to each of the $c_{buffer} - 1$ pages. This is repeated until all pages of $\mathcal{R}$ have been processed. Using the Nested-Block-Loop Join, the inner relation has to be loaded $\frac{|\mathcal{R}|}{(c_{buffer}-1)*c_{block}}$ times. Figure 9.2 shows the sequence in which tuples and blocks are being processed. Note that this approach shows the best performance if the join predicate Θ is not known in advance.

2.3 Sort-Merge-Loop Join

The next algorithm focuses on the evaluation of the *Equi-Join*, where Θ is equal to "=". First of all, both relations $\mathcal{R}$ and $\mathcal{S}$ are being sorted w.r.t. to the join attributes. In order to find the pairs of tuples $(r, s) \in \mathcal{R} \times \mathcal{S}$ that have equal join attributes, we can efficiently browse the tuples as shown in Figure 9.3. Here we exploit the property that, if the join attribute of r_i is larger than the join attribute of s_j, we do not have to check any pairs (r_k, s_j) where $k > i$. The reason is that the join attribute of r_k is equal or greater than the

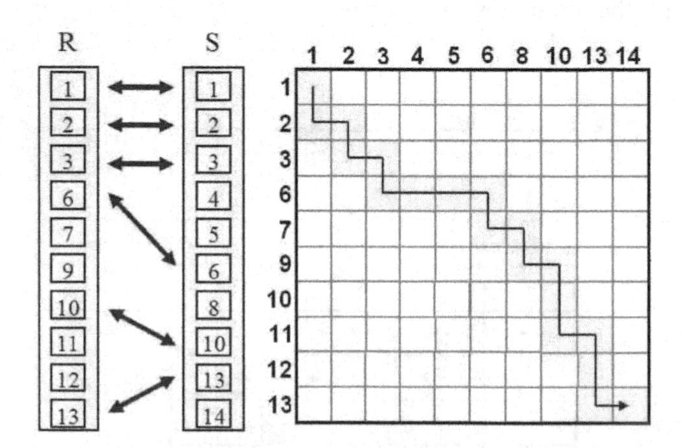

Figure 9.3. The figure to the right shows the order in which tuple pairs are joined using the Sort-Merge-Join for the two **sorted** relations $\mathcal{R}$ and $\mathcal{S}$ shown to the left.

join attribute of r_i and thus larger than the join attribute of s_j and, thus, cannot match the join predicate.

2.4 Other Join Methods

There are a lot of further standard join approaches which are not addressed here in detail. One class which should be mentioned here are join techniques which avoid comparing each pair of tuples by using a hash function used on the join attribute as a filter. These techniques exploit the fact that two join attributes can only be equal, if their hash key is equal. Finally, refinements of the filter results are required. Examples are the *Hashed-Loop Join* and the *Hash-Partitioned Join (GRACE)*.

2.5 Spatial Join Algorithms

In spatial databases, an important query is the *Spatial Join* [45]. In its basic form, a spatial join is a query which, given two sets $\mathcal{R}$ and $\mathcal{S}$ of spatially extended objects, finds all pairs consisting of one object of $\mathcal{R}$ and one object of $\mathcal{S}$ for which a spatial predicate is fulfilled. Examples are the distance predicate and the intersection predicate. Such queries have been studied extensively, and many efficient techniques exist for answering them [1, 6, 15, 16, 30, 47, 48]. Generally speaking, the spatial join operation is used to combine spatial objects of two sets according to some spatial property. For example, consider the spatial relations "Forests" and "Cities" where an attribute in each relation represents the borders of forests and cities respectively. The query "find all forests which are in a city" is an example of a spatial join.

Many algorithms designed for classical joins, such as hash-based join algorithms, cannot efficiently be applied to spatial joins. In particular hashing does not preserve the order of the data and therefore, objects that are close in the data space are not in the same bucket with high probability.

In spatial applications, the assumption is almost always true that a spatial index exists on a spatial relation. These indices can be exploited to support the efficient processing of spatial joins. On the basis of the availability of indices, spatial join methods operating on two relations can be classified into three classes:

- Class 1: Index on both relations
- Class 2: Index on one relation
- Class 3: No indices

The following sections give a broad overview of the three classes of join methods. Note that the following approaches are not limited to spatial data. If we consider objects with no extension, spatial joins can be adjusted to work for similarity joins that are applied to point objects which have been studied extensively in [10, 31, 11, 12]. This can be achieved by representing each object as a point in a feature space (vector space) where distance functions are used to measure the (dis-)similarity between objects.

2.6 Spatial Join using a spatial index structure for both relations

In [15] and [29], the R-tree [27] and particularly the R*-tree [7] are used as the underlying spatial access methods for processing spatial joins. Exploiting the R*-tree allows us to significantly reduce the number of pairs of spatial objects that need to be tested to check if they match the join condition. Here, it is assumed that both relations $\mathcal{R}$ and $\mathcal{S}$ to be joined consist of spatial objects or point objects of a d-dimensional vector space $\mathcal{R}^d$. Furthermore, it is assumed that each relation is separately indexed by an R*-tree or another member of the R-tree family. The central idea of performing the join with R*-trees is to use the property that directory rectangles form the minimum bounding hyper-rectangle of the data rectangles in the corresponding subtrees. Thus, if the rectangles of two directory entries, say $mbr_{\mathcal{R}}$ and $mbr_{\mathcal{S}}$, do not match the join predicate, there will be no pair (mbr_r, mbr_s) of entries in the corresponding subtrees matching the join predicate. Otherwise, there might be a pair of entries matching the join predicate in the corresponding subtrees and, thus, both subtrees corresponding to $mbr_{\mathcal{R}}$ and $mbr_{\mathcal{S}}$ have to be further evaluated. In its basic form, an efficient spatial join algorithm is presented in [15]. It starts from the roots of the trees and traverses simultaneously both of the trees in a depth-first order. For each qualifying pair of directory rectangles that match the join predicate, the algorithm follows the corresponding references to the nodes stored on the next lower level of the trees. Results are found when the leaf level is reached.

In the context of uncertain objects, this approach is adapted in [22]. Here, the page regions are used to conservatively approximate the uncertainty of point objects. In contrast to the original R-tree several uncertainty regions are assigned to a data/directory page which can be used to prune subtrees according to different join confidence thresholds. A more detailed description of this join technique is given in Section 4.2.0.

2.7 Spatial Join using a spatial index structure on one relation

In recent years, the main focus was on methods of Class 2 and Class 3. A simple Class 2 approach is the indexed nested-loop, where each tuple of the non-indexed relation is used as a query applied to the indexed relation. More efficient solutions are presented in [47, 46, 43]. These algorithms are based on building an index for the non-indexed relation in an efficient way in order to reduce the problem to the join where both relations are indexed (Class 1).

2.8 Spatial Join using no Spatial-Index Structure

For spatial join algorithms of Class 3, it is assumed that no indices are initially available. Several partitioning techniques have been proposed which partition the tuples into buckets and then use either hash-based or sweep-line techniques, e.g. the *Spatial Hash Join* [43], the *Partition-Based Spatial Merge Join (PBSM)* [47] or the *Scalable Sweeping-Based Spatial Join (SSSJ)* [6]. The latter approaches are adequate for relatively simply shaped 2D objects which can be well approximated by their minimal bounding boxes. In higher dimensions however, the minimal bounding box is a rather poor approximation. In order to use the *PBSM* or the *SSSJ* with decomposed objects, some modifications have to be done, as for instance duplicate elimination. Koudas and Sevcik solve in [32] the multidimensional join problem by introducing a generalization of the *SSSJ*, the *Multidimensional Spatial Join (MSJ)*. The sweep-line technique is adopted in [42] for join processing on uncertain spatial objects, details can be found in Section 4.3.

Methods for joining on complex highly resolved spatial data following the paradigm of multi-step query processing were proposed in [33], [34] and [35]. These join procedures are based on fast filter steps performed on object approximations. The core of these approaches is a cost-based decomposition algorithm, building the object approximations in a convenient way for the corresponding join process. The concept of using approximated object decompositions to accelerate a filter step of a multi-step join process was later transferred to solve the problem of similarity join for uncertain objects as proposed in [36] (cf. Section 4.2.0).

3. Uncertainty Models and Join Predicates

The problem of modeling uncertain data has been studied extensively in the literature [2, 25, 26, 41, 50]. In the following section, we show a variety of approaches that are utilized to model uncertain data, grouped by their data representation. In the literature three types of uncertainty models are prevalent:

- Continuous uncertainty (cf. [13, 14, 18, 19, 21, 22, 36–38]),
- discrete uncertainty (cf. [4, 5, 28, 49–53, 56]) and
- spatial uncertainty (cf. [17, 24, 39, 41, 55])

The uncertainty models, in particular, specify the representation of the uncertain data and define which comparison operators can be applied to the uncertain data in order to evaluate the join predicates. Additionally, we give a brief overview of join query predicates commonly used for uncertain databases, including the distance range predicate, the k-nearest-neighbor predicate and the spatial intersection predicate.

3.1 The Continuous Uncertainty Model

To capture the uncertainty of dynamic entities such as temperature, pressure and location, a data scheme known as *Probabilistic Uncertainty Model* was proposed in [19]. In many applications, uncertain attributes of objects are given by or can be approximated by a *Probabilistic Density Function (pdf)*. As an example imagine the situation where the position, direction and speed of a moving object o on a road network has been measured five minutes ago. Since the speed of o may have changed since the last measurement, the exact position of o is unknown. Nonetheless, we can incorporate the information about the last known position, direction and speed to specify a pdf of the current position of o. This pdf could be a uniform distribution over the area that o can reach from its last registered position at maximum speed.

Data Representation. Assume that each object of interest consists of at least one real-valued attribute a. In this model, a is treated as a continuous random variable. In [22], it is also assumed that each uncertain attribute value is mutually independent. The probabilistic representation of a consists of two components.

DEFINITION 9.2 (UNCERTAINTY INTERVAL) *An **Uncertainty Interval** of attribute a, denoted by $a.U$ is an interval $[a.l, a.r]$ where $a.l, a.r \in \mathcal{R}$, $a.r \geq a.l$ and $a \in a.U$.*

DEFINITION 9.3 (UNCERTAINTY PDF) *An **Uncertainty PDF** of attribute a denoted by $a.f(x)$, is a probability density function, such that $\int_{a.l}^{a.r} a.f(x)dx = 1$ and $a.f(x) = 0$ if $x \notin a.U$.*

Furthermore, we define the *Uncertainty Distribution Function* as follows:

DEFINITION 9.4 (UNCERTAINTY DISTRIBUTION FUNCTION) *An **Uncertainty Distribution Function** of an object attribute a denotes the probability distribution function $a.F(x) = \int_{-\infty}^{x} f(y)dy$.*

Note that $a.F(x) = 0$ if $x < a.l$ and $a.F(x) = 1$ if $x > a.r$. If a pdf is used to describe the value of an attribute, then the number of possible attribute values is infinite.

Comparison Operators. The evaluation of join predicates over the uncertain object attributes either requires certain predicates defined on the probabilistic similarity distance or requires the definition of probabilistic comparison

operators that can be directly applied to the uncertain objects. In [22] several probabilistic boolean comparison operators are defined on pairs (a, b) of uncertain objects.
The most common comparison comperator is the equality. Since a and b are represented by continuous functions, the probability of equality between a and b is zero, and a and b can never be equal. Given that the exact values for these data items are not known, a user is more likely to be interested in them being very close in value rather than exactly equal. Based upon this observation, equality is defined using a parameter, called $resolution$ c that denotes the distance that a and b may be apart in order to be considered equal.

DEFINITION 9.5 (EQUALITY OPERATOR $=_c$) *Given a resolution c, o_i is equal to o_j with probability*

$$P(a =_c b) = \int_{-\infty}^{\infty} a.f(x) \cdot (b.F(x+c) - b.F(x-c))dx$$

Using equality, inequality can be defined easily:

DEFINITION 9.6 (INEQUALITY OPERATOR $\neq_c$) *Given a resolution c, o_i is not equal to o_j with probability*

$$P(a \neq_c b) = 1 - \int_{-\infty}^{\infty} a.f(x) \cdot (b.F(x+c) - b.F(x-c))dx = 1 - P(a =_c b)$$

To address the question "Is a greater than b?", definitions can be found in [22] for the

- **Greater than** operator $>$: $o_i > o_j$ with probability $P(o_i > o_j)$ and the
- **Less than** operator $<$: $o_i < o_j$ with probability $P(o_i < o_j) = 1 - P(o_i < o_j)$.

These comparison operators return a boolean value in the case of certain objects and a probability value in the case of uncertain attributes.

Often, distance functions are used to compare objects to each other. A distance function assigns a distance value to a pair of uncertain objects. Often, this comparison operator is used to measure the (dis-)similarity between objects for similarity search applications. In the case of uncertain objects, the distance between two objects obviously is also uncertain. A frequently used distance function in the case of uncertainty is the probabilistic distance function as proposed in [36] that describes the probability distribution of all possible distances between two objects.

DEFINITION 9.7 (PROBABILISTIC DISTANCE FUNCTION (CONT.)) *Let $d : D \times D \rightarrow IR_0^+$ be a distance function defined on a pair of objects, and let*

$P(a \leq d(o,o') \leq b)$ denote the probability that $d(o,o')$ is between a and b. Then the probabilistic density function $f_d : D \times D \rightarrow (IR_0^+ \rightarrow IR_0^+ \cup \infty)$ is called probabilistic distance function.

Since the result of f_d is a probability density function, the probability that two objects have a certain distance $c \in \mathcal{R}$ to each other is zero. For this reason, we need the next equation in order to evaluate the distance between two objects. The probability that the distance between an object o and another object o' is at least a and at most b can be computed by

$$P(a \leq d(o,o') \leq b) = \int_{a \leq d \leq b} f_d(o,o') =$$

$$\int_{-\infty}^{\infty} a.f(x) \cdot (b.F(x+b) - b.F(x+a) + b.F(x-a) - b.F(x-b))dx \tag{9.1}$$

Note that this probability can also be expressed by means of the boolean comparison operator $=_c$ as follows:

$$P(a \leq d(o,o') \leq b) = (o =_b o') - (o =_a o')$$

Other function-based comparison operators can be defined as well. For example, uncertain spatial objects may require comparison operators that return the probability distribution of the spatial cover of two objects (cf. Section 3.2.0).

3.2 The Discrete Uncertainty Model

In this section, we consider the uncertainty of nominal variables such as weather conditions (e.g., sunshine, rain and snow), age in years or generally variables in situations where only discrete measurements, e.g. from a sensor, are given.

Data Representation. The most general approach to model discrete uncertainty is the *Possible Worlds Model*. In this model, every possible instance (or *possible world*) w of the database has a probability $c(w)$. The probability of the presence of a tuple in the database affects the probability of the presence or absence of any other tuple and vice-versa. An example of such a database can be given by considering positions of tigers. Let us assume that a set of possible positions of male tigers in a wildlife sanctuary is known for each tiger. Of course, each tiger may only exist at at most one position in a possible world. Note that possible worlds are mutually exclusive. Additionally, it is known that male tigers are territorial and the position of a tiger is affected by the presence of other tigers in its close vicinity. Consequently, there is a mutual dependency between the positions of the tigers. This full model respects the mutual dependency of tuples but is deficient in a computational point of view, because the number of possible worlds increases exponentially in the number of objects in the database.

For this reason, the *ULDB Model* (*Uncertainty-Lineage Database Model*, also called *X-Relation Model*), which was first introduced in the TRIO system [3, 8, 50, 54], is often used instead of the possible world model and extends the relational database model by utilizing lineage and uncertainty [9]. The x-relation model is a special form of the possible worlds model where independence between objects is assumed. The x-relation semantic allows us to represent the uncertainty of the attribute *values* (attribute level uncertainty) and the *presence* (existential uncertainty) of an object. Such an object is called an *x-tuple*. Each x-tuple consists of one or more *alternative tuples* representing the uncertainty of its contents. The uncertainty of the presence of an x-tuple is represented by the *maybe* annotation "?", as proposed in [3]. Independence is assumed among x-tuples. Each alternative tuple t is given a confidence value $c(t)$ with the probability that the respective alternative tuple exists. Note that for each x-tuple x consisting of alternative tuples $t \in x$ the following equation holds:

$$\sum_{t \in x} c(t) \leq 1$$

Table 9.1 shows an x-relation that contains information about the possible positions of tigers in a wildlife sanctuary. Here, the first x-tuple describes the tiger named "*Renzy*" who may be found at three possible (alternative) locations. He may be in his cave with a probability of 50% or located at the water hole and at the hunting grounds with a probability of 20% and 30%, respectively. This x-tuple logically yields three *possible instances*, one for each alternative location. Now, we know that a new tiger may have entered the wildlife sanctuary with a probability of 40%. In this case, it is not certain that the new tiger exists at all, which is an *existential* uncertainty, denoted by the "?" symbol to the right of the name of the tiger in the left column of the corresponding x-tuple. Taking into account the four alternatives (including the alternative of no new tiger) for the position of the new tiger, there are twelve possible instances of the tiger x-relation. In general, the possible instances of an x-relation $\mathcal{R}$ correspond to all combinations of alternatives for the x-tuples in $\mathcal{R}$. Note that in this model, the probability of the new tiger being at the water hole is not affected by the current position of *Renzy*, due to the independency assumption.

Comparison Operators. In order to evaluate join predicates on uncertain data using the possible worlds model, the definition of comparison operators

Tiger x-relation		
NAME	POSITION	PROBABILITY
Renzy	{ *Renzy*'s Den, Waterhole, Hunting Grounds}	{50%,20%,30% }
New Tiger ?	{Waterhole, Hunting Grounds, The Forest}	{10%,10%,20%}

Table 9.1. An x-relation containing x-tuples with possible positions of tigers.

is required. One of the most important comparison operators for the discrete uncertainty model is the *equality* operator.

DEFINITION 9.8 (EQUALITY OPERATOR $=_{PW}$ (POSSIBLE WORLDS)) *Let a and b be two uncertain objects with uncertain attributes $a.u$ and $b.u$. Let W be the set of possible worlds in the database and $c(w \in W)$ be the confidence of a possible world. The probability that $a.u$ is equal to $b.u$ is*

$$P(a.u =_{PW} b.u) = \sum_{w \in W, a.u = b.u} c(w).$$

As a consequence, all possible instances of the uncertain database have to be considered in order to compute the probability of equality between two objects. Other comparison operators can be treated in the same way. Even if each object has only two alternatives, the number of possible instances is 2^n where n is the number of objects in the database. Due to the dependency between objects, no possible world can be pruned. In fact, all possible worlds have to be enumerated, making the comparison of objects in the possible worlds model an NP-hard problem. Thus, the following comparison operators use the x-relation model, which only includes dependencies within, but not between, x-tuples.

DEFINITION 9.9 (EQUALITY OPERATOR $=_x$ (X-TUPLES)) *Two uncertain objects a and b represented by x-tuples with uncertain attributes $a.u$ and $b.v$ are equal with probability*

$$P(a =_x b) = \sum_{r \in a} \sum_{s \in b, r.u = s.v} c(r) \cdot c(s).$$

The probability of two objects being equal is the sum of the probabilities of the tuples to be equal. Unlike the possible worlds model, other objects do not have to be considered due to the independency between x-tuples.

Based upon the definition of equality, **inequality** can be defined as follows:

DEFINITION 9.10 (INEQUALITY OPERATOR $\neq_x$ (X-TUPLES)) *Two uncertain objects a and b represented by x-tuples with uncertain attributes $a.u$*

and b.v are not equal with probability

$$P(a \neq_x b) = 1 - \sum_{r \in a} \sum_{s \in b, r.u = s.v} c(r) \cdot c(s)$$

The *greater than* and *less than* relations can be defined in a similar fashion.

DEFINITION 9.11 (GREATER THAN $>_x$ (X-TUPLES)) *Let a and b be two uncertain objects represented by x-tuples with uncertain attributes a.u and b.v. The probability of a.u being greater than b.v is*

$$P(a.u >_x b.v) = \sum_{r \in a} \sum_{s \in b, r.u > s.v} c(r) \cdot c(s)$$

DEFINITION 9.12 (LESS THAN $<_x$ (X-TUPLES)) *Let a and b be two uncertain objects represented by x-tuples with uncertain attributes a.u and b.v. The probability of a.u being less than b.v is*

$$P(a.u <_x b.v) = \sum_{r \in a} \sum_{s \in b, r.u < s.v} c(r) \cdot c(s)$$

Similar to the continuous uncertainty model, distance-based comparison operators can be defined for the discrete uncertainty model. In the x-tuple model, a distance applied to two uncertain objects o_r and o_s returns a set of distance values, one for each pair of possible instances of o_r and o_s each associated with a probability value.

DEFINITION 9.13 (PROBABILISTIC DISTANCE SET (DISCRETE)) *Let o and o′ be two objects represented by x-tuples. Furthermore, let* $d : D \times D \rightarrow IR_0^+$ *be a distance function, and let* $P(a \leq d(o, o') \leq b)$ *denote the probability that the distance* $d(o, o')$ *is between a and b. Then, the probabilistic distance between two uncertain objects o and o′ is a set of distances associated with their probabilities* $s_d \subseteq IR \times [0 \ldots 1]$ *as defined as follows:*

$$s_d = \{(x, p) : IR_0^+ \times [0 \ldots 1] | \forall t \in a, \forall t' \in b, x = d(t, t') \wedge p = t.c \cdot t'.c\}$$

The Spatial Uncertainty Model. This model refers to objects which are assumed to have a spatial extension called spatial objects. This object representation is useful in biomedical and geographical applications. As an example, consider a satellite image with several buildings. All building objects are spatially extended because they cover a whole area, not just a single point. A possible task could be to automatically identify the type of buildings that can be seen there.

In the x-relation model, tuples belonging to an x-tuple are mutually exclusive. Thus an object contains at most one tuple in a possible world. In contrast to that, the *Spatial Uncertainty Model* assumes independence between possible tuples of an object, because one object may contain several tuples at the same time. A spatial uncertainty model is used in [42]. In this model, it is assumed that the number and the positions of the tuples are certain. Yet, there are two types of uncertainty:

- Uncertainty in the extent of an object: In the example above, which pixels in the satellite image belong to which building?
- Uncertainty in the class of an object: In the example above, a classifier could decide that there is a 70% probability of a given building being an airport.

First, a probabilistic description of the extents of the objects is required. In [40] a *probabilistic mask* is introduced which gives a probabilistic description of the uncertain extent of an object:

DEFINITION 9.14 (PROBABILISTIC MASK) *A* ***Probabilistic Mask*** M_a *for an object a is a set of tuples* $\{(\vec{x}, p_a)\}$ *such that each point* $\vec{x}$ *belongs to a with probability* p_a.

Additionally, the objects cannot be identified reliably. Therefore, *confidence values* are required to describe the classifier's estimate of the probability that an object belongs to a certain class.

DEFINITION 9.15 (PROBABILISTIC OBJECT) *A* ***Probabilistic Object*** *a is a pair* (M_a, p_{M_a})*, where* M_a *is a's probabilistic mask and* p_{M_a} *is the confidence value of a's class.*

In order to evaluate the join predicate, a construct which enables us to compare uncertain spatial objects is required. Similar to the boolean comparison operators defined in Section 3.2.0, *score* functions are introduced for this type of objects in [42], which already reflect the probability that the spatial join predicate is fulfilled.

DEFINITION 9.16 (POINT-LEVEL SCORE) *The score* s' *called* ***Point-Level Score*** *between two point objects* $a = (\{(\vec{x_a}, p_a)\}, p_{M_a})$ *and* $a = (\{(\vec{x_b}, p_b)\}, p_{M_b})$ *is*

$$s'(\vec{x_a}, p_a, \vec{x_b}, p_b) = p_a \cdot p_b \cdot \lambda \cdot e^{-\lambda \cdot d(\vec{x_a}, \vec{x_b})},$$

where λ *is a positive, domain-specific parameter that determines the relative importance of probability and distance, and d is a suitable distance function.*

DEFINITION 9.17 (OBJECT-LEVEL SCORE) *The* score s *also called* ***Object-Level Score*** *between two uncertain spatial objects* $a = (M_a, p_{M_a})$ *and*

$b = (M_b, p_{M_b})$ *is*

$$s(a,b) = \max_{\substack{(\vec{x_a}, p_a) \\ (\vec{x_b}, p_b)}} p_{M_a} \cdot p_{M_b} \cdot s'(\vec{x_a}, p_a, \vec{x_b}, p_b).$$

Note that the spatial comparison function that is required to evaluate the spatial join predicate is already included in the score functions. Thereby, the comparison mainly relates to the distance between the points of the spatial objects. However, there may exist other comparison operators for uncertain spatial objects, e.g. probabilistic spatial functions which, similar to the probabilistic distance functions, return a probability density function (pdf) according to a spatial comparison operator like the volume of the overlap between two objects.

The next section generalizes the concept of the *score* defined for a pair of uncertain objects which reflects the probability that a given join predicate in connection with a given comparison operator fulfills the join predicate for the object pair.

3.3 Join Predicates and Score

Given a comparison operator, the probability that a pair of objects satisfies a join predicate can be formalized as follows. The result of a boolean comparison operator ($=, \neq, <, >$ etc.) evaluated on uncertain objects directly returns a probability value which is called *score*. The same holds for the score defined for the spatial uncertainty model (cf. Section 3.2.0). If a distance function (probabilistic distance function, probabilistic spatial function, etc.) is used, a pdf is returned in the continuous case and a set of score values is returned in the discrete case. This pdf is used to derive the probability (*score*) that a given join predicate is satisfied.

The following section describes two types of join predicates. Here, we concentrate on the two most prominent join predicates. An important join predicate is the ε-*Range* join predicate.

DEFINITION 9.18 (UNCERTAIN ε-RANGE JOIN PRED. (CONT.)) *The probability that two uncertain objects a and b following the continuous uncertainty model have a score s greater than ε is given by*

$$s(a,b) = P(d(a,b) < \varepsilon) = P(0 \leq d(a,b) \leq \varepsilon) = \int_0^{\varepsilon} f_d(a,b)(x)dx$$

where f_d is a probabilistic distance function.

This definition can be easily adjusted to support discrete uncertain data.

DEFINITION 9.19 (UNCERTAIN ε-RANGE JOIN PREDICATE (DISCR.)) *The probability that two uncertain objects a and b (following the discrete uncertainty model) have a score s greater than ε is given by*

$$s(a,b) = P(d(a,b) < \epsilon) = P(0 \leq d(a,b) \leq \epsilon) = \sum_{(d,p) \in s_d(a,b)} \begin{cases} p & , if\, d < \varepsilon \\ 0 & else \end{cases}$$

where s_d is a probabilistic distance set.

Another important join query predicate is the k-NN predicate. The probability that an uncertain object b is one of the k nearest neighbors of an uncertain object a is:

$$s(a,b) = P(\text{for at least } (N-k) \text{ objects } o : d(o_r, o_s) \leq d(o_r, o))$$

The formal definition is given in [37].

There exists a large variety of other join predicates, that can be useful for certain applications (e.g. Reverse-k-NN, k-closest-pairs) that are not introduced here. Regardless of which join predicate is used, a score value is returned that describes the probability of o_r and o_s satisfying the join predicate.

3.4 Probabilistic Join Query Types

The comparison operators and join predicates assign to each pair of objects $(a,b) \in \mathcal{R} \times \mathcal{S}$ a score value. In general, the result of a *Probabilistic Join* $M(\mathcal{R},\mathcal{S})$ can be defined as follows:

DEFINITION 9.20 (PROBABILISTIC JOIN) *Given two relations $\mathcal{R}$ and $\mathcal{S}$ with uncertain objects. A* ***Probabilistic Join*** *$M(\mathcal{R},\mathcal{S})$ is a set*

$$M(\mathcal{R},\mathcal{S}) = (a,b,s(a,b)) \in \mathcal{R} \times \mathcal{S} \times [0\ldots1]$$

where $s(a,b)$ is the score between two objects a and b w.r.t. a given comparison operator and join predicate.

In the literature, various types of uncertain join queries are defined which differ in their result sets. This concerns the join pairs which are finally added to the result relation $\mathcal{J}$. In [22], two general types of join queries are proposed: The *Probabilistic Join Query* (PJQ) and the *Probabilistic Threshold Join Query* (PTJQ).

DEFINITION 9.21 (PROBABILISTIC JOIN QUERY (PJQ)) *Given two relations $\mathcal{R}$ and $\mathcal{S}$ with uncertain objects and a probabilistic join $M(\mathcal{R},\mathcal{S})$, a* ***Probabilistic Join Query*** *$PJQ(\mathcal{R},\mathcal{S})$ returns all triples $(a,b,s(a,b)) \in M(\mathcal{R},\mathcal{S})$, where $s(a,b) > 0$, i.e.*

$$PJQ(\mathcal{R},\mathcal{S}) = \{(a,b,s(a,b)) \in M(\mathcal{R},\mathcal{S}) | s(a,b) > 0\}.$$

Essentially, a PJQ returns join pairs with a non-zero probability of fulfilling the join predicate along with the associated probability.

The join condition can be further enforced by raising the confidence bound of the result candidates. This is realized by the probabilistic threshold join query (PTJQ), which additionally allows us to specify a confidence threshold τ. Only join results whose confidence exceeds τ are returned from the query.

DEFINITION 9.22 (PROBABILISTIC THRESH. JOIN QUERY (PTJQ)) *Given two relations $\mathcal{R}$ and $\mathcal{S}$ with uncertain objects, a confidence threshold $\tau \in [0\ldots1]$ and a probabilistic join $M(\mathcal{R},\mathcal{S})$, a* ***Probabilistic Threshold Join Query*** *$PTJQ(\mathcal{R},\mathcal{S})$ returns all triples $(a,b,s(a,b)) \in M(\mathcal{R},\mathcal{S})$, where $s(a,b) \geq \tau$, i.e.*

$$PTJQ(\mathcal{R},\mathcal{S}) = \{(a,b,s(a,b)) \in M(\mathcal{R},\mathcal{S}) | s(a,b) > \tau\}.$$

A PTJQ only returns join pairs that have probabilities higher than τ. Note, that the score $s(a,b)$ included in the result triple is optional. Since the probability threshold τ lower bounds the confidences of the results, the exact confidence values given by the score are often ommitted from the query output (cf. [22]).

Another important type of probabilistic query returns the k best join partners:

DEFINITION 9.23 (PROBABILISTIC TOP-k JOIN QUERY (PTOPkJQ)) *Given two relations $\mathcal{R}$ and $\mathcal{S}$ with uncertain objects and a probabilistic join $M(\mathcal{R},\mathcal{S})$, a* ***Probabilistic Top-k Join Query*** *PTopkJQ($\mathcal{R}$,$\mathcal{S}$) returns a set of k triples $(a,b,s(a,b)) \in M(\mathcal{R},\mathcal{S})$ with the highest score $s(a,b)$, i.e.*

$$PTopkJQ(\mathcal{R},\mathcal{S}) \subseteq M(\mathcal{R},\mathcal{S}), |PTopkJQ(\mathcal{R},\mathcal{S})| = k,$$

$$\forall (a,b,s(a,b)) \in PTopkJQ(\mathcal{R},\mathcal{S}),$$

$$\forall (a',b',s(a',b')) \in (M(\mathcal{R},\mathcal{S}) - PTopkJQ(\mathcal{R},\mathcal{S})) : s(a,b) \geq s(a',b').$$

Essentially, a PTop$k$JQ returns the k join pairs with the highest score.

The differences between the probabilistic join queries are illustrated in the following example.

3.5 Example

To improve the illustration, in this scenario one join relation $\mathcal{R}$ is assumed to have only one uncertain object which is called query object Q. It is joined with a set of uncertain objects of the other relation $\mathcal{S}$. The scenario is shown in Figure 9.4. $\mathcal{S}$ consists of seven uncertain objects A-G.

OBJECT	P(OBJECT IN ε-RANGE OF Q)	P(OBJECT IS 1-NN OF Q)
A	0	0
B	0	0.10
C	0.05	0.05
D	0.15	0.15
E	0.10	0.25
F	0.07	0.15
G	0.20	0.30

Table 9.2. Confidences of different join predicates w.r.t. the join between Q and the other objects. (cf. example shown in Figure 9.4)

First, we consider the ε-Range predicate. Table 9.2 shows for each uncertain object the score, which is the probability to be within the ε-Range of Q. These score values are also depicted in Figure 9.4. The results of the probabilistic join queries are shown in Table 9.3. First, a PJQ is performed: All objects except for A and B, which have a score of zero, are part of the result relation $\mathcal{J}$. If a PTJQ with $\tau = 10\%$ is issued, objects C and F are dropped, because their score is less than τ. If a PTop2JQ (PTop$k$JQ with $k = 2$) is performed, only objects D and G are returned, because D and G are the two objects with the highest score.

Now, we consider the 1-NN predicate. The probabilities of satisfying the join predicate and the query results are given in the corresponding tables 9.2 and 9.3. Figure 9.4 also shows, for each uncertain object, its score with respect to the query object Q. The score is now defined as the probability that an object is the nearest neighbor of Q. The result of a PJQ contains all objects except for A, because all objects except for A have a probability greater than zero of being the nearest neighbor of Q. Note that, even though B is further away than objects C-G, there exist possible instances of $\mathcal{S}$, in which B is the nearest neighbor of Q. A cannot be the nearest neighbor of Q, because the minimum distance between Q and A is larger than the maximum distance between Q and B. Thus, B is definitely closer to Q than A because it is closer to Q in any possible world. If a PTJQ with $\tau = 10\%$ is performed, then the result relation $\mathcal{J}$ contains objects B, D, E, F, G and a PTop2JQ results in the objects G and E.

3.6 Overview of Uncertainty Models and Probabilistic Join Queries

In the previous section, several uncertainty models are presented. This section gives a general overview of probabilistic join queries and the uncertainty models. First, let us summarize how the uncertainty models can be classified according to different criteria:

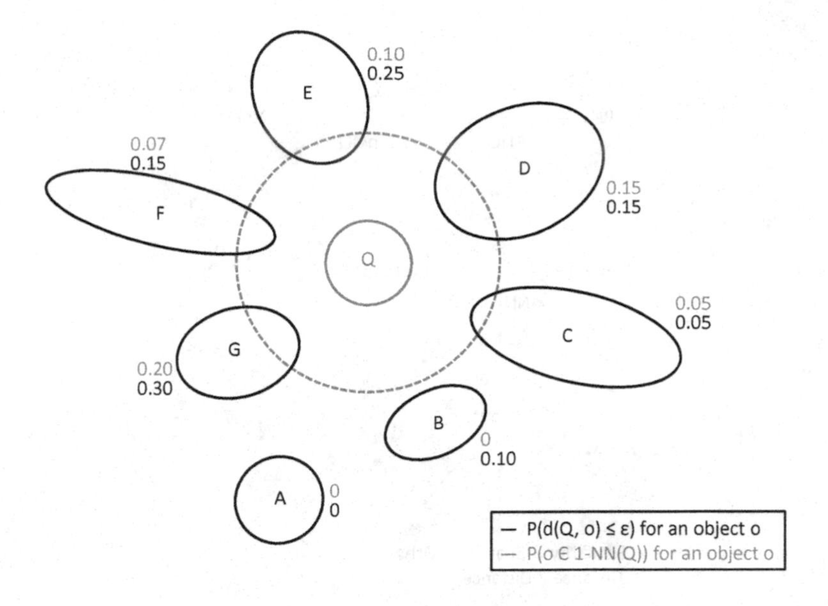

Figure 9.4. Example of two uncertain object relations $S = \{A, B, C, D, E, F, G\}$ and $\mathcal{R} = \{Q\}$ with given scores for the ε-Range join predicate and the 1-NN join predicate.

JOIN PREDICATE	ε-RANGE JOIN	k-NN JOIN
PJQ	$(Q, C, 0.05)$ $(Q, D, 0.15)$ $(Q, E, 0.10)$ $(Q, F, 0.07)$ $(Q, G, 0.20)$	$(Q, B, 0.10)$ $(Q, C, 0.05)$ $(Q, D, 0.15)$ $(Q, E, 0.25)$ $(Q, F, 0.15)$ $(Q, G, 0.30)$
PTJQ	(Q, D) (Q, E) (Q, G)	(Q, B) (Q, D) (Q, E) (Q, F) (Q, G)
PTop$k$JQ (k=2)	$(Q, G, 0.20)$ $(Q, D, 0.15)$	$(Q, G, 0.30)$ $(Q, E, 0.25)$

Table 9.3. Query results of different probabilistic join queries w.r.t. the different query predicates. (cf. example shown in Figure 9.4)

- Classification by data representation (continuous, discrete and spatial)
- Classification by the probabilistic query type (PJQ, PTJQ, PTop$k$JQ, etc.)

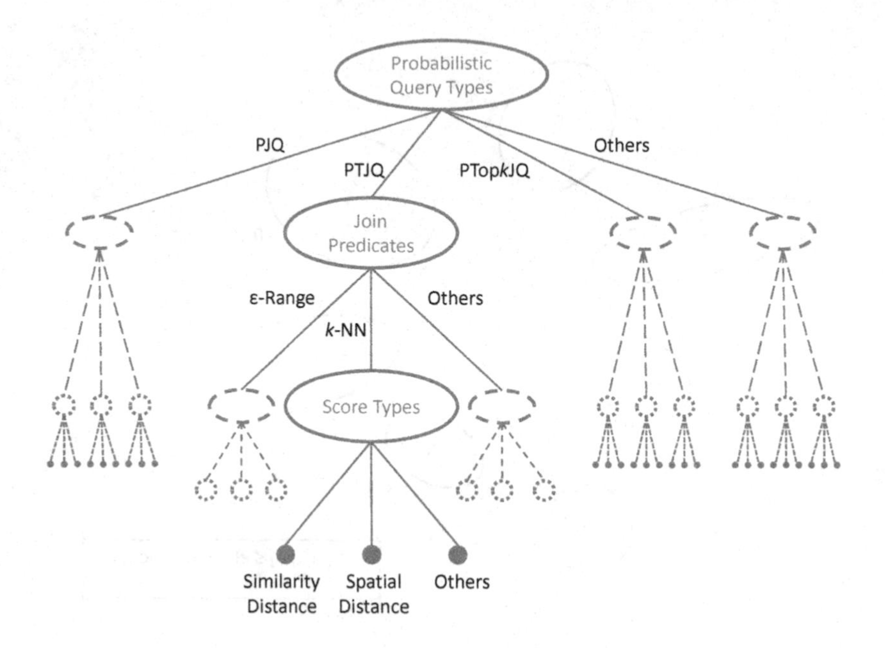

Figure 9.5. Overview of Uncertainty Models.

Source	Prob. Query Type	Join Predicate	Score Type
Ljosa [42]	PTJQ, PTop$k$JQ	ε-Range	Spatial Distance
Cheng [22]	PJQ, PTJQ	ε-Range	Similarity Distance
Kriegel [36, 37]	PJQ, PTJQ, PTop$k$JQ	ε-Range	Similarity Distance
Agrawal [4]	PTJQ, PTop$k$JQ, Other	Any	Any

Table 9.4. List of publications elaborated in the next section and respective classification of the uncertainty model. All of these approaches except for Cheng [22] 2006 use a discrete data representation. In Cheng [22] 2006 a continuous data representation is utilized.

- Classification by the join predicate (ε-Range, k-NN, etc.)

- Classification by the score function (boolean comparison operator, similarity distance, spatial distance etc.).

Figure 9.5 illustrates all possible combinations of the different probabilistic join queries with the different uncertainty models. Table 9.4 gives a brief overview of the approaches that are elaborated in the following section.

4. Approaches for Efficient Join Processing on Uncertain Data

While in the recent decade a lot of approaches that address the organization and efficient query processing of uncertain data have been published, only a few approaches directly address join queries on uncertain data. This section aims at surveying the currently followed research directions concerning joins on uncertain data. It exemplarily presents the most prominent representatives of the join categories which are currently of main interest in the research community. The approaches mainly differ in the representation of the uncertain data, the distance measures or other types of object comparisons, the types of queries and query predicates and the representation of the result. They can be classified into

- confidence-based join methods,
- probabilistic similarity join methods and
- probabilistic spatial join methods.

In the following, these categories are more closely illustrated by selecting concrete approaches for efficient join processing on uncertain data.

4.1 Confidence-Based Join Methods

Pure confidence-based join methods are mainly focused on reducing the search space w.r.t. the confidences of the result tuples of a join query. For the selection of the candidates to be joined, neither the *join-relevant attributes** ($\neq$ confidence values) of the objects, nor the join predicates are taken into account.

Let us assume that an object $o = \langle oid, pos \rangle$ is associated with a confidence value $c(o)$ denoting that the likelihood of the object identified by oid is currently located at position pos. If we join two relations with objects of this type, then the results obviously are also associated with confidence values. For example, if we try to join object $o_A = \langle A, coord(50, 14) \rangle \in \mathcal{R}$ with object $o_B = \langle B, coord(50, 14) \rangle \in \mathcal{S}$ with the given confidences $c(o_A)$ and $c(o_B)$, the result tuple $\langle A, B \rangle$ is associated with a confidence value $P((c_A, c_B) \in \mathcal{R} \bowtie \mathcal{S})$ which depends on the confidence values of both join partners o_A and o_B, i.e. $P((c_A, c_B) \in \mathcal{R} \bowtie \mathcal{S}) \leq c(o_A) \cdot c(o_B)$. Here, it is assumed that both objects o_A and o_B are independent of each other. Based on the confidences of the join results the probabilistic join queries PJQ, PTJQ and PTop$k$JQ can be directly applied (cf. Section 3.4). Furthermore, one can imagine the following types of join queries [4]:

*The object attributes that are used to evaluate the join predicate are called the *join-relevant attributes*. The confidence values assigned to the attributes are not included.

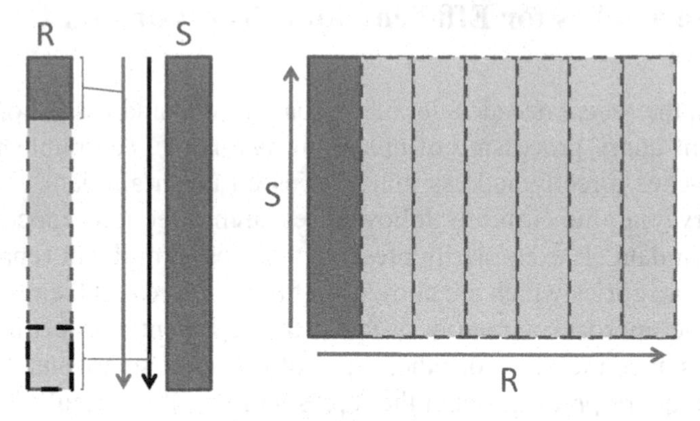

(a) Nested-loop-based join without confidence-based pruning

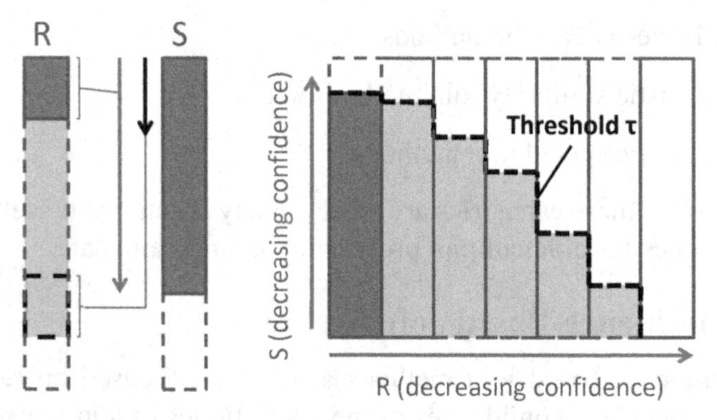

(b) Nested-loop-based join with confidence-based pruning (query type:*Threshold*)

Figure 9.6. Nested-loop-based Join Approaches.

- *Sorted*: Return result objects sorted by confidence.

- *Sorted-Threshold*: Combination of the *Top-k* and *Sorted* approach by returning result objects with confidence above a threshold τ, sorted by confidence in descending order.

Confidence-based Threshold Join Queries PTJQ. Agrawal and Widom propose in [4] efficient confidence-based join approaches for all query types mentioned above for the possible worlds model (cf. Section 3.2.0). They assume that stored relations provide efficient sorted access by confidence and that neither join relation fits into main memory. They also assume existential uncertainty of objects and independence between objects. Note that this approach can be applied regardless of the join-predicate and the type of score-function.

DEFINITION 9.24 (PROBLEM DEFINITION) *Let $\mathcal{R}$ and $\mathcal{S}$ be two relations and let $c(o) \in [0, 1]$ be a confidence value assigned to each object o in $\mathcal{R}$ and $\mathcal{S}$ which denotes the likelihood that o does exist in the corresponding relation $\mathcal{R}$ or $\mathcal{S}$. Furthermore, let $c(j) \in [0, 1]$ be a confidence value assigned to each tuple $j = (r, s) \in \mathcal{R} \times \mathcal{S}$ which denotes the likelihood that both objects r and s fulfill the join predicate. The problem is to find the join result $\mathcal{J} = \mathcal{R} \bowtie_{\Theta} \mathcal{S}$, whereas the confidence of each result tuple $j \in \mathcal{J}$ is above a given threshold value τ.*

The approach described in [4] sorts $\mathcal{R}$ and $\mathcal{S}$ by confidence in descending order and exploits the fact that, regardless of the join predicate, a tuple (r, s) can only exist in $\mathcal{J}$ if both r and s exist. With the assumption of object independence, the likelihood that two objects are joined is at most the product of their confidences. Note, that the assumption of the object independence can be relaxed if a combining function $f : \mathcal{R} \times \mathcal{S} \rightarrow [0, 1]$ is used instead of the product over the object confidences to compute the confidence of the result tuple. The combining function $f()$ can be any monotonic function, i.e. $f(r_i, s_j) \geq f(r_k, s_l)$ whenever $c(r_i) \geq c(r_k)$ and $c(s_j) \geq c(s_l)$. As a consequence, if the combining function applied to $r \in \mathcal{R}$ and $s \in \mathcal{S}$, e.g. the product of confidences $c(r) \cdot c(s)$, is less than the confidence threshold τ, then the tuple (r, s) cannot belong to the join result and, thus, can be excluded from the join in advance. With the assumption that the objects of both relations are sorted in descending order of their confidences and a monotonic combining function is used to determine the confidence of the corresponding result tuples then the probabilistic threshold join query can be accelerated by reducing the search space of a Nested-Block-Loop Join based query processing (cf. Section 2.2). This approach is illustrated in Figure 9.6. Figure 9.6(a) depicts the basic Nested-Block-Loop Join approach without confidence-based pruning. Here every pair of objects has to be tested in order to find all join result tuples. If confidence-based pruning is applied, then the number of pairs of objects can be reduced as shown in Figure 9.6(b), where only pairs of objects with a combined confidence greater than τ are explored.

Confidence-based Top-k Join Queries PTop$k$JQ. A similar pruning criterion as used for threshold join queries can be used for the probabilistic Top-k join queries. Remember, the problem of probabilistic Top-k join queries is to find join results $\mathcal{J} = \mathcal{R} \bowtie_k \mathcal{S}$, whereas J contains only the k result tuples with the highest confidence for a user-defined parameter k (cf. Section 3.4). The idea of this approach is again to prune the space to be explored using a threshold as in the previous algorithm. In this case, the threshold to be used is the confidence value of the kth tuple in the result, i.e., the minimum confidence value among the Top-k result tuples. Of course, this value is not known at the start of the algorithm. The approach of [4] uses a priority queue K to maintain the current Top-k result set. When a result tuple t is generated with a confidence greater than the confidence of the lowest-confidence result tuple t' in K, t' is replaced by t. When a result tuple t is generated with a confidence equal or less the lowest confidence result tuple in K, then the current iteration of the inner relation S may be aborted, because all subsequently generated result tuples in the current iteration of R have a lower confidence than t.

Confidence-Based Sorted-Threshold Join Queries. In a similar way as Top-k join queries, sorted and sorted-threshold join queries can be performed. *Sorted-Threshold Join Queries* try to find join results $\mathcal{J} = \mathcal{R} \bowtie^{s}_{\Theta} \mathcal{S}$, where the confidence of the existence of a tuple $j \in \mathcal{J}$ is above a given threshold value τ and the result is sorted by confidence in descending order. As proposed in [4] the algorithm explores the same pruned space as in the threshold join case, but in an order resembling Top-k. The sorted join query is a special case of a sorted-threshold join query with a threshold $\tau < 0$.

In addition to the join algorithms proposed in [4], bounds on the efficiency in comparison to other algorithms for the same problem in the same memory-constrained environment are proven.

4.2 Probabilistic Similarity Joins

The previous section focuses on join processing methods which only reduce the search space based on the confidence values of the input data, i.e. the attributes of the objects that the join predicate relates to are not considered. Knowledge about the attributes that are relevant for the join predicate, e.g. the positions of the objects, was not incorporated. For example, if we want to answer a distance range query on an uncertain database, then we want to find only those pairs of objects from $\mathcal{R}$ and $\mathcal{S}$ that are close to each other. However, the previous approach returns pairs of objects regardless of their distance, as long as their combined confidence is sufficient.

In general, similarity queries are very selective queries, i.e. only a very small portion of the candidates satisfies the query predicate. Thus, effective pruning strategies are very important for an efficient similarity query processing. Therefore, in contrast to the previous section, here we address methods which additionally take into account whether the join predicate is satisfiable. In particular, similarity join applications can significantly benefit from pruning those candidates whose attributes do not likely satisfy the join predicate. In this way, further exploration of candidates having a very low join probability can be avoided.

In the following, two approaches for a probabilistic similarity join are presented. They differ in the type of uncertain object representations used. The first approach assumes that the uncertainty is represented by continuous probabilistic density functions (pdf) while the second approach is based on objects represented by sets of discrete samples of alternative values (cf. Section 3).

Similarity Join Queries (Continuous Uncertainty Model). In the following, we assume that the query relevant attributes of each object are represented by *uncertainty intervals* and *uncertainty pdfs* (cf. Section 3.1.0). Cheng et al. study in [22] probabilistic similarity joins over uncertain data based on the continuous uncertainty model. To each uncertain object attribute an uncertainty interval accomplished by an uncertainty pdf is assigned. The score† of two uncertain objects, each represented by a continuous pdf, in turn leads to a continuous pdf representing the similarity probability distribution of both objects (cf. Section 3.3). In fact, the (probabilistic) similarity distance again consists of an uncertainty interval and an uncertainty pdf. This way, each join-pair $(a, b) \in \mathcal{R} \times \mathcal{S}$ (join candidate) is associated with a probability, indicating the likelihood that the two objects a and b are matched according to the given predicate. In [22], the boolean join predicates introduced in Section 3.1.0 are used.

The probabilistic predicates defined on pairs of uncertain objects allow to define probabilistic join queries. In [22], two join queries are proposed: The *Probabilistic Join Query (PJQ)* and the *Probabilistic Threshold Join Query (PTJQ)*. In the sequel, we will explain how the differences between PJQ and PTJQ are exploited for performance improvement. We note that the use of thresholds reduces the number of false positives, but it may also result in the introduction of false negatives. Thus, there is a tradeoff between the number of false positives and false negatives depending upon the threshold which is chosen. The reformulation of the join queries with thresholds is also helpful for improving the performance requirements on the method.
A number of pruning techniques are developed in order to improve the effectiveness of PTJQ processing. These pruning techniques are as follows:

- **Item-level Pruning:** In this case, two uncertain values are pruned without evaluating the probability.

- **Page-level Pruning:** In this case, two pages are pruned without probing into the data stored in each page.

- **Index-level Pruning:** In this case, the data which is stored in a subtree is pruned.

†In the context of a similarity join, the score between two objects obviously reflects the similarity between both objects.

Item-Level Pruning: The refinement process on item (object) level can be done by directly computing the join probability $P(a\Theta_u b)$ for every pair of $(a, b) \in \mathcal{R} \times \mathcal{S}$, where Θ_u is one of the boolean comparison operators introduced in Section 3.1.0. Only those pairs whose score is larger than τ are retained. For an arbitrary pdf, the join probability $P(a\Theta_u b)$ has to be computed with (relatively expensive) numerical integration methods. [22] shows a set of techniques to facilitate this computation. These methods do not compute $P(a\Theta_u b)$ directly. Instead, they establish pruning conditions that can be checked easily to decide whether $(a, b) \in \mathcal{R} \times \mathcal{S}$ satisfies the query predicate. They are applicable to any kind of uncertainty pdf, and do not require the knowledge of the specific form of $P(a\Theta_u b)$. These techniques are labelled "*item-level-pruning*", since pruning is performed based on testing a pair of data items. For **equality** and **inequality**, the following lemma holds:

LEMMA 9.25 *Suppose a and b are uncertain-valued variables, each described by an uncertainty interval $a.U = (a.l, a.r)$ and $b.U = (b.l, b.r)$ and an uncertainty pdf $a.f(x)$ and $b.f(x)$. Furthermore, let $a.U \cap b.U \neq \emptyset$ and let $l_{a,b,c}$ be $max(a.l - c, b.l - c)$ and $u_{a,b,c}$ be $min(a.r + c, b.r + c)$. Then, the following inequalities hold:*

$$P(a =_c b) \leq \\ min(a.F(u_{a,b,c}) - a.F(l_{a,b,c}), b.F(u_{a,b,c}) - b.F(l_{a,b,c})) \quad (9.2)$$

$$P(a \neq_c b) \geq \\ 1 - min(a.F(u_{a,b,c}) - a.F(l_{a,b,c}), b.F(u_{a,b,c}) - b.F(l_{a,b,c})) \quad (9.3)$$

For the proof of this lemma, see [22]. Figure 9.7 illustrates the upper bounding filter probability of the join predicate "$a =_c b$". Obviously, the minimum of both probabilities P_a and P_b build an upper bound of $P(a =_c b)$. It allows to quickly decide whether a candidate pair $(a, b) \in \mathcal{R} \times \mathcal{S}$ should be included into or excluded from the result, since the uncertainty cumulated density functions are known and Equations 9.2 and 9.3 can be computed easily. For equality, the lemma allows to prune away (a, b) if Equation 9.2 is less than the probability threshold τ and for inequality, (a, b) can immediately be identified as an answer if Equation 9.3 is larger than τ. For **Greater than** and **Less than** join predicates, similar lower and upper bounds of the probability that the join predicate is fulfilled can be efficiently computed.

Because the pdfs of the uncertain values are assumed to be known, the above estimations concerning the join probability allow us to perform a constant-time check to decide whether the exact value $P(a\Theta_u b)$ has to be computed. The

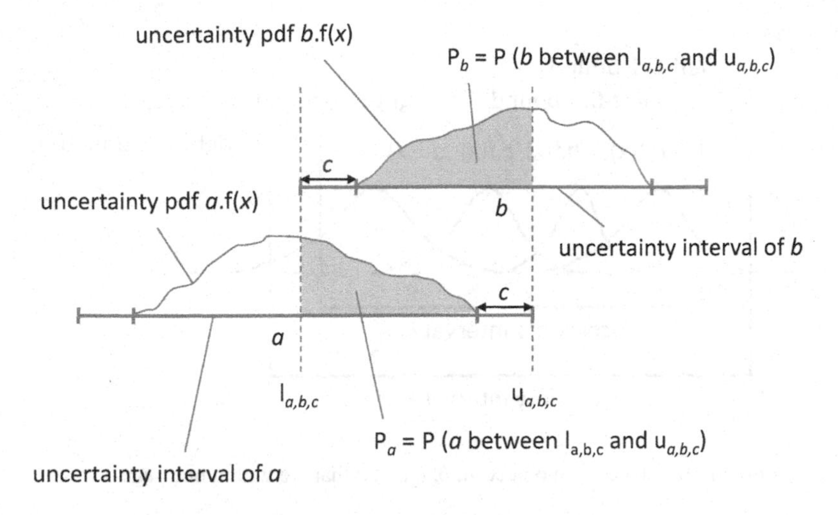

Figure 9.7. Upper bounding filter probability of the join predicate "$a =_c b$", i.e. $P(a =_c b) \leq \min\{P_a, P_b\}$.

problem of this approach is, that the uncertain intervals are used as the only pruning criterion. Thus, a lot of candidate pairs are generated that are not actually part of the answer (i.e. their probability is less than τ). The following pruning approach uses both uncertainty intervals and uncertainty pdfs for pruning, so that a smaller candidate set is produced.

Page-Level Pruning: The idea of this approach is to use a small overhead in order to facilitate the pruning of uncertain values as first proposed in [23]. The main idea is to augment some tighter bounds called *x-bounds* in each node in an interval R-tree. Each x-bound defines an uncertainty interval that is pre-calculated based on the properties of the uncertainty pdfs associated with the entries stored in that node. It describes the uncertainty interval for a page, such that every uncertain attribute stored in this page must have no more than a probability of x of being outside the interval. It is also assumed that x-bounds are "tight", i.e., the uncertainty interval associated with an x-bound is as small as possible. Figure 9.8 illustrates a page storing four uncertain attributes, a, b, c and d. As we can see, a has a probability less than 0.1 of lying to the left of the left-0.1-bound. Similarly, no object can have a probability above 0.3 of being outside of the uncertainty interval i_2, i.e. outside of the left 0.3-bound and right 0.3-bound. Finally, all the uncertainty intervals must be fully enclosed by the 0-bound, which is the same as the mbr of an index node.

Now given a page B with uncertain values stored in it and with respective x-bounds. B can be pruned from a range query q with uncertainty interval $[q.l, q.r]$ and probability threshold τ, if the following statements apply:

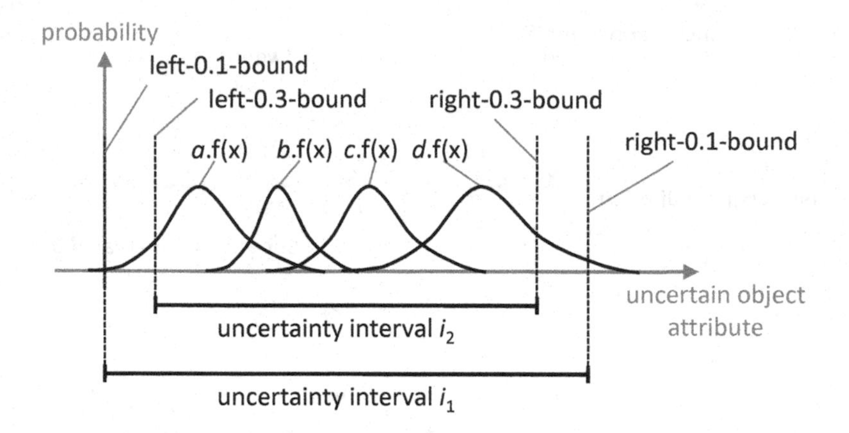

Figure 9.8. Uncertain objects a, b, c and d that are within one page $\mathcal{P}$.

- **CheckRight** There exists a right-x-bound $B.right(x)$ of B, such that $B.right(x) < q.l$ and $x < \tau$.
- **CheckLeft** There exists a left-x-bound $B.left(x)$ of B, such that $B.left(x) > q.r$ and $x < \tau$.

Two pages $B_{\mathcal{R}}$ and $B_{\mathcal{S}}$ of relations $\mathcal{R}$ and $\mathcal{S}$ are joined in two steps. The first step uses Checkleft and Checkright on page $B_{\mathcal{S}}$ using the 0-bound of page $B_{\mathcal{R}}$ extended with resolution c (cf. Section 3.1.0) to form a range query with the operator "$=_c$". In other words, the range query with the interval $[B_{\mathcal{R}}.l - c, B_{\mathcal{R}}.r + c]$ is checked against $B_{\mathcal{S}}$ using left- and right-x-bounds. If the first step does not result in pruning the pair $(B_{\mathcal{R}}, B_{\mathcal{S}})$, then another test is performed which exchanges the roles of $B_{\mathcal{R}}$ and $B_{\mathcal{S}}$. The range query is now constructed by using the 0-bound of $B_{\mathcal{S}}$, and tested against the uncertainty bounds in $B_{\mathcal{R}}$.

Index-Level Pruning: Although uncertainty tables can be used to improve the performance of page-based joins, they do not improve I/O performance, simply because the pages still have to be loaded in order to read the uncertainty tables. However, the idea of page-level pruning can be extended to improve I/O performance, by organizing the pages in a tree structure. An implementation of uncertainty relations in the index level is the Probability Threshold Index (PTI) [23], originally designed to answer probability threshold range queries. It is essentially an interval R-Tree, where each intermediate node is augmented with uncertainty tables. Specifically, for each child branch in a node, PTI stores both the mbr and the uncertainty table of each child. To perform the join, the 0-bound of each page from the outer relation is treated as a range query and tested against the PTI in the inner relation. All pages that are retrieved from

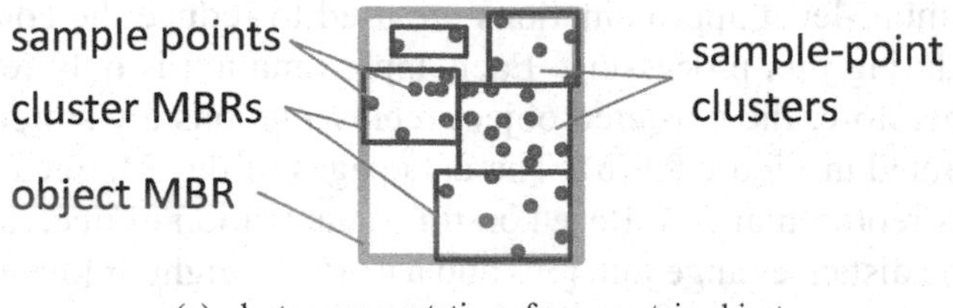

(a) cluster representation of an uncertain object

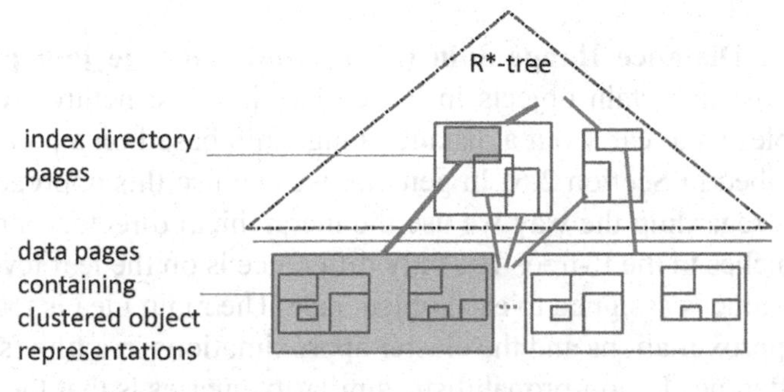

(b) Indexing uncertain objects

Figure 9.9. Representation and organization of uncertain objects (discrete uncertainty model)

the PTI are then individually compared with the page from where the range query is constructed, and the page-level pruning techniques (see the previous section) can then be used again to reduce computation efforts.

Similarity Join Queries (Discrete Uncertainty Model). Up to now, we assumed uncertain objects corresponding to the continuous probabilistic uncertainty model. In the following, we consider uncertain objects represented according to the discrete probabilistic uncertainty model (cf. Section 3.2).

The first probabilistic similarity join approach based on a discrete uncertain object representation was proposed by Kriegel et al. [36]. In particular, here, each uncertain object o_i is represented by a set of m points $\{o_{i,1}, .., o_{i,m}\}$ in a d-dimensional vector space $\mathcal{R}^d$. Thereby, each point $o_{i,j}$ represents an alternative position of o_i. Here, it is assumed that each sample point has the same probability that it matches with the position of the object. For efficiency reasons, they proposed to build k groups of sample points of each object by applying k-means clustering [44]. Each cluster is then approximated by a minimal bounding hyper-rectangle. Additionally, all clusters are again approximated by a minimal bounding hyper-rectangle. For each cluster c of an uncertain object o, it is assumed that the probability that o matches one of c's sample points is known. With the assumptions made above, this probability directly depends on the number of sample points contained in the cluster. This way, each uncertain object is approximated at different approximation levels as depicted in Figure

9.9(a). The multi-level approximations are used to reduce the computational complexity during join processing. Each approximation is only refined when required. Therefore, the clustered object representations are stored in an R*-tree as illustrated in Figure 9.9(b). The data pages of the R*-tree organize the object cluster representations. Based on this organization of uncertain objects, a probabilistic distance range join [36] and a nearest-neighbor join analogously to [37] can be efficiently supported.

Probabilistic Distance Range Join (PTJQ and ε-Range join predicate): Managing the uncertain objects in R-tree like index structures (cf. Figure 9.9(b)) enables us to carry out a distance-range join based on a parallel R-tree run as described in Section 2.6. In general, we can use this approach without any changes regarding the way we use the hierarchical directory structure for pruning branches in the R-tree. The only difference is on the leaf level where a probability value is assigned to each object pair. The main idea is to apply first the object approximations and the cluster approximations in a filter step. Here, the main difference to non-probabilistic similarity queries is that the filter does not directly relate to the similarity distances between objects but relates to the probability that the similarity distance between two objects matches the join predicate. In order to guarantee no false drops, in the filter step lower and upper bounds w.r.t. the join probability, i.e. the probability that a pair of objects fulfill the join predicate, have to be computed. If the minimal distance between two uncertain objects o_i and o_j matches the join predicate, i.e. is smaller or equal than a given ε-Range, then the probability $P_{filter}((o_i, o_j) \in \mathcal{R} \bowtie \mathcal{S})$ that both objects match the join predicate computed in the filter step is estimated to be 100%. This probability estimation can be refined by taking all pairs of object cluster representations into account. Assume that o_i is represented by n and o_j by m clusters. Let $c_{i,l}$ be a cluster of object $o_i \in \mathcal{R}$ with $1 \leq l \leq n$ and $c_{j,k}$ a cluster of object $o_j \in \mathcal{S}$ with $1 \leq k \leq m$. Furthermore, let $P(c_{i,l})$ and $P(c_{j,k})$ be the probabilities that the corresponding objects match one of the sample points contained in the clusters. Now, we get an upper bound estimation of the probability $P((o_i, o_j) \in \mathcal{R} \bowtie \mathcal{S})$ that both objects o_i and o_j match the join predicate if we sum up the products $P(c_{i,l}) \cdot P(c_{j,k})$ for all pairs of object clusters $(c_{i,l}, c_{j,k})$ such that the minimal distance between their representing approximations satisfies the join predicate.

An example is depicted in Figure 9.10. In this example, $P_{filter}(q, a) = \frac{2}{3} \geq P(q, a)$ and $P_{filter}(q, b) = \frac{1}{2} \geq P(q, b)$, where we assume that all object clusters contains the same number of object points. For a PTJQ with $\tau = 0.6$ the join candidate (q, b) cannot belong to the result set and, thus, can be pruned, while candidate (q, a) must be further refined.

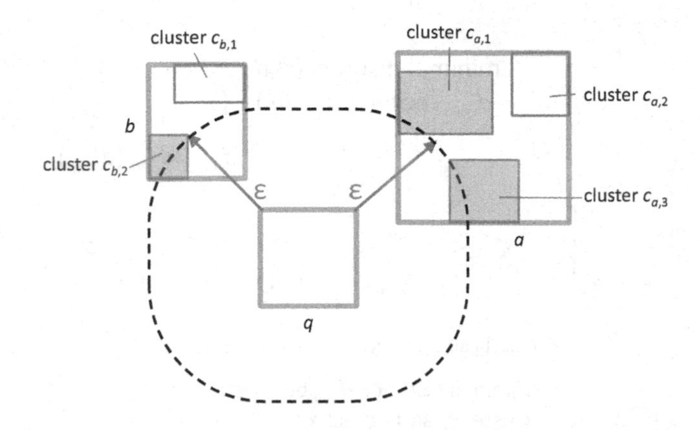

Figure 9.10. Example for a Probabilistic Distance Range Join Query.

Furthermore, in [36] it is shown how the join results with high probability can be reported very early. By means of a priority based refinement the results can be efficiently reported in the order of descending probabilities.

Probabilistic Nearest-Neighbor Join (PTJQ and 1-NN join predicate): Though, to the best of our knowledge, no approach for a probabilistic $(k-)$nearest-neighbor join exists, the basic solution is a nested-block-loop based join (cf. Section 2.2). However, the probabilistic (k-)nearest-neighbor query approaches, as proposed in [37], can be applied analogously to the index supported join approach as proposed for the probabilistic distance range join, because the representations of the uncertain objects and the index based organization is similar to that of the approach proposed in [36].

Similar to the probabilistic distance range join, the probabilistic nearest-neighbor join is based on the multi-step query paradigm. In a filter step, the object and cluster representations are used to compute upper and lower bounds for the join probability, i.e. the probability that two objects fulfill the join predicate. These bounds are then used to identify candidates which have to be refined, i.e. for which the uncertain distance and, thus, the probability that the candidate pair matches the query predicate, has to be refined. The filter strategy of a nearest-neighbor query is illustrated in Figure 9.11. In contrast to the probabilistic distance range join, the refinement w.r.t. an object depends on the location of the other objects. The distance between an object pair (q, o) does not have to be refined, if their maximal distance is lower than the minimal distance between q and all other objects, as illustrated in Figure 9.11(a). Similarly, this filter strategy can also be applied at the cluster approximation level as shown in Figure 9.11(b). Here, the refinement of the distance between the cluster pair (c_q, c_o) can be saved.

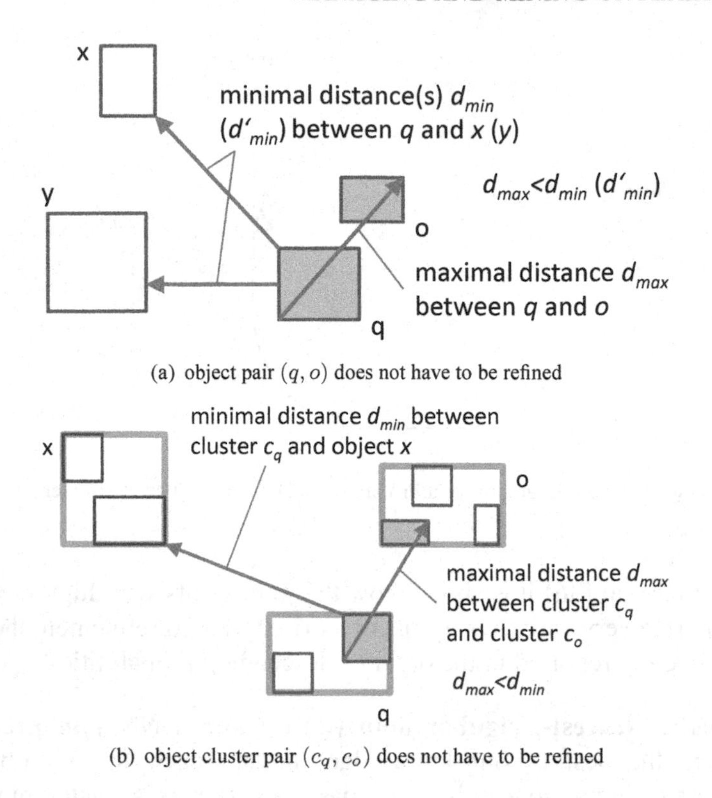

(a) object pair (q, o) does not have to be refined

(b) object cluster pair (c_q, c_o) does not have to be refined

Figure 9.11. Refinement criterions for uncertain object approximations

4.3 Probabilistic Spatial Join

In general, spatial joins are applied on spatial objects, i.e. objects having a certain position in space and a spatial extension. Spatial joins are based on spatial predicates that mainly refer to spatial topological predicates, e.g. intersect, volume of the overlap greater than ε, etc. and distance based predicates, e.g. distance range and nearest-neighbor predicates (cf. Section 3.3). For uncertain spatial objects with vague spatial attributes *Probabilistic Spatial Joins* (*PSpJ*) are required. A new look at spatial joins on probabilistic data is mandated by the large geographical and biomedical image datasets that are becoming available. For example, one might be interested in automatically identifying bodies of water and regions of dense population on satellite images. Note that in the literature often both probabilistic similarity joins and probabilistic spatial joins are abbreviated by the same term PSJ. To reduce confusion, here, this notation will remain for the probabilistic similarity join while the term *PSpJ* denotes probabilistic spatial join.

An uncertain spatial object consists of a set of points, each associated with a confidence value which reflects the likelihood that the point belongs to the corresponding object (cf. Section 3.2.0). An approach for the *PSpJ* based on this uncertainty model is proposed in [42]. Here, the uncertain spatial objects are composed of primitive volume elements with confidence values assigned to each of them. The score function is used to evaluate the join predicate for a pair of uncertain spatial objects. It is defined on a combination of the uncertain spatial extensions of the objects expressed by the confidences of the spatial primitives and the pairwise distances between the spatial primitives of both objects. In particular, the score function $s(a, b)$ defined on the object pair (a, b) is based on the distance between two volume elements that belong to objects a and b with high probability and it decays exponentially with increased distance of both volume elements.

Based on this score function, a *Probabilistic Threshold Join Query (PTJQ)* and a *Probabilistic Top-k Join Query (PTopkJQ)* were proposed (cf. Section 3.4). Here, the proposed join query methods are specific in such a way that the probabilistic query type is not based on the probability that a certain query predicate (e.g. the score between two objects is within a given ε-Range) is fulfilled. Rather, the score function itself returns a confidence value $s(a, b) \in [0, \lambda]$, where λ is a domain specific parameter. This confidence value reflects the likelihood that the spatial predicate is fulfilled without explicitly specifying the spatial predicate.

The principal idea of both proposed query algorithms is to transform d-dimensional object points (volume elements) into a $(d+1)$-dimensional space, where the $(d+1)^{th}$ dimension corresponds to a log-space of the object element confidences. This way, given an object point $a.p_i$ of object a, all object points $b.p_j$ of an object b that yield a score $s(a, b) > \tau$ above a given τ are within a triangle[‡] as illustrated in Figure 9.12. This geometric construct can be used to identify true hits and prune candidates that can be identified as true drops. The join is processed by applying a plane-sweep algorithm that works well for the one-dimensional case. Similar techniques applied for spatial join processing are proposed in [6, 47]. In the two-dimensional case, the algorithm can only be adapted to probabilistic spatial objects in the plane if the distance metric used for the score value is L_1. Like other plane-sweep algorithms, it is not likely to perform well on high-dimensional data.

5. Summary

In this chapter, a broad overview of diverse probabilistic join processing methods for uncertain objects is given. First, it formally comprises the most common uncertainty models used in the existing literature about probabilis-

[‡]For one-dimensional objects ($d = 1$) the corresponding geometry forms a triangle. In higher-dimensional spaces ($d > 1$) more complex geometries have to be considered, e.g. a pyramid for $d = 2$.

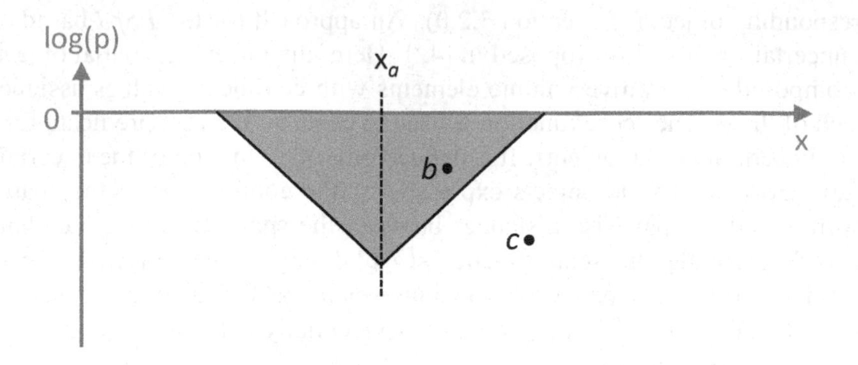

Figure 9.12. The score between two points $a = \langle x_a, p_a \rangle$ and $b = \langle x_b, p_b \rangle$ exceeds the threshold τ, as b is inside the triangle that is defined by a. Thus, the score between a and another point $c = \langle x_c, p_c \rangle$ does not exceed τ, as c is not inside the triangle.

tic query processing with a special focus on probabilistic joins. These models principally differ in the representation of the uncertain data. We distinguish between continuous uncertainty representation and discrete uncertainty representation. Additionally, this survey presents the spatial uncertainty model which is a special class of uncertainty models. It is defined for extended objects. Here, the location and extension of the objects are the uncertain attributes. Finally, the probabilistic evaluation of the join predicate is unified and called the score, which is defined for each join pair. Principally, the score of a join pair reflects the likelihood that a given join predicate is fulfilled. While the evaluation of the join predicate depends on the used object model or uncertainty model, the score can be applied independently to the probabilistic join evaluation. In this chapter, three probabilistic join query types are highlighted which are most commonly used in the literature: The probabilistic join query (PJQ), the probabilistic threshold join query (PTJQ) and the probabilistic Top-k join query (Top$k$JQ). Given two sets of objects, the PTJQ returns all pairs of objects from both sets which fulfill a given join predicate with a probability above a given threshold value. The first query type PJQ is a special form of the PTJQ, where the threshold value is set to zero. The PTop$k$JQ returns k pairs of objects with the highest probability that the pairs fulfill the join predicate.

The main part of this chapter exemplarily sketches existing probabilistic join approaches for uncertain data. These approaches are representatives of different probabilistic join categories. Since the core of the join approaches are the pruning strategies used to quickly reduce the candidate set in a filter step, this chapter concentrates on the different pruning criteria. The presented join approaches can be classified into confidence-based join, probabilistic similarity join and probabilistic spatial join. While the pruning criterion of the confidence-based join [4] uses only the confidences of the result tuples in order to prune the candidates, the other approaches additionally incorporate the join attributes, in particular the similarity and the spatial attributes. The probabilistic similarity join approaches are based on a multi-step query processing strategy. An index which organizes the uncertain data is exploited in order to evaluate the join at index level in a filter step. The approach which is based on continuous uncertainty representations [22] uses conservative approximations of the uncertain objects which are adjusted to specific uncertainty boundaries. Contrary, the approach [36], which is based on discrete uncertainty representations, decomposes the uncertain objects into multiple partitions which are approximated. These approximations are manageable in a better way and improve the filter selectivity in comparison to single object approximations. The probabilistic spatial join approach [42] reduces the probabilistic spatial join problem to a traditional spatial join problem by considering the spatial confidence attributes as additional spatial dimension. This way, plane-sweep techniques can be exploited to solve the probabilistic spatial join problem.

References

[1] D. J. Abel, V. Gaede, R. Power, and X. Zhou. Caching strategies for spatial joins. *GeoInformatica*, 3(1):33–59, 1999.

[2] S. Abiteboul, P. Kanellakis, and G. Grahne. On the representation and querying of sets of possible worlds. *SIGMOD Rec.*, 16(3):34–48, 1987.

[3] P. Agrawal, O. Benjelloun, A. D. Sarma, C. Hayworth, S. Nabar, T. Sugihara, and J. Widom. Trio: a system for data, uncertainty, and lineage. In *VLDB '06: Proceedings of the 32nd international conference on Very large data bases*, pages 1151–1154. VLDB Endowment, 2006.

[4] P. Agrawal and J. Widom. Confidence-aware joins in large uncertain databases. In *http://dbpubs.stanford.edu/pub/2007-14*, 2007.

[5] L. Antova, T. Jansen, C. Koch, and D. Olteanu. Fast and simple relational processing of uncertain data. In *Proceedings of the 24th International Conference on Data Engineering, ICDE 2008, April 7-12, 2008, Cancún, México*, pages 983–992, 2008.

[6] L. Arge, O. Procopiuc, S. Ramaswamy, T. Suel, and V. J. S. Scalable sweeping-based spatial join. In *In Proceedings of the 24rd International Conference on Very Large Data Bases (VLDB)*, page 570Ű581, 1998.

[7] N. Beckmann, H.-P. Kriegel, R. Schneider, and B. Seeger. The r*-tree: an efficient and robust access method for points and rectangles. *SIGMOD Rec.*, 19(2):322–331, 1990.

[8] O. Benjelloun, A. D. Sarma, A. Halevy, and J. Widom. Uldbs: databases with uncertainty and lineage. In *VLDB '06: Proceedings of the 32nd international conference on Very large data bases*, pages 953–964. VLDB Endowment, 2006.

[9] O. Benjelloun, A. D. Sarma, A. Y. Halevy, and J. Widom. Uldbs: Databases with uncertainty and lineage. In *Proceedings of the 32nd International Conference on Very Large Data Bases, Seoul, Korea, September 12-15, 2006*, pages 953–964, 2006.

[10] C. Böhm, B. Braunmüller, F. Krebs, and H.-P. Kriegel. Epsilon grid order: an algorithm for the similarity join on massive high-dimensional data. *SIGMOD Rec.*, 30(2):379–388, 2001.

[11] C. Böhm, F. Krebs, and H.-P. Kriegel. Optimal dimension order: A generic technique for the similarity join. In *DaWaK 2000: Proceedings of the 4th International Conference on Data Warehousing and Knowledge Discovery*, pages 135–149, London, UK, 2002. Springer-Verlag.

[12] C. Böhm and H.-P. Kriegel. A cost model and index architecture for the similarity join. In *ICDE '01: Proceedings of the 17th International Conference on Data Engineering*, page 411, Washington, DC, USA, 2001. IEEE Computer Society.

[13] C. Böhm, P. Kunath, A. Pryakhin, and M. Schubert. Querying objects modeled by arbitrary probability distributions. In *Advances in Spatial and Temporal Databases, 10th International Symposium, SSTD 2007, Boston, MA, USA, July 16-18, 2007, Proceedings*, pages 294–311, 2007.

[14] C. Böhm, A. Pryakhin, and M. Schubert. The gauss-tree: Efficient object identification in databases of probabilistic feature vectors. In *Proceedings of the 22nd International Conference on Data Engineering, ICDE 2006, 3-8 April 2006, Atlanta, GA, USA*, page 9, 2006.

[15] T. Brinkhoff, H.-P. Kriegel, and B. Seeger. Efficient processing of spatial joins using r-trees. *SIGMOD Rec.*, 22(2):237–246, 1993.

[16] T. Brinkhoff, H.-P. Kriegel, and B. Seeger. Parallel processing of spatial joins using r-trees. In *ICDE '96: Proceedings of the Twelfth International Conference on Data Engineering*, pages 258–265, Washington, DC, USA, 1996. IEEE Computer Society.

[17] J. Chen and R. Cheng. Efficient evaluation of imprecise location-dependent queries. In *Proceedings of the 23rd International Conference on Data Engineering, ICDE 2007, April 15-20, 2007, The Marmara Hotel, Istanbul, Turkey*, pages 586–595, 2007.

[18] R. Cheng, J. Chen, M. F. Mokbel, and C.-Y. Chow. Probabilistic verifiers: Evaluating constrained nearest-neighbor queries over uncertain data. In *Proceedings of the 24th International Conference on Data Engineering, ICDE 2008, April 7-12, 2008, Cancún, México*, pages 973–982, 2008.

[19] R. Cheng, D. V. Kalashnikov, and S. Prabhakar. Evaluating probabilistic queries over imprecise data. In *Proceedings of the 2003 ACM SIGMOD International Conference on Management of Data, San Diego, California, USA, June 9-12, 2003*, 2003.

[20] R. Cheng, D. V. Kalashnikov, and S. Prabhakar. Querying imprecise data in moving object environments. In *IEEE Transactions on Knowledge and Data Engineering*, 2004.

[21] R. Cheng, S. Singh, and S. Prabhakar. U-dbms: A database system for managing constantly-evolving data. In *Proceedings of the 31st International Conference on Very Large Data Bases, Trondheim, Norway, August 30 - September 2, 2005*, pages 1271–1274, 2005.

[22] R. Cheng, S. Singh, S. Prabhakar, R. Shah, J. S. Vitter, and Y. Xia. Efficient join processing over uncertain data. In *CIKM '06: Proceedings of the 15th ACM international conference on Information and knowledge management*, pages 738–747, New York, NY, USA, 2006. ACM.

[23] R. Cheng, Y. Xia, S. Prabhakar, R. Shah, and J. S. Vitter. Efficient indexing methods for probabilistic threshold queries over uncertain data. In *Proceedings of the Thirtieth International Conference on Very Large Data Bases, Toronto, Canada, August 31 - September 3 2004*, pages 876–887, 2004.

[24] X. Dai, M. L. Yiu, N. Mamoulis, Y. Tao, and M. Vaitis. Probabilistic spatial queries on existentially uncertain data. In *Advances in Spatial and Temporal Databases, 9th International Symposium, SSTD 2005, Angra dos Reis, Brazil, August 22-24, 2005, Proceedings*, pages 400–417, 2005.

[25] N. Dalvi and D. Suciu. Answering queries from statistics and probabilistic views. In *VLDB '05: Proceedings of the 31st international conference on Very large data bases*, pages 805–816. VLDB Endowment, 2005.

[26] X. Dong, A. Y. Halevy, and C. Yu. Data integration with uncertainty. In *VLDB '07: Proceedings of the 33rd international conference on Very large data bases*, pages 687–698. VLDB Endowment, 2007.

[27] A. Guttman. R-trees: a dynamic index structure for spatial searching. In *SIGMOD '84: Proceedings of the 1984 ACM SIGMOD international conference on Management of data*, pages 47–57, New York, NY, USA, 1984. ACM.

[28] M. Hua, J. Pei, W. Zhang, and X. Lin. Efficiently answering probabilistic threshold top-k queries on uncertain data. In *Proceedings of the 24th International Conference on Data Engineering, ICDE 2008, April 7-12, 2008, Cancún, México*, pages 1403–1405, 2008.

[29] Y.-W. Huang, N. Jing, and E. A. Rundensteiner. A cost model for estimating the performance of spatial joins using r-trees. In *SSDBM '97: Proceedings of the Ninth International Conference on Scientific and Statisti-*

cal Database Management, pages 30–38, Washington, DC, USA, 1997. IEEE Computer Society.

[30] Y.-W. Huang, N. Jing, and E. A. Rundensteiner. Spatial joins using r-trees: Breadth-first traversal with global optimizations. In *VLDB'97, Proceedings of 23rd International Conference on Very Large Data Bases, August 25-29, 1997, Athens, Greece*, pages 396–405, 1997.

[31] D. V. Kalashnikov and S. Prabhakar. Similarity join for low-and high-dimensional data. In *DASFAA '03: Proceedings of the Eighth International Conference on Database Systems for Advanced Applications*, page 7, Washington, DC, USA, 2003. IEEE Computer Society.

[32] N. Koudas and K. C. Sevcik. High dimensional similarity joins: Algorithms and performance evaluation. In *ICDE '98: Proceedings of the Fourteenth International Conference on Data Engineering*, pages 466–475, Washington, DC, USA, 1998. IEEE Computer Society.

[33] H.-P. Kriegel, P. Kunath, M. Pfeifle, and M. Renz. Spatial join for high-resolution objects. In *SSDBM '04: Proceedings of the 16th International Conference on Scientific and Statistical Database Management*, page 151, Washington, DC, USA, 2004. IEEE Computer Society.

[34] H.-P. Kriegel, P. Kunath, M. Pfeifle, and M. Renz. Distributed intersection join of complex interval sequences. In *Proc. 10th Int. Conf. on Database Systems for Advanced Applications (DASFAA'05), Bejing, China*, pages 748–760, 2005.

[35] H.-P. Kriegel, P. Kunath, M. Pfeifle, and M. Renz. Efficient join processing for complex rasterized objects. In *Proc. 7th Int. Conf. on Enterprise Information Systems (ICEIS'05), Miami, USA*, pages 20–30, 2005.

[36] H.-P. Kriegel, P. Kunath, M. Pfeifle, and M. Renz. Probabilistic similarity join on uncertain data. In *Database Systems for Advanced Applications, 11th International Conference, DASFAA 2006, Singapore, April 12-15, 2006, Proceedings*, pages 295–309, 2006.

[37] H.-P. Kriegel, P. Kunath, and M. Renz. Probabilistic nearest-neighbor query on uncertain objects. In *Advances in Databases: Concepts, Systems and Applications, 12th International Conference on Database Systems for Advanced Applications, DASFAA 2007, Bangkok, Thailand, April 9-12, 2007, Proceedings*, pages 337–348, 2007.

[38] H.-P. Kriegel and M. Pfeifle. Hierarchical density-based clustering of uncertain data. In *Proceedings of the 5th IEEE International Conference*

on Data Mining (ICDM 2005), 27-30 November 2005, Houston, Texas, USA, pages 689–692, 2005.

[39] X. Lian and L. C. 0002. Probabilistic ranked queries in uncertain databases. In *EDBT 2008, 11th International Conference on Extending Database Technology, Nantes, France, March 25-29, 2008, Proceedings*, pages 511–522, 2008.

[40] V. Ljosa and A. K. Singh. Probabilistic segmentation and analysis of horizontal cells. In *Proceedings of the 6th IEEE International Conference on Data Mining (ICDM 2006), 18-22 December 2006, Hong Kong, China*, pages 980–985, 2006.

[41] V. Ljosa and A. K. Singh. Apla: Indexing arbitrary probability distributions. In *In Proceedings of the 23rd International Conference on Data Engineering (ICDE)*, pages 946–955, 2007.

[42] V. Ljosa and A. K. Singh. Top-k spatial joins of probabilistic objects. In *Proceedings of the 24th International Conference on Data Engineering, ICDE 2008, April 7-12, 2008, Cancún, México*, pages 566–575, 2008.

[43] M.-L. Lo and C. V. Ravishankar. Spatial hash-joins. *SIGMOD Rec.*, 25(2):247–258, 1996.

[44] J. B. MacQueen. Some methods for classification and analysis of multivariate observations. In L. M. L. Cam and J. Neyman, editors, *Proc. of the fifth Berkeley Symposium on Mathematical Statistics and Probability*, volume 1, pages 281–297. University of California Press, 1967.

[45] J. A. Orenstein. Spatial query processing in an object-oriented database system. In *SIGMOD '86: Proceedings of the 1986 ACM SIGMOD international conference on Management of data*, pages 326–336, New York, NY, USA, 1986. ACM.

[46] A. Papadopoulos, P. Rigaux, and M. Scholl. A performance evaluation of spatial join processing strategies. In *SSD '99: Proceedings of the 6th International Symposium on Advances in Spatial Databases*, pages 286–307, London, UK, 1999. Springer-Verlag.

[47] J. M. Patel and D. J. DeWitt. Partition based spatial-merge join. In *Proceedings of the 1996 ACM SIGMOD International Conference on Management of Data, Montreal, Quebec, Canada, June 4-6, 1996*, pages 259–270, 1996.

[48] S. Ravada, S. Shekhar, C.-T. Lu, and S. Chawla. Optimizing join index based join processing: A graph partitioning approach. In *Symposium on Reliable Distributed Systems*, pages 302–308, 1998.

[49] C. Re, N. N. Dalvi, and D. Suciu. Efficient top-k query evaluation on probabilistic data. In *Proceedings of the 23rd International Conference on Data Engineering, ICDE 2007, April 15-20, 2007, The Marmara Hotel, Istanbul, Turkey*, pages 886–895, 2007.

[50] A. D. Sarma, O. Benjelloun, A. Halevy, and J. Widom. Working models for uncertain data. In *ICDE '06: Proceedings of the 22nd International Conference on Data Engineering*, page 7, Washington, DC, USA, 2006. IEEE Computer Society.

[51] P. Sen and A. Deshpande. Representing and querying correlated tuples in probabilistic databases. In *Proceedings of the 23rd International Conference on Data Engineering, ICDE 2007, April 15-20, 2007, The Marmara Hotel, Istanbul, Turkey*, pages 596–605, 2007.

[52] S. Singh, C. Mayfield, S. Prabhakar, R. Shah, and S. E. Hambrusch. Indexing uncertain categorical data. In *Proceedings of the 23rd International Conference on Data Engineering, ICDE 2007, April 15-20, 2007, The Marmara Hotel, Istanbul, Turkey*, pages 616–625, 2007.

[53] M. A. Soliman, I. F. Ilyas, and K. C.-C. Chang. Top-k query processing in uncertain databases. In *Proceedings of the 23rd International Conference on Data Engineering, ICDE 2007, April 15-20, 2007, The Marmara Hotel, Istanbul, Turkey*, pages 896–905, 2007.

[54] J. Widom. Trio: A system for integrated management of data, accuracy and lineage. In *CIDR'05: Proceedings of the 2nd International Conference Innovative Data Systems Research (CIDR)*, 2005.

[55] Z. Xu and H.-A. Jacobsen. Evaluating proximity relations under uncertainty. In *Proceedings of the 23rd International Conference on Data Engineering, ICDE 2007, April 15-20, 2007, The Marmara Hotel, Istanbul, Turkey*, pages 876–885, 2007.

[56] K. Yi, F. Li, G. Kollios, and D. Srivastava. Efficient processing of top-k queries in uncertain databases. In *Proceedings of the 24th International Conference on Data Engineering, ICDE 2008, April 7-12, 2008, Cancún, México*, pages 1406–1408, 2008.

[57] W. Zhao, R. Chellappa, P. Phillips, and A. Rosenfeld. Face recognition: A literature survey. In *ACM Computational Survey, 35(4)*, 2000.

Chapter 10

INDEXING UNCERTAIN DATA

Sunil Prabhakar
Department of Computer Science
Purdue University
sunil@cs.purdue.edu

Rahul Shah
Department of Computer Science
Louisiana State University
rahul@csc.lsu.edu

Sarvjeet Singh
Department of Computer Science
Purdue University
sarvjeet@cs.purdue.edu

Abstract As the volume of uncertain data increases, the cost of evaluating queries over this data will also increase. In order to scale uncertain databases to large data volumes, efficient query processing methods are needed. One of the key techniques for efficient query evaluation is indexing. Due to the nature of uncertain data and queries over this data, existing indexing solutions for precise data are often not directly portable to uncertain data. Even in situations where existing methods can be applied, it is often possible to build more effective indexes for uncertain data.

In this Chapter we discuss some of the recent ideas for indexing uncertain data in support of range, nearest-neighbor, and join queries. These indexes build on standard well-known indexes such as R-trees and/or signature trees. In some cases this involves augmenting the standard indexes with extra information. Sometimes more robust clustering criteria are required to make such indexes efficient.

Keywords: Indexing, PTI, x-bounds, Range queries, Nearest-neighbor queries

1. Introduction

As the size of a data collection grows, the cost of executing queries over the data also increases. In order to scale to large data collections, databases employ several techniques to curtail the computational cost (and effectively the time required to produce answers). One of the most effective, and ubiquitous, tools for reducing query execution cost in traditional databases is *indexing*. An index is a data structure that can significantly reduce the amount of data that needs to be processed when a query is executed. Virtually all existing database systems support a variety of index structures such as B-Trees, hash indexes, and R-Trees. Each index structure can be used for certain types of data (e.g., B-trees require that the data can be sorted), and query types (e.g., a hash index only supports exact equality match queries).

Similarly, when dealing with large collections of uncertain data, the cost of execution of queries is also a concern. Indexing can also be an effective mechanism for improving query processing cost, however, index structures for certain (precise) data cannot be directly used for uncertain data. To see why this is the case, consider for example, the case of a hash index which maps each data item to one of several buckets by applying a hash function to certain fields of the data item. In order to speed up an equality search using the hash index, we first hash the query fields using the same hash function to identify the bucket which must contain all the data items in our collection that match the query field. Hence, the query can limit its search to this bucket and ignore the rest of the buckets, thereby significantly reducing the number of data items to compare to the search key. With uncertain data, a given data item may have multiple possible values for some attributes. If we choose to build a hash index over this attribute how should we handle the multiple values – should a single data item hash to all the buckets corresponding to each possibility; or a single bucket? In either case, what happens at query time? What type of search are we expecting at query time (certainly not an exact match)?

This chapter discusses the following key question: How do we index uncertain data? Two related, and important preliminaries are: what is the nature of the uncertainty in the data; and what types of queries will be executed? Clearly, the answers to these preliminary questions impact what types of index structures are feasible. It should be noted that the availability of an index structure is only part of the solution to efficient query evaluation – an equally important component is a mechanism that will make use of the appropriate index structure for a given situation. This task is typically handled by the *Query Optimizer* in a database. This chapter does not discuss these issues. Furthermore, this chapter limits the discussion to probabilistic relational database models for handling uncertainty in data.

The nature of uncertainty in data is quite varied, and often depends on the application domain. Consider the following sample applications.

Example 1. Many data cleaning applications use automated methods to correct errors in data. Often, in such scenarios there is more than one reasonable alternative for the corrected value. In the standard relational model, one is forced to pick one among these alternative, which may lead to incorrectness. An uncertain model can allow multiple choices for an attribute value to be retained.

Example 2. Measured values in sensor data applications are notoriously imprecise. An example is an ongoing project at Purdue University that tracks the movement of nurses in order to study their behavior. Nurses carry RFID tags as they move around the hospital. Numerous readers located around the building report the presence of tags in their vicinity. The collected data is stored centrally in the form "Nurse2 in room6 at 10:10 am". Each nurse carries multiple tags. Difficulties arise due to the variability in the detection range of readers; multiple readers detecting the same tag; or a single tag being detected repeatedly between two readers (e.g., between room6 and the hallway – is the nurse in room6 all the time, just that the hallway sensor is detecting his tag or is she actually moving in and out?). Thus, the application may not be able to choose a single location for the nurse at all times with 100% certainty.

Example 3. Data collected from sensors (e.g., temperature sensors for weather, or GPS-based location data from cell phones), there is almost always some amount of inherent associated uncertainty. In addition, due to resource limitations such as battery power of sensor and network bandwidth, sensors only transmit data intermittently. Consequently, it is infeasible for a sensor database to contain exact value of each sensor at any given point in time. Thus the traditional model of a single value for a sensor reading is not a natural fit with this data. Instead, a more appropriate model is one where the sensor attribute can be represented as a probability distribution reflecting the inherent uncertainties and interpolation between measurements.

Overall, these kinds of emerging database applications require models which can handle uncertainty and semantics to define useful queries on such data.

In response to this need, several research projects have been undertaken recently (Orion [18], MayBMS [1], Mystiq [7], Trio [25], [21]). These projects represent a variety of data models. A major choice for each model is whether to incorporate probability values at the tuple or attribute level. This leads to two slightly different approaches in modelling and representing uncertain data. The two models are called *tuple uncertainty* model and *attribute uncertainty*

Car id	Problem	Probability
Car1	Brake	0.1
Car1	Tires	0.9
Car2	Trans	0.2
Car2	Suspension	0.8

Table 10.1. Example of a relation with x-tuples

Car id	Problem
Car1	{(Brake, 0.1), (Tires, 0.9)}
Car2	{(Trans, 0.2), (Suspension, 0.8)}

Table 10.2. Example of a relation with Attribute Uncertainty

model. In addition to capturing uncertainty in the data, the models must define the semantics of queries over the data. In this regard, virtually all models have adopted the standard *possible worlds semantics*. Some of these models are discussed below in Section 2. Some of the prominent index structures for uncertain data are discussed next. Section 3 presents index structures for data with continuous uncertainty. Section 4 discusses index structures for data with discrete uncertainty. Section 5 deals with indexes for supporting nearest-neighbor queries.

2. Data Models and Query Semantics

There are two main approaches for modeling uncertain relational data. One approach (Tuple uncertainty) is to attach a probability value with each tuple – the probability captures the likelihood of the given tuple being present in the given relation. The probability values for different tuples are assumed to be independent of each other, unless some dependency is explicitly given. These dependencies across tuples can be used to express mutually exclusive alternatives. Such tuples are called x-tuples.

Table 10.1 shows uncertainty information in a table expressed using tuple uncertainty. The tuples for Car id = Car1 are grouped together in a x-tuple, so they are mutually exclusive. Thus, Car1 has problems with either Brakes or Transmission with probability 0.1 and 0.9 respectively.

The second approach (Attribute uncertainty) allows for probability values at the attribute level. In this approach, a given tuple may have multiple alternatives for a given attribute. The alternatives may be a collection of discrete values with associate probabilities, or a continuous range(s) with a probability density function (pdf). The actual value of the attribute is a single value taken from this distribution. Multiple attributes of the same tuple are independent unless a joint distribution is provided. Joint distributions can be used to capture correlations. Table 10.2 shows the uncertain Table 10.1 expressed using Attribute Uncertainty.

Since, most database indexes are built on a single attribute, the discussion in this chapter assumes the attribute uncertainty model. It should be noted that both models are essentially the same, in that they use possible worlds semantics for probabilistic calculations and for verifying correctness of operations.

2.1 Uncertain Attribute types

Uncertain attributes can be broadly classified into the following types:

Discrete data :: The domain for the attribute is discrete (e.g., integers or city names). The domain may be ordered (e.g., integers) or not (e.g. colors). * An uncertain discrete value is represented as a set of values with probabilities. The values may or may not be contiguous or ordered. For example, the color of a car may be red with probability 0.7, or orange with a 0.3 probability: { (red, 0.7), (orange, 0.3) }.

Continuous data: : The domain for the attribute is the set of real numbers in a given range (e.g. length). An uncertain real-valued attribute is modelled as a pdf over a contiguous range of values. Apart from contiguity in values, this data is also ordered. A special case that is common, is that of multidimensional real-valued attributes (e.g., location). Such data can be represented as a pdf over a multidimensional region. For example, GPS devices are known to have a 2-dimensional Gaussian error in the reported longitude and latitude values, thus a GPS location is best modelled as a 2-dimensional pdf centered at the reported values.

3. Uncertainty Index for Continuous Domains

In this section, we consider index structures for continuous attributes in support of probabilistic threshold range queries (PTRQ). The dataset is a relation with a single uncertain attribute A. We are interested in the efficient evaluation of a range query given by the two end-points of the range: $[a, b]$ and threshold p. The query returns all tuples in the relation for which the probability of the tuple's value for attribute A falling in the range $[a, b]$ meets or exceeds the threshold p.

---

*Note that even though it may be possible to define an ordering over this domain (e.g., the alphabetic ordering of color names), the lack of order of interest is with respect to queries. If queries are not interested in retrieving colors based on their names, then the domain is effectively unordered for the purpose of indexing.

DEFINITION 10.1 *Given an uncertain relation R with uncertain attribute $R.u$, the probability threshold range query (PTRQ) with parameters $([a, b], p)$ returns all the tuples R_i such that the probability $Pr(a \leq R_i.u \leq b) > p$.*

Let $f_i(x)$ be the pdf that gives the uncertain value of attribute A for Tuple T_i of the relation. Let $[a_i, b_i]$ be the uncertainty interval for T_i. A straight-forward approach to evaluate PTRQ is to retrieve all tuples T_i which overlap with the query range, and for each compute its probability of falling in the query range [a,b]. This is given by

$$p_i = \int_a^b f_i(x)dx$$

where f_i is the PDF of A_i. Only those that exceed the threshold are returned as the answer to the query. Retrieving all the T_i's which overlap with the query interval can be efficiently (in fact optimally) done using interval B-trees [3].

While this solution is able to exploit existing index structures for improving query performance over uncertain data, it suffers from the following drawback: what if there are many intervals which overlap with the query interval but with probability lower than the threshold. In this case, the query unnecessarily retrieves these tuples (involves extra I/Os) and computes the corresponding probabilities (involves extra expensive integrations which may also be costly). Is it possible to prune away many, if not all, of these tuples that do not make it to the result without checking them individually? The *Probability Threshold Index* (PTI) provides one solution to this problem.

3.1 Probability Threshold Indexing

The above problems illustrate the inefficiency of using an interval index to answer a PTRQ. While the range search is being performed in the interval index, only uncertainty intervals are used for pruning out intervals which do not intersect $[a, b]$. Another piece of important uncertainty information, namely the uncertainty pdf, has not been utilized at all in this searching-and-pruning process.

The PTI is a modification of a one-dimensional R-tree, where probability information is augmented to its internal nodes to facilitate pruning.

To see how the PTI works, let us review how are range query is processed using a regular R-tree. Consider an R-tree that indexes the uncertain attribute A: each tuple is indexes using the range of its uncertainty interval $([a_i, b_i])$. The PTRQ is processed as follows. Starting with the root node, the query interval $[a, b]$ is compared with the minimum bounding rectangle (MBR) of each child in the node. If a child node's MBR does not overlap with the query range, the entire subtree rooted at that child is pruned from the search. In other words, any subtree with a zero probability of falling in the query range is pruned. The key idea of the PTI is to take the pruning one step further by pruning away

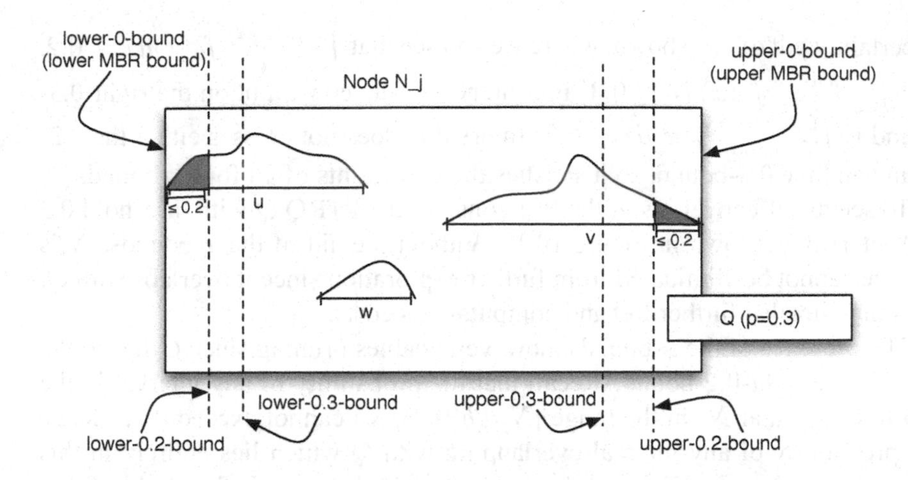

Figure 10.1. Inside an Node N_j, with a 0.2-bound and 0.3-bound. A PTRQ named Q is shown as an interval.

subtrees that may have a non-zero probability of overlap, but no larger than the threshold. This is achieved by generalizing the notion of the MBR and storing extra bounds (called lower-x-bounds and upper-x-bounds) along with the MBR entries in each node. the query range and the minimum bounding rectangle (MBR) of the root node of the subtree overlap. Let N_j denote the jth node of an R-tree, ordered by a pre-order traversal. Denote M_j to be the MBR of this node. The lower-x-bound and upper-x-bound of N_j is defined as follows.

DEFINITION 10.2 *A* **lower(upper)-x-bound** *of node N_j (denoted by $N_j.lb(x)$ ($N_j.ub(x)$)) is the largest value which guarantees that for any interval $[L_i, U_i]$ contained in the subtree rooted at N_j, the probability of being less that $N_j.lb(x(+)$ is at most x. This is to say, that the following must hold: $\int_{L_i}^{N_j.lb(x)} f_i(y)dy \leq x$ and $\int_{N_j.ub(x)}^{U_i} f_i(y)dy \leq x$.*

Using the definition of a x-bounds, the MBR of an internal node can be viewed as a pair of lower-0-bound and upper-0-bound, since all intervals in the node must be fully above the lower-0-bound (lower end of the MBR), and below the upper-0-bound (upper end of the MBR). Figure 10.1 that illustrates three children MBRs (u,v,w), in the form of one-dimensional intervals, contained in larger Node N_j. The domain increases from left to right. The 0.2-bounds and 0.3-bounds for N_j are also shown.

As Figure 10.1 shows, a lower(upper)-x-bound is a value below (above) which at most x% of any interval in the node can lie. For illustration, the uncertainty pdf of u is shown, where we can see that $\int_{L_u}^{N_j.lb(0.2)} f_u(x)dx \leq 0.2$, and $\int_{U_u}^{N_j.ub(0.3)} f_u(x)dx \leq 0.3$. For interval v, the constraint on the right-0.3-bound is $\int_{N_j.rb(0.3)}^{U_v} f_v(x)dx \leq 0.3$. Interval w does not crosses either the 0.2-bound and the 0.3-bound, so it satisfies the constraints of all four x-bounds.

To see the effectiveness of the PTI consider the PTRQ Q with threshold 0.3 and interval as shown in Figure 10.1. Without the aid of the x-bounds, N_j's subtree cannot be eliminated from further exploration since it overlaps with Q. This may involve further I/O and computations costs.

The presence of the x-bounds, however, enables pruning since Q lies to the right of the right-0.2-bound. Recall that the probability of any interval in the sub-tree rooted at N_j in the range $[N_j.ub(0.2), \infty]$ cannot exceed 0.2. Hence the probability of any interval overlapping with Q which lies entirely in this interval cannot exceed 0.2 and thus cannot meet the query's threshold of 0.3. Thus the node can be safely pruned away for this query. Compared with the case where no x-bounds are stored, this represents savings in terms of number of I/Os and computation time.

In general, node N_j can be pruned away for query Q with interval $[a, b]$ and threshold p, if the following two conditions hold for any of the lower or upper x-bounds of the node:

1 $[a, b]$ lies below (above) the lower(upper)-x-bound of N_j i.e., either $b < N_j.lb(x)$ or $a > N_j.ub(x)$ is true, and

2 $p \geq x$

If none of the x-bounds in N_j satisfies these two conditions, the node cannot be pruned and must be examined as with a regular R-tree.

Figure 10.2 illustrates an implementation of a PTI. Its framework is the same as R-tree, where each internal node stores the MBRs of its children and their corresponding pointers. In addition, the PTI stores a table $N_j.PT$ for each child N_j storing the lower and upper bounds for various values of x. Each entry of $M_j.PT$ is a tuple of the form <left-x-bound, right-x-bound>. To avoid repeated storage, a global table called T_G records the values of x for x-bounds. The i-th entry of $M_j.PT$ contains the x-bounds for the value of x stored in the i-th entry of T_G. The data items being indexed are uncertainty intervals and pdfs.

The PTI is a simple idea and is easy to implement. Although the fan-out of a PTI node is lower than an R-tree node because each entry is larger, the fan-out only logarithmically affects the height of the tree. Hence, in most cases this results in increase in height by an additive constant, which only has a minor

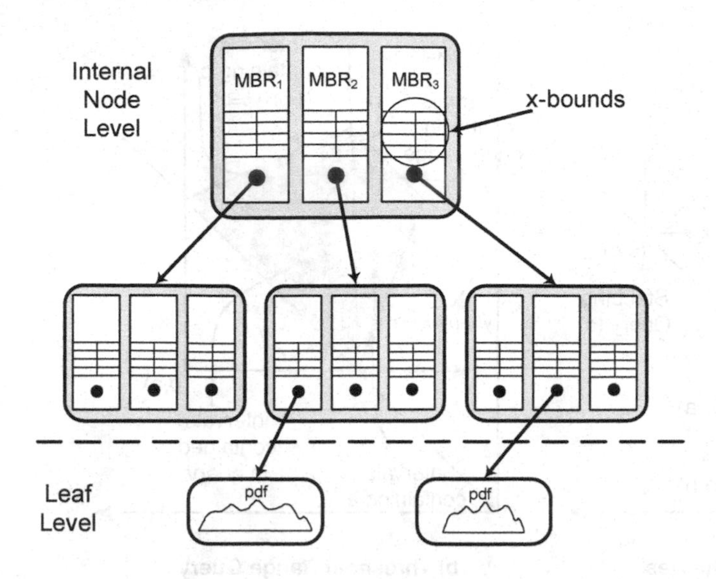

Figure 10.2. Structure of PTI

effect on PTI's performance. Indeed, its performance illustrates significant improvements over the R-tree. Although the discussion above was limited to a single dimension, the idea of the PTI can be easily extended to multiple dimensions.

3.2 Special Case: Uniform PDFS

For the special case where the pdf for each interval is a uniform distribution, it is possible to build an indexing using a mapping of intervals to the dual space. This version of the PTRQ query is called the PTQU problem. As it turns out, PTRQ is a hard problem to be provably solved even with uniform pdfs. However, good heuristics can be used for PTQU and they can be extended to PTQs when pdfs are arbitrary using the idea of PTI from the previous section. Furthermore, a provably good index for PTQU can exist if the threshold of probability p is a fixed constant.

3.3 2D mapping of intervals

Let us first explore the impact of the mapping on the data and query (PTQU) [3]. The mapping converts each interval $[x, y]$ to the point (x, y) in 2D space. Note that, for all intervals, $x < y$ and hence these points all lie in the region above (and to the left of) the line $x = y$. Figure 10.3(a) gives the illustration. A stabbing query is a particular kind of query associated with the notion of intervals. Given a point c, a stabbing query reports all the intervals containing point c. A stabbing query [3] for point c is converted to a two-sided

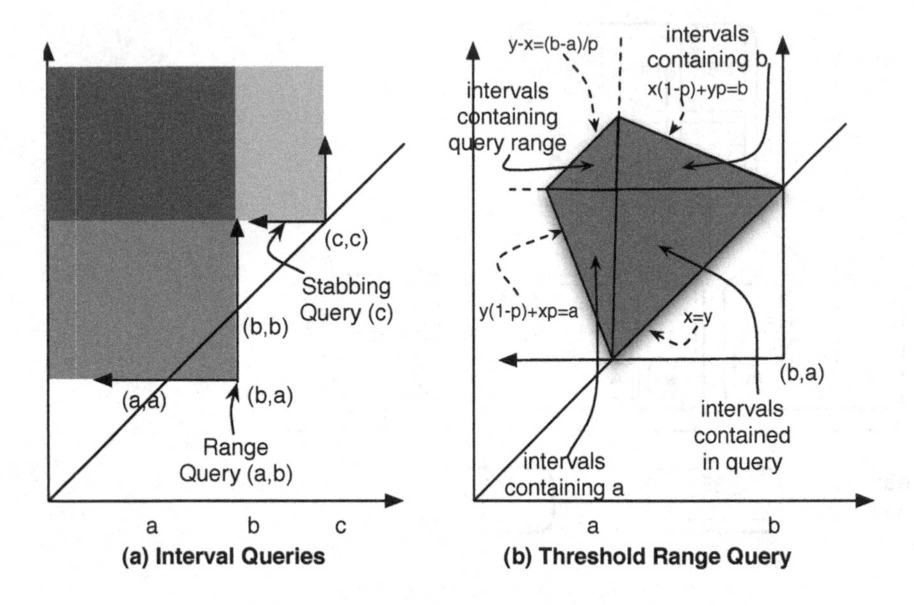

Figure 10.3. Probabilistic Threshold Queries with Uniform pdf

orthogonal query originating at point (c, c) in this mapping. A range query (a, b) is just a union of the stabbing queries for all the points from a to b. This is the same as a two-sided orthogonal query originating at point (b, a).

A PTQU (a, b, p) where $0 < p \leq 1$ now becomes a 3-sided trapezoidal query as shown in Figure 10.3(b). To see this, consider any point (x, y) (i.e. interval $[x, y]$) which satisfies PTQU (a, b, p). There are four main cases:

$x \leq a < b \leq y$:: The query lies within the interval. We simply require that the query covers a sufficient length of the interval, i.e., $b - a \geq p(y - x)$. Therefore point (x, y) is in the region below the line $y - x = (b - a)/p$. This line has slope 1.

$x \leq a < y \leq b$:: The query region is on the right of the interval. The amount of overlap is given by $y - a$. This condition translates to $y(1-p)+xp \geq a$. This corresponds to the region above the line $y(1 - p) + xp = a$ which has slope $-p/(1 - p)$.

$a \leq x < b \leq y$:: The query region is to the left of the interval. This is given by the region $x(1-p)+yp \leq b$. The separating line has slope $-(1-p)/p$.

$a < x < y < b$:: The entire interval lies within the query and hence it satisfies the PTQU for any p.

Thus, the region satisfying the query is given by the intersection of the three regions (first three) above, which corresponds to an isosceles trapezoid region.

The fourth side of the region given by line $x = y$ can be essentially considered redundant since there are no points below (or to the right) of this line. This as an open side of the trapezoid. Thus, under the 2D mapping, PTQU becomes a 3-sided trapezoidal query.

A simple index based on 2-D mapping. The above properties can be used to construct a simple R-tree based index for answering PTQU queries. The intervals are mapped to 2D points which are indexed using an R-tree. PTQU queries are converted to the corresponding trapezoids, which are evaluated as range queries against each page of R-tree (at any level). An entire subtree of the R-tree can be pruned if its bounding box has no overlap with the trapezoidal region for the query.

Even under this simplification of uniform pdfs, this problem is hard to solve theoretically. The problem is related to half-space queries, simplex queries or wedge queries in 2-D geometry. No efficient index exists and in fact some lower bounds on these problems suggest that the PTQU problem might also be hard to solve.

In any case, [10] does show that theoretical optimality is possible if the threshold of the query is fixed apriori before building the index. This can be obtained by novel use of 3-sided query index [2].

3.4 Join Queries

In this section, we briefly discuss how the concept of PTI indexing can be used to do answer join queries over uncertain data. First we present the definitions and semantics of join queries over uncertain data. Then we show how some of the standard join techniques can be combined with PTI indexing for efficient query processing.

Let us focus on one dimensional uncertain data. Later, we shall briefly discuss how this can be extended to multiple dimensions. For attribute a, let $a.f(x)$ denote its pdf function and $a.F(x)$ denote the corresponding cdf (cumulative density function). Let $a.l$ and $a.u$ denote the lower boundary and upper boundary of the uncertainty interval.

Given this, we can define:

DEFINITION 10.3 *The probability that uncertain attribute a is greater than b is given by*

$$\begin{aligned}
P(a > b) &= \int_{\max(a.l,b.l)}^{b.u} a.f(x)b.F(x)dx + 1 - a.F(b.u) \quad a.l \leq b.u < a.u \\
&= \int_{\max(a.l,b.l)}^{a.u} a.f(x)b.F(x)dx \quad b.l \leq a.u \leq b.u
\end{aligned}$$

DEFINITION 10.4 *Given a resolution parameter c, the probability a is equal to b (within distance c) is given by*

$$P(a =_c b) = \int_{-\infty}^{\infty} a.f(x).(b.F(x + c) - b.F(x - c))dx$$

.

The limits in the above definition can be made more restrictive than $-\infty$ to ∞ by observing the positions of their uncertainty intervals. Note that the resolution parameter c is needed to obtain a non-zero probability answer in the continuous case.

These definitions can also be used to compare uncertain attribute a to single certain value $v \in \mathcal{R}$. For example, $P(a =_c v) = a.F(v + c) - a.F(v - c)$ and $P(a > v) = 1 - a.F(v)$.

We now define the extension of probabilistic threshold queries to join queries. Let R and S be two relations having an uncertain attribute a.

DEFINITION 10.5 *Given an uncertainty comparator θ_u (for example $=_c, >, <, \neq_c$) a* **Probabilistic Threshold Join Query (PTRQ)** *returns all tuples (R_i, S_j) such that $P(R_i.a\theta_u S_j.a) > p$, where $p \in [0, 1]$ is the probability threshold.*

A simple way to evaluate such a join query is to take an interval join based approach [13]. We take each pair of uncertainty intervals and decide whether to evaluate the integral based on whether the uncertainty intervals of these two attributes overlap or not. Now, instead of evaluating a costly integral operation for each pair, we may be able to prune-in or prune-out a given pair based on the probability threshold p.

We take an example of Greater than queries. In this case, we can say that

1 If $a.l \leq b.r < a.r, P(a > b) \geq 1 - a.F(b.r)$.

2 If $a.l \leq b.l \leq a.r, P(a > b) \leq 1 - a.F(b.l)$.

It is much easier to check the above inequalities rather than evaluating the entire integral. Based on these bounds on probability we prune-out or prune-in certain pairs of attributes. These kinds of rules can be developed for other operators also. This kind of pruning is called item-level pruning.

The concept of item-level pruning can be taken further to page level pruning and also to index level pruning by using x-bounds as in PTI indexing. Say, one of the relations S in join is indexed using PTI. Now for each tuple $R.a$ from R, we can quickly prune-away the items in S by searching and traversing through the index on S. When we are at a certain node in the PTI based R-tree, we can in many cases conclude based on the x-bounds that no item in the subtree of that node will satisfy the join query with $R.a$

In many cases, it is worthwhile to index both the relations using PTI. The item-level pruning rule can be used to derive page-level pruning rules. Sometimes just by comparing the x-bounds of page r with page s we can conclude that no tuple indexed by page r will join with any tuple in page s. Details of these pruning rules can be found in [9].

Given item-level and page-level (and also index-level) pruning rules, these techniques can be applied to derive uncertain versions of standard Indexed Nested Loop Join (INLJ) and Block Nested Loop Join (BLNJ) algorithms. It is shown in [9] that different pruning strategies at various levels gives considerable performance gains in terms I/Os and number of integration operations.

3.5 Multi-dimensional Indexing

So far we have mainly focussed on continuous uncertain data in one dimension. The techniques there can be extended to handle multidimensional uncertain data also. One of the fundamental difference comes from the absence of well-defined cumulative density function (cdf) in this case. Here, the range queries involve a range defined by a hyper-rectangle. All objects that have more than a threshold probability mass within this hyper rectangle are to be reported.

The approach of [24] involves constructing CDF functions along each dimension. An R^*-tree like index is used, augmented with (left and right) x-bounds along each dimension similar to the PTI. These x-bounds define smaller hyper-rectangles compared to bounding boxes. Whenever the query region does not overlap with this smaller hyper-rectangle, the object can be pruned. Similar to the one dimensional case this pruning can be taken at page level and various levels in the index.

4. Uncertainty Index for discrete domains

Uncertainty in categorical data is commonplace in many applications, including data cleaning, database integration, and biological annotation. In such domains, the correct value of an attribute is often unknown, but may be selected from a reasonable number of alternatives. Current database management systems do not provide a convenient means for representing or manipulating this type of uncertainty. Two indexing structures for efficiently searching uncertain categorical data were proposed in [22] – one based on the R-tree and another based on an inverted index structure. Before we go into details of the indexing structures, we will discuss the data model and probabilistic queries supported by them.

Make	Location	Date	Text	Problem
Explorer	WA	2/3/06	...	{(Brake, 0.5), (Tires, 0.5)}
Camry	CA	3/5/05	...	{(Trans, 0.2, (Suspension, 0.8)}
Civic	TX	10/2/06	...	{(Exhaust, 0.4), (Brake, 0.6)}
Caravan	IN	7/2/06	...	{(Trans, 1.0)}

Table 10.3. Example of Uncertain Relation with an Uncertain Discrete Attribute

4.1 Data Model and Problem Definition

Under the categorical uncertainty model [5], a relation can have attributes that are allowed to take on uncertain values. For the sake of simplicity, we limit the discussion to relations with a single uncertain attribute, although the model makes no such restriction. The uncertain attributes are drawn from categorical domains. We shall call such an attribute an *uncertain discrete attribute* (UDA). Let $R.a$ be a particular attribute in relation R which is uncertain. $R.a$ takes values from the categorical domain D. Let $D = \{d_1, d_2, ..., d_N\}$, then $t.a$ is given by the probability distribution $Pr(t.a = d_i)$. Thus, $t.a$ can be represented by a probability vector $t.a = \langle p_1, p_2, ..., p_N \rangle$.

Table 10.3 is for a CRM application with UDA attribute `Problem`. The `Problem` field is derived from the `Text` field in the given tuple using a text classifier. A typical query on this data would be to report all the tuples which are highly likely to have a brake problem (i.e., `Problem = Brake`). Formally we define UDA as follows.

DEFINITION 10.6 *Given a discrete categorical domain* $D = \{d_1, .., d_N\}$, *an uncertain discrete attribute (UDA)* u *is a probability distribution over* D. *It can be represented by the probability vector* $u.P = \langle p_1, ..., p_N \rangle$ *such that* $Pr(u = d_i) = u.p_i$.

Given an element $d_i \in D$, the equality of $u = d_i$ is a probabilistic event. The probability of this equality is given by $Pr(u = d_i) = p_i$. The definition can be extended to equality between two UDAs u and v under the independence assumption as follows:

DEFINITION 10.7 *Given two UDAs* u *and* v, *the probability that they are equal is given by* $Pr(u = v) = \sum_{i=1}^{N} u.p_i \times v.p_i$.

This definition of equality is a natural extension of the usual equality operator for certain data. Analogous to the notion of equality of value is that of distributional similarity. Distribution similarity is the inverse of distributional divergence, which can be seen as a distance between two probability distributions. We consider the following distance functions between two distributions:

L_1:: $L_1(u,v) = \sum_{i=1}^{N} |u.p_i - v.p_i|$. This is the Manhattan distance between two distributions.

L_2:: $L_2(u,v) = \sqrt{\sum_{i=1}^{N}(u.p_i - v.p_i)^2}$. This is the Euclidean distance between two distributions.

$KL(u,v)$:: $KL(u,v) = \sum_{i=1}^{N} u.p_i \log(u.p_i/v.p_i)$. This is Kullback-Leibler (KL) divergence based on cross entropy measure. This measure comes from information theory. Unlike the above two, this is not a metric. Hence it is not directly usable for pruning search paths but can be used for clustering in an index [19].

We next define the queries over the UDAs:

DEFINITION 10.8 *Probabilistic equality query (PEQ): Given a UDA q, and a relation R with a UDA a, the query returns all tuples t from R, along with probability values, such that the probability value* $Pr(q = t.a) \geq 0$.

Often with PEQ there are many tuples qualifying with very low probabilities. In practice, only those tuples which qualify with sufficiently high probability are likely to be of interest. Hence the following queries are more meaningful: (1) equality queries which use probabilistic thresholds [5], and (2) equality queries which select k tuples with the highest probability values.

DEFINITION 10.9 *Probabilistic equality threshold query (PETQ): Given a UDA q, a relation R with UDA a, and a threshold* τ, $\tau \geq 0$. *The answer to the query is all tuples t from R such that* $Pr(q = t.a) \geq \tau$.

Analogous to PETQ, we define the top-k query PEQ-top-k, which returns the k tuples with the highest equality probability to the query UDA. Here a number k is specified (instead of the threshold τ and the answer to the query consists of k tuples from PEQ whose equality probability is the highest.

Having defined the data model and the queries over this data, we next discuss the two indexing structures we proposed in [22].

4.2 Probabilistic Inverted Index

Inverted indexes are popular structures in information retrieval [4]. The basic technique is to maintain a list of lists, where each element in the outer list corresponds to a domain element (i.e. the words). Each inner list stores the ids of documents in which the given word occurs, and for each document, the frequencies at which the word occurs.

Traditional applications assume these inner lists are sorted by document id. We introduce a probabilistic version of this structure, in which we store for

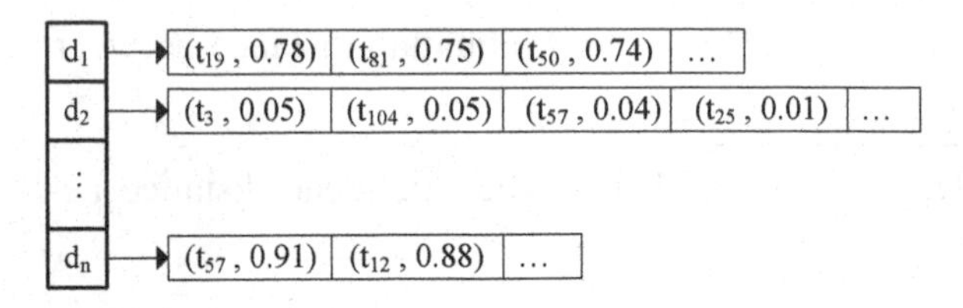

Figure 10.4. Probabilistic Inverted Index

each value in a categorical domain D a list of tuple-ids potentially belonging to D. Along with each tuple-id, we store the probability value that the tuple may belong to a given category. In contrast to the traditional structure, these inner lists are are sorted by descending probabilities. Depending on the type of data, the inner lists can be long. In practice, these lists (both inner or outer) are organized as dynamic structures such as B-trees, allowing efficient searches, insertions, and deletions.

Figure 10.4 shows an example of a probabilistic inverted index. At the base of the structure is a list of categories storing pointers to lists, corresponding to each item in D that occurs in the dataset. This is an inverted array storing, for each value in D, a pointer to a list of pairs. In the list $d_i.list$ corresponding to $d_i \in D$, the pairs (tid, p) store tuple-ids along with probabilities, indicating that tuple tid contains item d_i with probability p. That is, $d_i.list = \{(tid, p)|Pr(tid = d_i) = p > 0\}$. Again, we sort these lists in order of descending probabilities.

To insert/delete a tuple (UDA) tid in the index, we add/remove the tuple's information in tuple-list. To insert it in the inverted list, we dissect the tuple into the list of pairs. For each pair (d, p), we access the list of d and insert pair (tid, p) in the B-tree of this list. To delete, we search for tid in the list of d and delete it.

Next we describe search algorithms to answer the PETQ query given a UDA q and threshold τ. Let $q = \langle(d_{i_1}, p_{i_1}), (d_{i_2}, p_{i_2}), ..., (d_{i_l}, p_{i_l})\rangle$ such that $p_{i_1} \geq p_{i_2} \geq ... \geq p_{i_l}$. We first describe the brute force inverted index search which does not use probabilistic information to prune the search. Next we shall describe three heuristics by which the search can be concluded early. The three methods differ mainly in their stopping criteria and searching directions. Depending on the nature of queries and data, one may be preferable over others.

Inv-index-search:. This follows the brute-force inverted index based lookup. For all pairs (d_{i_j}, p_{i_j}) in q, we retrieve all the tuples in the list corresponding to each d. Now, from these candidate tuples we match with q to find out which of these qualify more than the threshold. This is a very simple method, and in many cases when these lists are not too big and the query

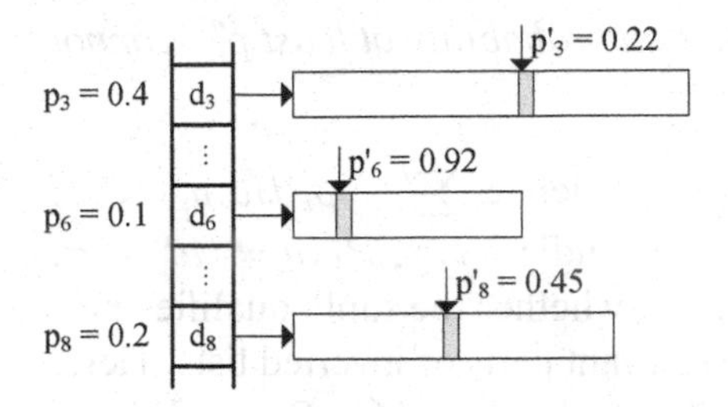

Figure 10.5. Highest-prob-first Search for $q = \langle (d_3, 0.4), (d_8, 0.2), (d_6, 0.1) \rangle$.

involves fewer d_{i_j}, this could be as good as any other method. However, the drawback of this method is that it reads the entire list for every query.

Highest-prob-first:. Here, we simultaneously search the lists for each d_{i_j}, maintaining in each $d_{i_j}.list$ a current pointer of the next item to process (see Figure 10.5). Let p'_{i_j} be the probability value of the pair pointed by the current pointer in this list. At each step, we consider the most promising tuple-id. That is, among all the tuples pointed by current pointers, move forward in that list of d_j where the next pair (tid, p'_{i_j}) maximizes the value $p'_{i_j} p_{i_j}$. The process stops when there are no more promising tuples. This happens when the sum of all current pointer probabilities scaled by their probability in query q falls below the threshold, i.e. when $\sum_{j=1}^{l} p'_{i_j} p_{i_j} < \tau$. This works very well for top-k queries when k is small.

Row Pruning:. In this approach, we employ the naive inverted index search but only consider lists of those items in D whose probability in query q is higher than threshold τ. It is easy to check that a tuple, all of whose items have probability less than τ in q, can never meet the threshold criteria. For processing top-k using this approach, we can start examining candidate tuples as we get them and update the threshold dynamically.

Column Pruning:. This approach is orthogonal to the row pruning. We retrieve all the lists which occur in the query. Each of these lists is pruned by probability τ. Thus, we ignore the part of the lists which have probability less than the threshold τ. This approach is more conducive to top-k queries.

The correctness of our stopping criteria is established by the following lemma. This applies to all three of the above cases.

LEMMA 10.10 *Let the query* $q = \{(d_{i_j}, p_{i_j}) | 1 \leq j \leq l\}$ *and threshold* τ. *Let* p'_{i_j} *be probability values such that* $\sum_{j=1}^{l} p_{i_j} p'_{i_j} < \tau$. *Then, any tuple tid which does not occur in any of the* $d_{i_j}.list$ *with probability at least* p'_{i_j}, *cannot satisfy the threshold query* (q, τ).

Proof: For any such tuple tid, $tid.p_{i_j} \leq p'_{i_j}$. Hence, $\sum_{j=1}^{l} p_{i_j} tid.p_{i_j} < \tau$. Since q only has positive probability values for indices i_j's, $Pr(q = tid) < \tau$.

In many cases, the random access to check whether the tuple qualifies performs poorly as against simply joining the relevant parts of inverted lists. Here, we use rank-join algorithms with early-out stopping [14, 16]. For each tuple so far encountered in our search, we maintain its lack parameter – the amount of probability value required for the tuple, and which lists it could come from. As soon as the probability values of required lists drop below a certain boundary such that a tuple can never qualify, we discard the tuple. If at any point the tuple's current probability value exceeds the threshold, we include it in the result set. The other tuples remain in the candidate set. A list can be discarded when no tuples in the candidate set reference it. Finally, once the size of this candidate set falls below some number (predetermined or determined by ratio to already selected result) we perform random accesses for these tuples.

4.3 Probabilistic Distribution R-tree

In this subsection, we describe an alternative indexing method based on the R-tree [15]. In this index, each UDA u is stored in a page with other similar UDAs which are organized as a tree.

We now describe our structure and operations by analogy to the R-tree. We design new definitions and methods for *Minimum Bounding Rectangles* (MBR), the area of an MBR, the MBR boundary, splitting criteria and insertion criteria. The concept of distributional clustering is central to this index. At the leaf level, each page contains several UDAs (as many as fit in one block) using the aforementioned *pairs* representation. Each list of pairs also stores the number of pairs in the list. The page stores the number of UDAs contained in it. Figure 10.6 shows an example of a PDR-tree index.

Each page can be described by its MBR boundaries. The MBR boundary for a page is a vector $v = \langle v_1, v_2, ..., v_N \rangle$ in R^N such that v_i is the maximum probability of item d_i in any of the UDA indexed in the subtree of the current page.

We maintain the essential pruning property of R-trees; if the MBR boundary does not qualify for the query, then we can be sure that none of the UDAs in the subtree of that page will qualify for the query. In this case, for good performance it is essential that we only insert a UDA in a given MBR if it is sufficiently tight with respect to its boundaries. There are several measures for the "area" of an MBR, the simplest one being the L_1 measure of the boundaries,

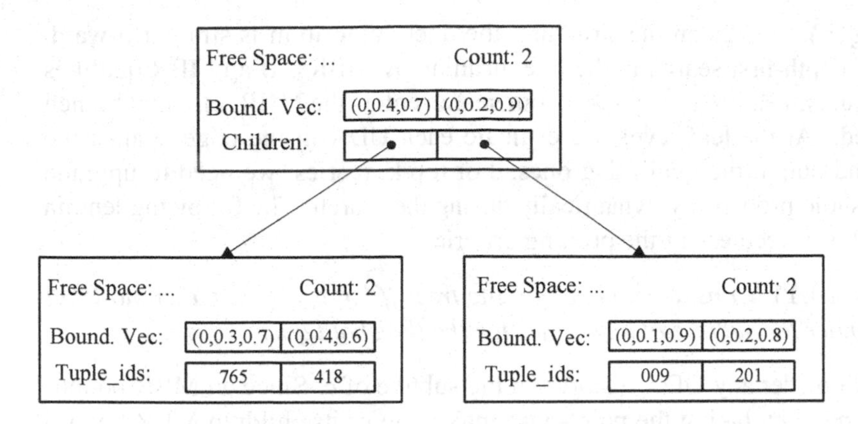

Figure 10.6. Probabilistic Distribution R-tree

which is $\sum_{i=1}^{N} v_i$. Our methods are designed to minimize the area of any MBR. Next, we describe how insert, split and PETQ are performed.

Insert(u). To insert a UDA into a page, we first update its MBR information according to u. Next, from the children of the current page we pick the best page to accommodate this new UDA. The following criteria (or their combination) are used to pick the best page: (1) Minimum area increase: we pick a page whose area increase is minimized after insertion of this new UDA; (2) Most similar MBR: we use distributional similarity measure of u with MBR boundary. This makes sure that even if a probability distribution fits in an MBR without causing an area increase, we may not end up having too many UDAs which are much smaller in probability values. Even though an MBR boundary is not a probability distribution in the strict sense, we can still apply most divergence measures described earlier.

Split( ). There are two alternative strategies to split an overfull page: *top-down* and *bottom-up*. In the top-down strategy, we pick two children MBRs whose boundaries are distributionally farthest from each other according to the divergence measures. With these two serving as the seeds for two clusters, all other UDAs are inserted into the closer cluster. In the bottom-up strategy, we begin with each element forming an independent cluster. In each step the closest pair of clusters (in terms of their distributional distance) are merged. This process stops when only two clusters remain. An additional consideration is to create a balanced split, so that two new nodes have a comparable number of objects. No cluster is allowed to contain more that $3/4$ of the total elements.

PETQ(q, τ). Given the structure, the query algorithm is straightforward. We do a depth-first search in the tree, pruning by MBRs. If an MBR qualifies for the query, i.e., if $\langle\langle c.v, q\rangle\rangle \geq \tau$, our search enters the MBR, else that branch is pruned. At the leaf level, we evaluate each UDA in the page against the query and output the qualifying ones. For top-k queries, we need to upgrade the threshold probability dynamically during the search. The following lemma proves the correctness of the pruning criteria.

LEMMA 10.11 *Consider a node c in the tree. If $\langle\langle c.v, q\rangle\rangle < \tau$ then no UDA stored under the subtree of c qualifies for the threshold query (q, τ).*

Proof: Consider any UDA u stored in the subtree of c. Since an MBR boundary is formed by taking the point-wise maximum of its children MBR boundaries, we can show by induction that $u.p_i \geq c.v.p_i$ and $q_i \geq 0$ for any i, $\langle\langle u, q\rangle\rangle < \langle\langle c.v, q\rangle\rangle < \tau$. Thus, u cannot qualify.

Both indexing techniques were implemented and the performance results for both real and synthetic datasets were presented in [22]. Both the index structures were shown to have good scalability with respect to dataset and domain size.

5. Indexing for Nearest Neighbor Queries

This section discusses indexing of uncertain data for another important class of queries: nearest-neighbor queries. Researchers have proposed several approaches for this problem. We will briefly review them here. The first definition of nearest neighbor queries was given by Cheng et. al [8]. It captures the probability that a given uncertain attribute is the nearest neighbor of a query point.

DEFINITION 10.12 *Given a query point q and a pdf-attribute a_i, a_i is a nearest neighbor of q if there exist x such that $dist(q, a_i) \leq x$ and for all other tuples in the relation with attributes a_j, $dist(q, a_j) > x$. For uncertain (pdf) attributes this is a probabilistic event. The probability P^i_{nn} that a_i is the nearest neighbor of q is equal to $\int_x Pr[(x \leq dist(q, a_i) \leq x + dx) \cap (\forall j \neq i,\ \ dist(q, a_j) \geq x + dx)]$*

In many scenarios, independence is assumed across different tuples. In such cases, the probability P^i_{nn} is given by $\int_x Pr[(x \leq dist(q, a_i) \leq x + dx)]\Pi_{j\neq i} Pr[dist(q, a_j) \geq x + dx]$.

Given this, we can define nearest neighbor queries as follows:

DEFINITION 10.13 *Given a query point q and a set of objects $O_1, O_2, ..., O_n$ with uncertain pdf attributes, a probabilistic nearest neighbor query (PNN) returns a pair (O_i, P^i_{nn}), i.e., tuples augmented with their probabilities of being the nearest neighbor.*

Two related definitions to PNN which are widely used are PNN-topk and PNNT:

DEFINITION 10.14 *Given a query point q and a number k, PNN-topk returns k objects* O_i*'s such that their corresponding* P_{nn}^i *are the k topmost probabilities (of being the nearest neighbor) among all the objects.*

DEFINITION 10.15 *Given a query point q and a probability threshold t, PNNT returns all the objects* O_i*'s such that their* $P_{nn}^i \geq t$.

In these definitions, P_{nn}^i for a given object, depends on all other objects in the database. Hence, this is very complicated to evaluate. This is the most natural definition on which many researchers have worked.

Cheng et al [11] present a simple pruning based approach for this problem. Given a query point q, the sphere of non-zero probability is first calculated. This is computed using the shortest radius x from q that encloses an entire object (along with its bounding intervals). Once this is calculated, only the objects whose uncertainty region overlaps with this sphere need to be considered. The rest can be pruned away. A regular R-tree based index can be used for this purpose. Once the candidate set of pdfs are obtained, brute-force processing is done to find out all the P_{nn}^i values for each of these objects. This simple pruning strategy can yield good results.

Ilyas et al [6] present a more holistic approach to the nearest-neighbor problem. With a focus on the top-k version of this problem, they first use an index to prune away irrelevant objects. The relevant candidates are evaluated in order of the shortest distance (in the bounding box) from the query point. As these objects are processed, the radius of interest grows. At each intermediate stage upper and/or lower bounds on the probability of each object being the nearest-neighbor are computed. These bounds are used to prune irrelevant objects. The stopping criteria used in this work are experimentally shown to effectively reduce the number of expensive probability computations performed. Their approach significantly gains over [11] in the number of integration operations, although not as much in the I/O cost of objects retrieved from the index. Thus their approach is more on search-optimization on top of the indexing cost.

A similar work has been done by Qi et al [20] where they consider PNNT, the threshold version of these queries. Their approach relies on augmenting an R-tree based index using two types of probability bounds stored within the entries for each node: (i) Absence Probability bounds (AP); and (ii) Maximum Probability bounds (MP). Their approach also handles the case of partial pdfs (objects which have a non-zero probability of not being in the database), which occur very often in uncertain database systems [18]. The AP bound captures the probability that there are none of the objects is present in a given region of the subtree. The MP bound captures the maximum probability for any one object being present in a given region of the subtree. These bounds enable effective pruning at various levels in the R-tree index. The probability bounds are calculated and updated as objects are encountered which yields effective pruning based on probabilities in addition to the distance based pruning. Thus the number of objects retrieved is less than with indexes which do not store this information. This directly leads to I/O savings and fewer objects being inspected.

Ljosa et al [17] gave a formal framework for summarizing the probability information at various levels of an index. They consider a slightly different definition of nearest neighbors. In most approaches to computing NN, it is costly to calculate probabilities since the probability for a particular object is dependent on all other overlapping objects. To overcome this drawback, they define the notion of *expected nearest neighbors (ENN)* to a query point q which is independent of other objects. The expected distance for each object O_i from query point q is defined as follows.

DEFINITION 10.16 *Given a query point q and an object O_i, the expected distance is obtained as* $Edist(q, O_i) = \int_x x.Pr[x \leq dist(q, O_i) \leq x + dx]$.

Given the expected distance $Edist$, it easy to define expected nearest neighbor (ENN) and also subsequently k-ENN as follows:

DEFINITION 10.17 *Given a query point q,the K-ENN query returns all k objects O_i's having smallest values of* $Edist(q, O_i)$.

Note that since this definition relies only on the distance function, it is possible to define k nearest neighbors easily. This differs from the notion of top-k probabilities for being the first nearest neighbors (1-NN) considered above.

The main contribution of [17] is a framework to summarize the probability histograms using piecewise approximations. Their approximation scheme also generalizes to (and can be summarized at) various levels in the index. They show how to use this information to efficiently answer the ENN queries.

Going back to our definition of PNNs, a simple but very elegant model has been proposed by Dai et al [12]. They limit objects to existentially uncertain data points. Here each uncertain object O_i is simply a point with existence probability p_i associated with it. Although this model is subsumed by Qi et. al (since they can also handle partial pdfs) the simplicity of their data model allows more efficient indexing. They show how to augment 2-D R-trees with probability information so that bounds can be computed during the index search. They also show how these simple probability bounds can be taken as the third dimension in R-tree indexing. As with all R-tree based approaches the insert/split heuristics have to account for the probability values to have meaningful clusters under a subtree note giving effective pruning.

A somewhat different definition of nearest neighbor queries is considered by Singh et al [23]. These are called nearest neighbor based on thresholds.

DEFINITION 10.18 *Given a query point q and a probability threshold p, the threshold distance* $dist_t(q, O_i)$ *is defined as radius* r *from* q *which overlaps with* O_i *to an extent that probability of* O_i *being within the sphere is p.*

Although this is not the focus of their paper, they show how their summary statistics can be used to compute a distance r from the query point q which will suffice to have k qualifying neighbors. Then, a PTRQ is executed with range parameters $(q - r, q + r)$. The output tuples can be ranked by their distances. This gives k-nearest neighbors similar to Ljosa et al.

References

[1] L. Antova, C. Koch, and D. Olteanu. 10^10^6 worlds and beyond: Efficient representation and processing of incomplete information. In *Proceedings of 23rd International Conference on Data Engineering (ICDE)*, 2007.

[2] L. Arge, V. Samoladas, and J. Vitter. On two-dimensional indexability and optimal range search indexing. In *Proc. ACM Symp. on Principles of Database Systems*, pages 346–357, 1999.

[3] L. Arge and J. Vitter. Optimal dynamic interval management in external memory. In *Proceedings of Annual Symposium on Foundations of Computer Science (FOCS)*, pages 560–569, 1996.

[4] R. Baeza-Yates and B. Ribeiro-Neto. *Modern Information Retrieval.* ACM Press / Addison-Wesley, 1999.

[5] Daniel Barbará, Hector Garcia-Molina, and Daryl Porter. The management of probabilistic data. *IEEE Transactions on Knowledge and Data Engineering*, 4(5):487–502, 1992.

[6] G. Beskales, M. Soliman, and I. Ilyas. Efficient search for the top-k probable nearest neighbors in uncertain databases. In *Proceedings of International Conference on Very Large Data Bases (VLDB)*, 2008.

[7] Jihad Boulos, Nilesh Dalvi, Bhushan Mandhani, Shobhit Mathur, Chris Re, and Dan Suciu. MYSTIQ: a system for finding more answers by using probabilities. In *Proceedings of ACM Special Interest Group on Management Of Data*, 2005.

[8] R. Cheng, D. V. Kalashnikov, and S. Prabhakar. Evaluating probabilistic queries over imprecise data. In *Proc. ACM SIGMOD Int. Conf. on Management of Data*, 2003.

[9] R. Cheng, S. Singh, S. Prabhakar, R. Shah, J. Vitter, and Y. Xia. Efficient join processing over uncertain data. In *Proceedings of ACM 15th Conference on Information and Knowledge Management (CIKM)*, 2006.

[10] R. Cheng, Y. Xia, S. Prabhakar, R. Shah, and J. Vitter. Efficient indexing methods for probabilistic threshold queries over uncertain data. In *Proceedings of International Conference on Very Large Data Bases (VLDB)*, 2004.

[11] Reynold Cheng, Sunil Prabhakar, and Dmitri V. Kalashnikov. Querying imprecise data in moving object environments. In *Proceedings of the International Conference on Data Engineering (ICDE'03)*, 2003.

[12] X. Dai, M. Yiu, N. Mamoulis, Y. Tao, and M. Vaitis. Probabilistic spatial queries on existentially uncertain data. In *Proceedings of International Symposium on Spatial and Temporal Databases (SSTD)*, pages 400–417, 2005.

[13] J. Enderle, M. Hampel, and T. Seidl. Joining interval data in relational databases. In *Proc. ACM SIGMOD Int. Conf. on Management of Data*, 2004.

[14] R. Fagin, A. Lotem, and M. Naor. Optimal aggregation algorithms for middleware. In *Proc. ACM Symp. on Principles of Database Systems*, 2001.

[15] A. Guttman. R-trees: A dynamic index structure for spatial searching. In *Proc. ACM SIGMOD Int. Conf. on Management of Data*, pages 47–57, 1984.

[16] I. Ilyas, W. Aref, and A. Elmagarmid. Supporting top-k join queries in relational databases. In *Proceedings of International Conference on Very Large Data Bases (VLDB)*, 2003.

[17] V. Ljosa and A. Singh. APLA: Indexing arbitrary probability distributions. In *Proceedings of 23rd International Conference on Data Engineering (ICDE)*, 2007.

[18] http://orion.cs.purdue.edu/, 2008.

[19] Fernando C. N. Pereira, Naftali Tishby, and Lillian Lee. Distributional clustering of english words. In *Meeting of the Association for Computational Linguistics*, 1993.

[20] Y. Qi, S. Singh, R. Shah, and S. Prabhakar. Indexing probabilistic nearest-neighbor threshold queries. In *Proceeding of workshop on Management of Uncertain Data (MUD)*, 2008.

[21] P. Sen and A. Deshpande. Representing and querying correlated tuples in probabilistic databases. In *Proceedings of 23rd International Conference on Data Engineering (ICDE)*, 2007.

[22] S. Singh, C. Mayfield, S. Prabhakar, R. Shah, and S. Hambrusch. Indexing uncertain categorical data. In *Proceedings of 23rd International Conference on Data Engineering (ICDE)*, 2007.

[23] S. Singh, C. Mayfield, R. Shah, S. Prabhakar, and S. Hambrusch. Query selectivity estimation for uncertain data. Technical Report TR–07–016, Purdue University, 2007.

[24] Y. Tao, R. Cheng, X. Xiao, W. Ngai, B. Kao, and S. Prabhakar. Indexing multi-dimensional uncertain data with arbitrary probability density functions. In *Proceedings of International Conference on Very Large Data Bases (VLDB)*, 2005.

[25] J. Widom. Trio: A system for integrated management of data, accuracy, and lineage. In *Proc. Conf. on Innovative Data Systems Research (CIDR)*, 2005.

Chapter 11

RANGE AND NEAREST NEIGHBOR QUERIES ON UNCERTAIN SPATIOTEMPORAL DATA

Yufei Tao
Department of Computer Science, Chinese University of Hong Kong
Sha Tin, New Territories, Hong Kong
taoyf@cse.cuhk.edu.hk

Abstract Uncertain data are abundant in numerous spatiotemporal applications providing location-based services. This chapter will first review these applications and explain why uncertainty is inherent, sometimes even intended, in their underlying data. Then, we will address two classical spatial queries: *range search* and *nearest neighbor retrieval*. We will see that the presence of uncertainty requires extending the traditional definitions of both queries, and thus, also invalidates the conventional algorithms designed for precise databases. We will survey the existing the variations of both queries specifically formulated on uncertain data. Finally, the algorithms for efficiently processing these queries will also be discussed.

Keywords: Spatiotemporal, uncertain, range search, nearest neighbor, probability thresholding

1. Introduction

A *spatiotemporal database* [15, 24, 29] manages a large number of moving objects such as human beings, vehicles, aircrafts, typhoons, wild-life animals, and so on. It plays an imperative role in numerous applications including location-based services, fleet/flight control, weather analysis and forecasting, zoology study, etc. In these contexts, the location of an object is captured using a positioning technology, e.g., GPS, cellular triangulation, location sensing, etc. These locations are sent to a database server, which organizes them in appropriate ways for supporting different analytical tasks efficiently.

Uncertainty is an inherent issue in spatiotemporal applications. Specifically, the database typically does not have the *exact* locations of the objects, but instead, must estimate their locations using probabilistic methods. Usually, uncertainty arises from three sources: *measurement imprecision*, *location-update policy*, and *privacy preservation*. Next, we will look at these issues in detail.

- *Measurement imprecision.* People are now able to track the locations of moving objects with much better precision, owing to the tremendous improvements in positioning techniques in the past decade. Nevertheless, all these techniques are still subject to errors, as nicely summarized in [34]. For instance, even in open air, using GPS to pinpoint a location may incur an error ranging from 10 to 300 meters. The latest WiFi technology has better accuracy but still suffers from an error up to 50 meters. As an example of commercial products, *Google Maps* allows a mobile phone user to identify her/his present location, but always associates the location with an error up to 200 meters.

  It is rather unlikely that the positioning error would be significantly reduced in near future. There are two reasons. First, upgrading the hardwares of positioning infrastructures (e.g., satellites for GPS) is extremely expensive, and lacks commercial motivation. Second, the current accuracy already satisfies the needs of most applications in practice, when combined with proper database techniques for tackling uncertainty, such as those discussed in this book.

- *Location-update policy.* In a location-based service such as a traffic control system, each moving object (e.g., a vehicle) is required to report its location periodically to a central server through wireless networks. Due to the bandwidth constraint, however, the system cannot afford to collect object locations at all timestamps. Hence, there is a tradeoff between how accurately the server can store objects' locations and the network overhead incurred.

  To achieve a good tradeoff, there have been significant research efforts [31–33] on developing effective *update policies*, which are a protocol determining when an object is supposed to send in its new location. One simple, yet effective, policy is *dead-reckoning* [33]. Specifically, an object must transmit its current location, only if it deviates from its last reported location by more than a distance ε, which is a system parameter. Accordingly, the database knows that currently an object o can be anywhere in an *uncertainty circle* that centers at its last updated location x, and has radius ε. A smaller ε promises higher precision in location recording, but entails larger transmission overhead.

- *Privacy preservation.* In the above scenarios, uncertainty is not desired, but is inevitable due to equipment limitations or economy concerns. In some applications, however, uncertainty may even be introduced intentionally to distort locations moderately in order to preserve individual privacy [2, 10, 11, 21]. For example, a mobile phone company is able to pinpoint its customers' locations using cellular triangulation. It is possible that the company may want to outsource these locations to an external agent to provide location-based services, such as environment tracking (e.g., update the UV index in the place of a customer to her/his phone), mobile advertising (e.g., alert a customer if s/he is near a cosmetic shop), friend connection (e.g., notify a customer whenever a friend is around), and so on. However, as the external agent may not be authorized to acquire customers' exact positions, the phone company must convert a location to a fuzzy form before outsourcing it.

One possible method to fulfill this purpose is *spatial cloaking* [21] based on *k-anonymity* [25, 27]. Specifically, for each customer, the phone company finds a *cloak rectangle* that covers her/his location, as well as the locations of $k-1$ other customers, where k is a system parameter. Then, the phone company outsources only the rectangle, instead of the customer's precise location. Location privacy is protected, because the external service provider only knows that the customer is inside a rectangle, but not her/his concrete position. Clearly, a greater k offers stronger privacy preservation, but leads to larger cloak rectangles that may in turn result in poorer location-based services.

The first step to cope with uncertainty is to formulate an *uncertainty region* $o.ur$ for each object o, which is guaranteed to cover the object's real location. For example, $o.ur$ can be the uncertainty circle under dead-reckoning, or the cloak rectangle in preserving location privacy. When $o.ur$ is large, however, it may be too fuzzy for the underlying applications. In this case, a common remedy is to model the object's current location with a probabilistic distribution function (pdf), denoted as $o.pdf(\cdot)$. More formally, $o.pdf(x)$ equals 0 for any location outside $o.ur$, and it satisfies

$$\int_{x \in o.ur} o.pdf(x)dx = 1 \tag{11.1}$$

for the locations in $o.ur$. The simplest $o.pdf(\cdot)$ describes the uniform distribution. This distribution is most reasonable when the database has no additional knowledge about the object. In some cases, the database may be able to justify other distributions. For example, in GPS, it would be reasonable to assume that $o.pdf(\cdot)$ follows a normal distribution whose mean is at the measured position.

Sometimes $o.pdf(\cdot)$ may not even be any regular probabilistic distribution, but instead, can be arbitrarily complex. For example, consider Figure 11.1 where x represents the last updated location of a vehicle o under

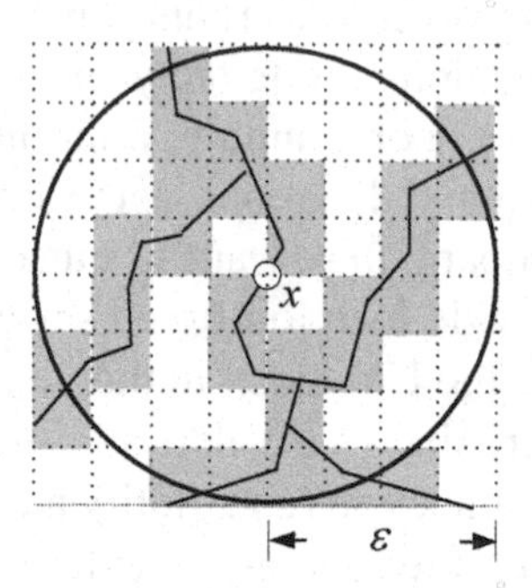

Figure 11.1. An example of irregular object pdf

dead-reckoning, and ε is the maximum permissible deviation from the vehicle's current position to x. The circle is the uncertainty circle $o.ur$ of o. The segments represent the road network where the vehicle is allowed to navigate. In other words, $o.pdf(x)$ has non-zero values only at positions x in the part of the network inside the circle. Accurately representing a complicated pdf may be space consuming. For example, to capture the $o.pdf(\cdot)$ in Figure 11.1, we would have to store the underlying road network as well. An alternative approach is to approximate $o.pdf(\cdot)$ using a histogram, which as shown in Figure 11.1 imposes a grid over the uncertainty circle $o.ur$. Shaded are the cells of the histogram that intersect at least one road segment. Each shaded cell is associated with a non-zero value equal to the probability that o appears in it. The white cells are associated with zeros. Apparently, the resolution of the grid provides an easy way to control the space consumption and quality of approximating $o.pdf(\cdot)$.

As in traditional spatial databases, *range search* and *nearest neighbor* (NN) *retrieval* are also the two most important queries on uncertain spatiotemporal objects. Given a search region r_q, a range query retrieves all the objects that fall in r_q. For example, *find all the vehicles in the downtown area.* Given a query point p_q, on the other hand, an NN query returns the object whose distance to p_q is the smallest. For instance, *find the taxi nearest to the Empire State building.* A direct extension is k *nearest neighbor search*, which, instead of retrieving the nearest object, extracts the k objects nearest to p_q.

In the presence of uncertainty, we need to interpret the query results more carefully. For example, given a range query r_q, an object o may have an uncertainty region that partially intersects r_q. In this case, we cannot claim for sure whether o satisfies or does not satisfy the query. Instead, o may qualify the query with a certain probability. Similarly, given an NN query p_q, each object also has a qualification probability, corresponding to the likelihood that it is the NN of p_q. In practice, a user may be interested in only those objects that may satisfy the query with high probability. This motivates *probability thresholding search* [7], which specifies a probability threshold t_q, and returns only objects with qualification probabilities at least t_q. Evidently, qualification probabilities depend on objects' pdfs, and are absent in traditional spatiotemporal databases dealing with precise data. As a result, the algorithms in precise spatiotemporal databases are no longer applicable to process probability thresholding queries.

The rest of the chapter will discuss the previous solutions to probability thresholding queries, focusing on range search and nearest neighbor retrieval in Sections 2 and 3, respectively. It is worth mentioning that although spatiotemporal applications provide the original motivation for this chapter, the techniques we cover are in fact applicable to general multidimensional uncertain data.

2. Range Search

Following the discussion earlier, we model an uncertain object o with a probability density function $o.pdf(\cdot)$, and an uncertainty region $o.ur$. We consider that the pdfs of different objects are mutually independent, and various objects can have totally different pdfs. For example, an object may have a pdf describing a uniform distribution, another could follow a Gaussian distribution, and yet another could possess an irregular distribution that can be described only by a histogram as in Figure 11.1. Furthermore, the uncertainty region of an object does not have to be convex, or can even be broken into multiple pieces. For instance, an object may appear inside two separate buildings, but not on the roads between the buildings.

2.1 Query Definitions

Let S be a set of uncertain objects. Given a region r_q, and a value $t_q \in (0, 1]$, a *nonfuzzy probability thresholding range query* returns all the objects $o \in S$ such that $Pr_{range}(o, r_q) \geq t_q$, where $Pr_{range}(o, r_q)$ is the appearance probability of o in r_q, and is computed as

$$Pr_{range}(o, r_q) = \int_{r_q \cap o.ur} o.pdf(x)dx. \tag{11.2}$$

To illustrate, the polygon in Figure 11.2a shows the uncertainty region $o.ur$ of an object o, and the rectangle corresponds to a query region r_q. If the possible location of o uniformly distributes inside $o.ur$, $Pr_{range}(o, r_q)$ equals the area of the intersection between $o.ur$ and r_q, i.e., the hatched region.

In nonfuzzy range search, the search region r_q is unique and has no uncertainty. Sometimes, we may want to explore the vicinity of an uncertain object. For example, a user may wish to *find all the cabs that are within 1 kilometers from the cab with license plate NY3056 with at least 50% probability*, where

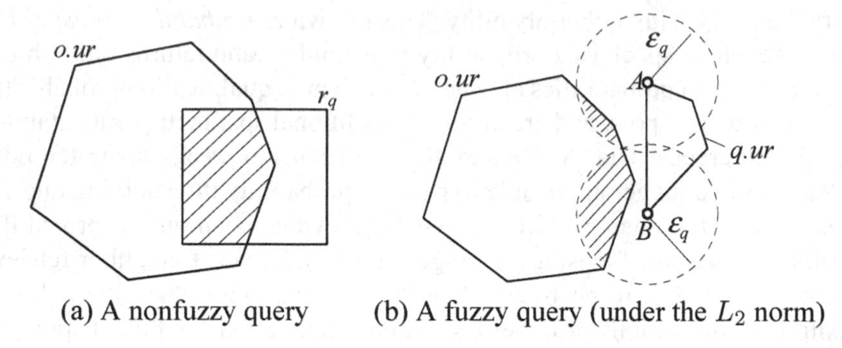

(a) A nonfuzzy query (b) A fuzzy query (under the L_2 norm)

Figure 11.2. Range search on uncertain data

the query object (i.e., the cab NY3056) is also uncertain. This motivates *fuzzy range search.*

Formally, let S be a set of uncertain objects, and q be an uncertain object that does not belong to S. Given a distance threshold ε_q, and a value $t_q \in (0, 1]$, a *fuzzy probability thresholding range query* returns all the objects $o \in S$ such that $Pr_{fuzzy}(o, q, \varepsilon_q) \geq t_q$, where $Pr_{fuzzy}(o, q, \varepsilon_q)$ is the probability that o and q have distance at most ε_q. If we regard o and q as random variables obeying pdfs $o.pdf(x)$ and $q.pdf(x)$ respectively, then

$$Pr_{fuzzy}(o, q, \varepsilon_q) = Pr\{dist(o, q) \leq \varepsilon_q\} \tag{11.3}$$

Since o and q are independent, Equation 11.3 can be re-written as

$$Pr_{fuzzy}(o, q, \varepsilon_q) = \int_{x \in q.ur} q.pdf(x) \cdot Pr_{range}(o, \odot(x, \varepsilon_q))dx. \tag{11.4}$$

where $Pr_{range}(\cdot, \cdot)$ is given in Equation 11.2, and $\odot(x, \varepsilon_q)$ is a circle that centers at point x and has radius ε_q.

As an example, the left and right polygons in Figure 11.2b demonstrate the uncertainty regions of a data object o and a query object q. The figure also shows two $\odot(x, \varepsilon_q)$, when x lies at point A and B, respectively. Again, for simplicity, assume that $o.pdf$ follows a uniform distribution inside $o.ur$. The area of the upper (lower) hatched region equals the probability $Pr_{range}(o, \odot(x, \varepsilon_q))$ for o and q to have a distance at most ε_q, when q is located at $x = A$ (B). In order to calculate $Pr_{fuzzy}(o, q, \varepsilon_q)$, (conceptually) we must examine the $Pr_{range}(o, \odot(x, \varepsilon_q))$ of all $x \in q.ur$. Note that the formulation of fuzzy search is independent of the distance metric employed. For example, under L_∞, $\odot(x, \varepsilon_q)$ is a square whose centroid falls at x and has a side length $2\varepsilon_q$.

In the sequel, we will refer to $Pr_{range}(o, r_q)$ and $Pr_{fuzzy}(o, q, \varepsilon_q)$ simply as the *qualification probability* of o, when the query type is clear from the context.

2.2 Filter and Refinement

For both nonfuzzy and fuzzy range search, it can be rather expensive to calculate the qualification probability of an object. There are two reasons. First, as shown in Equations 11.2 and 11.4, such calculation requires solving multidimensional integrals, which is known to be a time consuming process, especially when the integral region and integrated function are complex. Note that this problem is particularly serious for fuzzy queries, because Equation 11.4 essentially has two layers of multidimensional integrals, noticing that $Pr_{range}(o, \odot(x, \varepsilon_q))$ needs to be unfolded into an integral similar to Equation 11.2. The second reason behind the large cost of assessing qualification probabilities is that an object's pdf may be in the form of a histogram (c.f. Figure 11.1), which must reside in the disk, and hence, its retrieval incurs I/Os.

A similar situation was also encountered in range search in a spatial database of polygonal objects. In that case, verifying whether an object intersects the search region is expensive as it needs to fetch the object's polygon from the disk, which can occupy a large number of pages. The *filter refinement* framework is introduced to reduce the query cost dramatically in this situation. The framework contains a *filter step* followed by a *refinement step*. The filter step first uses efficient operations to quickly compute a *candidate set*, which is much smaller than the underlying database, and is guaranteed to be a super set of the final result. In other words, any object outside the candidate set can be safely ignored. Then, the refinement step examines each object in the candidate set to verify whether it indeed satisfies the query.

Range search on uncertain data [3, 28, 30] also follows the filter refinement framework, in order to minimize the number of qualification probability evaluations. To achieve this, the filter step has two missions. First, it needs to prune as many nonqualifying objects as possible. Second, it also needs to validate as many qualifying objects as possible. The refinement step performs the expensive qualification probability evaluation, only if an object can be neither pruned nor validated.

The rest of this section will explain the rationales behind the algorithms in [3, 28, 30] for solving nonfuzzy and fuzzy range search. As we will see, a crucial concept underlying this technique is *probabilistically constrained rectangle* (PCR). We note that in the literature PCR has also appeared under a different name of *p-bound* [3].

2.3 Nonfuzzy Range Search

A *probabilistically constrained region* (PCR) of an object o depends on a parameter $c \in [0, 0.5]$. We denote it as $o.pcr(c)$. It is a d-dimensional rectangle, obtained by pushing, respectively, each face of $o.mbr$ inward, until the appearance probability of o in the area swept by the face equals c. Figure 11.3a illustrates the construction of a 2D $o.pcr(c)$, where the polygon represents the uncertainty region $o.ur$ of o, and the dashed rectangle is the MBR of o, denoted as $o.mbr$. The $o.pcr(c)$, which is the grey area, is decided by 4 lines $l_{[1]+}$, $l_{[1]-}$, $l_{[2]+}$, and $l_{[2]-}$. Line $l_{[1]+}$ has the property that, the appearance probability of o on the right of $l_{[1]+}$ (i.e., the hatched area) is c. Similarly, $l_{[1]-}$ is obtained in such a way that the appearance likelihood of o on the left of $l_{[1]-}$ equals c. It follows that the probability that o lies between $l_{[1]-}$ and $l_{[1]+}$ is $1 - 2c$. Lines $l_{[2]+}$ and $l_{[2]-}$ are obtained in the same way, except that they horizontally partition $o.ur$.

PCRs can be used to prune or validate an object, without computing its accurate qualification probability. Let us assume that the grey box in Figure 11.3a is the $o.pcr(0.1)$ of o. Figure 11.3b shows the same PCR and $o.mbr$ again, together with the search region r_{q_1} of a nonfuzzy range query q_1 whose probability threshold t_{q_1} equals 0.9. As r_{q_1} does not fully contain $o.pcr(0.1)$, we can immediately assert that o cannot qualify q_1. Indeed, since o falls in the hatched region with probability 0.1, the appearance probability of o in r_{q_1} must be smaller than $1 - 0.1 = 0.9$. Figure 11.3c illustrates pruning the same object with respect to another query q_2 having $t_{q_2} = 0.1$. This time, o is disqualified because r_{q_2} does not intersect $o.pcr(0.1)$ (the pruning conditions are different for q_1 and q_2). In fact, since r_{q_2} lies entirely on the right of $l_{[1]+}$, the appearance probability of o in r_{q_2} is definitely smaller than 0.1.

The second row of Figure 11.3 presents three situations where o can be validated using $o.pcr(0.1)$, with respect to queries q_3, q_4, q_5 having probability thresholds $t_{q_3} = 0.9$, $t_{q_4} = 0.8$, and $t_{q_5} = 0.1$, respectively. In Figure 11.3d (or Figure 11.3f), o must satisfy q_3 (or q_5) due to the fact that r_{q_3} (or r_{q_5}) fully covers the part of $o.mbr$ on the right (or left) of $l_{[1]-}$, which implies that the appearance probability of o in the query region must be at least $1 - 0.1 = 0.9$ (or 0.1), where 0.1 is the likelihood for o to fall in the hatched area. Similarly, in Figure 11.3e, o definitely qualifies q_4, since r_{q_4} contains the portion of $o.mbr$ between $l_{[1]-}$ and $l_{[1]+}$, where the appearance probability of o equals $1 - 0.1 - 0.1 = 0.8$.

The queries in Figures 11.3d-11.3f share a common property: the projection of the search region contains that of $o.mbr$ along one (specifically, the vertical) dimension. Accordingly, we say that those queries *1-cover o.mbr*. In fact, validation is also possible, even if a query *0-covers o.mbr*, namely, the projection of the query area does not contain that of $o.mbr$ on any dimension. Next, we

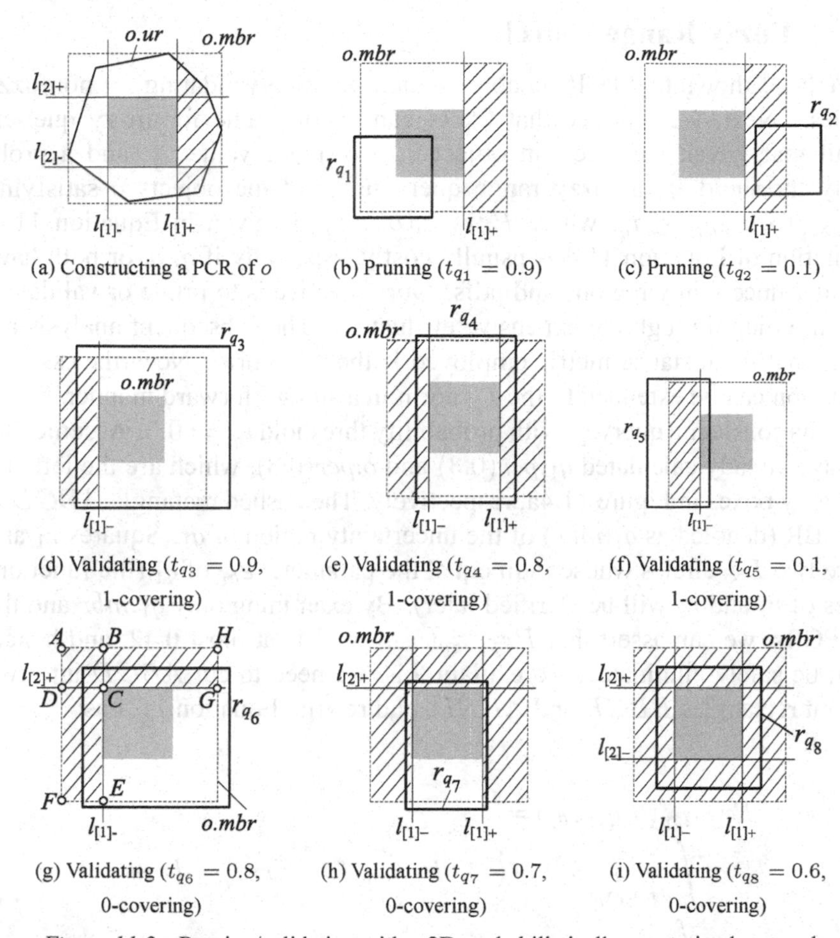

(a) Constructing a PCR of o (b) Pruning ($t_{q1} = 0.9$) (c) Pruning ($t_{q2} = 0.1$)

(d) Validating ($t_{q3} = 0.9$, 1-covering) (e) Validating ($t_{q4} = 0.8$, 1-covering) (f) Validating ($t_{q5} = 0.1$, 1-covering)

(g) Validating ($t_{q6} = 0.8$, 0-covering) (h) Validating ($t_{q7} = 0.7$, 0-covering) (i) Validating ($t_{q8} = 0.6$, 0-covering)

Figure 11.3. Pruning/validating with a 2D probabilistically constrained rectangle

illustrate this using the third row of Figure 11.3, where the queries q_6, q_7, q_8 have probability thresholds $t_{q6} = 0.8$, $t_{q7} = 0.7$, and $t_{q8} = 0.6$, respectively.

In Figure 11.3g, o is guaranteed to qualify q_6, since r_{q6} covers entirely the part of $o.mbr$ outside the hatched area. Observe that the appearance probability of o in the hatched area is *at most* 0.2. To explain this, we decompose the area into three rectangles $ABCD$, $DCEF$, $BCGH$, and denote the probabilities for o to lie in them as ρ_{ABCD}, ρ_{DCEF}, and ρ_{BCGH}, respectively. By the definition of $l_{[1]-}$, we know that $\rho_{ABCD} + \rho_{DCEF} = 0.1$, whereas, by $l_{[2]+}$, we have $\rho_{ABCD} + \rho_{BCGH} = 0.1$. Since ρ_{ABCD}, ρ_{DCEF}, and ρ_{BCGH} are nonnegative, it holds that $\rho_{ABCD} + \rho_{DCEF} + \rho_{BCGH} \leq 0.2$. This, in turn, indicates that o falls in r_{q6} with probability at least 0.8. With similar reasoning, it is not hard to verify that, in Figure 11.3h (Figure 11.3i), the appearance probability of o in the hatched area is at most 0.3 (0.4), meaning that o definitely satisfies q_7 (q_8).

2.4 Fuzzy Range Search

We have shown that PCRs enable efficient pruning/validating for nonfuzzy queries. Next, we will see that PCRs can do the same for fuzzy queries. Recall that, given an uncertain object q, a distance value ε_q, and a probability threshold t_q, a fuzzy range query finds all the objects o satisfying $Pr_{fuzzy}(o, q, \varepsilon_q) \geq t_q$, where $Pr_{fuzzy}(o, q, \varepsilon_q)$ is given in Equation 11.4. Evaluation of Equation 11.4 is usually costly, especially if q, o, or both have irregular uncertainty regions and pdfs. Our objective is to prune or validate o without going through the expensive evaluation. The subsequent analysis assumes that the distance metric employed is the L_∞ norm. Nevertheless, our discussion can be extended to the L_2 norm in a straightforward manner.

Let us consider a query q_1 with probability threshold $t_{q_1} = 0.5$. Assume that we have already calculated $q_1.pcr(0.3)$ and $o.pcr(0.3)$, which are the left and right grey boxes in Figure 11.4a, respectively. The dashed rectangle $ABCD$ is the MBR (denoted as $q_1.mbr$) of the uncertainty region of q_1. Squares r_1 and r_2 are two L_∞ circles whose radii equal the parameter ε_{q_1} of q_1 (the functionalities of r_1 and r_2 will be clarified later). By examining only $q_1.mbr$ and the two PCRs, we can assert that $Pr_{fuzzy}(o, q_1, \varepsilon_{q_1})$ is at most 0.42, and hence, o can be safely eliminated. To explain this, we need to cut $ABCD$ into two disjoint rectangles $EBCF$ and $AEFD$, and rewrite Equation 11.4 as:

$$
\begin{aligned}
&Pr_{fuzzy}(o, q_1, \varepsilon_{q_1}) = \\
&\quad \int_{x \in EBCF} q_1.pdf(x) \cdot Pr_{range}(o, \odot(x, \varepsilon_{q_1}))dx + \\
&\quad \int_{x \in AEFD} q_1.pdf(x) \cdot Pr_{range}(o, \odot(x, \varepsilon_{q_1}))dx.
\end{aligned} \tag{11.5}
$$

where $\odot(x, \varepsilon_{q_1})$ is a square that centers at point x, and has a side length of $2\varepsilon_{q_1}$. Observe that, for any $x \in EBCF$, $Pr_{range}(o, \odot(x, \varepsilon_{q_1}))$ must be bounded by 0.7, due to the fact that $\odot(x, \varepsilon_{q_1})$ does not fully cover $o.pcr(0.3)$. For example, r_1, which is the $\odot(x, \varepsilon_{q_1})$ for $x = B$, does not fully cover $o.pcr(0.3)$. On the other hand, for any $x \in AEFD$, $Pr_{range}(o, \odot(x, \varepsilon_{q_1}))$ never exceeds 0.3, because $\odot(x, \varepsilon_{q_1})$ does not intersect $o.pcr(0.3)$. For instance, r_2, which is the $\odot(x, \varepsilon_{q_1})$ for $x = G$, is disjoint with $o.pcr(0.3)$. As a result,

$$
\begin{aligned}
&Pr_{fuzzy}(o, q_1, \varepsilon_{q_1}) \\
&\leq \; 0.7 \int_{x \in EBCF} q_1.pdf(x)dx + 0.3 \int_{x \in AEFD} q_1.pdf(x)dx \\
&= \; 0.7 \times 0.3 + 0.3 \times 0.7 = 0.42.
\end{aligned} \tag{11.6}
$$

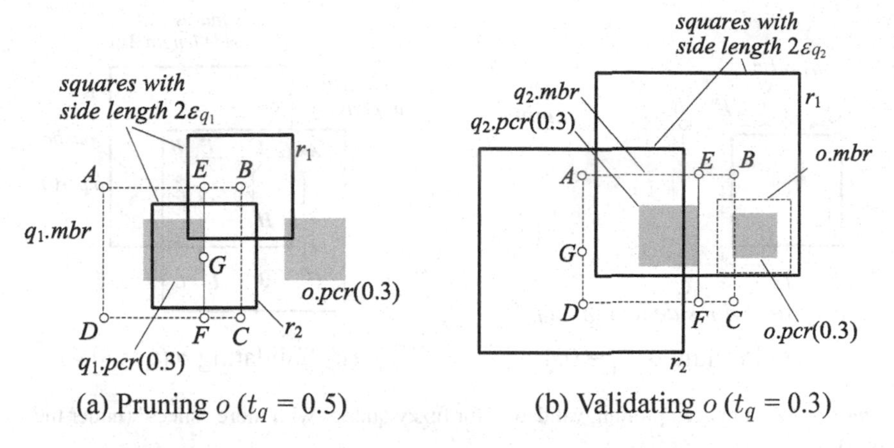

Figure 11.4. Pruning/validating with PCRs for fuzzy queries (under the L_∞ norm)

Let q_2 be another query with probability threshold $t_{q2} = 0.3$. The left and right grey boxes in Figure 11.4b demonstrate $q_2.pcr(0.3)$ and $o.pcr(0.3)$, respectively, whereas the larger and smaller dashed rectangles capture $q_2.mbr$ and $o.mbr$, respectively. Squares r_1 and r_2 are again two L_∞ circles whose radii equal the parameter ε_{q2} of q_2. Based on the above information, we can claim that $Pr_{fuzzy}(o, q_2, \varepsilon_{q2}) \geq 0.3$, and hence, o can be validated. To clarify this, we again divide $q_2.mbr$ into rectangles $EBCF$ and $AEFD$, and scrutinize Equation 11.4. Here, for any $x \in EBCF$, $Pr_{range}(o, \odot(x, \varepsilon_{q2}))$ is 1, because $\odot(x, \varepsilon_{q2})$ necessarily contains $o.mbr$ (r_1 illustrates an example of $\odot(x, \varepsilon_{q2})$ for $x = E$). However, when x distributes in $AEFD$, $Pr_{range}(o, \odot(x, \varepsilon_{q2}))$ may drop to 0, as is exemplified by r_2, which is the $\odot(x, \varepsilon_{q2})$ for $x = G$. It follows that

$$
\begin{aligned}
Pr_{fuzzy}(o, q_1, \varepsilon_{q1}) &\geq 1 \cdot \int_{x \in EBCF} q_1.pdf(x)dx + 0 \int_{x \in AEFD} q_1.pdf(x)dx \\
&= 1 \times 0.3 + 0 \times 0.7 = 0.3. \qquad (11.7)
\end{aligned}
$$

In the above examples, we "sliced" $q.mbr$ into two rectangles for pruning and validating. In fact, stronger pruning/validation effects are possible by performing the slicing more aggressively. Assume that, instead of 0.5, the query q_1 in Figure 11.4a has a lower $t_{q1} = 0.4$. Hence, o can no longer be disqualified as described with Inequality 11.5 (as $0.42 > t_{q1}$). However, we can actually derive a tighter upper bound 0.33 of $Pr_{fuzzy}(o, q_1, \varepsilon_{q1})$, and thus, still eliminate o. For this purpose, we should divide $q.mbr$ into three rectangles $EBCF$, $IEFJ$, and $AIJD$ as in Figure 11.5a, which repeats the content of Figure 11.4a, except for including $o.mbr$ (i.e., the right dashed box). Accordingly:

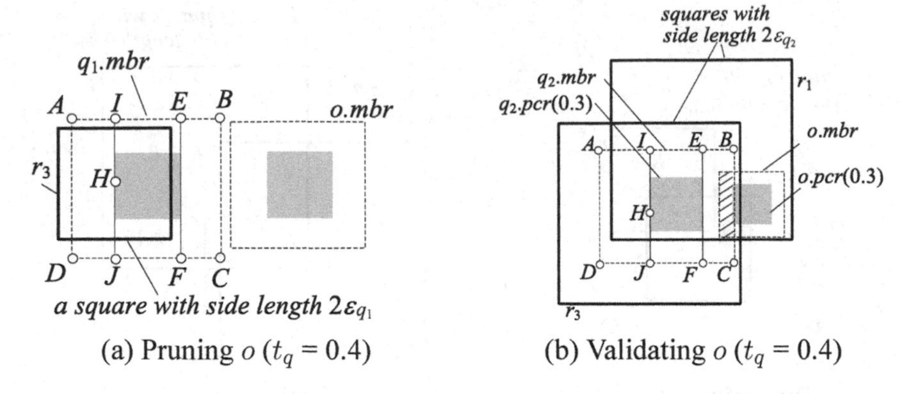

(a) Pruning o ($t_q = 0.4$) (b) Validating o ($t_q = 0.4$)

Figure 11.5. Enhanced pruning/validating for fuzzy queries with more "slices" (under the L_∞ norm)

$$
\begin{aligned}
&Pr_{fuzzy}(o, q_1, \varepsilon_{q_1}) = \\
&\quad \int_{x \in EBCF} q_1.pdf(x) \cdot Pr_{range}(o, \odot(x, \varepsilon_{q_1}))dx + \\
&\quad \int_{x \in IEFJ} q_1.pdf(x) \cdot Pr_{range}(o, \odot(x, \varepsilon_{q_1}))dx + \\
&\quad \int_{x \in AIJD} q_1.pdf(x) \cdot Pr_{range}(o, \odot(x, \varepsilon_{q_1}))dx. \qquad (11.8)
\end{aligned}
$$

As analyzed earlier with Figure 11.4a, for any point $x \in EBCF$, $Pr_{range}(o, \odot(x, \varepsilon_{q_1})) \leq 0.7$, whereas, for any point $x \in IEFJ \subset ABCD$, $Pr_{range}(o, \odot(x, \varepsilon_{q_1})) \leq 0.3$. Furthermore, notice that, given any point $x \in AIJD$, $Pr_{fuzzy}(o, q_1, \varepsilon_{q_1})$ is always 0, because $\odot(x, \varepsilon_{q_1})$ is disjoint with $o.mbr$. For instance, rectangle r_3 is the $\odot(x, \varepsilon_{q_1})$ when x lies at H; evidently, it is impossible for o to appear in r_3. Therefore,

$$
\begin{aligned}
&\text{Equation 11.7} \\
&\quad \geq 0.7 \int_{x \in EBCF} q_1.pdf(x)dx + 0.3 \int_{x \in IEFJ} q_1.pdf(x)dx + \\
&\quad\quad 0 \int_{x \in AIJD} q_1.pdf(x)dx \\
&\quad = 0.7 \times 0.3 + 0.3 \times 0.4 + 0 \times 0.3 = 0.33. \qquad (11.9)
\end{aligned}
$$

Similarly, suppose that the query q_2 in Figure 11.4b has a probability threshold $t_{q_2} = 0.4$, in which case o cannot be confirmed as a qualifying object with Inequality 11.6. Next, we will use Figure 11.5b, where the grey and dashed rectangles have the same meaning as in Figure 11.4b, to derive a new lower bound 0.42 of $Pr_{fuzzy}(o, q_2, \varepsilon_{q2})$, which thus validates o.

Let us break $q_2.mbr$ into rectangles $EBCF$, $IEFJ$, and $AIJD$. Then, $Pr_{fuzzy}(o, q_2, \varepsilon_{q2})$ can be represented as Equation 11.7. Following the analysis that led to Inequality 11.6, we know that, for $x \in EBCF$, $Pr_{fuzzy}(o, q_2, \varepsilon_{q2}) = 1$, and, for $x \in AIJD$, (obviously) $Pr_{fuzzy}(o, q_2, \varepsilon_{q2}) \geq 0$. The new observation here is that, for $x \in IEFJ$, $Pr_{fuzzy}(o, q_2, \varepsilon_{q2}) \geq 0.3$, since $\odot(x, \varepsilon_{q_2})$ always fully covers the hatched area in Figure 11.5b, which is the part of $o.mbr$ on the left of $o.pcr(0.3)$. Rectangle r_3 shows an example of $\odot(x, \varepsilon_{q_2})$ when $x = H$. By the above reasoning, o has a probability of at least 0.3 to lie in r_3. Therefore,

$$
\begin{aligned}
&\text{Equation 11.7} \\
&\geq 1 \int_{x \in EBCF} q_1.pdf(x)dx + 0.3 \int_{x \in IEFJ} q_1.pdf(x)dx + \\
&\quad 0 \int_{x \in AIJD} q_1.pdf(x)dx \\
&= 1 \times 0.3 + 0.3 \times 0.4 + 0 \times 0.3 = 0.42. \qquad (11.10)
\end{aligned}
$$

2.5 Indexing

The above discussion provides the basic intuition as to how PCRs can be utilized to prune and validate an object. Based on this idea, Tao et al. [28, 30] propose the *U-tree* for indexing multidimensional uncertain objects. Specifically, for each object o, the U-tree stores a small number of PCRs $o.pcr(p)$ at several values of p, called the *catalog values*. The same set of catalog values are used for all objects. These PCRs are then organized in an R-tree manner into a balanced structure. Given a query, it may be possible to prune the entire subtree of an intermediate entry by utilizing the information stored in the entry [30], thus saving considerable I/Os. In [30], Tao et al. propose an analytical model that can be applied to choose the optimal number of catalog values to achieve the best query performance.

Besides PCRs and the U-tree, nonfuzzy range search has also been addressed using other techniques. Ljosa and Singh [20] suggest that higher efficiency of range search may be possible by working with objects *cumulative probability functions* (cdf), as opposed to their pdfs. Motivated by this, they develop a method to approximate an arbitrary cdf with a piecewise linear function. This method leads to *the APLA-tree*, which is able to solve 1D range queries efficiently. In 2D space, Ljosa and Singh [20] propose to approximate a cdf as the product of two 1D cdfs, one for each dimension. This may work well for axis-independent pdfs such as uniform and normal distributions. However, for the other pdfs, especially irregular pdfs as in Figure 11.1, the approxima-

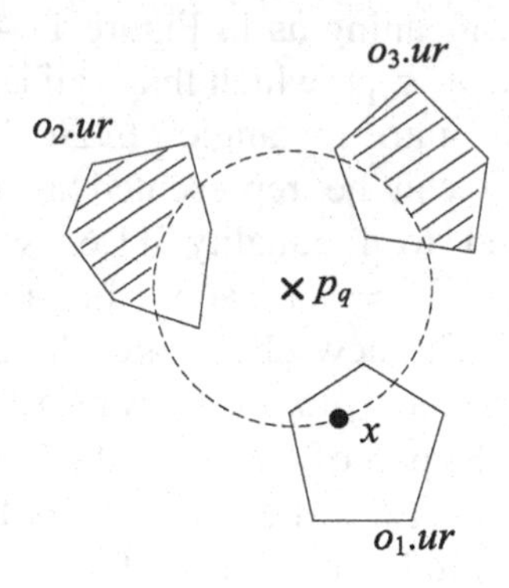

Figure 11.6. Illustration of calculating an NN probability

tion may have poor quality, which in turn may significantly compromise the efficiency of the APLA-tree.

Bohm et al. [1] propose *the Gause-tree* to index uncertain objects whose pdfs are all normal distributions. Each normal distribution is described by a mean μ and a variance σ. Treating (μ, σ) as a 2D point, the Gause-tree can be regarded as a 2D R-tree on the resulting (μ, σ) of all objects. Bohm et al. [1] show that effective pruning is possible at the intermediate levels of the Gause-tree to quickly discard the subtrees that cannot contain any qualifying object.

3. Nearest Neighbor Retrieval

In this section, we discuss the existing solutions to NN search on uncertain multidimensional data. Our analysis will use the same modeling of uncertain objects as in the previous section. Specifically, each object o is represented by an uncertainty region $o.ur$ and a probability density function $o.pdf(\cdot)$.

3.1 Query Definition

Let S be a set of uncertain objects. Given a query point p_q, a probability threshold $t_q \in (0, 1]$, a *probability thresholding nearest neighbor query* finds all objects $o \in S$ satisfying $Pr_{nn}(o, p_q) \geq t_q$, where $Pr_{nn}(o, p_q)$ is the *NN probability* of o. Specifically, $Pr_{nn}(o, p_q)$ is calculated as:

$$Pr_{nn}(o, p_q) = \int_{x \in o.ur} o.pdf(x) \cdot Pr_{nn}(o, p_q | o = x) dx \tag{11.11}$$

where $Pr_{nn}(o, p_q | o = x)$ is the probability that o is the NN of p_q by being at point x. This is also the probability that *all* the other objects in S are outside the circle $\odot(p_q, \|x, p_q\|)$ that centers at p_q and has radius equal to the distance $\|x, p_q\|$ between x and p_q. We refer to $Pr_{nn}(o, p_q)$ as the *qualification probability* of object o.

To illustrate, consider Figure 11.6 where the polygons represent the uncertainty regions of three objects o_1, o_2, o_3. To calculate the NN probability $Pr_{nn}(o_1, p_q)$ of o_1, conceptually we need to compute the $Pr_{nn}(o, p_q|o = x)$ for all locations x in the uncertainty region of o_1. Assume that x is the black dot as shown in Figure 11.6. The circle is $\odot(p_q, \|x, p_q\|)$. Let u (v) be the probability that o_2 (o_3) appears in the left (right) shaded area. Then, $Pr_{nn}(o, p_q|o = x)$ equals $u \cdot v$.

3.2 Query Processing

Cheng et al. [5] develop an algorithm for solving probability thresholiding NN search following the filter refinement framework. The filter step retrieves a *candidate set* where every object has a non-zero NN-probability. In other words, all objects outside the candidate set have no chance of being the NN of the query point p_q, and therefore, do not need to be considered. The refinement step simply calculates the NN-probabilities of all the objects in the candidate set according to Equation 11.11.

To illustrate the filter step, let us define two metrics. Given an uncertainty region ur, let $mindist(p_q, ur)$ be the shortest distance between the query point p_q and any point on the boundary of ur. Conversely, let $maxdist(p_q, u_r)$ be the longest distance. For example, given point p_q and the uncertainty region $o.ur$ as shown in Figure 11.6a, $mindist(p_q, o.ur)$ equals the length of segment p_qA, and $maxdist(p_q, o.ur)$ equals that of segment p_qB.

The filter step identifies a *minmax circle* and selects all objects whose uncertainty regions intersect the circle. Specifically, the minmax circle is a circle that centers at the query point p_q, and its radius r equals the smallest maxdist from p_q to the uncertainty regions of all objects, namely:

$$r = \min_{o \in S} maxdist(p_q, o.ur). \tag{11.12}$$

As an example, consider that the dataset S has 4 objects o_1, o_2, ..., o_4 whose uncertainty regions are presented in Figure 11.7b. Then, the radius r of the minmax circle is determined by the maxdist between p_q and $o_2.ur$, as all other objects have greater maxdist to p_q. Figure 11.7b shows the minmax circle in dashed line. The uncertainty regions of all the objects except o_4 intersect the circle. Hence, the candidate set is $\{o_1, o_2, o_3\}$. It is easy to see that o_4 cannot be the NN of p_q, because o_2 definitely is closer to p_q, regardless of where o_2 and o_4 are located in their uncertainty regions. We refer to this method as the *minmax algorithm*.

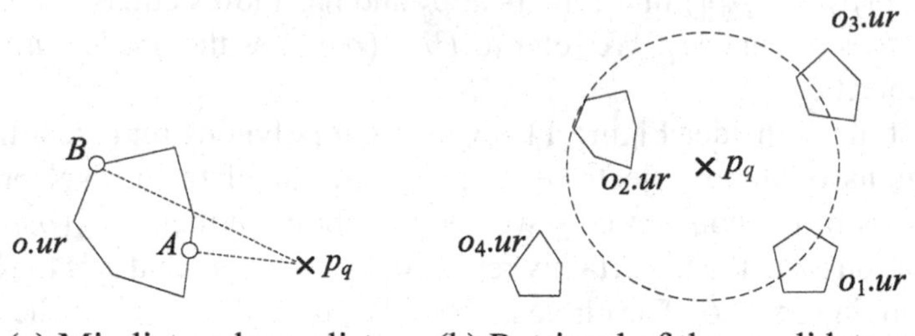

Figure 11.7. Illustration of the filter step

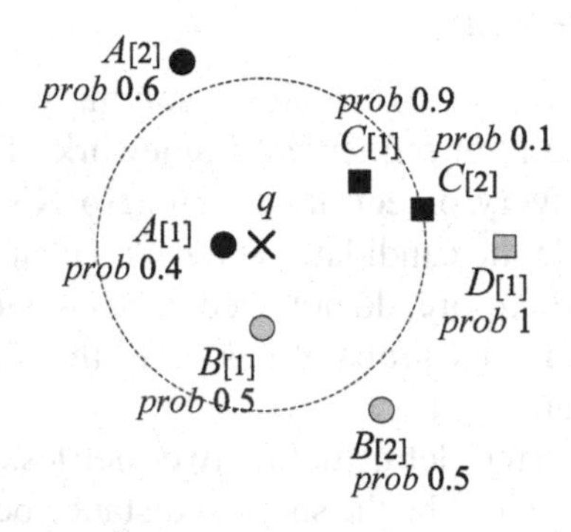

Figure 11.8. Illustration of calculating an NN probability

Kriegel et al. [17] develop another solution which assumes *discrete pdfs*. Specifically, the probability density function $o.pdf(\cdot)$ of an object o is defined by a set of *instances*, each of which is a point, and is associated with a probability. Formally, let $o.s$ be the number of instances of o, and denote them as $o[1]$, $o[2]$, ..., $o[o.s]$. The probability associated with instance $o[i]$ $(1 \leq i \leq s)$ is $o.pdf(o[i])$. The probabilities of all the instances must sum up to 1, namely:

$$\sum_{i=1}^{o.s} o.pdf(o[i]) = 1. \tag{11.13}$$

Figure 11.8 provides an example with four objects A, B, C, and D. Except D, which has only one instance, all other objects have two instances each, i.e., $A.s = B.s = C.s = 2$ and $D.s = 1$. For instance, the first (second) instance $A[1]$ ($A[2]$) of A carries a probability 0.4 (0.6), indicating that A may be located at $A[1]$ ($A[2]$) with 40% (60%) chance.

Note that discrete pdfs still fit into the modeling of uncertain objects we have been using. In particular, for an object o with a discrete $o.pdf(\cdot)$, its uncertainty region $o.ur$ is not a continuous area, but instead, a set of $o.s$ points corresponding to the instances of o. The *minmax algorithm* described earlier with Figure 11.7 can also be applied to perform NN search on discrete pdfs. Specifically, the filter step first decides the minmax circle as the circle in Figure 11.8, and retrieves the candidate set $\{A, B, C\}$. D is not a candidate because it has no instance in the circle. Then, the refinement step calculates the NN-probabilities of A, B, C from their instances.

In terms of indexing, discrete pdfs provide an important flexibility that continuous pdfs do not have. That is, the instances of an object can be stored in different parts of a data structure, as opposed to a continuous pdf that must be stored as a whole. It turns out that such a flexibility permits the development of an *incremental* algorithm [8, 17] that is faster than the *minmax algorithm*.

Incremental assumes that the instances of all objects are indexed by an R-tree. It invokes the *best first* algorithm [12] to retrieve the instances in ascending order of their distances to the query point p_q. During the retrieval, for every object such that at least one of its instances has been seen, *Incremental* maintains both a lower bound and an upper bound of the NN-probability of the object. Once the lower bound reaches the probability threshold t_q, the object is returned as a result. On the other hand, as soon as the upper bound drops below t_q, the object is pruned.

Let us use the example in Figure 11.8 to demonstrate the algorithm, assuming that the probability threshold t_q equals 0.4. $A[1]$ is the first instance retrieved, since it is nearest to the query point p_q. At this moment, we know that A has at least 40% probability to be the NN of p_q. This is because A has 40% chance to be located at $A[1]$, and when it does, no other object can be closer to p_q. Hence, *incremental* reports A as a result. The next two instances fetched are $B[1]$ and $C[1]$. The algorithm terminates here by claiming that there is no more result. In particular, now we can assert that B has at most 0.33 probability to be the NN. To see this, note that B can be the NN only under two events: (i) A is not at $A[1]$, and B is at $B[1]$; (ii) A is not at $A[1]$, C is not at $C[2]$, and B is at an instance that has not been found yet. Event (i) occurs with 0.3 probability, which is the product of the probability 0.6 of $A \neq A[1]$ and the probability 0.5 of $B = B[1]$. Event (ii) occurs with 0.03 probability, which is the product of the probability 0.6 of $A \neq A[1]$, the probability 0.1 of $C \neq C[1]$, and the probability 0.5 of $B \neq B[1]$. Similarly, we can also claim that the NN-probability of C is at most 0.2, and the NN-probability of any object other than A, B, C is at most 0.18. Therefore, *incremental* concludes that no more result is possible.

3.3 Variations of Nearest Neighbor Retrieval

Cheng et al. [4] observe that, in practice, a user may tolerate some small error in objects' qualification probabilities. Specifically, even if an object's qualification probability may be lower than the probability threshold t_q, it is still acceptable, as long as it is *not much* lower. Motivated by this, Cheng et al. [4] proposed the *constrained nearest neighbor query*. Specifically, given a query point p_q, a probability threshold t_q, and a *tolerance threshold* Δ_q, a constrained nearest neighbor query returns a set of objects o whose NN-probabilities must be at least $t_q - \Delta_q$. Such a query can be processed faster than a conventional probability thresholding NN query, because once we know that an object's NN-probability has a lower bound at least $t_q - \Delta_q$, we can immediately return the object, without having to calculate the NN-probability more precisely.

Recall that our discussion of range search addresses both a nonfuzzy and a fuzzy version. Adapting the taxonomy to NN retrieval, the NN query we have been analyzing is *nonfuzzy* because the query point p_q is a precise location. Similarly, we can also define fuzzy NN search. Specifically, let S be a set of uncertain objects. Given an uncertain object q and a probability threshold t_q, a *fuzzy NN query* retrieves all objects $o \in S$ satisfying $Pr_{fnn}(o, q) \geq t_q$. Here, $Pr_{fnn}(o, q)$ is the probability that o is the NN of q, and can be obtained as:

$$Pr_{fnn}(o, q) = \int_{p_q \in q.ur} q.pdf(p_q) \cdot Pr_{nn}(o, p_q) dp_q \tag{11.14}$$

where $Pr_{nn}(\cdot, \cdot)$ is given in Equation 11.11. In [17], Kriegel et al. extend the *incremental* algorithm described earlier to solve fuzzy NN queries, utilizing an interesting idea of clustering the instances of an object into small groups.

Furthermore, the *minmax algorithm* [5] can also be easily extended to fuzzy NN retrieval. Towards this, given an uncertain object o, let $mindist(o.ur, q.ur)$ be the shortest distance of any two points in $o.ur$ and $q.ur$, respectively. Conversely, let $maxdist(o.ur, q.ur)$ be the longest distance. In the filter step, we first decide a value r, which is the minimum of the maxdist of all objects:

$$r = \min_{o \in S} maxdist(o.ur, q.ur). \tag{11.15}$$

Then, we create a candidate set including all the objects o satisfying $mindist(o.ur, q.ur) \leq r$. Finally, the refinement step calculates the qualification probability of each object by Equation 11.14.

So far our discussion has focused on single nearest neighbor search. It appears that k nearest neighbor retrieval on uncertain multidimensional data has not been specifically studied yet. Nevertheless, there has been considerable work [9, 13, 19, 26, 35] on *top-k search* on uncertain data. Conventionally, when the dataset contains only precise objects, a top-k query retrieves the k objects that minimize a certain preference function. Interestingly, a kNN query can be regarded as a form of top-k search, where the preference function is the distance between an object and the query point. On uncertain data, several variations of top-k retrieval have been proposed. Next, we adapt those variations to kNN search on uncertain objects.

For simplicity, we first consider discrete pdfs. Let S be a set of n objects o_1, o_2, ..., o_n. The probability density function $o_i.pdf(\cdot)$ of each object o_i $(1 \leq i \leq n)$ has non-zero values at $o_i.s$ instances $o_i[1]$, $o_i[2]$, ..., $o_i[o_i.s]$. This is equivalent to say that the location of o_i has $o_i.s$ *options*, each chosen with a probability determined by $o_i.pdf(\cdot)$. The options of different objects are independent. Imagine we randomly choose an option for every object following its pdf. As each option is a point, the n options translate to a set W of n points. This W is a possible *configuration* of the objects' actual locations. This configuration occurs with a probability $Pr(W)$ that equals the product of the probabilities of all options.

Reasoning with W is much easier. In particular, it is a *precise* dataset with n points, because we have fixed the location of every object to one of its instances. Hence, it is unambiguous which k points from W are the k NNs of the query point p_q. Use $kNN(W)$ to denote an ordered set of those k NN, sorted in ascending order of their distances to p_q. Call $kNN(W)$ the *kNN-set of* W. Apparently, as W occurs with probability $Pr(W)$, $kNN(W)$ is the real kNN result also with probability $Pr(W)$.

The above discussion gives an *enumeration* approach to tackle uncertainty. Essentially we simply enumerate all the possible configurations, and derive the kNN result for every configuration. Next, we can decide what we want by summarizing all these configurations in a statistical manner. Many decisions are possible, and they lead to different versions of uncertain kNN search. Next, we describe three of them, corresponding to three formulations of uncertain top-k retrieval proposed in [13, 26].

- *U-topk-NN.* It is possible for two different configurations W_1 and W_2 to have the same kNN-set, i.e., $kNN(W_1) = kNN(W_2)$. In other words, the *aggregated probability* of a kNN-set S_{knn} should be the sum of the probabilities of all configurations W such that $kNN(W) = S_{knn}$. *U-topk-NN* returns the kNN-set with the largest aggregated probability.

- *U-kRanks-NN.* Sometimes a user may want to know which object has the highest probability of being the NN of p_q. Also, independently, which object has the highest probability of being the second NN of p_q. Similarly for the 3rd, ..., and up to the k-th NN. *U-topk-NN* returns exactly this information. Specifically, given an object o and a value of $i \in [1, k]$, its *iNN-probability* equals the sum of the probabilities of all configurations W where o is the i-th NN of p_q. *U-topk-NN* returns the objects with the greatest 1NN-, 2NN-, ..., kNN-probabilities, respectively. Note

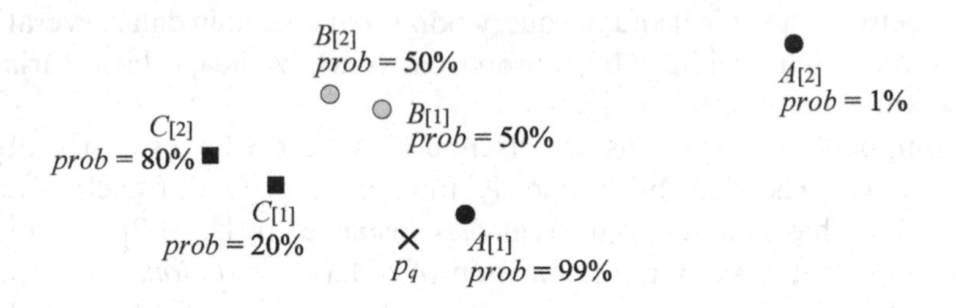

Figure 11.9. NN retrieval by expected distances

that these objects are not necessarily different. It is possible for the same object to have the greatest iNN-probability for multiple values of i.

- *Probabilistic topk-NN.* In some cases, a user may not be interested in the ordering of the k NNs. Instead, s/he may care about only *whether* an object is in the kNN set. Motivated by this, given an object o, we can define its *kNN-probability* to be the sum of the probabilities of all configurations W where o belongs to the kNN-set $kNN(W)$. Given a probability threshold t_q, *probabilistic topk-NN search* returns all objects whose kNN-probabilities are at least t_q.

The above queries are adapted from *U-topk* [26], *U-kRanks* [26], and *probabilistic top-k* [13] respectively. In [13, 26], the authors also provide algorithms for solving these top-k problems. Those algorithms, as well as their improved versions [35], can be modified to solve the above kNN problems as well. It is worth mentioning that, in the literature, a configuration W in our earlier discussion is often termed a *possible world*.

Finally, it is worth mentioning that NN search on uncertain data can also be performed based on objects' *expected distances* to the query point p_q. Specifically, given an uncertain object o with uncertainty region $o.ur$ and probability density function $o.pdf(\cdot)$, its expected distance to p_q equals

$$\int_{x \in o.ur} o.pdf(x) \cdot \|x, p_q\| dx. \tag{11.16}$$

Thus, we may simply return the object with the smallest expected distance [20]. This approach has two advantages. First, for many regular pdfs such as uniform and normal distributions, expected distances effectively reflect the relative superiority of objects. Second, extension to kNN search is trivial.

Not surprisingly, for non-regular pdfs, the expected distance is not a reliable indicator of the quality of an object. To understand this, consider Figure 11.9, which shows the possible instances of objects A, B, and C. Intuitively, A is the best object, because it is almost sure (i.e., with 99% probability) to be the

NN of p_q (A is at $A[1]$ with 99% likelihood, and $A[1]$ is nearer to p_q than all the instances of B and C). However, A has a large expected distance, because its instance $A[2]$ is faraway from q. In fact, without affecting the NN-probability of A, we can *arbitrarily* increase the expected distance of A, by pushing $A[2]$ sufficiently away from p_q.

4. Summary

We started this chapter by reviewing the applications that require management of uncertain spatiotemporal data. It is clear that the presence of uncertainty introduces numerous issues that do not exist in conventional precise spatiotemporal databases. This creates the opportunity of formulating novel query types that extend the traditional queries by taking uncertainty into account. We discussed the existing formulations of uncertain range search and nearest neighbor retrieval. A major difference between these queries and their counterparts on precise data is that, a query on uncertain data should avoid retrieving objects with low qualification probabilities. This is usually achieved by either specifying a probability threshold, or returning only the few objects with the largest qualification probabilities. We also surveyed the known algorithms for answering uncertain queries, and elaborated their underlying ideas.

Although our discussion focuses on range search and nearest neighbor retrieval due to their vast importance in practice, we note that other types of queries have also been studied on uncertain multidimensional data. For example, in the context of privacy preservation, Mokbel et al. [21] and Kalnis et al. [14] analyze how to use cloaked rectangles to answer range and NN queries conservatively. Lian and Chen [18] investigate the *group nearest neighbor query* [22]. Kriegel et al. [16] and Cheng et al. [6] propose algorithms for *probabilistic spatial joins*. Xu and Jacobsen [34] consider the problem of continuously monitoring of *n-body constraints* on uncertain objects. Pei et al. [23] study *probabilistic skylines*, and Lian and Chen [19] discuss the related problem of *bichromatic skylines* [9].

References

[1] C. Bohm, A. Pryakhin, and M. Schubert. The gauss-tree: Efficient object identification in databases of probabilistic feature vectors. In *ICDE*, 2006.

[2] F. Bonchi, O. Abul, and M. Nanni. Never walk alone: Uncertainty for anonymity in moving objects databases. In *ICDE*, pages 376–385, 2008.

[3] J. Chen and R. Cheng. Efficient evaluation of imprecise location-dependent queries. In *ICDE*, pages 586–595, 2007.

[4] R. Cheng, J. Chen, M. Mokbel, and C.-Y. Chow. Probabilistic verifiers: Evaluating constrained nearest-neighbor queries over uncertain data. In *ICDE*, 2008.

[5] R. Cheng, D. V. Kalashnikov, and S. Prabhakar. Querying imprecise data in moving object environments. *TKDE*, 16(9):1112–1127, 2004.

[6] R. Cheng, S. Singh, S. Prabhakar, R. Shah, J. S. Vitter, and Y. Xia. Efficient join processing over uncertain data. In *CIKM*, pages 738–747, 2006.

[7] R. Cheng, Y. Xia, S. Prabhakar, R. Shah, and J. S. Vitter. Efficient indexing methods for probabilistic threshold queries over uncertain data. In *VLDB*, pages 876–887, 2004.

[8] X. Dai, M. L. Yiu, N. Mamoulis, Y. Tao, and M. Vaitis. Probabilistic spatial queries on existentially uncertain data. In *SSTD*, pages 400–417, 2005.

[9] E. Dellis and B. Seeger. Efficient computation of reverse skyline queries. In *VLDB*, pages 291–302, 2007.

[10] B. Gedik and L. Liu. Location privacy in mobile systems: A personalized anonymization model. In *ICDCS*, pages 620–629, 2005.

[11] G. Ghinita, P. Kalnis, and S. Skiadopoulos. Prive: Anonymous location-based queries in distributed mobile systems. In *WWW*, pages 371–380, 2007.

[12] G. R. Hjaltason and H. Samet. Distance browsing in spatial databases. *TODS*, 24(2):265–318, 1999.

[13] M. Hua, J. Pei, W. Zhang, and X. Lin. Ranking queries on uncertain data: A probabilistic threshold approach. In *SIGMOD*, 2008.

[14] P. Kalnis, G. Ghinita, K. Mouratidis, and D. Papadias. Preventing location-based identity inference in anonymous spatial queries. *IEEE Trans. Knowl. Data Eng.*, 19(12):1719–1733, 2007.

[15] G. Kollios, D. Gunopulos, and V. J. Tsotras. On indexing mobile objects. In *PODS*, pages 261–272, 1999.

[16] H.-P. Kriegel, P. Kunath, M. Pfeifle, and M. Renz. Probabilistic similarity join on uncertain data. pages 295–309, 2006.

[17] H.-P. Kriegel, P. Kunath, and M. Renz. Probabilistic nearest-neighbor query on uncertain objects. pages 337–348, 2007.

[18] X. Lian and L. Chen. Probabilistic group nearest neighbor queries in uncertain databases. *TKDE*.

[19] X. Lian and L. Chen. Probabilistic ranked queries in uncertain databases. In *EDBT*, 2008.

[20] V. Ljosa and A. K. Singh. Apla: Indexing arbitrary probability distributions. In *ICDE*, pages 946–955, 2007.

[21] M. F. Mokbel, C.-Y. Chow, and W. G. Aref. The new casper: Query processing for location services without compromising privacy. In *VLDB*, pages 763–774, 2006.

[22] D. Papadias, Y. Tao, K. Mouratidis, and C. K. Hui. Aggregate nearest neighbor queries in spatial databases. *TODS*, 30(2):529–576, 2005.

[23] J. Pei, B. Jiang, X. Lin, and Y. Yuan. Probabilistic skylines on uncertain data. In *VLDB*, pages 15–26, 2007.

[24] S. Saltenis, C. S. Jensen, S. T. Leutenegger, and M. A. Lopez. Indexing the positions of continuously moving objects. In *SIGMOD*, pages 331–342, 2000.

[25] P. Samarati. Protecting respondents' identities in microdata release. *TKDE*, 13(6):1010–1027, 2001.

[26] M. A. Soliman, I. F. Ilyas, and K. C.-C. Chang. Top-k query processing in uncertain databases. In *ICDE*, pages 896–905, 2007.

[27] L. Sweeney. Achieving k-anonymity privacy protection using generalization and suppression. *International Journal on Uncertainty, Fuzziness and Knowledge-based Systems*, 10(5):571–588, 2002.

[28] Y. Tao, R. Cheng, X. Xiao, W. K. Ngai, B. Kao, and S. Prabhakar. Indexing multi-dimensional uncertain data with arbitrary probability density functions. In *VLDB*, pages 922–933, 2005.

[29] Y. Tao and D. Papadias. Mv3r-tree: A spatio-temporal access method for timestamp and interval queries. In *VLDB*, pages 431–440, 2001.

[30] Y. Tao, X. Xiao, and R. Cheng. Range search on multidimensional uncertain data. *TODS*, 32(3), 2007.

[31] G. Trajcevski, O. Wolfson, K. Hinrichs, and S. Chamberlain. Managing uncertainty in moving objects databases. *TODS*, 29(3):463–507, 2004.

[32] O. Wolfson, S. Chamberlain, S. Dao, L. Jiang, and G. Mendez. Cost and imprecision in modeling the position of moving objects. In *ICDE*, pages 588–596, 1998.

[33] O. Wolfson, A. P. Sistla, S. Chamberlain, and Y. Yesha. Updating and querying databases that track mobile units. *Distributed and Parallel Databases*, 7(3):257–387, 1999.

[34] Z. Xu and H.-A. Jacobsen. Evaluating proximity relations under uncertainty. In *ICDE*, pages 876–885, 2007.

[35] K. Yi, F. Li, G. Kollios, and D. Srivastava. Efficient processing of top-k queries in uncertain databases. In *ICDE*, pages 1406–1408, 2008.

Chapter 12

PROBABILISTIC XML

Edward Hung
Department of Computing,
The Hong Kong Polytechnic University, Hong Kong
csehung@comp.polyu.edu.hk

Abstract Interest in XML databases has been expanding rapidly over the last few years. In this chapter, we will study the problem of incorporating probabilistic information into XML databases. First we will consider the sources of probabilistic XML data. We will then describe the initial attempts in using XML tags to represent simple uncertainty in text and XML data. After that, we will describe Zhao et al's framework of Semi-structured Probabilistic Objects (SPOs), which uses semi-structured data to describe probabilistic data. Next, we will describe Nierman and Jagadish's proposal of ProTDB which considers uncertainty in the structure of XML data with limited probabilistic distribution in point probability. Finally we will see how Hung et al.'s PXML and PIXML models handle uncertain structures with arbitrary distributions (in point probability and interval probability), which also provide a set of algebraic operations, queries, and aggregate operators.

Keywords: Probabilistic XML, probabilistic semistructure database, algebra, aggregate, interval probability

1. Introduction

Over the last few years, there has been considerable interest in Extensible Markup Language (XML) databases. A proliferation of semi-structured data models has been proposed [1–4], along with associated query languages [5, 6] and algebras [7, 8]. XML is a simple but very flexible markup language derived from SGML, which is now mainly used for exchange, transmission and manipulation of data on the web [9]. XML tags are not predefined, which means that it gives users flexibility to define their own tags according to their domains and applications.

XML has the advantage of not placing hard constraints on the *structure* of the data. Nowadays, data is more often generated in XML for easier data transmission and manipulation over the web. Since an XML instance specifies deterministic relationships between objects, in cases where we would also like to avoid hard constraints on the object-level structure, it is necessary to have a model for applications in some domains to represent, store and manipulate uncertainty over the relationships between objects.

This uncertainty is necessary when relationships between objects and values for attributes of objects are not known with absolute certainty. For example, this occurs when the sensor inputs are noisy. This provides new challenges on how to manipulate and maintain such new kinds of database systems.

2. Sources of Uncertainty in XML Data

A common source for this uncertainty comes when a semi-structured representation is constructed from a noisy input source: uncertainty in sensor readings, information extraction using probabilistic parsing of input sources and image processing all may result in a semi-structured instance in which there is uncertainty. Another source for this uncertainty comes from the need to represent nondeterministic processes using a semi-structured model. In this case, it may be desirable to represent the distribution over possible substructures explicitly, rather than forcing a particular choice. Examples where this may hold include biological domains, manufacturing processes and financial applications.

There are numerous applications for which a probabilistic XML data model is quite natural and for which a query language that supports probabilistic inference provides important functionality. Probabilistic inference supports capabilities for predictive and 'what-if' types of analysis. For example, consider the use of a variety of predictive programs[10] for the stock market. Such programs usually return probabilistic information. If a company wanted to export this data into an XML format, they would need methods to store probabilistic data in XML. The financial marketplace is a hotbed of both predictive and XML activity (e.g. the FIX standard for financial data is XML based). There is the same need to store probabilistic data in XML for programs that predict expected energy usage and cost, expected failure rates for machine parts, and in general, for any predictive program. Another useful class of applications where there is a need for probabilistic XML data is image processing programs that process images (automatically) using image identification methods and store the results in an XML database. Such image processing algorithms often use statistical classifiers[11] and often yield uncertain data as output. If such information is to be stored in an XML database, then it would be very useful to have the ability to automatically query this uncertain information. Another important application is in automated manufacturing monitoring and diagnosis. A corporate manufacturing floor may use sensors to track what happens on the manufacturing floor. The results of the sensor readings may be automatically piped to a fault diagnosis program that may identify zero, one, or many possible faults with a variety of probabilities on the space of faults. When such analysis is stored in a database, there is a natural need for probabilities.

In addition to these types of applications, the NSIR system for searching documents at the University of Michigan[12] returns documents based along with probabilities. Search engines such as Google return documents relevant to a user query but not return answers to user questions. NSIR is an architecture that augments existing search engines to support natural language question answering. NSIR takes some probabilistic approaches (such as probabilistic phrase reranking (PPR) using proximity and question type features) to the stages of passage extraction, phrase extraction, and answer ranking. Queries with appropriate follow-up questions are repeatedly submitted to NSIR in order to populate a probabilistic XML databases used in [13].

Likewise, Nierman, et al. point out the use of probabilistic semi-structured databases in scientific areas such as protein chemistry[13]. Experimental errors in scientific data are common due to the measurement impreciseness of the equipments. One challenge in proteomics is to analyze the production of different proteins under different conditions. To identify a protein, different tools with varying degrees of reliability may be used. It is crucial to the analysts how to effectively and efficiently model the identity of a protein (and its degree of certainty) in a model.

3. Modeling Uncertainty using Tags

The initiative of the guidelines of Text Encoding Initiative (TEI)[14] is to provide an effective representation of features in a text which need to be identified explicitly for facilitating the processing of the text by computer programs. The guidelines include a set of markers (or tags) which may be inserted in the electronic representation of the text so as to mark the text structure and other textual features. Following the ratification of XML recommendation in 1998 and its rapid adoption, the TEI Consortium was formed in 2001 to revise both the text and the DTDs of the scheme of TEI in a way which supported XML unambiguously. One particular item addressed is to indicate that some encoded text are uncertain, and to indicate who is responsible for the markup of the electronic text. The guidelines provide three methods of recording this uncertainty.[15]

The first method to record uncertainty is to attach a note (using tag `<note>`) to the element or location about which is uncertain. In the following paragraph, for example, it might be uncertain whether "Essex" should be marked as a place name or a personal name, since both could be possible in the given context:

```
  Elizabeth went to Essex.  She had always liked
Essex.
```

The following shows how uncertainty is recorded using `<note>`.

```
  <persName>Elizabeth</persName> went to
  <placeName id="p1"> Essex</placeName>.  She had
  always liked <placeName id="p2">
Essex</placeName>.
  <note type="uncertainty" resp="MSM" target="p1
p2">
  It is not clear here whether <mentioned> Essex
  </mentioned> refers to the place or to the
  nobleman.  If the latter, it should be tagged as
  a personal name. -MSM</note>
```

In spite of its relative simplicity, this technique cannot convey the nature and degree of uncertainty systematically and thus is not suitable for automatic processing. The `<certainty>` element may be used to record uncertainty in a more structured way for at least simple automatic processing,

The following example uses `<certainty>` element to indicate the element in question (`target = "p1"`), where (i) the `locus` attribute indicates what aspect of the markup we are uncertain about (in this case, whether we have used the correct element type), (ii) the `degree` attribute records the degree of confidence, and (iii) the `assertedValue` attribute provides an alternative choice of generic identifier (in this case `<persName>`).

```
  Elizabeth went to
  <placeName id="p1">Essex</placeName>.
  <!- 60% chance that P1 is a placename,
  40% chance a personal name.  ->
  <certainty target="p1" locus="#gi"
  desc="probably a placename, but possibly not"
  degree="0.6"/>
  <certainty target="p1" locus="#gi"
  assertedValue="persName"
  desc="may refer to the Earl of Essex"
degree="0.4"/>
```

Furthermore, conditional probability can be represented using `given` attribute. For example, in the sentence “Elizabeth went to Essex; she had always liked Essex,” we may feel there is a 60 percent chance it means the county, and a 40 percent chance it means the earl. Additionally, we think that the two occurrences of the word should agree with each other, i.e., there is no chance at all that one occurrence refers to the county and one to the earl.

```
Elizabeth went to <placeName id="p1">
Essex</placeName>.  She had always liked
<placeName id="p2">Essex</placeName>.
<!- 60% chance that P1 is a placename,
40% chance a personal name.  ->
<certainty id="cert-1" target="p1" locus="#gi"
desc="probably a placename, but possibly not"
degree="0.6"/>
<certainty id="cert-2" target="p1" locus="#gi"
desc="may refer to the Earl of Essex"
assertedValue="persName" degree="0.4"/>
<!- 60% chance that P2 is a placename,
40% chance a personal name.
100% chance that it agrees with P1.  ->
<certainty target="p2" locus="#gi" given="cert-1"
desc="if P1 is a placename, P2 certainly is"
degree="1.0"/>
<certainty target="p2" locus="#gi"
assertedValue="persName" given="cert-2"
desc="if p1 refers to the Earl of Essex,
so does P2" degree="1.0"/>
```

The usage of `<certainty>` tag in TEI Guidelines is an early initiative to bring uncertainty and XML together. However, their motivation is to mark the text structure. Although XML tags are used, the underlying data structure is still a plain text. Thus, it does not make a good use of the flexibility and power of XML to represent semi-structured data. A better example is the work of Ustunkaya et al.[16]. They proposed to represent uncertainty using attribute `FuzzyPredicate` to relate the fuzzy (possible) values of XML elements. For example, the following represents an image to have a color of red or green:

```
<ProductInfo>
  <book>
    <image>
      <colors>
        <color FuzzyPredicate="OR">red</color>
        <color FuzzyPredicate="OR">blue</color>
      </colors>
    </image>
    ...
  </book>
</ProductInfo>
```

A range of values can be represented by some specific tags to denote the minimum and maximum values like the following:

ω: S1	
DA	P
A	0.4
B	0.1
C	0.2
F	0.3

ω: S2		
DA	LY	P
A	A	0.01
A	B	0.02
A	C	0.2
A	F	0.01
B	A	0.05
B	B	0.12
...	...	...
F	C	0.03
F	F	0.01

ω: S3		
city: Lexington job: Managerial		
DA	LY	P
A	A	0.01
A	B	0.02
A	C	0.2
A	F	0.01
B	A	0.05
B	B	0.12
...	...	...
F	C	0.03
F	F	0.01

ω: S4		
city: Lexington job: Managerial		
DA	LY	P
A	A	0.01
A	B	0.02
A	C	0.2
A	F	0.01
B	A	0.05
B	B	0.12
...	...	...
F	C	0.03
F	F	0.01
SE = B DR $\in$ { A, B }		

Figure 12.1. Different types of probabilistic information to be stored in the database for risk analysis applications (from left to right: single variable (Driver Age (DA)) probability distribution, joint probability distribution of two variables (Drive Age (DA) and License Years (LY)), joint probability distribution with context (city and job), and conditional joint probability distribution with context (given the condition specified on variables SE and DR).[17]

```
<ProductInfo>
  <book>
    <minPrice>$12.92</minPrice>
    <maxPrice>$80.00</maxPrice>
  </book>
</ProductInfo>
```

Using XML tags to represent semi-structured data with uncertainty will be described in later section.

4. Modeling Uncertainty using Semi-structured Data

In the work of Zhao et al.[17], a model was proposed that allows probabilistic information to be stored using semi-structured databases. It was the first to deal with probabilities and semi-structured data. They pioneered the integration of probabilities and semi-structured data by introducing a semi-structured model to support storage and querying of probabilistic information in flexible forms such as (i) a simple interval probability distribution, (ii) a joint interval probability distribution, or (iii) a simple or joint conditional interval probability distribution. Their model allows us to use an object (semi-structured probabilistic object or SPO) to represent the probability table of one or more random variables, the extended context and the extended conditionals. Intuitively, contexts provide information about when a probability distribution is applicable. The formal definition of a SPO is as follow:

DEFINITION 12.1 *A Semi-structured Probabilistic Object (SPO) S is defined as a tuple* $S =< T, V, P, C, \omega >$, *where*

- *T is the context of S and provides supporting information for a probability distribution. It includes the known values of certain parameters, which are not considered to be random variables by the application.*
- *V is a set of random variables that participate in S and determine the probability distribution described in S. We require that* $V \neq \emptyset$.
- *P is the probability table of S. If only one random variable participates, it is a simple probability distribution table; otherwise the distribution will be joint. A probability table may be complete, when the information about the probability of every instance is supplied, or incomplete.*
- *C is the conditional of S. A probability table may represent a distribution, conditioned by some prior information. The conditional part of its SPO stores the prior information in one of two forms: "random variable* u *has value* x*" or "the value of random variable* u *is restricted to a subset X of its values".*
- ω *is a unique identifier when S is inserted into the database.*

For example, in Figure 12.1, the rightmost table illustrates a SPO where (from top to bottom) (i) T states the context (city is Lexington and race is Asian), (ii) V specifies the two random variables (Drive Age (DA) and License Years (LY)), (iii) P shows the joint probability table with the last column P containing the probability of each possible combination of values of DA and LY, (iv) C indicates the conditional ($SE = B, DR \in \{A, B\}$).

Zhao et al. also developed an elegant algebra and a prototype implementation to query databases of such SPOs. The resulting algebra called Semi-structured Probabilistic Algebra (SP-Algebra) contains three standard set operations (union, intersection, difference) and extends the definitions of standard relational operations (selection, projection, Cartesian product and join). A new operation called conditionalization[18] is also defined which returns the conditional probability distributions of input SPOs.

The SPO model appears to be similar to PXML described in the later section but in fact it is quite different. An SPO itself can be represented in a semi-structured way, but its main body is just a flat table. It cannot show the semi-structured relationship among variables. Only contexts (but not random variables) are represented in a semi-structured form. Contexts are "regular relational attributes", i.e., the context provides already known information when the probability distribution is given on real "random variables". Detailed comparison will be found in later section.

5. Modeling Uncertainty in XML with Independent or Mutually Exclusive Distribution in Point Probability

More recently, Nierman and Jagadish developed a framework called ProTDB to extend the XML data model to include probabilities[13]. Their approach addresses the several modeling challenges of XML data: due to its structure, due to the possiblility of uncertainty association at multiple granularities, and due to the possibility of missing and repeated sub-elements. They demonstrated to manage probabilistic XML data extracted from the web using a natural language analysis system NSIR. Since there are many possible sources of error in the query processing, NSIR returns multiple possible answers, each with a probability. A probabilistic XML database was generated by repeatedly submitting queries. Examples of queries and answers include:

- What is the name of the President of the United States?
  - George Bush (0.7); George W. Bush (0.4); Bill Clinton (0.2)
- How old is Bill Clinton?
  - 55 (0.3)

They consider several issues in the representation of uncertainty in XML data. Instead of a probability associated with a tuple as in a probabilistic relational model, a probability is associated with an element or even an attribute value in ProTDB model. A probability associated with an element is treated as the existential probability that the state of the world includes this element and the sub-tree rooted at it. More formally, each node is dependent upon its root to node chain. Each probability p on an element in the XML document is assigned conditioned on the fact that the parent element exists, i.e., if the parent exists, then the element's probability equals p; when the parent does not exist, then the element's probability equals 0. Consider a chain $A \leftarrow B \leftarrow C$ from root A. The following probabilities are assigned to nodes A, B and C: $Prob(A), Prob(B|A), Prob(C|B)$. To obtain $Prob(B)$, the probability that B exists in some state of the world, we can use Bayes' formula: $Prob(B|A) = \dfrac{Prob(A|B) \times Prob(B)}{Prob(A)}$. However, a parent must exist if its children exist, so $Prob(A|B) = 1$. Therefore,

$Prob(B) = Prob(B|A) \times Prob(A)$. By default, all probabilities between nodes not in an ancestor-descendant chain are independent. However, it also allows sibling nodes to be specified as "mutually-exclusive".

XML DTD's are extended to use a `Prob` attribute for each element. To enable the use of this attribute, for each element:

`<!ELEMENT elementName ...>`, the DTD should contain the following:

`<!ATTLIST elementName Prob CDATA "1.0">`.

The DTD should also define the distribution (`Dist`) and value (`Val`) elements as follows:

```
<!ELEMENT Dist (Val+)>
<!ATTLIST Dist type (independent
| mutually-exclusive) "independent">
<!ELEMENT Val (#PCDATA)>
<!ATTLIST Val Prob CDATA "1">
```

For a leaf element `curElement` (i.e., it contains only text, or #PCDATA), its original definition in the DTD:

`<!ELEMENT curElement (#PCDATA)>`

will be changed to:

`<!ELEMENT curElement (#PCDATA|Dist)>`

For non-leaf element `curElement`, two changes will be made to the DTD:

1. Change the element definition from: `<!ELEMENT curElement (prev-def)>` to: `<!ELEMENT curElement ((prev-def)| Dist)>` (where `prev-def` is the original definition of `curElement`)

2. and add this element's previous definition to the `Val` construct: `<!ELEMENT Val (X)>` to be: `<!ELEMENT Val (X | (prev-def))>`

The following shows a fragment of an XML document for input to ProTDB system[13].

```
1.<countries>
2.   <country Prob='.9'>
3.    <countryName>United States</countryName>
4.    <coordinates Prob='.9'>
5.     <latitude>
6.      <direction>North</direction>
7.      <degrees Prob='.8'>38</degrees>
8.      <minutes>00</minutes>
9.     </latitude>
10.     <longitude>
11.      <direction>West</direction>
12.      <degrees>97</degrees>
```

```
13.     </longitude>
14.    </coordinates>
15.    <government>
16.     <independenceDay Prob='.85'>07/04/1776
17.     </independenceDay>
18.     <chiefOfState>
19.      <Dist type="mutually-exclusive">
20.       <Val Prob='.5'>
21.        <title Prob='.75'>President</title>
22.        <name>
23.         <Dist>
24.          <Val Prob='.4'>George W. Bush</Val>
25.          <Val Prob='.7'>George Bush</Val>
26.         </Dist>
27.        </name>
28.        <age>
29.         <Dist type="mutually-exclusive">
30.          <Val Prob='.2'>54</Val>
31.          <Val Prob='.35'>55</Val>
32.          <Val Prob='.1'>56</Val>
33.          <Val Prob='.15'>77</Val>
34.         </Dist>
35.        </age>
36.        <spouse>
37.         <Dist type="mutually-exclusive">
38.          <Val Prob='.5'>Laura Welch</Val>
39.          <Val Prob='.2'>Barbara Pierce</Val>
40.         </Dist>
41.        </spouse>
42.       </Val>
43.       <Val Prob='.2'>
44.        <title Prob='.65'>President</title>
45.        <name>Bill Clinton</name>
46.        <age Prob='.3'>55</age>
47.       </Val>
48.      </Dist>
49.     </chiefOfState>
50.    </government>
51.   </country>
52.
53.   <country>
54.    <countryName>Uruguay</countryName>
```

```
55.      ...
56.    </country>
57.</countries>
```

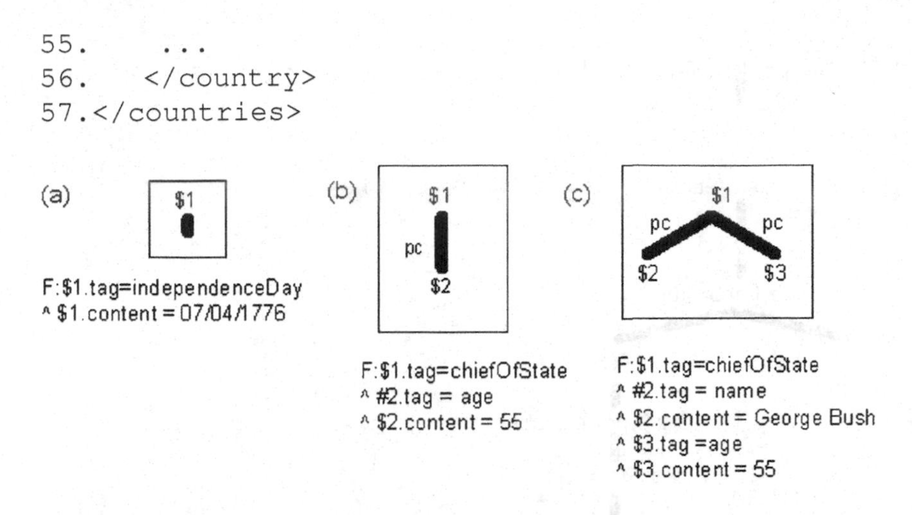

Figure 12.2. Three pattern trees[13]

As a query language, they use a variation of their earlier work on the TAX algebra for XML[8] and use pattern trees. Figure 12.2 shows three examples of pattern trees[13]. The "pc" on the edge means "parent-child" relationship. TAX queries include single node queries (e.g., Figure 12.2(a)), conjunctive queries (e.g., Figure 12.2(b, c)) and disjunctive queries. The output of a query is a set of subtrees (matching data trees), each with a global probability (the probability that this subtree exists). Figure 12.3 shows the matching data trees and associated global probabilities for the queries in Figure 12.2[13].

6. Formal Model of Probabilistic Semi-structured Data with Arbitrary Probabilistic Distributions

In this section, we describe a probabilistic XML (PXML) model for probabilistic semi-structured data[19]. The advantage of this approach is that it supports a flexible representation that allows the specification of a wide class of distributions over semi-structured instances. There are two semantics provided for the model, where the semantics are probabilistically coherent. Next, the relational algebra was extended to handle probabilistic semi-structured data. Efficient algorithms were developed for answering queries that use this algebra. Furthermore, aggregate operators were considered for PXML data and provide two semantics[20]. First, in the ordinary semantics, answers to PXML aggregate queries are defined using the potentially huge number of such compatible instances. It is shown how probabilistic XML instances can be directly manipulated without the exponential blowup. This method is shown to be correct for most aggregate queries which are important and useful. The second semantics is the expected semantics, which returns the expected value of the answer in

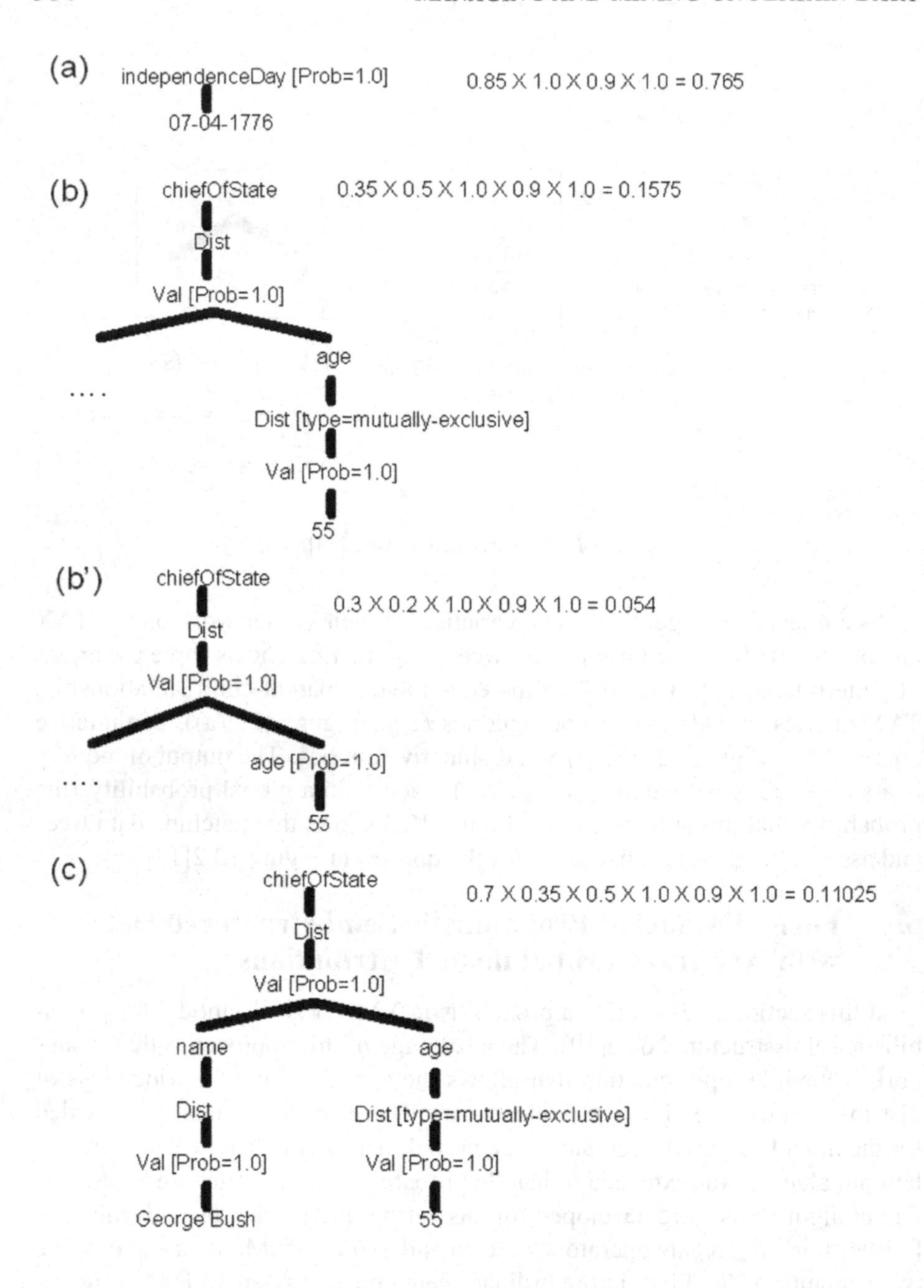

Figure 12.3. Data trees matching the query pattern trees in Figure 12.2[13]

the ordinary semantics. In [20], a series of experiments were described that implement the algebraic operations, and both exact and approximate aggregate operations – these experiments validate the utility of this approach. We will

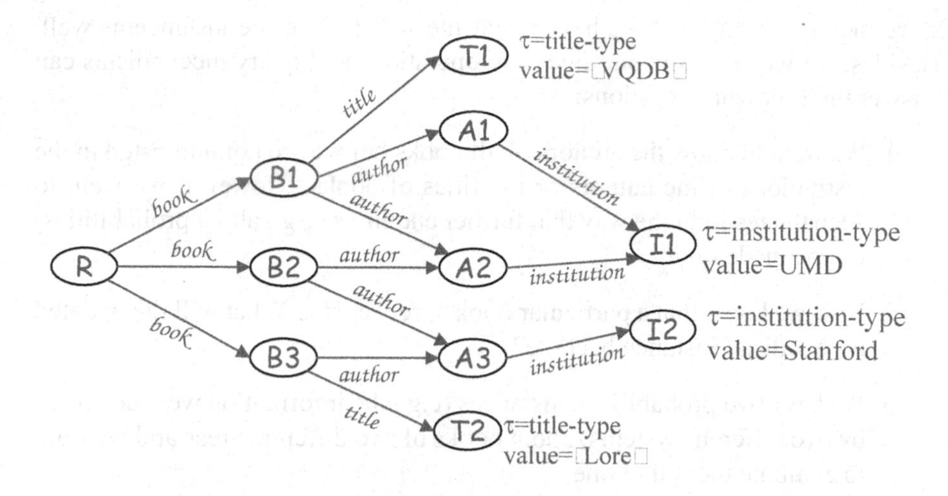

Figure 12.4. A semi-structured instance for a bibliographic domain.

also have a comparison among PXML, SPO and ProTDB and an introduction of PIXML (an interval probability version of PXML).

6.1 Motivating Examples

Here we provide two applications as our motivating examples used to illustrate the PXML model, semantics, algebra, query and aggregate operators.

A Bibliographical Application. As our first running example, we will use a bibliographic application. This example is rather simple, but we assume it will be accessible to all readers. In this case, we assume that the uncertainty arises from the information extraction techniques used to construct the bibliography. Consider a citation index such as Citeseer [21] or DBLP [22]. In Citeseer, the indexes are created by crawling the web, and operations include parsing postscript and PDF documents. Often, there will be uncertainty over the existence of a reference (have we correctly identified a bibliographic reference?), the type of the reference (is the reference a conference paper, a journal article or a book?), the existence of subfields of the reference such as author, title and year, the identity of the author (does Hung refer to Edward Hung or Sheung-lun Hung or many other tens of authors with "Hung" as their last names or first names?). In such environments, uncertainty abounds.

Semi-structured data is a natural way to store such data because for an application of this kind, we have some idea of what the structure of data looks like (e.g. the general hierarchical structure alluded to above). However, semi-structured data models do not provide support for uncertainty over the relationships in an instance. In this section, we will extend this model to naturally store the uncertainty that we have about the structure of the instance as well. Besides, we will see how our algebraic operations and query mechanisms can answer the following questions:

1 We want to know the authors of all books but we are not interested in the institutions of the authors or the titles of books. However, we want to keep the result in the way that further enquiries (e.g., about probabilities) can be made on it.

2 Now we know that a particular book surely exists. What will the updated probabilistic instance become?

3 We have two probabilistic instances (e.g., the information were collected by two different systems) about books of two different areas and we want to combine them into one.

4 We want to know the probability that a particular author exists.

A Surveillance Application. Consider a surveillance application where a battlefield is being monitored. Image processing methods are used to classify objects appearing in images. Some objects are classified as vehicle convoys or refugee groups. Vehicle convoys may be further classified into individual vehicles, which may be further classified into categories such as tanks, cars, armored personnel carriers. However, there may be uncertainty over the number of vehicles in a convoy as well as the categorization of a vehicle. For example, image processing methods often use Bayesian statistical models[11] to capture uncertainty in their identification of image objects. Further uncertainty may arise because image processing methods may not explicitly extract the identity of the objects. Semi-structured data is a natural way to store such data because for a surveillance application of this kind, we have some idea of what the structure of data looks like (e.g. the general structure described above). However, the above example demonstrates the need for a semi-structured model to store uncertain information in uncertain environments.

Aggregate queries are natural queries for users to ask in such applications. To date, we are aware of no formal model of aggregate computations in probabilistic XML databases. Examples of queries that users may wish to ask include: *How many convoys are there (in some collection of images)? How many tanks are there in total? On the average, how many tanks are there in a convoy? What is the ratio of the total number of tanks to the total number of trucks?* In more complex examples, there are many other important queries. If convoys include an estimate of the number of soldiers per vehicle, we may be interested in the total number (sum) of soldiers. We may also be interested in the average number of soldiers per convoy, the average number of soldiers per tank, etc.

6.2 Probabilistic Semi-structured Data Model

In this section, we introduce the probabilistic semi-structured data (PXML) model. We first review the definition of a semi-structured data model. We then introduce the syntax of PXML followed by the semantics of PXML.

Semi-structured Data Model. We start by recalling some simple graph concepts.

DEFINITION 12.2 *Let V be a finite set (of vertices), $E \subseteq V \times V$ be a set (of edges) and $\ell : E \to \mathcal{L}$ be a mapping from edges to a set $\mathcal{L}$ of strings called* labels. *The triple $G = (V, E, \ell)$ is an* **edge labeled directed graph**.

DEFINITION 12.3 *Suppose $G = (V, E, \ell)$ is any rooted, edge-labeled directed graph. For $o \in V$:*

- *The* **children** $\mathsf{children}(o)$ *of o is the set $\{o' \mid (o, o') \in E\}$.*
- *The parents of o, $\mathsf{parents}(o)$, is the set $\{o' \mid (o', o) \in E\}$.*
- *The descendants of o is the set $\mathsf{des}(o) = \{o' \mid$ there is a directed path from o to o' in $G\}$, i.e., o's descendants include o's children as well as children of o's descendants.*
- *The non-descendants of o is the set $\mathsf{non\text{-}des}(o) = \{o' | o' \in V \wedge o' \notin \mathsf{des}(o) \cup \{o\}\}$, i.e., all vertices except o's descendants are o's non-descendants.*
- *We use $\mathsf{lch}(o, l)$ to denote the set of children of o with label l. More formally,*
$$\mathsf{lch}(o, l) = \{o' \mid (o, o') \in E \wedge \ell(o, o') = l\}.$$
- *A vertex o is called a* **leaf** *iff $\mathsf{children}(o) = \emptyset$.*

It is important to note that our graphs are not restricted to trees— in fact, the above definition allows cycles. However, in our probabilistic semi-structured data model, we will restrict attention to directed acyclic graphs.

DEFINITION 12.4 *A* **semi-structured instance** *$\mathcal{S}$ over a set of objects $\mathcal{O}$, a set of labels $\mathcal{L}$, and a set of types $\mathcal{T}$, is a 5-tuple $\mathcal{S} = (V, E, \ell, \tau, \mathsf{val})$ where:*

1. *$G = (V, E, \ell)$ is a rooted, directed graph where $V \subseteq \mathcal{O}$, $E \subseteq V \times V$ and $\ell : E \to \mathcal{L}$;*
2. *τ associates a type in $\mathcal{T}$ with each leaf object o in G.*
3. *val associates a value in the domain $\mathsf{dom}(\tau(o))$ with each leaf object o.*

We illustrate the above definition through an example from the bibliographic domain.

EXAMPLE 12.5 *Figure 12.4 shows a graph representing a part of the bibliographic domain. The instance is defined over the set of objects $\mathcal{O} =$*

$\{R, B1, B2, B3, T1, T2, A1, A2, A3, I1, I2\}$. *The set of labels is* $\mathcal{L} = \{book, title, author, institution\}$. *There are two types,* title-type *and* instition-type, *with domains given by:* $\text{dom(title-type)} = \{\text{VQDB}, \text{Lore}\}$ *and* $\text{dom(institution-type)} = \{\text{Stanford}, \text{UMD}\}$. *The graph shows that the relationships between the objects in the domain and the types and values of the leaves.*

The PXML Probabilistic Data Model. In this section, we develop the basic syntax of the PXML probabilistic data model. However, before defining the important concept of a probabilistic instance, we need some intermediate concepts.

A central notion that allows us to provide coherent probabilistic semantics is that of a weak instance. A weak instance describes the objects that can occur in a semi-structured instance, the labels that can occur on the edges in an instance and constraints on the number of children an object might have. We will later define a probabilistic instance to be a weak instance with some probabilistic attributes.

DEFINITION 12.6 *A* **weak instance** $\mathcal{W}$ *with respect to* $\mathcal{O}$, $\mathcal{L}$ *and* $\mathcal{T}$ *is a 5-tuple* $\mathcal{W} = (V, \mathsf{lch}, \tau, \mathsf{val}, \mathsf{card})$ *where:*

1. $V \subseteq \mathcal{O}$.
2. *For each object* $o \in V$ *and each label* $l \in \mathcal{L}$, $\mathsf{lch}(o, l)$ *specifies the set of objects that* **may** *be children of* o *with label* l. *We assume that for each object* o *and distinct labels* l_1, l_2, $\mathsf{lch}(o, l_1) \cap \mathsf{lch}(o, l_2) = \emptyset$. *(This condition says that two edges with different labels cannot lead to the same child).*
3. τ *associates a type in* $\mathcal{T}$ *with each leaf vertex.*
4. val *associates a value in* $\text{dom}(\tau(o))$ *with each leaf object* o.
5. card *is mapping which constrains the number of children with a given label* l. card *associates with each object* $o \in V$ *and each label* $l \in \mathcal{L}$, *an integer-valued interval* $\mathsf{card}(o, l) = [min, max]$, *where* $min \geq 0$, *and* $max \geq min$. *We use* $\mathsf{card}(object, l).min$ *and* $\mathsf{card}(object, l).max$ *to refer to the lower and upper bounds respectively.*

A weak instance implicitly defines, for each object and each label, a set of potential sets of children. Consider the following example.

EXAMPLE 12.7 *Consider a weak instance with* $V = \{R, B1, B2, B3, T1, T2, A1, A2, A3, I1, I2\}$. *We may have* $\mathsf{lch}(R, book) = \{B1, B2, B3\}$ *indicating that B1 and B2 are possible book-children of R. Likewise, we may have* $\mathsf{lch}(B1, author) = \{A1, A2\}$. *If* $\mathsf{card}(B1, author) = [1, 2]$, *then B1 can have between one and two authors. The set of possible* author-*children of B1 is thus* $\{\{A1\}, \{A2\}, \{A1, A2\}\}$.

Likewise, if $\mathsf{card}(A1, institution) = [1, 1]$ *then* $A1$ *must have exactly one (primary) institution.*

We formalize the reasoning in the above example below.

DEFINITION 12.8 *Suppose* $\mathcal{W} = (V, \mathsf{lch}, \tau, \mathsf{val}, \mathsf{card})$ *is a weak instance and* $o \in V$ *and* l *is a label. A set* c *of objects in* V *is a* **potential** *l-child set of o w.r.t. the above weak instance iff:*

1 *If* $o' \in \mathsf{c}$ *then* $o' \in \mathsf{lch}(o, l)$ *and*
2 *The cardinality of* c *lies in the closed interval* $\mathsf{card}(o, l)$.

We use the notation $\mathsf{PL}(o, l)$ *to denote the set of all potential l-child sets of o.*

As $\mathsf{PL}(o, l)$ denotes the set of all potential child sets of o with labels l, we can define the set of all potential child sets of o with *any* labels as the following.

DEFINITION 12.9 *Suppose* $\mathcal{W} = (V, \mathsf{lch}, \tau, \mathsf{val}, \mathsf{card})$ *is a weak instance and* $o \in V$. *A* **potential child set** *of o is any set Q of subsets of V such that* $Q = \bigcup H$ *where H is a hitting set* of* $\{\mathsf{PL}(o, l) \mid (\exists o')o' \in \mathsf{lch}(o, l)\}$. *We use* $\mathsf{potchildren}(o)$ *to denote the set of all potential child sets of o w.r.t. a weak instance.*

Once a weak instance is fixed, $\mathsf{potchildren}(o)$ is well defined for each o. We will use this to define the *weak instance graph* below. We will need this in our definition of a probabilistic instance.

DEFINITION 12.10 *Given a weak instance* $\mathcal{W} = (V, \mathsf{lch}, \tau, \mathsf{val}, \mathsf{card})$, *the* **weak instance graph**, $\mathcal{G}_{\mathcal{W}} = (V, E)$, *is a graph over the same set of nodes V, and for each pair of nodes o and o', there is an edge from o to o' iff* $\exists c \in \mathsf{potchildren}(o)$ *such that* $o' \in c$.

Before we define a probabilistic instance, let us first introduce the notion of a local probability model for a set of children of an object. We adopt the framework of classical probability theory so that the sum of the probabilities of all potential child sets equals 1.

DEFINITION 12.11 *Suppose* $\mathcal{W} = (V, \mathsf{lch}, \tau, \mathsf{val}, \mathsf{card})$ *is a weak instance. Let* $o \in V$ *be a non-leaf object. An* **object probability function** *(OPF for short) for o w.r.t. $\mathcal{W}$ is a mapping* $\omega : \mathsf{potchildren}(o) \rightarrow [0, 1]$ *such that OPF is a legal probability distribution, i.e.,* $\Sigma_{\mathsf{c} \in \mathsf{potchildren}(o)} \omega(\mathsf{c}) = 1$.

*Suppose $\mathbf{S} = \{S_1, \ldots, S_n\}$ where each S_i is a set. A ***hitting set*** for $\mathbf{S}$ is a set H such that (i) for all $1 \leq i \leq n$, $H \cap S_i \neq \emptyset$ and (ii) there is no $H' \subset H$ satisfying condition (i).

DEFINITION 12.12 *Suppose* $\mathcal{W} = (V, \mathsf{lch}, \tau, \mathsf{val}, \mathsf{card})$ *is a weak instance. Let* $o \in V$ *be a leaf object. A* **value probability function** *(VPF for short) for* o *w.r.t.* $\mathcal{W}$ *is a mapping* $\omega : \mathsf{dom}(\tau(o)) \to [0,1]$ *such that VPF is a legal probability distribution, i.e.,* $\Sigma_{\mathsf{v} \in \mathsf{dom}(\tau(o))} \omega(\mathsf{v}) = 1$.

An object probability function provides the model theory needed to study a single non-leaf object (and its children) in a probabilistic instance to be defined later. It defines the probability of a set of children of an object existing *given* that the parent object exists. Thus it is the conditional probability for a set of children to exist, under the condition that their parent exists in the semi-structured instance. Similarly, the value probability function provides the model theory needed to study a leaf object, and defines a distribution over values for the object.

DEFINITION 12.13 *Suppose* $\mathcal{W} = (V, \mathsf{lch}, \tau, \mathsf{val}, \mathsf{card})$ *is a weak instance. A* **local interpretation** *is a mapping* $\wp$ *from the set of objects* $o \in V$ *to local probability functions. For non-leaf objects,* $\wp(o)$ *returns an OPF, and for leaf objects,* $\wp(o)$ *returns a VPF.*

Intuitively, a local interpretation specifies, for each object in the weak instance, a local probability function.

DEFINITION 12.14 *A* **probabilistic instance** $\mathcal{I}$ *is a 6-tuple* $\mathcal{I} = (V, \mathsf{lch}, \tau, \mathsf{val}, \mathsf{card}, \wp)$ *where:*

1. $\mathcal{W} = (V, \mathsf{lch}, \tau, \mathsf{val}, \mathsf{card})$ *is a weak instance and*
2. $\wp$ *is a local interpretation.*

A probabilistic instance consists of a weak instance, together with a probability associated with each potential child of each object in the weak instance.

EXAMPLE 12.15 *Figure 12.5 shows a very simple probabilistic instance. The set* $\mathcal{O}$ *of objects is the same as in our earlier* PXML *example. The figure shows the potential* lch *of each object; for example,* $\mathsf{lch}(B1, author) = \{A1, A2\}$. *The cardinality constraints are also shown in the figure; for example, object B1 can have 1 to 2 authors and 0 to 1 titles. The tables on the right of Figure 12.5 show the local probability models for each of the objects. The tables show the probability of each potential child of an object. For example, if B2 exists, the probability A1 is one of its authors is* 0.8.

The components $\mathcal{O}, \mathcal{L}, \mathcal{T}$ of a probabilistic instance are identical to those in a semi-structured instance. However, in a probabilistic instance, there is uncertainty over:

- The number of sub-objects of an object o;

o	l	lch(o, l)
R	book	{B1, B2, B3}
B1	title	{T1}
B1	author	{A1, A2}
B2	author	{ A1, A2, A3}
B3	title	{T2}
B3	author	{A3}
A1	institution	{I1}
A2	institution	{I1, I2}
A3	institution	{I2}

o	l	card(o, l)
R	book	[2,3]
B1	author	[1,2]
B1	title	[0,1]
B2	author	[2,2]
B3	author	[1,1]
B3	title	[1,1]
A1	institution	[0,1]
A2	institution	[1,1]
A3	institution	[1,1]

c ∈ potchildren(R)	$\wp(R)$(c)
{B1, B2}	0.2
{B1, B3}	0.2
{B2, B3}	0.2
{B1, B2, B3}	0.4

c ∈ potchildren($B1$)	$\wp(B1)$(c)
{A1}	0.3
{A1, T1}	0.35
{A2}	0.1
{A2, T1}	0.15
{A1, A2}	0.05
{A1, A2, T1}	0.05

c ∈ potchildren($B2$)	$\wp(B2)$(c)
{A1, A2}	0.4
{A1, A3}	0.4
{A2, A3}	0.2

c ∈ potchildren($B3$)	$\wp(B3)$(c)
{A3, T2}	1.0

c ∈ potchildren($A1$)	$\wp(A1)$(c)
{}	0.2
{I1}	0.8

c ∈ potchildren($A2$)	$\wp(A2)$(c)
{I1}	0.5
{I2}	0.5

c ∈ potchildren($A3$)	$\wp(A3)$(c)
{I2}	1.0

Figure 12.5. A probabilistic instance for the bibliographic domain.

- The identity of the sub-objects.
- The values of the leaf objects.

This uncertainty is captured through the function $\wp(o)$. We may define $\wp(o)$ more compactly, in the case where there are some symmetries or independence constraints that can be exploited in the representation. For example, if the occurrence of each category of labeled objects is independent, then we can simply specify a probability for each subset of objects with the same label and compute the joint probability as the product of the individual probabilities. For instance, if the existence of author and title objects is independent, then we only need to specify a distribution over authors and a distribution over titles. Furthermore, in some domains it may be the case that some objects are indistiguishable. For example in an object recognition system, we may not be able to distinguish between vehicles. Then if we have two vehicles, vehicle1 and vehicle2, and a bridge bridge1 in a scene S1, we may not be able to distinguish between a scene that has a bridge1 and vehicle1 in it from a scene that has bridge1 and vehicle2 in it. In this case, $\wp(S1)(\{bridge1, vehicle1\}) = \wp(S1)(\{bridge1, vehicle2\})$. The semantics of the model we have proposed is fully general, in that we can have arbitrary distributions over the sets of children of an object. However in the case where there is additional structure that can be exploited, we plan to allow compact representations of the distributions and make use of the additional structure effectively when answering queries.

6.3 Semantics

In this section, we develop a semantics for probabilistic semi-structured databases. We can use a PXML model to represent our uncertainty about the world as a distribution over possible semi-structured instances. A probabilistic instance *implicitly* is shorthand for a set of (possible) semi-structured instances—these are the only instances that are *compatible* with the information we do have about the actual world state which is defined by our weak instance. We begin by defining the notion of the set of semi-structured instances that are compatible with a weak instance.

DEFINITION 12.16 *Let* $\mathcal{S} = (V_\mathcal{S}, E, \ell, \tau_\mathcal{S}, \mathsf{val}_\mathcal{S})$ *be a semi-structured instance over a set of objects* $\mathcal{O}$*, a set of labels* $\mathcal{L}$ *and a set of types* $\mathcal{T}$ *and let* $\mathcal{W} = (V_\mathcal{W}, \mathsf{lch}_\mathcal{W}, \tau_\mathcal{W}, \mathsf{val}_\mathcal{W}, \mathsf{card})$ *be a weak instance.* $\mathcal{S}$ *is compatible with* $\mathcal{W}$ *if the root of* $\mathcal{W}$ *is in* $\mathcal{S}$ *and for each* o *in* $V_\mathcal{S}$*:*

- *o is also in $V_\mathcal{W}$.*
- *If o is a leaf in $\mathcal{S}$, then o is also a leaf in $\mathcal{W}$, $\tau_\mathcal{S}(o) = \tau_\mathcal{W}(o)$ and and $\mathsf{val}_\mathcal{S}(o) \in \tau_\mathcal{S}(o)$.*
- *If o is not a leaf in $\mathcal{S}$ then*
  - *For each edge (o, o') with label l in $\mathcal{S}$, $o' \in \mathsf{lch}_\mathcal{W}(o, l)$,*
  - *For each label $l \in \mathcal{L}$, let $k = |\{o' | (o, o') \in E \wedge \ell(E) = l\}|$, then $card(o, l).min \leq k \leq card(o, l).max$.*

We use $\mathcal{D}(\mathcal{W})$ to denote the set of all semi-structured instances that are compatible with a weak instance $\mathcal{W}$. Similarly, for a probabilistic instance $\mathcal{I} = (V, \mathsf{lch}_\mathcal{I}, \tau_\mathcal{I}, \mathsf{val}_\mathcal{I}, \mathsf{card}, \wp)$, we use $\mathcal{D}(\mathcal{I})$ to denote the set of all semi-structured instances that are compatible with $\mathcal{I}$'s associated weak instance $\mathcal{W} = (V, \mathsf{lch}_\mathcal{I}, \tau_\mathcal{I}, \mathsf{val}_\mathcal{I}, \mathsf{card})$.

We now define a *global interpretation* based on the set of a compatible instances of a weak instance.

DEFINITION 12.17 *Consider a weak instance* $\mathcal{W} = (V, \mathsf{lch}, \tau, \mathsf{val}, \mathsf{card})$. *A* **global interpretation** $\mathcal{P}$ *is a mapping from* $\mathcal{D}(\mathcal{W})$ *to* $[0, 1]$ *such that* $\Sigma_{S \in \mathcal{D}(\mathcal{W})} \mathcal{P}(S) = 1$.

Intuitively, a global interpretation is a distribution over the set of semi-structured instances compatible with a weak instance.

Recall that a local interpretation defines the local semantics of an object. In addition, it enables us to define the global semantics for our model. First we must impose an acyclicity requirement on the weak instance graph. This is required to ensure that our probabilistic model is coherent.

DEFINITION 12.18 *Let* $\mathcal{W} = (V_{\mathcal{W}}, \mathsf{lch}_{\mathcal{W}}, \tau_{\mathcal{W}}, \mathsf{val}_{\mathcal{W}}, \mathsf{card})$ *be a weak instance.* $\mathcal{W}$ *is* **acyclic** *if its associated weak instance graph* $\mathcal{G}_{\mathcal{W}}$ *is acyclic.*

We do not restrict our probabilistic instance to be trees. For example, the probabilistic instance in Figure 12.5 whose weak instance graph shown in Figure 12.4 is an acyclic graph.

Given a probabilistic instance $\mathcal{I}$ over an acyclic weak instance $\mathcal{W}$, the probability of any particular instance can be computed from the OPF and VPF entries corresponding to each object in the instance and its children. We are now going to define the relationship between the local interpretation and the global interpretation.

DEFINITION 12.19 *Let* $\wp$ *be local interpretation for a weak instance* $\mathcal{W} = (V, \mathsf{lch}, \tau, \mathsf{val}, \mathsf{card})$. *Then* $\mathcal{P}_{\wp}$ *returns a function defined as follows: for any instance* $S \in \mathcal{D}(\mathcal{W})$, $\mathcal{P}_{\wp}(S) = \prod_{o \in S} \wp(o)(\mathsf{children}_{\mathcal{S}}(o))$, *where if o is not a leaf in* $\mathcal{W}$, *then* $\mathsf{children}_{\mathcal{S}}(o) = \{o' | (o, o') \in E\}$, *i.e., the set of children of o in instance* $\mathcal{S}$*; otherwise,* $\mathsf{children}_{\mathcal{S}}(o) = \mathsf{val}_{\mathcal{S}}(o)$, *i.e., the value of o in instance* $\mathcal{S}$.

In order to use this definition for the semantics of our model, we must first show that the above function is in fact a legal global interpretation.

THEOREM 12.20 *Suppose* $\wp$ *is a local interpretation for a weak instance* $\mathcal{W} = (V, \mathsf{lch}, \tau, \mathsf{val}, \mathsf{card})$. *Then* $\mathcal{P}_{\wp}$ *is a global interpretation for* $\mathcal{W}$.

EXAMPLE 12.21 *Consider* S_1 *in Figure 12.6 and the probabilistic semi-structured instance from Figure 12.5.*

$$\begin{aligned} P(S_1) &= P(B1, B2 \mid R)\, P(A1, T1 \mid B1) P(A1, A2 \mid B2)\, P(I1 \mid A1)\, P(I1 \mid A2) \\ &= 0.2 \cdot 0.35 \cdot 0.4 \cdot 0.8 \cdot 0.5 = 0.00448 \end{aligned}$$

An important question is whether we can go the other way: from a global interpretation, can we find a local interpretation for a weak instance $\mathcal{W}(V, \mathsf{lch}, \tau, \mathsf{val}, \mathsf{card})$? It turns out that we can **if** the global interpretation can be factored in a manner consistent with the structure constraints imposed by $\mathcal{W}(V, \mathsf{lch}, \tau, \mathsf{val}, \mathsf{card})$. One way to ensure this is to impose a set of independence constraints on the distribution $\mathcal{P}$.

DEFINITION 12.22 *Suppose* $\mathcal{P}$ *is a global interpretation and* $\mathcal{W} = (V, \mathsf{lch}, \tau, \mathsf{val}, \mathsf{card})$ *is a weak instance.* $\mathcal{P}$ **satisfies** $\mathcal{W}$ *iff for every non-leaf object*

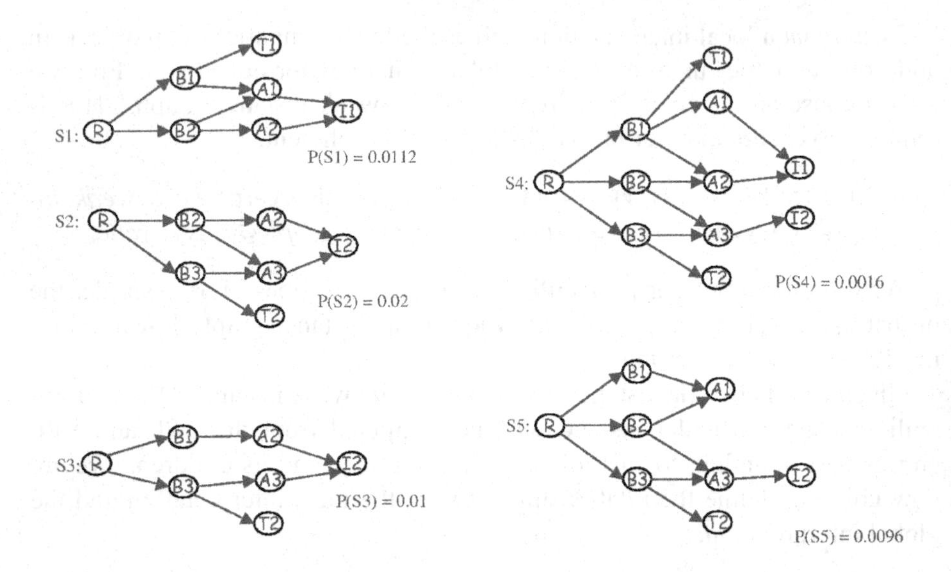

Figure 12.6. Some of semi-structured instances compatible with the probabilistic instance in Figure 12.5.

$o \in V$ and each $c \in \mathsf{potchildren}(o)$ (and for every leaf object $o \in V$ and each $c \in \mathsf{dom}(\tau(o))$), it is the case that $\mathcal{P}(c|\mathsf{non\text{-}des}_{\mathcal{W}}(o)) = \mathcal{P}(c)$ where $\mathsf{non\text{-}des}_{\mathcal{W}}(o)$ are the nondescendants of o in the weak instance graph $\mathcal{G}_{\mathcal{W}}$.

In other words, given that o occurs in the instance, the probability of any potential children c of o is independent of the nondescendants of o in the instance.

Furthermore, given a global interpretation that satisfies a weak instance, we can find a local interpretation associated with it in the following manner:

DEFINITION 12.23 ($\tilde{D}$ OPERATOR) *Suppose $\mathsf{c} \in \mathsf{potchildren}(o)$ for some non-leaf object o*[†] *and suppose $\mathcal{P}$ is a global interpretation. $\omega_{\mathcal{P},o}$, is defined as follows.*

$$\omega_{\mathcal{P},o}(\mathsf{c}) \quad = \quad \frac{\Sigma_{S\in\mathcal{D}(\mathcal{W})\,\wedge o\in S\,\wedge\,\mathsf{children}_S(o)=\mathsf{c}}\mathcal{P}(S)}{\Sigma_{S\in\mathcal{D}(\mathcal{W})\,\wedge o\in S}\mathcal{P}(S)}.$$

Then, $\tilde{D}(\mathcal{P})$ returns a function defined as follows: for any non-leaf object o, $\tilde{D}(\mathcal{P})(o) = \omega_{\mathcal{P},o}$.

Intuitively, we construct $\omega_{\mathcal{P},o}(\mathsf{c})$ as follows. Find all semi-structured instances S that are compatible with $\mathcal{W}$ and eliminate those for which o's set of

[†] For leaf objects, $\mathsf{c} \in \mathsf{dom}(\tau(o))$ and $\mathsf{children}_S(o) = \mathsf{val}(o)$ in the formula.

children is not c. The sum of the (normalized) probabilities assigned to the remaining semi-structured instances by $\mathcal{P}$ is assigned to c by the OPF[‡] $\omega_{\mathcal{P},o}(\mathsf{c})$. By doing this for each object o and each of its potential child sets, we get a local interpretation.

THEOREM 12.24 *Suppose $\mathcal{P}$ is a global interpretation for a weak instance $\mathcal{W} = (V, \mathsf{lch}, \tau, \mathsf{val}, \mathsf{card})$. Then $\tilde{D}(\mathcal{P})$ is a local interpretation for $\mathcal{W}$.*

6.4 PXML Algebra and Comparison with Previous Work

In [19], an algebra was also proposed to support querying probabilistic semi-structured data. In addition, efficient algorithms for answering these queries were described and implemented for evaluation. Relational algebra is based on relation names and attribute names while PXML algebra is based on probabilistic instance names and *path expressions*. The definition of path expressions is a variation of the standard definition [5].

DEFINITION 12.25 *An* **edge sequence** *is a sequence $l_1.\ldots.l_n$, where the l_i's are labels of edges. A* **path expression** $p = r.l_1.\ldots.l_n$ *is an object (oid) r, followed by a (possibly empty) edge sequence $l_1.\ldots._n$; p denotes the set of objects that can be reached via the sequence of edges with labels $l_1.\ldots._n$.*

A path expression is used to locate objects in an instance. We say $o \in p$ iff there is a path p to reach o. For example, in the example instance in Figure 12.4, $A2 \in R.book.author$ because there is a path from R to reach $A2$ through a path that is labeled $book.author$.

In the PXML algebra, a set of operators were defined: projection (ancestor projection, descendant projection, and single projection), selection, and cross product (join can be defined in terms of these operations in the standard way). Finally a probabilistic point query was also proposed, similar to the ProTDB query in [13].

The SPO model in [17] appears to be similar to PXML described in the later section but in fact it is quite different. An SPO itself can be represented in a semi-structured way, but its main body is just a flat table. It cannot show the semi-structured relationship among variables. Only contexts (but not random variables) are represented in a semi-structured form. Contexts are "regular relational attributes", i.e., the context provides already known information when the probability distribution is given on real "random variables". In contrast, PXML model allows data to be represented in a truly semi-structured manner. The syntax and semantics of the model were modified by introducing cardinality and object probability functions to demonstrate the uncertainty of the

‡VPF for leaf objects; note that for the rest of this section, when we mention OPF, it is also true for the case of VPF.

number and the identity of objects existing in possible worlds. Every possible world is a semi-structured instance compatible with the probabilistic instance. The representation of a possible world (semi-structured instance) is the same as the one widely accepted nowadays. However, the model of Dekhtyar et al. cannot do this. Their model also requires random variables to have distinct variable names (or edge labels) (in PXML model, they are the children connected to their parents with the same edge label). Consequently, their model cannot allow two or more variables with the same variable names (no matter their values are the same or different) in a single possible world. Their model also cannot capture the uncertainty of cardinality. On the other hand, PXML model can represent their table. For each random variable, define a set of children (with the possible variable values) connected to their parent with the same edge label (set as the variable name). The cardinality associates with the parent object with each label is set to $[1, 1]$ so that each random variable can have exactly one value in each possible world. The extended context and extended conditionals in SPO can be represented by two subtrees with corresponding edge labels and values connected to the parent object.

ProTDB proposed by Nierman and Jagadish[13] is similar in spirit to PXML – however there are a few important differences. In ProTDB, independent conditional probabilities are assigned to each individual child of an object (i.e., independent of the other children of a node); PXML supports arbitrary distributions over sets of children. Furthermore, dependencies are required to be tree-structured in ProTDB, whereas PXML allows arbitrary acyclic dependency models. In the other words, their answers are correct under the assumption of conditional independence and under the condition that the underlying graph is tree-structured. Thus the PXML data model subsumes the ProTDB data model. In addition, it was proved that the semantics of PXML is probabilistically coherent. Another important difference is in the queries supported. There is no direct mapping among PXML algebra and ProTDB query language. For example, in ProTDB's conjunctive query, given a query pattern tree, they return a set of subtrees (with some modified node probabilities) from the given instance, each with a global probability. There is no direct mapping between their conjunctive query and PXML's ancestor projection because the former finds subtrees matching the pattern tree, while the latter uses a path expression. Each of former subtrees is restricted to match the query pattern tree and has a fixed structure while the output of PXML is a probabilistic instance which implicitly includes many possible structures. Strictly speaking, ProTDB's output (a set of instances, each with a global probability) is not the same kind as the input (one instance, with a global probability = 1 by definition). Besides, they don't have operators like Cartesian product in PXML model, which is fundamental for other important operators like join.

o	l	$\mathsf{lch}(o,l)$
I1	convoy	{ convoy1, convoy2 }
convoy1	tank	{ tank1, tank2 }
convoy2	truck	{ truck1 }

o	$\tau(o)$	$\mathsf{val}(o)$
tank1	tank-type	T-80
tank2	tank-type	T-72
truck1	truck-type	rover

o	l	$\mathsf{card}(o,l)$
I1	convoy	[1,2]
convoy1	tank	[1,1]
convoy2	truck	[1,1]

$c \in \mathsf{potchildren}(I1)$	$\wp(I1)(c)$
{ convoy1}	0.3
{ convoy2}	0.2
{ convoy1, convoy2}	0.5

$c \in \mathsf{potchildren}(convoy1)$	$\wp(convoy1)(c)$
{ tank1}	0.4
{ tank2}	0.6

$c \in \mathsf{potchildren}(convoy2)$	$\wp(convoy2)(c)$
{ truck1}	1

Figure 12.7. A probabilistic instance for the surveillance domain.

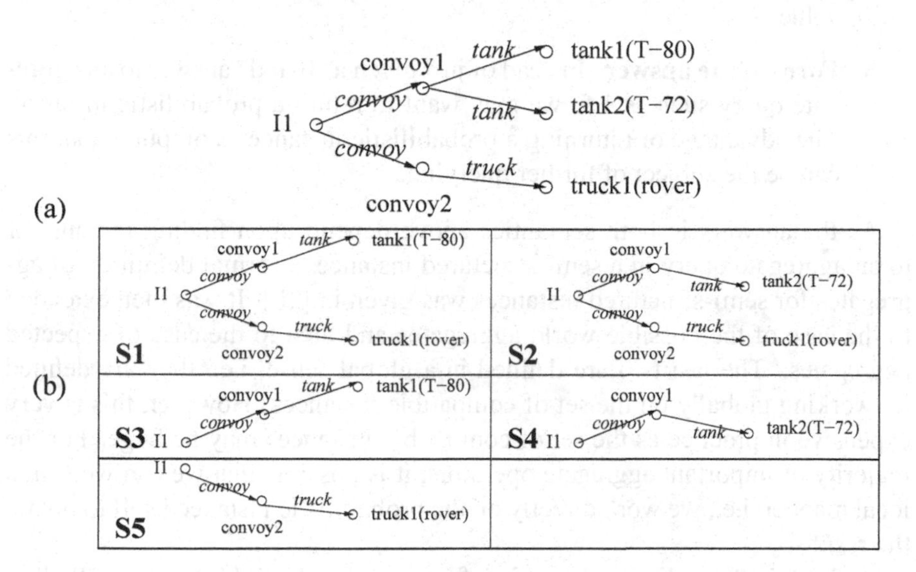

Figure 12.8. (a) The graph structure of the probabilistic instance in Figure 12.7. (b) The set of semi-structured instances compatible with the probabilistic instance in Figure 12.7.

6.5 Probabilistic Aggregate Operations

In this section, we consider another useful class of PXML operations, operations that use aggregates. We will use as a running example the surveillance application introduced earlier. Consider the probabilistic instance shown in

Figure 12.7 and its associated graph shown in Figure 12.8(a). In the following we will define the declarative semantics of aggregate queries. Answering aggregate queries in PXML raises three important issues:

- **Possible-worlds answer:** Consider a query that wishes to count the number of objects in all convoys in probabilistic instance $I1$. This probabilistic instance has five compatible semi-structured instances marked as $S1, \ldots, S5$ in Figure 12.8. Each of these instances has between 1 and 2 objects - as a consequence, we may want to return the set $\{1, 2\}$ indicating that the answer to the count query is not known precisely, but is either 1 or 2, together with the probability that it is one (0.5 in the example) and the probability that it is 2 (also 0.5 in the example).

- **Expected answer:** Alternatively, we could use the statistical notion of *expected value*. In this case, we always return one count for any count query. We multiply the number of objects in $S1$ (i.e. 2) by the probability of $S1$ (i.e. $0.5 \times 0.4 \times 1 = 0.2$) and add this to the number of objects in $S2$ (i.e. 2) by the probability of $S2$ (i.e., $0.5 \times 0.6 \times 1 = 0.3$) and so on. In the above example, we would return the answer 1.5 as the expected value.

- **Form of the answer:** Instead of just given a "bland" answer to an aggregate query such as 1.5, we may want to return a probabilistic instance. The advantage of returning a probabilistic instance as output is that this can be the subject of further querying.

As the answers in both semantics above depend upon finding the answer to an aggregate query in a semi-structured instance, a formal definition of aggregates for semi-structured instances was given in [20]. It was then extended to the case of the possible world aggregates and then to the case of expected aggregates. The last two are defined in a global sense, i.e., they are defined by working globally on the set of compatible instances. However, this is very expensive in practice as the set of compatible instances may be huge. For the majority of important aggregate operators, it is possible that we can work in a local manner, i.e., we work directly on the probabilistic instance itself to obtain the results.

In short, in the ordinary semantics, for every non-leaf object o, the CP algorithm is proposed to compute the results of aggregate function applied on every possible combination of possible aggregate values computed at o's children and also computes the corresponding probabilities of such results. However, in practice, we do not want to execute a probabilistic aggregate operation by applying the aggregate operator on all compatible instances. The idempotent-distributive property of a large class of aggregate operators allows us to execute the probabilistic aggregate operation locally on every non-leaf objects in a bottom-up manner. Examples of such aggregate operators include count, sum, min and max. The worse case time complexity increases with the actual total number of possible aggregate values. In practice, it may be large or small, depending on the application domain, the instance and the aggregate operator used. However, if it is unreasonably large, we can use pruning techniques to reduce the size. For example, we can set a threshold such that when a probability table T is passed up, values with probability smaller than the threshold are pruned. Another method is to keep the h most probable values in the probability table. In addition, a hybrid of the above two and other pruning methods can be used. Since the effect of pruning techniques depends on the application domain, the instance, the aggregate operator used and the thresholds used, users are advised to fine tune their pruning methods according to the desired performance of their applications.

In the expectation semantics, for every non-leaf object o, the SE algorithm was proposed to compute the expected result of aggregate function applied on every possible combination of expected values computed at o's children.

6.6 Modeling Interval Uncertainty in Semi-structured Data

In [23, 24], PXML model was extended so that interval probabilities are used instead of point probabilities to represent uncertainty. Two alternative formal semantics for the Probabilistic Interval XML (PIXML) model were provided: a declarative (model-theoretic) semantics, and an operational semantics that can be used for computation. In the W3C formal specification of XML, an instance is considered as an ordered rooted tree in which cycles can possibly appear[25]. In [23, 24], an instance is assumed to be an acyclic graph - this assumption will be needed to provide a coherent semantics to PIXML databases. An operational semantics was also provided that is provably correct for a queries over a large class of probabilistic instances called tree-structured instances.

Interval Probabilities. An extension to handle interval probabilities is useful because almost all statistical evidence involves margins of error. For instance, when a statistical estimate says that something is true with probability 95% with a $\pm 2\%$ margin of error, then this really corresponds to saying the event's probability lies in the interval $[0.93, 0.97]$. Likewise, using intervals is valuable when one does not know the relationship between different events. For example, if we know the probabilities of events e_1, e_2 and want to know the probability of both of them holding, then we can, in general, only infer an interval for the conjunction of e_1, e_2 ([26, 27]) *unless* we know something more about the dependencies or lack thereof between the events. Furthermore, it is also natural for a human judgement to be expressed as an interval probability rather than an exact point probability. For example, a human expert may say that the vehicle in a picture is *likely* a tank. If he or she is asked to indicate a probability, he or she may feel difficulty to give a point probability (say, 60%), but he or she may feel more natural to give an interval probability (say, 40% to 70%), which also reflects the nature of uncertainty. An extreme case is $[0,1]$ (i.e., "0% to 100%") which indicates that we have no information about the probability or likeliness of an event.

Below I quickly review definitions and give some important theorems for interval probabilities. Given an interval $I = [x, y]$ I will often use the notation $I.lb$ to denote x and $I.ub$ to denote y.

An ***interval function*** ι w.r.t. a set S associates, with each $s \in S$, a closed subinterval $[lb(s), ub(s)] \subseteq [0,1]$. ι is called an ***interval probability function*** if $\sum_{s\in S} lb(s) \leq 1$ and $\sum_{s\in S} ub(s) \geq 1$. A ***probability distribution*** w.r.t. a set S over an interval probability function ι is a mapping $\mathcal{P} : S \to [0,1]$ where

1 $\forall s \in S, lb(s) \leq \mathcal{P}(s) \leq ub(s)$, and

2 $\Sigma_{s\in S}\mathcal{P}(s) = 1$.

LEMMA 12.26 *For any set S and any interval probability function ι w.r.t. S, there exists a probability distribution P(S) which is compatible with ι.*

It may be noted that among the possible distributions, there has been work such as [28] to find the one with maximum entropy. An interval probability function ι w.r.t. S is ***tight*** iff for any interval probability function ι' w.r.t. S such that every probability distribution $\mathcal{P}$ over ι is also a probability distribution over ι', $\iota(s).lb \geq \iota'(s).lb$ and $\iota(s).ub \leq \iota'(s).ub$ where $s \in S$. If every probability distribution P over ι' is also a probability distribution over ι, then we say that ι is the **tight equivalent** of ι'. A ***tightening operator***, tight, is a mapping from interval probability functions to interval probability functions such that tight(ι) produces a tight equivalent of ι. The following result (Theorem 2 of [29]) tells us that we can always tighten an interval probability function.

THEOREM 12.27 *[29, Theorem 2] Suppose ι, ι' are interval probability functions over S and $tight(\iota') = \iota$. Let $s \in S$. Then:*

$$\iota(s) = \left[max\left(\iota'(s).lb, 1 - \sum_{s' \in S \wedge s' \neq s} \iota'(s').ub\right), min\left(\iota'(s).ub, 1 - \sum_{s' \in S \wedge s' \neq s} \iota'(s').lb\right)\right].$$

For example, we can use the above formula to check that the interval probability functions in Figure 12.5 are tight. Throughout the rest of this section, unless explicitly specified otherwise, it will be assumed that all interval probability functions are tight.

The PIXML Data Model. In probabilistic XML, we have uncertainty because we do not know which of various possible semi-structured instances is "correct." Rather than defining a point probability for each instance, we will use interval probabilities to give bounds on the probabilities for structure. In this section, we will first define a probabilistic *interval* semi-structured instance. Details of its model theoretic semantics can be found in [23, 24].

Recall the definitions of a weak instance, a potential l-child set, a potential child set, a weak instance graph, an object probability function (OPF) and a local interpretation. A probabilistic semi-structured instance defined in previous section uses a local interpretation to map a set of OPFs to non-leaf objects for the *point* probabilities of children sets. Here, a probabilistic interval semi-structured instance uses ipf for a similar purpose; however, instead of point probabilities, *interval* probabilities are used in ipf.

DEFINITION 12.28 *A* ***probabilistic instance*** $\mathcal{I}$ *is a 6-tuple* $\mathcal{I} = (V, \mathsf{lch}, \tau, \mathsf{val}, \mathsf{card}, \mathsf{ipf})$ *where:*

1 $\mathcal{W} = (V, \mathsf{lch}, \tau, \mathsf{val}, \mathsf{card})$ *is a weak instance and*

2 ipf *is a mapping which associates with each non-leaf object* $o \in V$*, an interval probability function* ipf *w.r.t.* $\mathsf{potchildren}(o)$*, where* $\mathsf{c} \in \mathsf{potchildren}(o)$ *and* $\mathsf{ipf}(o, \mathsf{c}) = [lb, ub]$.

Intuitively, a probabilistic instance consists of a weak instance, together with probability intervals associated with each potential child set of each object in the weak instance. Similarly, given a probabilistic instance, we can obtain its weak instance graph from its corresponding weak instance.

EXAMPLE 12.29 *Figure 12.9 shows a very simple probabilistic instance.*[§] *The set* $\mathcal{O}$ *of objects is* $\{I1, convoy1, convoy2, tank1, tank2, truck1\}$*. The first table shows the legal children of each of the objects, along with their labels. The cardinality constraints are shown in the third table; for example object* $I1$ *can have from one to two convoy-children. The tables on the right of Figure 12.9 shows the* ipf *of each potential child of I1, convoy1 and convoy2. Intuitively,* $\mathsf{ipf}(I1, \{convoy1\}) = [0.2, 0.4]$ *says that the probability of having only convoy1 is between* 0.2 *and* 0.4.

In [24], we also proposed a query language to query such probabilistic instances. We then provided an operational semantics that is proven to be sound and complete.

[§] Here we only show objects with non-empty set of children.

o	l	$\mathsf{lch}(o,l)$
I1	convoy	{ convoy1, convoy2 }
convoy1	tank	{ tank1, tank2 }
convoy2	truck	{ truck1 }

o	$\tau(o)$	$\mathsf{val}(o)$
tank1	tank-type	T-80
tank2	tank-type	T-72
truck1	truck-type	rover

o	l	$\mathsf{card}(o,l)$
I1	convoy	[1,2]
convoy1	tank	[1,1]
convoy2	truck	[1,1]

$\mathsf{c} \in \mathsf{potchildren}(I1)$	$\mathsf{ipf}(I1, \mathsf{c})$
{ convoy1}	[0.2, 0.4]
{ convoy2}	[0.1, 0.4]
{ convoy1, convoy2}	[0.4, 0.7]

$\mathsf{c} \in \mathsf{potchildren}(convoy1)$	$\mathsf{ipf}(convoy1, \mathsf{c})$
{ tank1}	[0.2, 0.7]
{ tank2}	[0.3, 0.8]

$\mathsf{c} \in \mathsf{potchildren}(convoy2)$	$\mathsf{ipf}(convoy2, \mathsf{c})$
{ truck1}	[1, 1]

Figure 12.9. A probabilistic instance for the surveillance domain.

7. Summary

As XML is used more widely to represent textual sources, multimedia sources, biological and chemical applications, Nierman and Jagadish[13] have argued eloquently for the need to represent uncertainty and handle probabilistic reasoning in semi-structured data of applications like information retrieval and protein chemistry applications. There are many other applications ranging from image surveillance applications to the management of predictive information.

After the initial attempt of using XML tages to model uncertainty, researchers began to develop different models to combine uncertainty and semistructure data, e.g., SPOs (Zhao et al.[17]), ProTDB (Nierman and Jagadish[13]), PXML (Hung et al.[19]) and PIXML (Hung et al.[23, 24]).

For the last two new probabilistic semi-structured data models, PXML and PIXML models, a formal theory was developed for probabilistic semi-structured data. While graph models of semi-structured data are augmented to include probabilistic information, point probabilities are used in PXML and interval probabilities are used in PIXML. Algebraic operations and queries were proposed, with efficient processing techniques implemented. The declarative semantics and the soundness and completeness results are the first of their kind.

Two formal models for probabilistic aggregates (the *possible-worlds* semantics and the *expectation* semantics) were then introduced[20]. Though these semantics are declaratively defined over a large space of “compatible” semi-structured instances, it is able to find a succinct way of representing them and manipulating them.

From using XML tags to represent uncertainty in text, to developing models in representing uncertain structures in XML databases, it has been shown the on-going efforts of researchers taken in advancing the power and flexibility of modeling and querying such probabilistic semi-structured data.

Acknowledgements

This work has been partially supported by grant PolyU 5174/04E from Hong Kong RGC.

References

[1] Papakonstantinou, Y., Garcia-Molina, H., Widom, J.: Object exchange across heterogeneous information sources. In: Proc. of the Eleventh International Conference on Data Engineering, Taipei, Taiwan (1995) 251–260

[2] McHugh, J., Abiteboul, S., Goldman, R., Quass, D., Widom, J.: Lore: A database management system for semi-structured data. SIGMOD Record **26**(3) (1997) 54–66

[3] Thompson, H., Beech, D., Maloney, M., Mendelsohn, N.: XML schema part 1: Structures. W3C Recommendation (2004) Available at `http://www.w3.org/TR/xmlschema-1/`.

[4] Biron, P., Malhotra, A.: XML schema part 2: Datatypes. W3C Recommendation (2004) Available at `http://www.w3.org/TR/xmlschema-2/`.

[5] Abiteboul, S., Quass, D., McHugh, J., Widom, J., Wiener, J.: The Lorel query language for semi-structured data. Journal of Digital Libraries **1**(1) (1996) 68–88

[6] Deutsch, A., Fernandez, M., Florescu, D., Levy, A., Suciu, D.: XML-QL: A query language for XML. Available at `http://www.w3.org/xml/` (1998)

[7] Beeri, C., Tzaban, Y.: SAL: An algebra for semi-structured data and XML. In: Informal Proceedings of the ACM International Workshop on the Web and Databases (WebDB'99), Philadelphia, Pennsylvania, USA (1999)

[8] Jagadish, H., Lakshmanan, V., Srivastava, D., Thompson, K.: TAX: A tree algebra for XML. In: Proc. of Int. Workshop on Database Programming Languages (DBPL'01), Roma, Italy (2001)

[9] Bray, T., Paoli, J., Sperberg-McQueen, C.M., Maler, E., Yergeau, F.: Extensible markup language (XML) 1.0

(third edition). W3C Recommendation (2004) Available at http://www.w3c.org/TR/2004/REC-xml-20040204/.

[10] Bouwerman, B., O'Connell, R.: Forecasting and Time Series: An Applied Approach. Brooks/Cole Publishing, Florence, Kentucky, USA (2000)

[11] Kamberova, G., Bajcsy, R.: Stereo depth estimation: the confidence interval approach. In: Proc. of Intl. Conf. Computer Vision (ICCV98), Bombay, India (1998)

[12] Radev, D., Fan, W., Qi, H.: Probabilistic question answering from the web. In: Proceedings of the 11th International World Wide Web Conference, Honolulu, Haiwaii (2002) 408–419

[13] Nierman, A., Jagadish, H.: ProTDB: Probabilistic data in XML. In: Proc. of the 28th VLDB Conference, Hong Kong, China (2002)

[14] : The xml version of the tei guidelines. (Text Encoding Initiative (TEI) Consortium. 2002. http://www.tei-c.org/release/doc/tei-p4-doc/html/FM1.html)

[15] : The xml version of the tei guidelines, chapter 17: Certainty and responsibility. (Text Encoding Initiative (TEI) Consortium. 2002. http://www.tei-c.org/release/doc/tei-p4-doc/html/CE.html)

[16] Ustunkaya, E., Yazici, A., George, R.: Fuzzy data representation and querying in xml database. International Journal of Uncertainty, Fuzziness and Knowledge-Based Systems **15** (2007) 43–57

[17] Zhao, W., Dekhtyar, A., Goldsmith, J.: A framework for management of semi-structured probabilistic data. Journal of Intelligent Information Systems **25**(3) (2005) 293–332

[18] Dey, D., Sarkar, S.: A probabilistic relational model and algebra. ACM Transactions on Database Systems **21**(3) (1996) 339 – 369

[19] Hung, E., Getoor, L., Subrahmanian, V.: PXML: A probabilistic semi-structured data model and algebra. In: Proc. of 19th International Conference on Data Engineering (ICDE), Bangalore, India (2003)

[20] Hung, E.: Managing Uncertainty and Ontologies in Databases. Ph.D Dissertation, University of Maryland, College Park (2005)

[21] : Citeseer (NEC researchindex). (Available at http://citeseer.nj.nec.com/cs/)

[22] : DBLP (computer science bibliography). (Available at http://www.informatik.uni-trier.de/ ley/db/)

[23] Hung, E., Getoor, L., Subrahmanian, V.: Probabilistic Interval XML. In: Proc. of International Conference on Database Theory (ICDT), Siena, Italy (2003)

[24] Hung, E., Getoor, L., Subrahmanian, V.: Probabilistic Interval XML. ACM Transactions on Computational Logic (TOCL) **8**(4) (2007) 1–38

[25] : Extensible Markup Language (XML). (Available at `http://www.w3.org/XML`)

[26] Boole, G.: The Laws of Thought. Macmillan (1954)

[27] Fagin, R., Halpern, J., Megiddo, N.: A logic for reasoning about probabilities. Information and Computation (1990) 78–128

[28] Goldman, S., Rivest, R.: A non-iterative maximum entropy algorithm. In: Proceedings of the 2nd Annual Conference on Uncertainty in Artificial Intelligence (UAI-86), New York, NY, Elsevier Science Publishing Comapny, Inc. (1986) 133–148

[29] Dekhtyar, A., Goldsmith, J., Hawkes, S.: Semi-structured probabilistic databases. In: Proceedings of 2001 Conference on Statisitcal and Scientific Database Management (SSDBM), Fairfax, VA, USA (2001)

Chapter 13

ON CLUSTERING ALGORITHMS FOR UNCERTAIN DATA

Charu C. Aggarwal
IBM T. J. Watson Research Center
Hawthorne, NY 10532
charu@us.ibm.com

Abstract
In this chapter, we will study the clustering problem for uncertain data. When information about the data uncertainty is available, it can be leveraged in order to improve the quality of the underlying results. We will provide a survey of the different algorithms for clustering uncertain data. This includes recent algorithms for clustering static data sets, as well as the algorithms for clustering uncertain data streams.

Keywords: Clustering, Stream Clustering, k-means, density-based clustering

1. Introduction

Many data sets which are collected often have uncertainty built into them. In many cases, the underlying uncertainty can be easily measured and collected. When this is the case, it is possible to use the uncertainty in order to improve the results of data mining algorithms. This is because the uncertainty provides a probabilistic measure of the relative importance of different attributes in data mining algorithms. The use of such information can enhance the effectiveness of data mining algorithms, because the uncertainty provides a guidance in the use of different attributes during the mining process. Some examples of real applications in which the uncertainty may be used are as follows:

- In many cases, imprecise instruments may be used in order to collect the data. In such cases, the level of uncertainty can be measured by prior experimentation.

- In many cases, the data may be imputed by statistical methods. In other cases, parts of the data may be generated by using statistical methods such as forecasting. In such cases, the uncertainty may be inferred from the methodology used in order to perform the forecasting.

- Many privacy-preserving data mining techniques use agglomeration of underlying records in order to create pseudo-records. For example, in many surveys, the data is presented as a collection in order to preserve the underlying uncertainty. In other cases, perturbations [7] may be added to the data from a known probability distribution. In such cases, the uncertainty may be available as an end result of the privacy-preservation process. Recent work [5] has explicitly connected the problem of privacy-preservation with that of uncertain data mining.

The problem of uncertain data has been studied in the traditional database literature [9, 26], though the issue has seen a revival in recent years [2, 5, 10, 13, 15, 16, 24, 30–32]. The driving force behind this revival has been the evolution of new hardware technologies such as sensors which cannot collect the data in a completely accurate way. In many cases, it has become increasingly possible to collect the uncertainty along with the underlying data values. Many data mining and management techniques need to be carefully re-designed in order to work effectively with uncertain data. This is because the uncertainty in the data can change the results in a subtle way, so that deterministic algorithms may often create misleading results [2]. While the raw values of the data can always be used in conjunction with data mining algorithms, the uncertainty provides additional insights which are not otherwise available. A survey of recent techniques for uncertain data mining may be found in [1].

The problem of clustering is a well known and important one in the data mining and management communities. Details of a variety of clustering algorithms may be found in [21, 20]. The clustering problem has been widely studied in the traditional database literature [19, 22, 33] because of its applications to a variety of customer segmentation and data mining problems. The presence of uncertainty significantly affects the behavior of the underlying clusters. This is because the presence of uncertainty along a particular attribute may affect the expected distance between the data point and that particular attribute. In most real applications, there is considerable skew in the uncertainty behavior across different attributes. The incorporation of uncertainty into the clustering behavior can significantly affect the quality of the underlying results. An example is illustrated in [23] in which uncertainty was incorporated into the clustering process in an application of sales merchandising. It was shown that this approach significantly improves the quality of the underlying results.

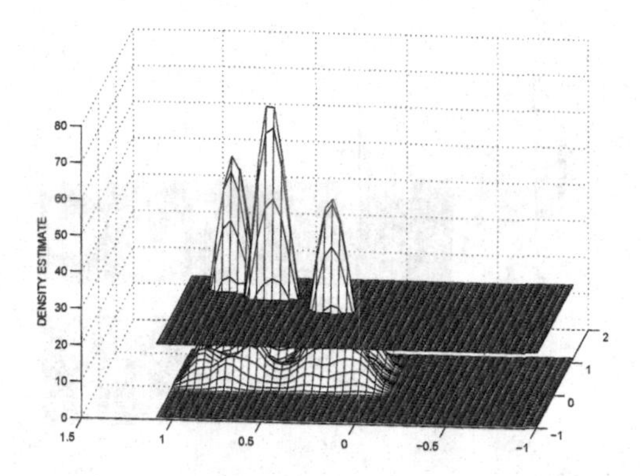

Figure 13.1. Density Based Profile with Lower Density Threshold

In this chapter, we will provide a survey of clustering algorithms for uncertain data. The main algorithms known for the case of uncertain data are as follows:

- **Density-based Method:** A density-based method for uncertain data was proposed in [24]. This is referred to as the FDBSCAN algorithm. This approach modifies the DBSCAN algorithm to the case of uncertain data. An alternative method modifies the OPTICS algorithm to the case of uncertain data [25]. This is referred to as the FOPTICS algorithm.

- The K-means algorithm has been modified for the case of uncertain data [29]. This is referred to as the UK-means algorithm.

- The problem of clustering uncertain data has been extended to the case of data streams [3]. For this purpose, we extend the micro-clustering approach [6] to the case of data streams.

In this chapter, we will provide a detailed discussion of each of the above algorithms for uncertain data. This chapter is organized as follows. In the next section, we will discuss density-based clustering algorithms for uncertain data. In section 3, we will discuss the UK-means algorithm for clustering uncertain data. Section 13.3 discusses streaming algorithms for clustering uncertain data. Section 5 discusses approximation algorithms for clustering uncertain data. Section 6 contains the conclusions and summary.

2. Density Based Clustering Algorithms

The presence of uncertainty changes the nature of the underlying clusters, since it affects the distance function computations between different data points. A technique has been proposed in [24] in order to find density based

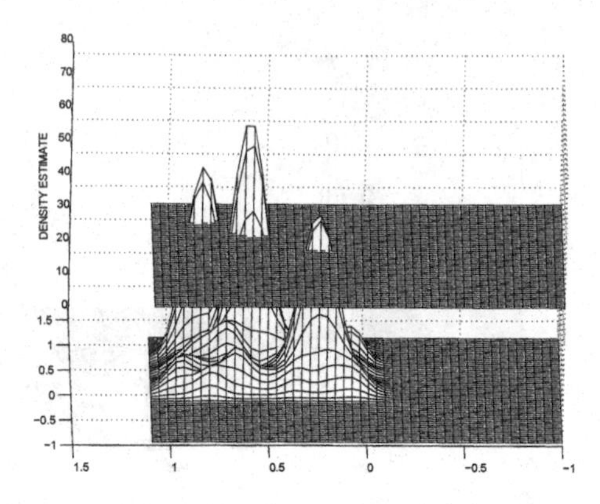

Figure 13.2. Density Based Profile with Higher Density Threshold

clusters from uncertain data. The key idea in this approach is to compute uncertain distances effectively between objects which are probabilistically specified. The fuzzy distance is defined in terms of the distance distribution function. This distance distribution function encodes the probability that the distances between two uncertain objects lie within a certain user-defined range. Let $d(\overline{X}, \overline{Y})$ be the random variable representing the distance between $\overline{X}$ and $\overline{Y}$. The distance distribution function is formally defined as follows:

DEFINITION 13.1 *Let $\overline{X}$ and $\overline{Y}$ be two uncertain records, and let $p(\overline{X}, \overline{Y})$ represent the distance density function between these objects. Then, the probability that the distance lies within the range (a, b) is given by the following relationship:*

$$P(a \leq d(\overline{X}, \overline{Y}) \leq b) = \int_a^b p(\overline{X}, \overline{Y})(z)dz \qquad (13.1)$$

Based on this technique and the distance density function, the method in [24] defines a *reachability probability* between two data points. This defines the probability that one data point is directly reachable from another with the use of a path, such that each point on it has density greater than a particular threshold. We note that this is a direct probabilistic extension of the deterministic reachability concept which is defined in the DBSCAN algorithm [17]. In the deterministic version of the algorithm [17], data points are grouped into clusters when they are reachable from one another by a path which is such that every point on this path has a minimum threshold data density. To this effect, the algorithm uses the condition that the ϵ-neighborhood of a data point should contain at least $MinPts$ data points. The algorithm starts off at a given data point and checks if the ϵ neighborhood contains $MinPts$ data points. If this is the case, the algorithm repeats the process for each point in this cluster and keeps adding points until no more points can be added. One can plot the density profile of a data set by plotting the number of data points in the ϵ-neighborhood of various regions, and plotting a smoothed version of the curve. This is similar to the concept of probabilistic density estimation. Intuitively, this approach corresponds to the continuous contours of intersection between the density thresholds of Figures 13.1 and 13.2 with the corresponding density profiles. The density threshold depends upon the value of $MinPts$. Note that the data points in any contiguous region will have density greater than the threshold. Note that the use of a higher density threshold (Figure 13.2) results in 3 clusters, whereas the use of a lower density threshold results in 2 clusters. The fuzzy version of the DBSCAN algorithm (referred to as FDBSCAN) works in a similar way as the DBSCAN algorithm, except that the density at a given point is uncertain because of the underling uncertainty of the data points. This corresponds to the fact that the number of data points within the ϵ-neighborhood of a given data point can be estimated only probabilistically, and is essentially an uncertain variable. Correspondingly, the reachability from one point to another is no longer deterministic, since other data points may lie within the ϵ-neighborhood of a given point with a certain probability, which may be less than 1. Therefore, the additional constraint that the computed reachability probability must be greater than 0.5 is added. Thus, this is a generalization of the deterministic version of the algorithm in which the reachability probability is always set to 1.

Another related technique discussed in [25] is that of hierarchical density based clustering. An effective (deterministic) density based hierarchical clustering algorithm is OPTICS [8]. We note that the core idea in OPTICS is quite similar to DBSCAN and is based on the concept of *reachability distance* between data points. While the method in DBSCAN defines a *global density parameter* which is used as a threshold in order to define reachability, the work in [25] points out that different regions in the data may have different data density, as a result of which it may not be possible to define the clusters effectively with a single density parameter. Rather, many different values of the density parameter define different (hierarchical) insights about the underlying clusters. The goal is to define an implicit output in terms of ordering data points, so that when the DBSCAN is applied with this ordering, once can obtain the hierarchical clustering at any level for different values of the density parameter. The key is to ensure that the clusters at different levels of the hierarchy are consistent with one another. One observation is that clusters defined over a lower value of ϵ are completely contained in clusters defined over a higher value of ϵ, if the value of $MinPts$ is not varied. Therefore, the data points are ordered based on the value of ϵ required in order to obtain $MinPts$ in the ϵ-neighborhood. If the data points with smaller values of ϵ are processed first, then it is assured that higher density regions are always processed before lower density regions.

This ensures that if the DBSCAN algorithm is used for different values of ϵ with this ordering, then a consistent result is obtained. Thus, the output of the OPTICS algorithm is not the cluster membership, but it is the order in which the data points are processed. We note that that since the OPTICS algorithm shares so many characteristics with the DBSCAN algorithm, it is fairly easy to extend the OPTICS algorithm to the uncertain case using the same approach as that was used for extending the DBSCAN algorithm. This is referred to as the FOPTICS algorithm. Note that one of the core-concepts needed to order to data points is to determine the value of ϵ which is needed in order to obtain $MinPts$ in the corresponding neighborhood. In the uncertain case, this value is defined probabilistically, and the corresponding expected values are used to order the data points.

3. The UK-means and CK-means Algorithms

A common approach to clustering is the k-means algorithm. In the k-means algorithm, we construct clusters around a pre-defined number of cluster centers. A variety of distance functions may be used in order to map the points to the different clusters. A k-means approach to clustering uncertain data was studied in the context of moving object data [29]. In the case of moving objects, the actual locations of the objects may change over time as the data is reported intermittently. Thus, the position of a vehicle could be an arbitrary or circle region which uses the reported location as its center and has a size which is dependent upon the speed and direction of the vehicle. A probability density function could be used to model the probability of presence of the vehicle at a given location at a particular time.

The UK-means clustering approach is very similar to the K-means clustering approach, except that we use the *expected distance* from the data's uncertainty region to the representative of the candidate cluster to which it is assigned. Clearly, a key challenge is the computation of the expected distances between the data points and the centroids for the k-means algorithm. A natural technique for computing these expected distances is to use Monte-carlo sampling, in which samples for the data points are used in order to compute the uncertain distances. This approach can be very expensive because a large number of samples may be required in order to compute the distances accurately. Clearly, some kind of pruning is required in order to improve the efficiency of the approach.

The idea here is to use branch-and-bound techniques in order to minimize the number of expected distance computations between data points and cluster representatives. The broad idea is that once an upper bound on the minimum distance of a particular data point to some cluster representative has been quantified, it is necessary to to perform the computation between this point and another cluster representative, if it can be proved that the corresponding distance is greater than this bound. In order to compute the bounds, the minimum bounding rectangle for the representative point for a cluster region is computed. The uncertain data point also represents a region over which the object may be distributed. For each representative cluster, its minimum bounding rectangle is used to compute the following two quantities with respect to the uncertain data point:

- The minimum limit on the expected distance between the MBR of the representative point and the uncertain region for the data point itself.

- The maximum limit on the expected distance between the MBR of the representative point and the uncertain region for the data point itself.

These upper and lower bound computations are facilitated by the use of the Minimum Bounding Rectangles in conjunction with the triangle inequality. We note that a cluster representative can be pruned, if its maximum limit is less than the minimum limit for some other representative. The approach is [29] constructs a k-d tree on the cluster representatives in order to promote an orderly pruning strategy and minimize the number of representatives which need to be accessed. This approach is used to design an efficient algorithm for clustering uncertain location data.

A different approach called the CK-means algorithm was presented in [27]. It was observed in [27] that the pruning effectiveness of the technique in [29] is not guaranteed. Therefore, the technique is [27] proposes a simple formula for expected distance computations, so that the cost of the computations can be considerably reduced. The effective result is that it is possible to perform the clustering on the uncertain data with the use of the traditional k-means algorithm, while producing the same result as the UK-means algorithm.

4. UMicro: Streaming Algorithms for Clustering Uncertain Data

In this section, we will introduce $UMicro$, the Uncertain MICROclustering algorithm for data streams. We will first introduce some additional notations and definitions. We assume that we have a data stream which contains d dimensions. The actual records in the data are denoted by $\overline{X_1}, \overline{X_2}, \ldots \overline{X_N} \ldots$. We assume that the estimated error associated with the jth dimension for data point $\overline{X_i}$ is denoted by $\psi_j(\overline{X_i})$. This error is defined in terms of the standard deviation of the error associated with the value of the jth dimension of $\overline{X_i}$. The corresponding d-dimensional error vector is denoted by $\overline{\psi(\overline{X_i})}$. Thus, the input to the algorithm is a data stream in which the ith pair is denoted by $(\overline{X_i}, \overline{\psi(\overline{X_i})})$.

We note that many techniques in the uncertain data management literature [12] work with the assumption that the entire probability density function is available. We make the more modest assumption that the standard error of individual entries is available. In many real applications, this is a more realistic assumption for data mining purposes, since complete probability distributions are rarely available, and are usually inserted only as a modeling assumption. The interpretation of this error value can vary with the nature of the data mining application. For example, in a scientific application in which the measurements can vary from one observation to another, the error value is the standard deviation of the observations over a large number of measurements. In a k-anonymity based data (or incomplete data) mining application, this is the standard deviation of the partially specified (or imputed) fields in the data.

We will develop a method for clustering uncertain data streams with the use of a micro-clustering model. The micro-clustering model was first proposed in [33] for large data sets, and subsequently adapted in [6] for the case of deterministic data streams. We will see that the uncertainty in the underlying data significantly affects the quality of the clusters with methods that use such error information.

In order to incorporate the uncertainty into the clustering process, we need a method to incorporate and leverage the error information into the micro-clustering statistics and algorithms. As discussed earlier, it is assumed that the data stream consists of a set of multi-dimensional records $\overline{X_1} \ldots \overline{X_k} \ldots$ arriving at time stamps $T_1 \ldots T_k \ldots$. Each $\overline{X_i}$ is a multi-dimensional record containing d dimensions which are denoted by $\overline{X_i} = (x_i^1 \ldots x_i^d)$. In order to apply the micro-clustering method to the uncertain data mining problem, we need to also define the concept of error-based micro-clusters. We define such micro-clusters as follows:

DEFINITION 13.2 *An uncertain micro-cluster for a set of d-dimensional points $\overline{X_{i_1}} \ldots \overline{X_{i_n}}$ with time stamps $T_{i_1} \ldots T_{i_n}$ and error vectors $\overline{\psi(X_{i_1})} \ldots \overline{\psi(X_{i_n})}$ is defined as the $(3 \cdot d + 2)$ tuple $(\overline{CF2^x}(\mathcal{C}), \overline{EF2^x}(\mathcal{C}), \overline{CF1^x}(\mathcal{C}), t(\mathcal{C}), n(\mathcal{C}))$, wherein $\overline{CF2^x}(\mathcal{C})$, $\overline{EF2^x}(\mathcal{C})$, and $\overline{CF1^x}(\mathcal{C})$ each correspond to a vector of d entries. The entries in $\overline{EF2^x}(\mathcal{C})$ correspond to the error-based entries. The definition of each of these entries is as follows:*

• *For each dimension, the sum of the squares of the data values is maintained in $\overline{CF2^x}(\mathcal{C})$. Thus, $\overline{CF2^x}(\mathcal{C})$ contains d values. The p-th entry of $\overline{CF2^x}(\mathcal{C})$ is equal to $\sum_{j=1}^{n}(x_{i_j}^p)^2$. This corresponds to the second moment of the data values along the p-th dimension.*

• *For each dimension, the sum of the squares of the errors in the data values is maintained in $\overline{EF2^x}(\mathcal{C})$. Thus, $\overline{EF2^x}(\mathcal{C})$ contains d values. The p-th entry of $\overline{EF2^x}(\mathcal{C})$ is equal to $\sum_{j=1}^{n}\psi_p(\overline{X_{i_j}})^2$. This corresponds to the sum of squares of the errors in the records along the p-th dimension.*

• *For each dimension, the sum of the data values is maintained in* $\overline{CF1^x}(\mathcal{C})$. *Thus,* $\overline{CF1^x}(\mathcal{C})$ *contains d values. The p-th entry of* $\overline{CF1^x}(\mathcal{C})$ *is equal to* $\sum_{j=1}^{n} x_{i_j}^p$. *This corresponds to the first moment of the values along the p-th dimension.*

• *The number of points in the data is maintained in* $n(\mathcal{C})$.

• *The time stamp of the last update to the micro-cluster is maintained in* $t(\mathcal{C})$.

We note that the uncertain definition of micro-clusters differs from the deterministic definition, since we have added an additional d values corresponding to the error information in the records. We will refer to the uncertain micro-cluster for a set of points $\mathcal{C}$ by $\overline{ECF}(\mathcal{C})$. We note that error based micro-clusters maintain the important *additive property* [6] which is critical to its use in the clustering process. We restate the additive property as follows:

PROPERTY 4.1 *Let* $\mathcal{C}_1$ *and* $\mathcal{C}_2$ *be two sets of points. Then all non-temporal components of the error-based cluster feature vector* $\overline{ECF}(\mathcal{C}_1 \cup \mathcal{C}_2)$ *are given by the sum of* $\overline{ECF}(\mathcal{C}_1)$ *and* $\overline{ECF}(\mathcal{C}_2)$.

The additive property follows from the fact that the statistics in the individual micro-clusters are expressed as a separable additive sum of the statistics over individual data points. We note that the single temporal component $t(\mathcal{C}_1 \cup \mathcal{C}_2)$ is given by $\max\{t(\mathcal{C}_1), t(\mathcal{C}_2)\}$. We note that the additive property is an important one, since it ensures that it is easy to keep track of the cluster statistics as new data points arrive. Next we will discuss the process of uncertain micro-clustering.

4.1 The UMicro Algorithm: Overview

The $UMicro$ algorithm works using an iterative approach which maintains a number of micro-cluster centroids around which the clusters are built. It is assumed that one of the inputs to the algorithm is n_{micro}, which is the number of micro-clusters to be constructed. The algorithm starts off with a number of null clusters and initially creates new singleton clusters, to which new points are added subsequently. For any incoming data point, the closest cluster centroid is determined. The closest cluster centroid is determined by using the *expected distance* of the uncertain data point to the *uncertain micro-clusters*. The process of expected distance computation for the closest centroid is tricky, and will be subsequently discussed. Furthermore, for the incoming data point, it is determined whether it lies within a *critical uncertainty boundary* of the micro-cluster. If it lies within this critical uncertainty boundary, then the data point is added to the micro-cluster, otherwise a new micro-cluster needs to be created containing the singleton data point. In order to create a new micro-cluster, it must either be added to the current set of micro-clusters, or it needs

Algorithm *UMicro*(Number of Clusters: n_{micro})
begin
$S = \{\}$;
{ S is the current set of micro-cluster statistics. S contains at most n_{micro} elements }
repeat
Receive the next stream point $\overline{X}$;
{ Initially, when S is null, the computations below cannot be performed, and $\overline{X}$ is simply added as a singleton micro-cluster to S }
Compute the *expected similarity* of $\overline{X}$ to the closest micro-cluster $\mathcal{M}$ in S;
Compute critical uncertainty boundary of $\mathcal{M}$;
if $\overline{X}$ lies inside uncertainty boundary
add $\overline{X}$ to statistics of $\mathcal{M}$
else
add a new micro-cluster to S containing singleton point $\overline{X}$;
if $|S| = n_{micro} + 1$ remove the least recently updated micro-cluster from S;
until data stream ends;
end

Figure 13.3. The UMicro Algorithm

to replace one of the older micro-clusters. In the initial stages of the algorithm, the current number of micro-clusters is less than n_{micro}. If this is the case, then the new data point is added to the current set of micro-clusters as a separate micro-cluster with a singleton point in it. Otherwise, the new data point needs to replace one of the older micro-clusters. For this purpose, we always replace the least recently updated micro-cluster from the data set. This information is available from the temporal time stamp in the different micro-clusters. The overall framework for the uncertain stream clustering algorithm is illustrated in Figure 13.3. Next, we will discuss the process of computation of individual subroutines such as the expected distance or the uncertain boundary.

4.2 Computing Expected Similarity

In order to compute the expected similarity of the data point $\overline{X}$ to the centroid of the cluster $\mathcal{C}$, we need to determine a closed form expression which is expressed only in terms of $\overline{X}$ and $ECF(\mathcal{C})$. We note that just as the individual data points are essential random variables with a given error, the centroid $\overline{Z}$ of a cluster $\mathcal{C}$ is also a random variable. We make the following observation about the centroid of a cluster:

LEMMA 13.3 *Let $\overline{Z}$ be the random variable representing the centroid of cluster $\mathcal{C}$. Then, the following result holds true:*

$$E[||Z||^2] = \sum_{j=1}^{d} CF1(\mathcal{C})_j^2/n(\mathcal{C})^2 + \sum_{j=1}^{d} EF2(\mathcal{C})_j/n(\mathcal{C})^2 \tag{13.2}$$

Proof: We note that the random variable Z_j is given by the current instantiation of the centroid and the mean of $n(\mathcal{C})$ different error terms for the points in cluster $\mathcal{C}$. Therefore, we have:

$$Z_j = CF1(\mathcal{C})_j/n(\mathcal{C}) + \sum_{\overline{X}\in\mathcal{C}} e_j(\overline{X})/n(\mathcal{C}) \tag{13.3}$$

Then, by squaring $\overline{Z_j}$ and taking the expected value, we obtain the following:

$$\begin{aligned} E[Z_j^2] = CF1(\mathcal{C})_j^2/n(\mathcal{C})^2 + \\ +2\cdot\sum_{\overline{X}\in\mathcal{C}} E[e_j(\overline{X})]\cdot CF1(\mathcal{C})_j/n(\mathcal{C})^2 + \\ +E[(\sum_{\overline{X}\in\mathcal{C}} e_j(\overline{X}))^2]/n(\mathcal{C})^2 \end{aligned}$$

Now, we note that the error term is a random variable with standard deviation $\psi_j(\cdot)$ and zero mean. Therefore $E[e_j]=0$. Further, since it is assumed that the random variables corresponding to the errors of different records are independent of one another, we have $E[e_j(\overline{X})\cdot e_j(\overline{Y})] = E[e_j(\overline{X})]\cdot E[e_j(\overline{Y})] = 0$. By using these relationships in the expansion of the above equation we get:

$$\begin{aligned} E[Z_j^2] &= CF1(\mathcal{C})_j^2/n(\mathcal{C})^2 + \sum_{\overline{X}\in\mathcal{C}} E[e_j(\overline{X})^2]/n(\mathcal{C})^2 \\ &= CF1(\mathcal{C})_j^2/n(\mathcal{C})^2 + \sum_{\overline{X}\in\mathcal{C}} \psi_j(\overline{X})^2/n(\mathcal{C})^2 \\ &= CF1(\mathcal{C})_j^2/n(\mathcal{C})^2 + EF2(\mathcal{C})_j/n(\mathcal{C})^2 \end{aligned}$$

By adding the value of $E[Z_j^2]$ over different values of j, we get:

$$E[||Z||^2] = \sum_{j=1}^{d} CF1(\mathcal{C})_j^2/n(\mathcal{C})^2 + \sum_{j=1}^{d} EF2(\mathcal{C})_j/n(\mathcal{C})^2 \tag{13.4}$$

This proves the desired result.

Next, we will use the above result to directly estimate the expected distance between the centroid of cluster $\mathcal{C}$ and the data point $\overline{X}$. We will prove the following result:

LEMMA 13.4 *Let v denote the expected value of the square of the distance between the uncertain data point $\overline{X} = (x_1 \ldots x_d)$ (with instantiation $(x_1 \ldots x_d)$ and error vector $(\psi_1(\overline{X}) \ldots \psi_d(\overline{X}))$ and the centroid of cluster $\mathcal{C}$. Then, v is given by the following expression:*

$$v = \sum_{j=1}^{d} CF1(\mathcal{C})_j^2/n(\mathcal{C})^2 + \sum_{j=1}^{d} EF2(\mathcal{C})_j/n(\mathcal{C})^2 + \sum_{j=1}^{d} x_j^2 + \\ + \sum_{j=1}^{d} (\psi_j(\overline{X}))^2 - 2\sum_{j=1}^{d} x_j \cdot CF1(\mathcal{C})_j/n(\mathcal{C})$$

Proof: Let $\overline{Z}$ represent the centroid of cluster $\mathcal{C}$. Then, we have:

$$\begin{aligned} v &= E[||\overline{X} - \overline{Z}||^2] \\ &= E[||\overline{X}||^2] + E[||\overline{Z}||^2] - 2E[\overline{X} \cdot \overline{Z}] \\ &= E[||\overline{X}||^2] + E[||\overline{Z}||^2] - 2E[\overline{X}] \cdot E[\overline{Z}] (\text{indep. of } \overline{X} \text{ and } \overline{Z}) \end{aligned}$$

Next, we will analyze the individual terms in the above expression. We note that the value of X is a random variable, whose expected value is equal to its current instantiation, and it has an error along the jth dimension which is equal to $\psi_j(\overline{X})$. Therefore, the expected value of $E[||\overline{X}||^2]$ is given by:

$$\begin{aligned} E[||\overline{X}||^2] &= (E[X])^2 + \sum_{j=1}^{d} (\psi_j(\overline{X}))^2 \\ &= \sum_{j=1}^{d} x_j^2 + \sum_{j=1}^{d} (\psi_j(\overline{X}))^2 \end{aligned}$$

Now, we note that the jth term of $E[Z]$ is equal to the jth dimension of the centroid of cluster $\mathcal{C}$. This is given by the expression $CF1(\mathcal{C})_j/n(\mathcal{C})$, where $CF1_j(\mathcal{C})$ is the jth term of the first order cluster component $CF1(\mathcal{C})$. Therefore, the value of $E[X] \cdot E[Z]$ is given by the following expression:

$$E[X] \cdot E[Z] = \sum_{j=1}^{d} x_j \cdot CF1(\mathcal{C})_j/n(\mathcal{C}) \tag{13.5}$$

The results above and Lemma 13.3 define the values of $E[||X||^2]$, $E[||Z||^2]$, and $E[X \cdot Z]$. Note that all of these values occur in the right hand side of the following relationship:

$$v = E[||\overline{X}||^2] + E[||\overline{Z}||^2] - 2E[\overline{X}] \cdot E[\overline{Z}] \tag{13.6}$$

By substituting the corresponding values in the right hand side of the above relationship, we get:

$$v = \sum_{j=1}^{d} CF1(\mathcal{C})_j^2/n(\mathcal{C})^2 + \sum_{j=1}^{d} EF2(\mathcal{C})_j/n(\mathcal{C})^2 + \sum_{j=1}^{d} x_j^2 + \\ + \sum_{j=1}^{d} (\psi_j(\overline{X}))^2 - 2\sum_{j=1}^{d} x_j \cdot CF1(\mathcal{C})_j/n(\mathcal{C})$$

The result follows.

The result of Lemma 13.4 establishes how the square of the distance may be computed (in expected value) using the error information in the data point $\overline{X}$ and the micro-cluster statistics of $\mathcal{C}$. Note that this is an efficient computation which requires $O(d)$ operations, which is asymptotically the same as the deterministic case. This is important since distance function computation is the most repetitive of all operations in the clustering algorithm, and we would want it to be as efficient as possible.

While the expected distances can be directly used as a distance function, the uncertainty adds a lot of noise to the computation. We would like to remove as much noise as possible in order to determine the most accurate clusters. Therefore, we design a dimension counting similarity function which prunes the uncertain dimensions during the similarity calculations. This is done by computing the variance σ_j^2 along each dimension j. The computation of the variance can be done by using the cluster feature statistics of the different micro-clusters. The cluster feature statistics of all micro-clusters are added to create one global cluster feature vector. The variance of the data points along each dimension can then be computed from this vector by using the method discussed in [33]. For each dimension j and threshold value $thresh$, we add the *similarity value* $\max\{0, 1 - E[||X - Z||_j^2]/(thresh * \sigma_j^2)\}$ to the computation. We note that this is a similarity value rather than a distance value, since larger values imply greater similarity. Furthermore, dimensions which have a large amount of uncertainty are also likely to have greater values of $E[||X - Z||_j^2]$, and are often pruned from the computation. This improves the quality of the similarity computation.

4.3 Computing the Uncertain Boundary

In this section, we will describe the process of computing the uncertain boundary of a micro-cluster. Once the closest micro-cluster for an incoming point has been determined, we need to decide whether it should be added to the corresponding micro-clustering statistics, or whether a new micro-cluster containing a singleton point should be created. We create a new micro-cluster, if the incoming point lies outside the uncertainty boundary of the micro-cluster. The uncertainty boundary of a micro-cluster is defined in terms of the standard deviation of the distances of the data points about the centroid of the micro-cluster. Specifically, we use t standard deviations from the centroid of the cluster as a boundary for the decision of whether to include that particular point in the micro-cluster. A choice of $t = 3$ ensures a high level of certainty that the point does not belong to that cluster with the use of the normal distribution assumption. Let $\overline{W}$ be the centroid of the cluster $\mathcal{C}$, and let the set of points in it be denoted by $\overline{Y_1} \ldots \overline{Y_r}$. Then, the uncertain radius U is denoted as follows:

$$U = \sum_{i=1}^{r} \sum_{j=1}^{d} E[||Y_i - W||_j^2] \tag{13.7}$$

The expression on the right hand side of the above Equation can be evaluated by using the relationship of Lemma 13.4.

4.4 Further Enhancements

The method for clustering uncertain data streams can be further enhanced in several ways:

- In many applications, it is desirable to examine the clusters over a specific time horizon rather than the entire history of the data stream. In order to achieve this goal, a pyramidal time frame [6] can be used for stream classification. In this time-frame, snapshots are stored in different orders depending upon the level of recency. This can be used in order to retrieve clusters over a particular horizon with very high accuracy.

- In some cases, the behavior of the data stream may evolve over time. In such cases, it is useful to apply a *decay-weighted* approach. In the decay-weighted approach, each point in the stream is a weighted by a factor which decays over time. Such an approach can be useful in a number of scenarios in which the behavior of the data stream changes considerably over time. In order to use the decay-weighted approach, the key modification is to define the micro-clusters with a weighted sum of the data points, as opposed to the explicit sums. It can be shown that such an approach can be combined with a lazy-update method in order to effectively maintain the micro-clusters.

Recently, this method has also been extended to the case of projected clustering of uncertain data streams [4]. The *UPStream* algorithm simultaneously computes the projected dimensions and the assignment of data points to clusters during the process. The case of high dimensional data is much more challenging, since it requires us to determine the relevant projected dimensions in the presence of uncertainty. Details may be found in [4].

5. Approximation Algorithms for Clustering Uncertain Data

Recently, techniques have been designed for approximation algorithms for uncertain clustering in [14]. The work in [14] discusses extensions of the k-mean and k-median version of the problems. Bi-criteria algorithms are designed for each of these cases. One algorithm achieves a $(1+\epsilon)$-approximation to the best uncertain k-centers with the use of $O(k \cdot \epsilon^{-1} \cdot \log^2(n))$ centers. The second algorithm picks $2k$ centers and achieves a constant-factor approximation.

A key approach proposed in the paper [14] is the use of a transformation from the uncertain case to a weighted version of the deterministic case. We note that solutions to the weighted version of the deterministic clustering problem are well known, and require only a polynomial blow-up in the problem size. The key assumption in solving the weighted deterministic case is that the ratio of the largest to smallest weights is polynomial. This assumption is assumed to be maintained in the transformation. This approach can be used in order to solve both the uncertain k-means and k-median version of the problem.

6. Conclusions and Summary

In this chapter, we discussed new techniques for clustering uncertain data. The uncertainty in the data may either be specified in the form of a probability density function or in the form of variances of the attributes. The specification of the variance requires less modeling effort, but is more challenging from a clustering point of view. The problem of clustering is significantly affected by the uncertainty, because different attributes may have different levels of uncertainty embedded in them. Therefore, treating all attributes evenly may not provide the best clustering results. This chapter provides a survey of the different algorithms for clustering uncertain data. We discuss extensions of the density-based approach as well as a K-means approach for clustering. We also discussed a new method for clustering uncertain data streams.

Acknowledgements

Research was sponsored in part by the US Army Research laboratory and the UK ministry of Defense under Agreement Number W911NF-06-3-0001. The views and conclusions contained in this document are those of the author and should not be interpreted as representing the official policies of the US Government, the US Army Research Laboratory, the UK Ministry of Defense, or the UK Government. The US and UK governments are authorized to reproduce and distribute reprints for Government purposes notwithstanding any copyright notice hereon.

References

[1] C.C. Aggarwal, P. S. Yu. " A Survey of Uncertain Data Algorithms and Applications," in *IEEE Transactions on Knowledge and Data Engineering*, to appear, 2009.

[2] C. C. Aggarwal, "On Density Based Transforms for Uncertain Data Mining," in *ICDE Conference Proceedings*, 2007.

[3] C. C. Aggarwal and P. S. Yu, "A Framework for Clustering Uncertain Data Streams," in *ICDE Conference*, 2008.

[4] C. C. Aggarwal, "On High-Dimensional Projected Clustering of Uncertain Data Streams," in *ICDE Conference*, 2009 (poster version). Full version in *IBM Research Report*, 2009.

[5] C. C. Aggarwal, "On Unifying Privacy and Uncertain Data Models," in *ICDE Conference Proceedings*, 2008.

[6] C. C. Aggarwal, J. Han, J. Wang, and P. Yu, "A Framework for Clustering Evolving Data Streams," in *VLDB Conference*, 2003.

[7] R. Agrawal, and R. Srikant, "Privacy-Preserving Data Mining," in *ACM SIGMOD Conference*, 2000.

[8] M. Ankerst, M. M. Breunig, H.-P. Kriegel, and J. Sander, "OPTICS: Ordering Points to Identify the Clustering Structure," in *ACM SIGMOD Conference*, 1999.

[9] D. Barbara, H. Garcia-Molina, D. Porter: The management of probabilistic data. *IEEE Transactions on Knowledge and Data Engineering*, 4(5), pp. 487–502, 1992.

[10] D. Burdick, P. Deshpande, T. Jayram, R. Ramakrishnan, and S. Vaithyanathan, "OLAP Over Uncertain and Imprecise Data," in *VLDB Conference Proceedings*, 2005.

[11] A. L. P. Chen, J.-S. Chiu, and F. S.-C. Tseng, "Evaluating Aggregate Operations over Imprecise Data," in *IEEE Transactions on Knowledge and Data Engineering*, vol. 8, no. 2, pp. 273–294, 1996.

[12] R. Cheng, Y. Xia, S. Prabhakar, R. Shah, and J. Vitter, "Efficient Indexing Methods for Probabilistic Threshold Queries over Uncertain Data," in *VLDB Conference Proceedings*, 2004.

[13] R. Cheng, D. Kalashnikov, and S. Prabhakar, "Evaluating Probabilistic Queries over Imprecise Data," in *SIGMOD Conference*, 2003.

[14] G. Cormode, and A. McGregor, "Approximation algorithms for clustering uncertain data," in *PODS Conference*, pp. 191-200, 2008.

[15] N. Dalvi, and D. Suciu, "Efficient Query Evaluation on Probabilistic Databases," in *VLDB Conference Proceedings*, 2004.

[16] A. Das Sarma, O. Benjelloun, A. Halevy, and J. Widom, "Working Models for Uncertain Data," in *ICDE Conference Proceedings*, 2006.

[17] M. Ester, H.-P. Kriegel, J. Sander, and X. Xu, "A Density Based Algorithm for Discovcering Clusters in Large Spatial Databases with Noise," in *KDD Conference*, 1996.

[18] H. Garcia-Molina, and D. Porter, "The Management of Probabilistic Data," in *IEEE Transactions on Knowledge and Data Engineering*, vol. 4, pp. 487–501, 1992.

[19] S. Guha, R. Rastogi, and K. Shim, CURE: An Efficient Clustering Algorithm for Large Databases, in *ACM SIGMOD Conference*, 1998.

[20] A. Jain, and R. Dubes, Algorithms for Clustering Data, *Prentice Hall,* New Jersey, 1998.

[21] L. Kaufman, P. Rousseeuw.Finding Groups in Data: An Intrduction to Cluster Analysis,*Wiley Interscience*, 1990.

[22] R. Ng and J. Han, "Efficient and Effective Clustering Algorithms for Spatial Data Mining," in *VLDB Conference*, 1994.

[23] M. Kumar, N. Patel, and J. Woo, "Clustering seasonality patterns in the presence of errors," in *ACM KDD Conference Proceedings*, pp. 557-563, 2002.

[24] H.-P. Kriegel, and M. Pfeifle, "Density-Based Clustering of Uncertain Data," in *ACM KDD Conference Proceedings*, 2005.

[25] H.-P. Kriegel, and M. Pfeifle, "Hierarchical Density Based Clustering of Uncertain Data," in *ICDM Conference*, 2005.

[26] L. V. S. Lakshmanan, N. Leone, R. Ross, and V. S. Subrahmanian, "ProbView: A Flexible Probabilistic Database System," in *ACM Transactions on Database Systems*, vol. 22, no. 3, pp. 419–469, 1997.

[27] S. D. Lee, B. Kao, and R. Cheng, "Reducing UK-means to K-means," in *ICDM Workshops*, 2006.

[28] S. I. McClean, B. W. Scotney, and M. Shapcott, "Aggregation of Imprecise and Uncertain Information in Databases," in *IEEE Transactions on Knowledge and Data Engineering*, vol. 13, no. 6, pp. 902–912, 2001.

[29] W. Ngai, B. Kao, C. Chui, R. Cheng, M. Chau, and K. Y. Yip, "Efficient Clustering of Uncertain Data," in *ICDM Conference Proceedings*, 2006.

[30] D. Pfozer, and C. Jensen, "Capturing the uncertainty of moving object representations," in *SSDM Conference*, 1999.

[31] S. Singh, C. Mayfield, S. Prabhakar, R. Shah, and S. Hambrusch, "Indexing Uncertain Categorical Data," in *ICDE Conference*, 2007.

[32] Y. Tao, R. Cheng, X. Xiao, W. Ngai, B. Kao, and S. Prabhakar, "Indexing Multi-dimensional Uncertain Data with Arbitrary Probabality Density Functions," in *VLDB Conference*, 2005.

[33] T. Zhang, R. Ramakrishnan, and M. Livny, "BIRCH: An Efficient Data Clustering Method for Very Large Databases," in *ACM SIGMOD Conference Proceedings,* 1996.

Chapter 14

ON APPLICATIONS OF DENSITY TRANSFORMS FOR UNCERTAIN DATA MINING

Applications to Classification and Outlier Detection

Charu C. Aggarwal
IBM T. J. Watson Research Center
Hawthorne, NY 10532
charu@us.ibm.com

Abstract In this chapter, we will examine a general density-based approach for handling uncertain data. The broad idea is that implicit information about the errors can be indirectly incorporated into the density estimate. We discuss methods for constructing error-adjusted densities of data sets, and using these densities as intermediate representations in order to perform more accurate mining. We discuss the mathematical foundations behind the method and establish ways of extending it to very large scale data mining problems. As concrete examples of our technique, we show how to apply the intermediate density representation in order to accurately solve the classification and outlier detection problems. This approach has the potential in constructing intermediate representations as a broad platform for data mining applications.

Keywords: Density Transforms, Uncertain Data, Classification, Outlier Detection

1. Introduction

While data collection methodologies have become increasingly sophisticated in recent years, the problem of inaccurate data continues to be a challenge for many data mining problems. This is because data collection methodologies are often inaccurate and are based on incomplete or inaccurate information. For example, the information collected from surveys is highly incomplete and either needs to be imputed or ignored altogether. In other cases, the base data for the data mining process may itself be only an estimation from other underlying phenomena. In many cases, a quantitative estimation of the noise in different fields is available. An example is illustrated in [12], in which error-driven methods are used to improve the quality of retail sales merchandising. Many scientific methods for data collection are known to have error-estimation methodologies built into the data collection and feature extraction process. Such data with error estimations or probabilistic representations of the underlying data are referred to as *uncertain data* [2]. We summarize a number of real applications, in which such error information can be known or estimated a-priori:

- When the inaccuracy arises out of the limitations of data collection equipment, the statistical error of data collection can be estimated by prior experimentation. In such cases, different features of observation may be collected to a different level of approximation.

- In the case of missing data, imputation procedures can be used [15] to estimate the missing values. If such procedures are used, then the statistical error of imputation for a given entry is often known a-priori.

- Many data mining methods are often applied to *derived* data sets which are generated by statistical methods such as forecasting. In such cases, the error of the data can be derived from the methodology used to construct the data.

- In many applications, the data is available only on a partially aggregated basis. For example, many demographic data sets only include the statistics of household income over different localities rather than the precise income for individuals.

The results of data mining are often subtly dependent upon the errors in the data. For example, consider the case illustrated in Figure 14.1. In this case, we have illustrated a two dimensional binary classification problem. It is clear that the errors along dimension 1 are higher than the errors along dimension 2. In addition to a test example X, we have illustrated two training examples Y and Z, whose errors are illustrated in the same figure with the use of oval shapes. For a given test example X, a nearest neighbor classifier would pick the class label of data point Y. However, the data point Z may have a much higher probability of being the nearest neighbor to X than the data point Y. This is because the data point X lies within the error boundary of Z. It is important to design a technique which can use the relative errors of the different data points over the different dimensions in order to improve the accuracy of the data mining process.

Thus, it is clear that the failure to use the error information in data mining models can result in a loss of accuracy. In this chapter, we will discuss a

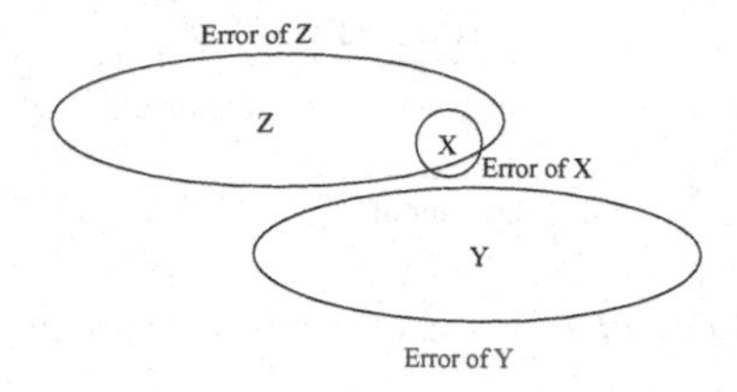

Figure 14.1. Effect of Errors on Classification

general and scalable approach to uncertain data mining with the use of multivariate density estimation. We will show that density estimation methods provide an effective intermediate representation, which captures information about the noise in the underlying data. The density estimate can be leveraged by designing an algorithm which works with error-adjusted densities rather than individual data points. As a specific example, we will discuss the case of the classification problem. As we will see, only a few minor modifications to existing methods are required in order to apply the method a density based representation of the data. In general, we expect that our approach can be used for a wide variety of data mining applications which use density estimation as an intermediate representation during the analytical process.

In order to improve the scalability of our technique, we show how to use a compression approach in order to generalize it for very large data sets. This goal is achieved by developing a density estimation process which works with (error-adjusted) micro-clusters in the data. The statistics of these error-adjusted micro-clusters are used to compute a kernel function which can be used to estimate the density in a time proportional to the number of micro-clusters. Furthermore, since the micro-clusters are stored in main memory, the procedure can be efficiently applied to very large data sets, even when the probability densities need to re-computed repeatedly in different subspaces during the mining process.

This chapter is organized as follows. In the next section, we will discuss the method of density based estimation of uncertain data sets. In section 3, we will discuss the application of the procedure to the problem of classification. We will show how to use the error estimates in order to gain additional information about the data mining process. In section 4, we will discuss the application of the method to the outlier detection problem. Section 5 contains the conclusions and summary.

2. Kernel Density Estimation with Errors

We will first introduce some additional notations and definitions. We assume that we have a data set $\mathcal{D}$, containing N points and d dimensions. The

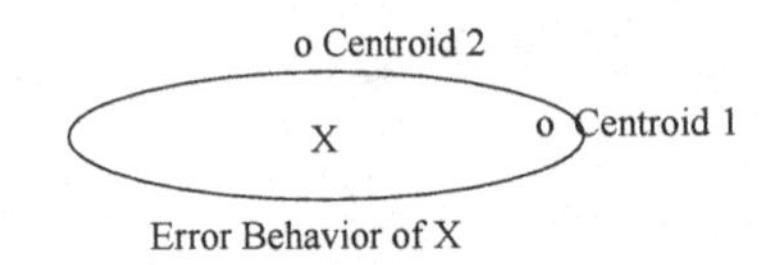

Figure 14.2. Effect of Errors on Clustering

actual records in the data are denoted by $\overline{X_1}, \overline{X_2}, \ldots \overline{X_N}$. We assume that the estimated error associated with the jth dimension for data point $\overline{X_i}$ is denoted by $\psi_j(\overline{X_i})$. The interpretation of this error value can vary with the nature of the data mining application. For example, in a scientific application in which the measurements can vary from one observation to another, the error value is the standard deviation of the observations over a large number of measurements. In a k-anonymity based data (or incomplete data) mining application, this is the standard deviation of the partially specified (or imputed) fields in the data. Even though the error may be defined as a function of the field (or dimension) in most applications, we have made the most general assumption in which the error is defined by both the row and the field. This can be the case in many applications in which different parts of the data are derived from heterogeneous sources.

The idea in kernel density estimation [17] is to provide a continuous estimate of the density of the data at a given point. The value of the density at a given point is estimated as the sum of the smoothed values of kernel functions $K'_h(\cdot)$ associated with each point in the data set. Each kernel function is associated with a kernel width h which determines the level of smoothing created by the function. The kernel estimation $\overline{f}(x)$ based on N data points and kernel function $K'_h(\cdot)$ is defined as follows:

$$\overline{f}(x) = (1/N) \cdot \sum_{i=1}^{N} K'_h(x - \overline{X_i}) \tag{14.1}$$

Thus, each discrete point $\overline{X_i}$ in the data set is replaced by a continuous function $K'_h(\cdot)$ which peaks at X_i and has a variance which is determined by the smoothing parameter h. An example of such a distribution would be a gaussian kernel with width h.

$$K'_h(x - \overline{X_i}) = (1/\sqrt{2\pi} \cdot h) \cdot e^{-(x-\overline{X_i})^2/(2h^2)} \tag{14.2}$$

The overall effect of kernel density estimation is to convert the (discrete) data set into a continuous density estimate by replacing each data point with a smoothed “bump”, whose width is determined by h. The density distribution at a given coordinate is equal to the sum of the contributions of all the “bumps” represented by the data points. The result is a continuous distribution in which the random artifacts are suppressed and the density behavior provides a global overview of the dense as well as sparsely populated regions of the data. The estimation error depends upon the kernel width h which is chosen in a data driven manner. A widely used rule for approximating the bandwidth is the Silverman approximation rule [17] for which h may be chosen to be $1.06 \cdot \sigma \cdot N^{-1/5}$, where σ^2 is the variance of the N data points. It has been shown [17] that for most smooth functions $K'_h(\cdot)$, when the number of data points goes to infinity, the estimator $\overline{f}(x)$ asymptotically converges to the true density function $f(x)$, provided that the width h is chosen using the above relationship. For the d-dimensional case, the kernel function is chosen to be the product of d identical kernels $K_i(\cdot)$, each with its own smoothing parameter h_i.

The presence of errors can change the density estimates because of the different levels of error in different entries or fields. For example a data point or field with very large error should affect the density estimation of its locality to a smaller extent than one which contains small errors. When estimations of such errors are available, it is desirable to incorporate them into the estimation process. A direct way of doing so is to adapt the kernel function so that the measurement errors are taken into account during the calculation of the density distribution. Correspondingly, we define the following error-based kernel $Q'_h(x - X_i, \psi(\overline{X_i})$ function, which depends both upon the error as well as the values of the underlying data points.

$$Q'_h(x - X_i, \psi(\overline{X_i}) = (1/\sqrt{2\pi} \cdot (h + \psi(\overline{X_i}))) \cdot e^{\frac{-(x-X_i)^2}{(2\cdot(h^2+\psi(\overline{X_i})^2))}} \tag{14.3}$$

The overall effect of changing the kernel function is that the width of the bandwidth along the corresponding dimension is increased by $\psi(\overline{X_i})$. The intuition behind this choice of modifying the kernel is to adjust the contributions of the kernel function for a point depending upon its (error-based) probability density. Note that in the limiting case, when there are a large number of data points N, the value of the bandwidth h goes to zero, and this kernel function has a gaussian distribution with standard error exactly equal to the standard error of the data point. Conversely, the error-based kernel function converges to the standard kernel function when the value of the error $\psi(\overline{X_i})$ is 0. Therefore, in these boundary cases, the direct error-based generalization of the kernel function has a probability distribution with the same standard error as the data point. It is also clear that in the limiting case of a large number of data points, (when the bandwidth h tends to zero by the Silverman rule) the kernel function reflects the errors in each data point accurately. As in the previous case, the error-based density at a given data point is defined as the sum of the error-based kernels over different data points. Therefore, we define the error based density $\overline{f^Q}$ at point x as follows:

$$\overline{f^Q}(x, \psi(\overline{X_i})) = (1/N) \cdot \sum_{i=1}^{N} Q'_h(x - \overline{X_i}, \psi(\overline{X_i})) \quad (14.4)$$

As in the previous case, we can easily generalize the definition to the multi-dimensional case. Specifically, the error for the jth dimension is denoted by $\psi_j(\overline{X_i})$. The overall kernel function is defined as the product of the kernel function for the different dimensions.

Our aim is to use the joint probability densities over different subspaces in order to design data mining algorithms. This is much more general than the *univariate* approach for privacy preserving data mining discussed in [6], and is applicable to a much wider range of error-based and privacy-preserving approaches, since it does not require the probability distribution of the noise in the data. This joint probability density may need to be repeatedly computed over different subsets of dimensions for particular data mining problems. If the data set is too large to maintain in main memory, we need to perform repeated passes over the data for the computation of the density over different subsets of dimensions.

Since this option does not scale very well for large data sets, we need to develop methods to condense the data points into a smaller number of *pseudo-points*, but with slightly larger error. Such an approach is easily implementable for larger data sets, since the pseudo-points can be maintained in main memory for the computation of the density over different subsets of dimensions. However, this requires us to modify the method of computation of the error-based densities using micro-clusters [5] instead of individual data points.

2.1 Scalability for Large Data Sets

The method can be generalized to very large data sets and data streams. In order to generate the approach to very large data sets, we condense the data into a smaller number of micro-clusters. While the concept of micro-clustering has been discussed in [5, 18], our aim in this chapter is to modify and leverage it for error-based density estimation. In order to achieve this goal, we need to discuss how the error-based density may be computed using micro-cluster statistics instead of individual data points. Our first step is to define the micro-clusters in the data as suggested in [5]. However, since the work in [5] does not use errors, the definition of a micro-cluster needs to be modified correspondingly.

It is assumed that the data stream consists of a set of multi-dimensional records $\overline{X_1} \ldots \overline{X_k} \ldots$ arriving at time stamps $T_1 \ldots T_k \ldots$. Each $\overline{X_i}$ is a multi-dimensional record containing d dimensions which are denoted by $\overline{X_i} = (x_i^1 \ldots x_i^d)$.

We will first begin by defining the concept of error-based micro-clusters more precisely.

DEFINITION 14.1 *A micro-cluster for a set of d-dimensional points* $X_{i_1} \ldots X_{i_n}$ *with time stamps* $T_{i_1} \ldots T_{i_n}$ *is defined as the* $(3 \cdot d + 1)$ *tuple* $(\overline{CF2^x}(\mathcal{C}), \overline{EF2^x}(\mathcal{C}), \overline{CF1^x}(\mathcal{C}), n(\mathcal{C}))$, *wherein* $\overline{CF2^x}(\mathcal{C})$, $\overline{EF2^x}(\mathcal{C})$, *and* $\overline{CF1^x}(\mathcal{C})$ *each correspond to a vector of d entries. The definition of each of these entries is as follows:*

- *For each dimension, the sum of the squares of the data values is maintained in* $\overline{CF2^x}(\mathcal{C})$. *Thus,* $\overline{CF2^x}(\mathcal{C})$ *contains d values. The p-th entry of* $\overline{CF2^x}(\mathcal{C})$ *is equal to* $\sum_{j=1}^{n}(x_{i_j}^p)^2$.
- *For each dimension, the sum of the squares of the errors in the data values is maintained in* $\overline{EF2^x}(\mathcal{C})$. *Thus,* $\overline{EF2^x}(\mathcal{C})$ *contains d values. The p-th entry of* $\overline{EF2^x}(\mathcal{C})$ *is equal to* $\sum_{j=1}^{n}\psi_p(X_{i_j})^2$.
- *For each dimension, the sum of the data values is maintained in* $\overline{CF1^x}(\mathcal{C})$. *Thus,* $\overline{CF1^x}(\mathcal{C})$ *contains d values. The p-th entry of* $\overline{CF1^x}(\mathcal{C})$ *is equal to* $\sum_{j=1}^{n} x_{i_j}^p$.
- *The number of points in the data is maintained in* $n(\mathcal{C})$.

We will refer to the micro-cluster for a set of points $\mathcal{C}$ by $\overline{CFT}(\mathcal{C})$. As in [18], this summary information can be expressed in an additive way over the different data points. This makes it very easy to create and maintain the clusters using a single pass of the data. The actual maintenance of the micro-clusters is a variation of the approach discussed in [5]. In this variation, we maintain the micro-cluster statistics for the q different centroids. These q centroids are chosen randomly. Each incoming data point is always assigned to its closest micro-cluster centroid using a nearest neighbor algorithm, and is never allowed to create a new micro-cluster. This is different from [5] in which a new micro-cluster is created whenever the incoming data point does not naturally fit in a micro-cluster. Similarly, clusters are never discarded as in [5]. This is required to ensure that all data points are reflected in the micro-cluster statistics. In addition, it is necessary to take the errors into account during the computation of the micro-cluster statistics. Consider the example illustrated in Figure 14.2 in which we have shown the error behavior of data point X by an elliptical error region. While the data point X is closer to centroid 2, it is more likely to belong to the cluster corresponding to centroid 1. This is because the error behavior of the data point is skewed in such a way that it would have been more likely to coincide with centroid 1 simply because of an error in measurement. Thus, we need to adjust for the errors corresponding to the different dimensions during the distance calculations. Thus, let us consider the data point $\overline{X}$ and centroid $\overline{c} = (c_1 \ldots c_d)$. Then, the distance $dist(\overline{Y}, \overline{c})$ between data point

$\overline{Y} = (Y_1 \ldots Y_d)$ and $\overline{c}$ is given by the following relationship:

$$dist(\overline{Y}, \overline{c}) = \sum_{j=1}^{d} \max\{0, (Y_j - c_j)^2 - \psi_j(\overline{Y})^2\} \quad (14.5)$$

We note that this is an error-adjusted variation of the Euclidean distance metric. The error adjustment ensures that the dimensions which have large errors do not contribute significantly to the distance function. If the true distance along a particular dimension j is less than the average error $\psi_j(\overline{Y})$, then the value of the error-adjusted distance is defined to be zero. Therefore, we use a distance function which reflects the best-case scenario along each dimension. Such an approach turns out to be more effective for noisy data sets in high dimensionality [3].

The aim of the micro-clustering process is to compress the data so that the resulting statistics can be held in main memory for repeated passes during the density estimation process over different subspaces. Therefore, the number of micro-clusters q is defined by the amount of main memory available. Given the large memory sizes available even on modest desktop hardware today, this corresponds to thousands of micro-clusters for data sets containing hundreds of dimensions. This means that a high level of granularity in data representation can be maintained. This level of granularity is critical in using the micro-clustering as a surrogate for the original data.

Each micro-cluster is subsequently used as a summary representation in order to compute the densities over different subspaces. The key is to design a kernel function which adapts to the variance of the data points within a cluster as well as their errors. At the same time, the kernel function should be computable *using only the micro-cluster statistics of the individual clusters.* The first step is to recognize that each micro-cluster is treated as a pseudo-point with a new error defined both by the variance of the points in it as well as their individual errors. Thus, we need to compute the error of the pseudo-point corresponding to a micro-cluster. For this purpose, we assume that each data point $\overline{X}$ in a micro-cluster is a mutually independent observation with bias equal to its distance from the centroid and a variance equal to the error $\psi(\overline{X})$. Therefore, the true error $\phi(\overline{X}, \mathcal{C})^2$ of the data point $\overline{X}$ assuming that it is an instantiation of the pseudo-observation corresponding to the micro-cluster $\mathcal{C}$ is given by:

$$\phi_j(\overline{X}, \mathcal{C})^2 = bias_j(\overline{X}, \mathcal{C})^2 + \psi_j(\overline{X})^2 \quad (14.6)$$

Here $\phi_j(\overline{X}, \overline{C})$ refers to the error in the jth dimension. We note that this result follows from the well known statistical relationship that the true error is defined by the squared sum of the bias and the variance. Once we have established the above relationship, we can use the independence assumption to derive the overall error for the micro-cluster pseudo-point in terms of the micro-cluster summary statistics:

LEMMA 14.2 *The true error $\Delta(\mathcal{C})$ for the pseudo-observation for a micro-cluster $\mathcal{C} = \{Y_1 \ldots Y_r\}$ is given by the relationship:*

$$\Delta_j(\mathcal{C}) = \sum_{i=1}^{r} \frac{\phi_j(\overline{Y_i}, \mathcal{C})^2}{r} = \frac{CF2_j^x(\mathcal{C})}{r} - \frac{CF1_j^x(\mathcal{C})^2}{r^2} + \frac{EF2_j(\mathcal{C})}{r} \tag{14.7}$$

Proof: These results easily follow from Equation 14.6. By averaging the results of Equation 14.6 over different data points, we get:

$$\sum_{i=1}^{r} \phi_j(\overline{Y_i}, \mathcal{C})^2/r = \sum_{i=1}^{r} bias_j(\overline{Y_i}, \mathcal{C})^2/r + \sum_{j=1}^{r} \psi_j(\overline{Y_i})^2/r \tag{14.8}$$

We will now examine the individual terms on the right hand side of Equation 14.8. We first note that the value of $\sum_{i=1}^{r} bias_j(\overline{Y_i}, \mathcal{C})^2/r$ corresponds to the variance of the data points in cluster $\mathcal{C}$. From [18], we know that this corresponds to $CF2_j^x(\mathcal{C})/r - CF1_j^x(\mathcal{C})^2/r^2$. It further follows from the definition of a micro-cluster that the expression $\sum_{j=1}^{r} \psi_j(\overline{Y_i})^2/r$ evaluates to $EF2_j(\mathcal{C})/r$. By substituting the corresponding values in Equation 14.8, the result follows.

The above result on the true error of a micro-cluster pseudo-observation can then be used in order to compute the error-based kernel function for a micro-cluster. We denote the centroid of the cluster $\mathcal{C}$ by $c(\mathcal{C})$. Correspondingly, we define the kernel function for the micro-cluster $\mathcal{C}$ in an analogous way to the error-based definition of a data point:

$$Q'_h(x - c(\mathcal{C}), \Delta(\mathcal{C}) = (1/\sqrt{2\pi} \cdot (h + \Delta(\mathcal{C}))) \cdot e^{-\frac{(x - X_i)^2}{(2 \cdot (h^2 + \Delta(\mathcal{C})^2))}} \tag{14.9}$$

As in the previous case, we need to define the overall density as a sum of the densities of the corresponding micro-clusters. The only difference is that we need to define the overall density as the weighted sum of the corresponding kernel functions. Therefore, if the data contains the clusters $\mathcal{C}_1 \ldots \mathcal{C}_m$, then we define the density estimate at x as follows:

$$\overline{f^Q}(x, \Delta(\mathcal{C})) = (1/N) \cdot \sum_{i=1}^{m} n(\mathcal{C}_i) \cdot Q'_h(x - c(\mathcal{C}_i), \Delta(\overline{X_i})) \tag{14.10}$$

The density estimate is defined by the weighted estimate of the contributions from the different micro-clusters. This estimate can be used for a variety of data mining purposes. In the next section, we will describe one such application.

Algorithm *DensityBasedClassification*(Test Point: x,
Accuracy Threshold: a);
begin
  $C_1 = \{1, \ldots d\}$;
  for each dimension S in C_1 and label l_j
      compute $\mathcal{A}(x, S, l_j)$;
  L_1 is the set of dimensions in C_1
    for which for some $j \in \{1 \ldots k\}$
    $\mathcal{A}(x, S, l_j) > a$;
  i=1;
  while L_i is not empty
  begin
    Generate C_{i+1} by joining L_i with L_1;
    for each subset S of dimensions in C_{i+1}
        and class label l_j compute $\mathcal{A}(x, S, l_j)$;
    L_{i+1} is the set of dimensions in C_{i+1}
      for which for some $j \in \{1 \ldots k\}$ we
        have $\mathcal{A}(x, S, l_j) > a$;
    $i = i + 1$;
  end;
  $\mathcal{L} = \cup_i L_i$;
  $N = \{\}$;
  while $\mathcal{L}$ is not empty
  begin
    Add set with highest local accuracy in $\mathcal{L}$ to N;
    Remove all sets in $\mathcal{L}$ which overlap with sets in N;
  end;
  report majority class label in N;
end

Figure 14.3. Density Based Classification of Data

3. Leveraging Density Estimation for Classification

We note that error-based densities can be used for a variety of data mining purposes. This is because the density distribution of the data set is a surrogate for the actual points in it. When joint distributions over different subspaces are available, then many data mining algorithms can be re-designed to work with density based methods. Thus, the key approach to apply the method to an arbitrary data mining problem is to design a *density based algorithm* for the problem. We further note that clustering algorithms such as DBSCAN [9] and many other data mining approaches work with joint probability densities as intermediate representations. In all these cases, our approach provides a direct (and scalable) solution to the corresponding problem. We further note that unlike the work in [6], which is inherently designed to work with univariate re-construction, our results (which are derived from only simple error statistics rather than entire distributions) are tailored to a far wider class of methods. The joint probability distribution makes it easier to directly generate error-based generalizations of more complex methods such as multi-variate classifiers.

In this chapter, we will discuss such a generalization for the classification problem [8, 11]. We define the notations for the classification problem as follows: We have a data set $\mathcal{D}$ containing a set of d-dimensional records, and k class labels which are denoted by $l_1 \ldots l_k$. The subset of the data corresponding to the class label l_i is denoted by $\mathcal{D}_i$. In this chapter, we will design a density based adaptation of rule-based classifiers. Therefore, in order to perform the classification, we need to find relevant classification rules for a particular test instance. The challenge is to use the density based approach in order to find the particular subsets of dimensions that are most discriminatory for a particular test instance. Therefore, we need to find those subsets of dimensions in which the instance-specific local density of the data for a particular class is significantly higher than its density in the overall data. The first step is to compute the density over a subset of dimensions S. We denote the density at a given point x, subset of dimensions S, and data set $\mathcal{D}$ by $g(x, S, \mathcal{D})$. The process of computation of the density over a given subset of dimensions is exactly similar to our discussion of the full dimensional case, except that we use only the subset of dimensions S rather than the full dimensionality.

The first step in the classification process is to pre-compute the error-based micro-clusters together with the corresponding statistics. Furthermore, the micro-clusters are computed separately for each of the data sets $\mathcal{D}_1 \ldots \mathcal{D}_k$ and $\mathcal{D}$. We note that this process is performed only *once* as a pre-processing step. Subsequently, the compressed micro-clusters for the different classes are used in order to generate the accuracy density estimates over the different subspaces. The statistics of these micro-clusters are used in order to compute the densities using the relations discussed in the previous section. However, since the densities need to be computed in an example-specific way, the density calculation is performed during the classification process itself. The process of data compression makes the density computation particularly efficient for large data sets. We also note that in order to calculate the density over a particular subspace, we can use Equations 14.9, and 14.10, except that we only need to apply them to a subspace of the data rather than the entire data dimensionality.

In order to construct the final set of rules, we use an iterative approach in which we find the most relevant subspaces for the classification process. In order to define the relevance of a particular subspace, we define its density based local accuracy $\mathcal{A}(x, S, l_i)$ as follows:

$$\mathcal{A}(x, S, l_i) = \frac{|\mathcal{D}_i| \cdot g(x, S, \mathcal{D}_i)}{|\mathcal{D}| \cdot g(x, S, \mathcal{D})} \quad (14.11)$$

This dominant class $dom(x, S)$ at point x is defined as the class label with the highest accuracy. Correspondingly, we have:

$$dom(x, S) = \text{argmax}_i \mathcal{A}(x, S, l_i) \quad (14.12)$$

In order to determine the most relevant subspaces, we perform a bottom up methodology to find those combinations of dimensions which have high accuracy for a given test example. We also impose the requirement that in order for a subspace to be considered in the set of $(i+1)$-dimensional variations, at least one subset of it needs to satisfy the accuracy threshold requirements. We impose this constraint in order to facilitate the use of a roll-up technique in our algorithm. In most cases, this does not affect the use of the technique when only lower dimensional projections of the data are used for the classification process. In the roll-up process, we assume that C_i is a set of candidate i-dimensional subspaces, and L_i is a subset of C_i, which have sufficient discriminatory power for that test instance. We iterate over increasing values of i, and join the candidate set L_i with the set L_1 in order to determine C_{i+1}. We find the subspaces in C_{i+1} which have accuracy above a certain threshold (subject to the additional constraints discussed). In order to find such subspaces, we use the relationships discussed in Equations 14.11 and 14.12. Thus, we need to compute the local density accuracy over each set of dimensions in C_{i+1} in order to determine the set L_{i+1}. We note that this can be achieved quite efficiently because of the fact that the computations in Figure 14.11 can be performed directly over the micro-clusters rather than the original data points. Since there are significantly fewer number of micro-clusters (which reside in main memory), the density calculation can be performed very efficiently. Once the accuracy density for each subspace in C_{i+1} has been computed, we retain only the dimension subsets for which the accuracy density is above the threshold a. This process is repeated for higher dimensionalities until the set C_{i+1} is empty. We assume that the final set of discriminatory dimensions are stored in $\mathcal{L} = \cup_i L_i$.

Finally, the non-overlapping sets of dimensions with the highest accuracy above the pre-defined threshold of a are used in order to predict the class label. In order to achieve this goal, we first determine the subspace of dimensions with the highest local accuracy. Then, we repeatedly find the next non-overlapping subset of dimensions with the highest level of accuracy until all possibilities are exhausted. The majority class label from these non-overlapping subsets of dimensions are reported as the relevant class label. We note that we do not necessarily need to continue with the process of finding

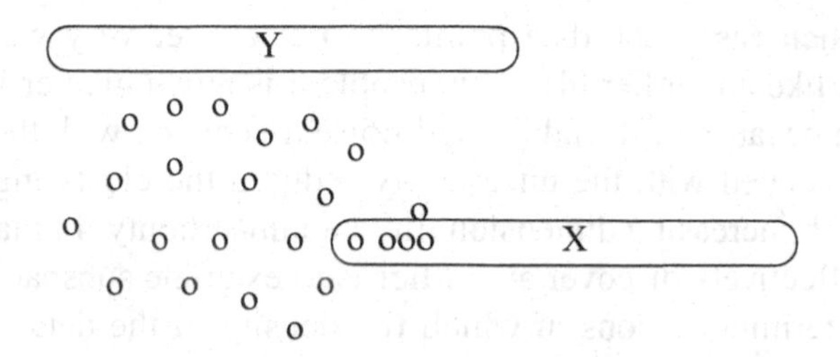

Figure 14.4. Outlier Detection Issues with Uncertain Data

the next overlapping subspace repeatedly until all possibilities are exhausted. Rather, it is possible to terminate the process after finding at most p non-overlapping subsets of dimensions.

This kind of approach can in fact be generalized to any data mining approach which use probability densities instead of individual data points. Many data mining algorithms can be naturally extended from point based processing to density based processing in order to apply this kind of approach. This is useful in order to leverage this methodology in general and practical settings. The overall algorithm is illustrated in Figure 14.3.

4. Application of Density Based Approach to Outlier Detection

A key problem in data mining is that of outlier detection. The problem of outlier detection has been widely studied in the data mining community [4, 7, 16, 14]. The addition of noise to the data makes the problem far more difficult from the perspective of uncertainty. In order to explore this point further, we have illustrated a 2-dimensional data set in Figure 14.4. We have also illustrated two points in the data set along with the corresponding contours of uncertainty. If we did not use the uncertainty behavior across the different attributes, we would be more likely to label X as an outlier, since it seems further away from the data set. However, in reality, the data point Y is much closer to being an outlier, since the corresponding contours do not overlap with the dense regions in the data. The important point to be noted here is that the relative uncertainty in the different attributes over a given record is important to use in the process of determination of the outlier-like behavior of a given record. In general, a higher level of uncertainty of a given record can result in it behaving like an outlier, even though the record itself may not be an outlier.

Another problem with outlier detection techniques is that an increase in dimensionality results in difficulty in identification of outliers in the full dimensional case. This is because a full dimensional data set is inherently sparse, and therefore every pair of points may be approximately equidistant from one another. In such cases, the data points do not cluster very well, and every point behaves like an outlier [4]. This problem is much greater in the case of uncertain data because of the additional noise associated with the uncertainty. The noise associated with the uncertainty reduces the clustering behavior of the data set with increasing dimensionality and uncertainty. In many cases, the only way to effectively discover an outlier is to examine subspaces of the data in order to determine regions in which the density of the data is abnormally low. We will use such a subspace exploration process in combination with density estimation in order to determine the outliers in the underlying data.

4.1 Outlier Detection Approach

We will first define some notations and definitions in order to facilitate further description. Let us assume that the database $\mathcal{D}$ contains N records. The ith record in the database contains d dimensions, and is denoted by $\overline{X_i}$. It is also assumed that each dimension of a record has a probability density function (pdf) associated with it. The probability density function for the record $\overline{X_i}$ along the jth dimension is denoted by $h_i^j(\cdot)$. Thus, the mean value of $h_i(\cdot)$ is $\overline{X_i}$. It is also assumed that the standard deviation of $h_i^j(\cdot)$ along the jth dimension is $\psi_j(\overline{X_i})$.

The outlier detection algorithm relies on the fact that an outlier in high dimensional space shows up in a region of abnormally low density in at least one subspace of the data. Therefore, the outlier detection algorithm constructs a density estimate of the underlying data in various subspaces and uses these density estimates in order to determine the outliers in the data. We note that the presence of the uncertainty in the data affects how the density estimate is constructed. Data points and attributes with less uncertainty tend to be more concentrated in their contribution to the density estimate of the data. In a later section, we will explain how the density estimate of the data is constructed efficiently. First, we will discuss the outlier algorithm assuming that the density estimate is available. Since the uncertainty affects the approach to outlier detection, we need a definition of the outlier detection problem, which takes such uncertainty into account. In order to achieve this goal, we will define a sequence of terms and notations.

First, we will define the η-probability of a given data point. This defines the probability that a data point lies in a region with (overall data) density at least η. Since the data point $\overline{X_i}$ is itself uncertain, the probability density of the overall data set will vary along the contour of its uncertainty. The *η-probability* may be estimated by integrating the probability density function of the data point over those regions of overall data density which are greater than η.

DEFINITION 14.3 *The η-probability of a data point $\overline{X_i}$ is defined as the probability that the uncertain data point lies in a region with (overall data) density at least η.*

We note that the η-probability of a data point can be computed in any subspace of the data. Clearly, the lower the η-probability, the less likely that the data point lies in a dense region. Such a point is an outlier. We note that the probability that a given data point lies in a region with density estimate at least η can be determined by integrating the density estimate of the probability density function of the data point over regions with probability density at least η. Consider the subspace S which is defined* by a subset of dimensions $J = \{1 \dots r\}$. Let the density function of the overall data set over the subspace J and coordinates $(x_1 \dots x_r)$ be given by $G(x_1 \dots x_r)$. Then, the probability p_i that the uncertain data point $\overline{X_i}$ lies in region with data density at least η is given by the following:

$$p_i = \int_{G(x_1 \dots x_r) \geq \eta} \pi_{j=1}^{r} h_i^j(x_j) dx_j \tag{14.13}$$

We note that the value of p_i is hard to compute exactly, but we will show that it can be estimated quite accurately. In order for the data point $\overline{X_i}$ to be an outlier, this value of p_i needs to be less than a user-defined parameter δ. This ensures that the data point has low probability of lying in a dense region of the data, even after the uncertainty in the position of the point is taken into account. Such a point is referred to as a (δ, η)-outlier.

DEFINITION 14.4 *We will define an uncertain data point $\overline{X_i}$ to be a (δ, η)-outlier, if the η-probability of $\overline{X_i}$ in some subspace is less than δ.*

Next, we will discuss the overall algorithm for outlier detection.

4.2 Subspace Exploration for Outlier Detection

The overall algorithm for outlier detection uses a roll-up approach in which we test the outlier-like behavior of data points in different subspaces. The algorithm uses as input the parameters δ and η which define the outlier-like behavior of the data points. In addition, the maximum dimensionality r of the subspaces is input to the algorithm. In order for the parameter η to have the same significance across different subspaces, we normalize the data, so that the standard deviation along each dimension is one unit.

The overall algorithm for outlier detection is illustrated in Figure 14.5. The algorithm works by using a bottom up enumeration approach in which it tests

*We can assume without loss of generality that $J = \{1 \dots r\}$ by re-ordering the dimensions appropriately. We have used $J = \{1, \dots r\}$ for notational convenience only.

```
Algorithm DetectOutlier(Data: D, Parameters: η, δ
          MaximumDim: r);
begin
  O = {};
  C_1 = All 1-dimensional candidates;
  i=1;
  while (C_i is not null) and (i <= r)
    begin
    Determine if any data point in D − O
      is a (δ, η)-outlier with respect to
      subspaces in C_i and add corresponding point
      to O;
    C_{i+1} = C_i ⊕ C_1;
    { The sign ⊕ corresponds to dimensional
     concatenation of each element of C_i with
     those dimensions of C_1 which are not
     already included in the corresponding subspace
     element in C_i; }
    i = i + 1;
    end
end
```

Figure 14.5. Outlier Detection Algorithm

the possibility that the different points in the data are outliers. In order to do so, it successively appends dimensions to candidate subspaces, and tests whether different data points are outliers in these subspaces. The algorithm starts off with the outlier set $\mathcal{O}$ set to null. The set $\mathcal{C}_i$ denotes the candidate set of subspaces which are used for testing the possibility of a given data point being an outlier. The algorithm uses the concept of (δ, η)-outlier in order to test whether a given data point is an outlier. In the event that a data point is indeed an outlier it is added to $\mathcal{O}$, and we move on to testing subspaces of the next higher dimension. This process continues until either $\mathcal{C}_i$ is null, or we have exhausted all subspaces of dimensionality up to r. We note that this approach requires two subroutines which can be significantly difficult to implement:

- We need to determine whether a given data point is a (δ, η)-outlier. Note that such a determination requires the computation of the integral in Equation 14.13, which can be quite difficult to estimate. Therefore, we will design a technique to estimate the integral in an efficient and accurate way with the use of probabilistic sampling.

- We need repeated density estimation in different subspaces over the entire data set. This can be extremely expensive, if it is not implemented carefully. We will use a cluster-based compression approach in order to improve the efficiency of the density estimation process.

5. Conclusions

This chapter presents a framework for performing density based mining of uncertain data. The density-based framework constructs an intermediate representation of the data which can be leveraged for a variety of problems. In this chapter, we discuss the application of this technique to the problems of clustering and outlier detection. The method can typically be used for any problem which use intermediate density representations for the underlying data set. A natural direction of research would be generalize this technique to a wide variety of closely related problems in the data mining field.

Acknowledgements

Research was sponsored in part by the US Army Research laboratory and the UK ministry of Defense under Agreement Number W911NF-06-3-0001. The views and conclusions contained in this document are those of the author and should not be interpreted as representing the official policies of the US Government, the US Army Research Laboratory, the UK Ministry of Defense, or the UK Government. The US and UK governments are authorized to reproduce and distribute reprints for Government purposes notwithstanding any copyright notice hereon.

References

[1] C. C. Aggarwal. On Density Based Transforms for Uncertain Data Mining. *ICDE Conference*, 2007.

[2] C. C. Aggarwal, P. S. Yu. A Survey of Uncertain Data Algorithms and Applications, *IEEE Transactions on Knowledge and Data Engineering*, to appear, 2009.

[3] C. C. Aggarwal, A. Hinneburg, D. A. Keim. On the surprising behavior of distance metrics in high dimensional space. *ICDT Conference*, 2001.

[4] C. C. Aggarwal, P. S. Yu. Outlier Detection for High Dimensional Data. *ACM SIGMOD Conference*, 2001.

[5] C. C. Aggarwal, J. Han, J. Wang, P. Yu. A Framework for Clustering Evolving Data Streams. *VLDB Conference*, 2003.

[6] R. Agrawal, R. Srikant. Privacy Preserving Data Mining. *ACM SIGMOD Conference*, 2000.

[7] M. Breunig, H.-P. Kriegel, R. Ng, and J. Sander, " LOF: Identifying Density-Based Local Outliers",*ACM SIGMOD Conference on Management of Data*, 2000.

[8] R. Duda, P. Hart. *Pattern Classification and Scene Analysis*, Wiley, New York, 1973.

[9] M. Ester, H.-P. Kriegel, J. Sander, X. Xu: A Density-Based Algorithm for Discovering Clusters in Large Spatial Databases with Noise, *KDD Conference*, 1996.

[10] D. Burdick, P. Deshpande, T. Jayram, R. Ramakrishnan, S. Vaithyanathan. OLAP Over Uncertain and Imprecise Data. *VLDB Conference*, 2005.

[11] M. James. Classification Algorithms, *Wiley*, 1985.

[12] M. Kumar, N. Patel, J. Woo. Clustering Seasonality Patterns in the presence of Errors. *KDD Conference*, 2002.

[13] H.-P. Kriegel, M. Pfeifle. Density-Based Clustering of Uncertain Data. *ACM KDD Conference*, 2005.

[14] E. Knorr, and R. Ng. Algorithms for Mining Distance-based Outliers in Very Large Databases, *VLDB Conference*, 1998.

[15] R. Little, D. Rubin. Statistical Analysis with Missing Data Values. *Wiley Series in Prob. and Stats.*, 1987.

[16] S. Ramaswamy, R. Rastogi, and K. Shim, Efficient Algorithms for Mining Outliers from Large Data Sets, *ACM SIGMOD Conference on Management of Data*, 2000.

[17] B. W. Silverman. *Density Estimation for Statistics and Data Analysis.* Chapman and Hall, 1986.

[18] T. Zhang, R. Ramakrishnan, M. Livny. BIRCH: An Efficient Data Clustering Method for Very Large Databases. *ACM SIGMOD Conference,* 1996.

[19] T. Zhang, R. Ramakrishnan, M. Livny. Fast Density Estimation Using CF-Kernel for Very Large Databases. *ACM KDD Conference,* 1999.

Chapter 15

FREQUENT PATTERN MINING ALGORITHMS WITH UNCERTAIN DATA

Charu C. Aggarwal
IBM T. J. Watson Research Center
Hawthorne, NY 10532, USA
charu@us.ibm.com

Yan Li
Tsinghua University
Beijing, China
liyan06@mails.tsinghua.edu.cn

Jianyong Wang
Tsinghua University
Beijing, China
jianyong@tsinghua.edu.cn

Jing Wang
Stern School of Business, NYU
New York, NY, USA
jwang5@stern.nyu.edu

Abstract Uncertain data sets have become popular in recent years because of advances in recent years in hardware data collection technology. In uncertain data sets, the values of the underlying data sets may not be fully specified. In this chapter, we will discuss the frequent pattern mining for uncertain data sets. We will show how the broad classes of algorithms can be extended to the uncertain data setting. In particular, we will discuss the candidate generate-and-test algorithms, hyper-structure algorithms and the pattern growth based algorithms. One of our insightful and interesting observations is that the experimental behavior of different classes of algorithms is very different in the uncertain case as compared to the deterministic case. In particular, the hyper-structure and the candidate generate-and-test algorithms perform much better than the tree-based algorithms. This counter-intuitive behavior compared to the case of deterministic data is an important observation from the perspective of frequent pattern mining algorithm design in the case of uncertain data. We will test the approach on a number of real and synthetic data sets, and show the effectiveness of two of our approaches over competitive techniques.

Keywords: Uncertain Data, Association Rule Mining, Frequent Pattern Mining

1. Introduction

In recent years, the problem of mining uncertain data sets has gained importance because of its numerous applications to a wide variety of problems [3, 13–16, 11, 19, 22, 24, 27]. This is because data collection methodologies are often inaccurate and are based on incomplete or inaccurate information. For example, sensor data sets are usually noisy and lead to a number of challenges in processing, cleaning and mining the data. Some techniques for adaptive cleaning of such data streams may be found in [20]. In many cases, estimations of the uncertainty in the data are available from the methodology used to measure or reconstruct the data. Such estimates may either be specified in the form of error variances [3] or in the form of probability density functions [14].

The problem of frequent pattern mining of uncertain data is often encountered in privacy-preserving data mining applications [17] in which the presence or absence of different items is tailored with the use of a probabilistic model. In other applications, the transactions may be constructed based on future expected behavior, and may therefore be probabilistic in nature. In some cases, the transaction data may be constructed from surveys which are inherently incomplete, and the corresponding values may be probabilistically modeled. In general, a variety of online methods exist [26] to estimate missing data values along with the corresponding errors.

In recent years, a number of data mining problems have been studied in the context of uncertain data. In particular, the problems of indexing, clustering, classification and outlier detection [3–5, 7, 14, 22, 23, 28, 33] have been studied extensively in the context of uncertain data. While traditional data mining algorithms can be applied with the use of *mean values* of the underlying data, it has been observed [3] that the incorporation of uncertainty information into the mining algorithm significantly improves the quality of the underlying results. The work in [34] studies the problem of frequent item mining, though the work is restricted to the problem of mining 1-items in the data. The general problem of frequent pattern mining has also been studied in a limited way in [32, 33], and a variety of pruning strategies are proposed in order to speed up the Apriori algorithm for the uncertain case. A number of algorithms have also been proposed for extending the FP-growth algorithm [25, 6] to the uncertain case. This chapter will provide a comparative study of the study of the problem of frequent pattern mining in a comprehensive way. The chapter will study the different classes of algorithms for this case, and will provide a comprehensive study of the behavior of different algorithms in different scenarios. One observation from our extensions to the uncertain case is that the respective algorithms do not show similar trends to the deterministic case. For example, in the deterministic case, the FP-growth algorithm is well known to be an extremely efficient approach. However, in our tests, we found that the extensions of the candidate generate-and-test as well as the hyper-structure based algorithms are much more effective in cases in which the uncertainty probabilities are high. In such cases, many pruning methods, which work well for the case of deterministic algorithms introduce greater overheads from their implementation than the advantages gained from using them. This is because the extensions of some of the algorithms to the uncertain case are significantly more complex, and require different kinds of trade-offs in the underlying computations. For example, the maintenance of probabilistic information can introduce additional overhead.

This chapter is organized as follows. The next section defines the uncertain version of the problem. In Section 3 we will discuss the extension of Apriori-style algorithms to the uncertain version of the problem. In Section 4, we will discuss the extension of set-enumeration based algorithms. In Section 5, we explore how to extend two popular pattern growth based algorithms, H-mine and FP-growth to the uncertain case. Section 6 presents the experimental results. Section 7 studies the problem of frequent item mining with a more general probabilistic model: the *possible worlds model*. The conclusions and summary are discussed in section 8.

2. Frequent Pattern Mining of Uncertain Data Sets

In this section, we will discuss frequent pattern mining for uncertain data sets. We will first introduce some additional notations and definitions. We assume that we have a database $\mathcal{D}$ containing N transactions. We assume that the total number of unique items is d, and each item is denoted by a unique index in the range of $\{1 \dots d\}$. In sparse databases, only a small number of items have a nonzero probability of appearing in a given transaction. Let us assume that the ith transaction in database $\mathcal{D}$ contains n_i items with non-zero probability. Let us assume that the items in the ith transaction are denoted by

$j_1^i \ldots j_{n_i}^i$. Without loss of generality, we can assume that these items are in sorted order. We assume that the probability of the ith item being present in transaction T is given by $p(i, T)$. Thus, in the uncertain version of the problem, the item may be present in the transaction T with the above-mentioned probability.

First, we will define the frequent pattern mining problem. Since the transactions are probabilistic in nature, it is impossible to count the frequency of itemsets deterministically. Therefore, we count the frequent itemsets only in expected value. In order to do so, we need to count the expected probability of presence of an itemset in a given transaction. Let $s(I)$ be the support of the itemset I. This support can only be counted in probabilistic value. The expected support of an itemset I is defined as follows:

DEFINITION 15.1 *The expected support of itemset I is denoted by $E[s(I)]$, and is defined as the sum of the expected probabilities of presence of I in each of the transactions in the database.*

The problem of frequent itemset mining is defined in the context of uncertain databases as follows:

DEFINITION 15.2 *An itemset I is said to be frequent when the expected support of the itemset is larger than the user-defined threshold $minsup$.*

Note that the expected number of occurrences of the itemset I can be counted by summing the probability of presence of the itemsets in the different transactions in the database. The probability of the presence of itemset I in a given transaction can be computed using the relationship below.

OBSERVATION 2.1 *The expected probability of the itemset I occurring in a given transaction T is denoted by $p(I, T)$ and is the product of the corresponding probabilities. Therefore, we have the following relationship:*

$$p(I, T) = \prod_{i \in I} p(i, T) \tag{15.1}$$

Next, we will discuss how broad classes of algorithms can be generalized to the uncertain version of the problem. First, we will discuss the candidate generate-and-test algorithms.

3. Apriori-style Algorithms

These are algorithms which use the candidate generate-and-test paradigm for frequent pattern mining. These can be *join-based* [8] or *set-enumerations based* [1]. The conventional *Apriori* algorithm [8] belongs to this category.

The *Apriori* algorithm uses a candidate generate-and-test approach which uses repeated joins on frequent itemsets in order to construct candidates with one more item. A key property for the correctness of *Apriori*-like algorithms is the downward closure property. We will see that the downward-closure property is true in the uncertain version of the problem as well.

LEMMA 15.3 *If a pattern I is frequent in expected support, then all subsets of the pattern are also frequent in expected support.*

Proof: Let J be a subset of I. We will first show that for any transaction T, $p(J,T) \geq p(I,T)$. Since J is a subset of I, we have:

$$\frac{p(I,T)}{p(J,T)} = \prod_{i \in I-J} p(i,T) \leq 1 \tag{15.2}$$

This implies that $p(J,T) \geq p(I,T)$. Summing this over the entire database $\mathcal{D}$, we get:

$$\sum_{T \in \mathcal{D}} p(J,T) \geq \sum_{T \in \mathcal{D}} p(I,T) \tag{15.3}$$

$$E[s(J)] \geq E[s(I)] \tag{15.4}$$

Equation 15.4 above can be derived from Equation 15.3 by using the fact that the values on the left hand and right hand side correspond to the expected support values of J and I respectively. ■

The maintenance of the downward closure property means that we can continue to use the candidate-generate-and-test algorithms without the risk of losing true frequent patterns during the counting process. In addition, pruning tricks (such as those discussed in *Apriori*) which use the downward closure property can be used directly. Therefore the major steps in generalizing candidate generate-and-test algorithms are as follows:

- All steps for candidate generation using joins and in pruning with the downward closure property remain the same.
- The counting procedure needs to be modified using Observation 2.1.

3.1 Pruning Methods for Apriori-Style Algorithms

We note that a variety of techniques can be used to further improve the pruning power of Apriori-style algorithms. A key feature that can be used in order to improve the pruning power is the presence of items with *low existential probabilities*. We note that since the support counts on the transactions use the product of the probabilities, items with low existential probabilities add very little to the support count. For example, consider the case when the support threshold is 1%, and an item has existential probability of 0.01% in a given transaction. The use of this item during the support counting is not very useful for mining purposes.

As the percentage of items with low existential probabilities increases, The fraction of such insignificant increments to the support counts of candidate itemsets increases. Therefore, the work in [32] trims these items during the support counting process. A specific probability threshold is used in order to trim the items. This trimming threshold is set to be marginally above the minimum support threshold. It uses a smart error estimation method in order to determine the potential error caused by this trimming process. Thus, at the end of the process, the method can prune many candidate itemsets, but there are also some *potentially frequent itemsets*, in which it is not clear whether or not they are frequent as a result of the errors caused by the trimming. Therefore, the method in [32] uses a final *patch up phase* in which the true support counts of these potentially frequent itemsets are computed in order to determine which of the itemsets are truly frequent.

The technique in [32] works well when the frequencies of the underlying items follow a *bimodal distribution.* In a bimodal distribution, the items fall into one of two categories. The first category of items have very high existential probability, and the second category of items have very low existential probability. In such cases, it is possible to pick the trimming threshold effectively. However, it is much more difficult to find an appropriate trimming threshold in the case where the existential probabilities of the different items are more uniform. In order to deal with such situations, a *decremental pruning technique* was proposed in [33].

The decremental pruning technique exploits the statistical characteristics of existential probabilities to gradually reduce the set of candidate itemsets. The key idea is to estimate the upper bounds of candidate itemset supports progressively after each database transaction is processed. If a candidate itemset's upper bound falls below the support threshold, then it is immediately pruned. Consider an itemset X for which the support is being estimated and let X' be any subset of X. We denote the *decremental counter* $S_e(X, X', k)$ as an upper bound on the value of the support after k transactions $t_1 \dots t_k$ have been processed. We denote the probability of item x in transaction t_i by $P_{t_i}(x)$. The counter $S_e(X, X', k)$ is defined as follows:

DEFINITION 15.4 *Consider a database D with $|D$ transactions $t_1 \dots t_{|D|}$. Let the support of item x in transaction t_i be denoted by $P_{t_i}(x)$. For any non-empty $X' \subset X$, $k \geq 0$, the value of the decremental counter $S_e(X, X', k)$ is defined as an upper bound on the expected support of X after k transactions*

$t_1 \ldots t_k$ have been processed. Specifically, we have:

$$S_e(X, X', k) = \sum_{i=1}^{k} \pi_{xinX} P_{t_i}(X) + \sum_{i=k+1}^{|D|} \pi_{xinX} P_{t_i}(X) \quad (15.5)$$

The fact that $S_e(X, X', k)$ is an upper bound is easy to observe, because the first expression $\sum_{i=1}^{k} \pi_{xinX} P_{t_i}(X)$ is an exact expected support for the first k transactions, whereas the second expression is an upper bound on the expected support by setting the probabilities of all items in $X - X'$ to 1. One observation is that as more transactions are processed (or the value of k increases), the upper bound $S_e(X, X', k)$ reduces. Once this upper bound falls below the threshold support, the itemset X can be dropped from consideration.

We note that any subset X' can be used for pruning purposes. The use of a larger subsets provides better pruning power, but also increases the space- and time-overhead in maintaining these decremental counters. Of course there are an exponential number $2^{|X|} - 2$ possibilities for the counters, and it is not possible to maintain all of them. Therefore, a number of techniques are proposed in [33] for picking and using these counters effectively. The two methods for picking and leveraging the decremental counters are as follows:

- **Aggregate by Singletons:** In this case, only the aggregate counts of frequent singletons are maintained. In [33] an additional inductive relationship has been proposed to efficiently update the decremental counters.

- **Common Prefix Method:** In this case, the decremental counts of itemsets with a common prefix are aggregated. It is assumed that the items follow and certain ordering and prefixes are computed with respect to this ordering. Only decremental counters are maintained for those subsets X' of X, which are also prefixes of X. As in the case of singletons, an inductive relationship has been proposed to efficiently update the decremental counters.

For more details of the two techniques discussed above and the inductive relationship to update the counters, we refer the reader to [33]. It has been shown in [33] that this technique improves the pruning effectiveness of the Apriori algorithm. We note however, that such techniques are effective only when there are a substantial number of items with low existential probabilities. When this is not the case, the use of such an approach can be counter-productive, since the additional overhead incurred for implementing the pruning approach can be greater than the advantages gained from pruning. In [6], the effect of using such pruning strategies has been studied for the case of high existential probabilities. It has been shown in [6] that in such cases, the more straightforward extension of the Apriori algorithm is more efficient than the variations [33] with such pruning strategies. Some of these results will be presented in the comparison study in section 6 of this chapter.

4. Set-Enumeration Methods

Similar techniques can be used in order to extend set-enumeration based methods. In set-enumeration based methods, we construct the candidates by building the set enumeration-tree [30] or lexicographic tree. A number of recent algorithms belong to this category. These include the *MaxMiner* algorithm [9], the *DepthProject* algorithm [1], the *TreeProjection* algorithm [2] and MAFIA [12]. These methods typically use top-down tree-extension in conjunction with branch validation and pruning using the downward closure property. Different algorithms use different strategies for generation of the tree in order to obtain the most optimum results. Since the set-enumeration based algorithms are also based on the downward closure property, they can be easily extended to the uncertain version of the problem. The key modifications to set-enumeration based candidate generate-and-test algorithms are as follows:

- The tree-extension phase uses the ordering of the different items in order to construct it in top-down fashion. The tree extension phase is exactly the same as in candidate generate-and-test algorithms.
- The counting of frequent patterns uses Observation 2.1.
- In the event that transactions are projected on specific branches of the tree (as in [1, 2]), we can perform the projection process, except that we need to retain the probabilities of presence of specific items along with the transactions. Also note that the probabilities for the items across different transactions need to be maintained respectively, even if the transactions are identical after projection. This is because the probabilities of the individual items will not be identical after projection. Therefore each projected transaction needs to be maintained separately.
- The pruning of the branches of the tree remains identical because of the downward closure property.

We note that set-enumeration algorithms are conceptually quite similar to the join-based algorithms, except that the candidates are enumerated differently.

5. Pattern Growth based Mining Algorithms

There are also some popular algorithms for mining frequent patterns which are based on the pattern growth paradigm. Among these methods, the H-mine [29] and FP-growth [18] algorithms are two representative ones. Their main difference lies in the data representation structures. FP-growth adopts a prefix tree structure while H-mine uses a hyper-linked array based structure. We will see that the use of such different structures have a substantially different impact in the uncertain case as compared to the deterministic case. Next, we will discuss the extension of each of these algorithms to the uncertain case in some detail.

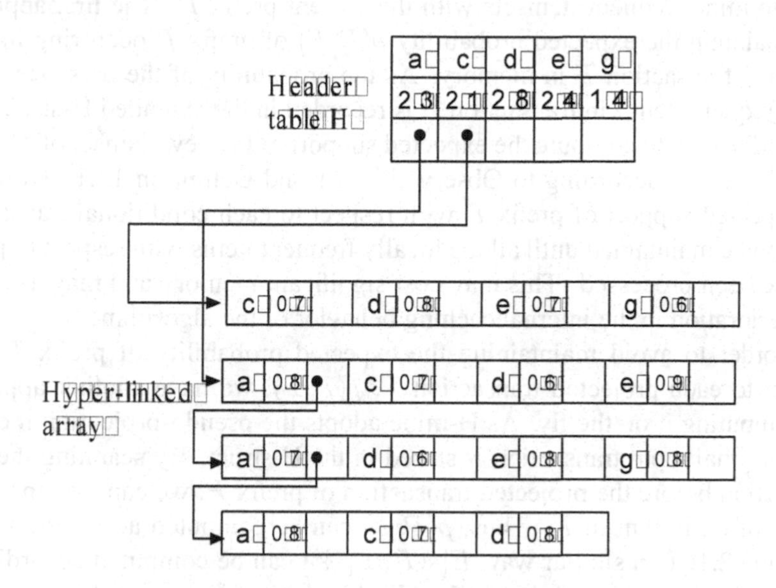

Figure 15.1. H-Struct

5.1 Extending the H-mine algorithm

The H-mine algorithm proposed in [29] adopts a hyper-linked data structure called H-struct. Similar to FP-growth, it is a partition-based divide-and-conquer method. Initially, H-mine scans the input database once to find the frequent items. The infrequent items are removed from the database. The frequent items left in each input transaction are sorted according to a certain global ordering scheme. The transformed database is stored in an array structure, where each row corresponds to one transaction. During the mining process, there always exists a prefix itemset (denoted by P, which is initially empty). H-mine needs to construct a header table which records the starting places of the projected transactions. By following the links in the header table, H-mine can locate all the projected transactions and find the locally frequent items by scanning the projected transactions. The locally frequent items with respect to the prefix P can be used to extend P to longer prefix itemsets.

As the hyper-linked array structure used in H-mine is not in a compressed form, it is relatively easy to extend the H-struct for mining frequent itemsets from uncertain data. As described in [29], each frequent item in a transaction is stored in an entry of the H-struct structure with two fields: an item_id and a hyper-link. In addition, the probability $p(i, T)$ of the presence of item i in transaction T is maintained. Figure 15.1 shows an example of an extended H-struct structure*. With the extended H-struct structure there are two ways to mine frequent itemsets with the current prefix P. The first approach is to maintain the expected probability $p(P, T)$ of prefix P occurring in each projected transaction T in memory. As the probability of the presence of locally frequent item i in transaction T is recorded in the extended H-struct, it is straightforward to compute the expected support of the new itemset of $P \cup \{i\}$, $E[s(P \cup \{i\})]$, according to Observation 2.1 and Definition 15.1. However, the expected support of prefix P with respect to each conditional transaction needs to be maintained until all the locally frequent items with respect to prefix P have been processed. This may cost significant memory and may also lead to deterioration in the internal caching behavior of the algorithm.

In order to avoid maintaining the expected probability of prefix P with respect to each projected transaction T, $p(P, T)$, we have another approach for computing it on the fly. As H-mine adopts the pseudo-projection method, each original input transaction is stored in the H-struct. By scanning the sub-transaction before the projected transaction of prefix P, we can find the probability of each item in P. Thus, $p(P, T)$ can be computed according to Observation 2.1. In a similar way, $E[s(P \cup \{i\})]$ can be computed according to Observation 2.1 and Definition 15.1. In this chapter, we adopt the second approach for computing the expected support of the current prefix itemset P. This is because the use of on-the-fly computations reduces the space-requirements of the technique. The reduced space-requirements also indirectly improve the locality of the caching behavior of the underlying computations. This leads to improved efficiency of the overall algorithm.

5.2 Extending the FP-growth Algorithm

FP-growth [18] is one of the most popular frequent itemset mining algorithms which are based on the pattern growth paradigm. It adopts a prefix tree structure, FP-tree, to represent the database (or conditional databases). As FP-tree is a compressed structure, it poses several challenges when we try to adapt the FP-growth algorithm for uncertain data sets. These challenges are as follows:

- In the original FP-tree structure, each node has a 'count' entry which records the number of transactions containing the prefix path from the

*Note that the second row of the header table in the H-struct structure stores the sum of item probabilities for each locally frequent item.

root node to this node. For uncertain data, if we just store in a node the sum of item probabilities with respect to the transactions containing the prefix path, we will no longer be able to determine the probability of the presence of an item in each transaction. Thus, there is an irreversible loss of information in the uncertain case with the use of a compressed structure. Thus, we need to find a different and efficient way to store the item probabilities without losing too much information.

- The original FP-growth algorithm mines frequent itemsets by searching the tree structure in a bottom up manner. The computation of the support of a given prefix path is quite straightforward. Its support is simply the support of the lowest node of the path. However, for uncertain data, the expected support of a given prefix path should be computed according to Definition 15.1. As we no longer know the mapping among the item probabilities and the transactions, it is impossible to compute the exact expected support of a given prefix according to Definition 15.1.

- Since it is impossible to determine the exact expected support of each frequent itemset based on the FP-tree structure, we may need to first mine all the candidate itemsets, and then remove the infrequent itemsets by checking the original database. The process of determining such infrequent itemsets efficiently can be quite difficult in the uncertain case.

There are two extreme solutions to adapt the FP-tree structure for uncertain data mining. Let us denote the FP-tee built from uncertain data by UFP-tree. The first one is to store (in each node) the sum of item probabilities with respect to the transactions containing the prefix path from the root to it. The UFP-tree built in this way is as compact as the original FP-tree. However, it cannot even be used to compute the the upper bound or lower bound of the expected support of an itemset, because it loses information with respect to the distinct probability values for different transactions. Another extreme solution is to split a node into m nodes if the item in this node has m distinct probability values. In this case, we can compute the exact expected support. On the other hand, the UFP-tree built in this way consumes a lot of memory.

In this work, we adopt a compromise by storing a subset of the probabilities for the item in each node. These probabilities are selected using clustering, and are stored as floating point numbers. This method does not consume too much memory, and we will show that it allows us to compute an upper bound on the expected support of any itemset. We then compute a set of candidate frequent itemsets based on this upper bound. This set of candidates provides us a superset of the complete set of real frequent itemsets. Any remaining false positives will then need to be removed by scanning the input database. Next, we will discuss the adaptation of different steps of the FP-growth algorithm to the uncertain case.

Construction of the UFP-tree. The process of constructing the UFP-tree for uncertain data is very similar to the construction of the FP-tree for deterministic data sets. The main difference lies in the information stored in each node. The UFP-tree is built using the following steps. First, the database is scanned to find the frequent items and to generate a support descending item list. Then, the transactions are read one by one, and the infrequent items are pruned. The remaining frequent items are sorted according to the frequent item list. The re-ordered transactions are inserted into the UFP-tree.

As discussed earlier in this chapter, each node of the UFP-tree stores a summary of the probabilities of the non-zero probability items in those transactions which share the same prefix path in clusters. We partition the probabilities into a set of k clusters. The corresponding parameters created for the ith cluster by the partitioning are represented by c_i and m_i ($1 \leq i \leq k$), where c_i denotes the maximum probability value in the ith cluster and m_i is the number of item probabilities in the ith cluster. We assume that $c_1 > c_2 > \ldots > c_k$. The reason that we store the maximum probability value in each cluster instead of the center of the cluster (i.e., the average value of all the item probabilities in this cluster) is to make sure that the support computed from the summary is no less than the true support. In Section 5.2.0, we will introduce a method to compute an upper bound on the true support based on the cluster information stored in each node. Besides the global UFP-tree construction from the database, conditional UFP-trees are generated from the global UFP-tree. Therefore, there are two different situations which need the data summarization in the construction process. We will discuss the solutions separately under the two situations.

There are several clustering and data summarization methods available for our task. The choice of the proper method should consider two factors. The first is memory usage. This also indirectly affects the performance since lower memory consumption results in better internal caching behavior on most machines. Since there could be a large number of nodes in the UFP-tree, the summarization of the probabilities in each node should be as concise as possible in order to reduce memory consumption. The trade-off is that greater conciseness leads to lower precision. In the mining process, we compute the upper bound of the support of each itemset according to the summarization of probabilities stored in each node. We use this to compute the candidate frequent itemsets. If the precision of the summarization is too low, the difference between the upper bound and the true support will be large and a large number of false positives may be generated. This will increase the memory and space requirements for the elimination process of the false positives. Clearly, the tradeoff needs to be carefully exploited in order to optimize the performance of the underlying algorithm.

This problem is closely related to that of building V-optimal histograms [21] for time-series data. It is however not natural to apply the V-optimal technique to this situation. During the construction of the UFP-tree, the transactions are read sequentially, and the item probabilities corresponding to a certain node will typically arrive in neither ascending nor descending order. In order to apply the V-optimal algorithm to this set of probabilities (which are floating point numbers) in each node, we would need to sort these numbers in ascending or descending order, and this is time consuming. Furthermore, the time and space complexities of the complete V-optimal method are $O(n^2 \cdot k)$ and $O(n \cdot k)$ respectively. Because of the expensive behavior of the V-optimal method, we decided to use k-means clustering instead. However, if we store all the item probabilities associated with each node before applying k-means clustering during the UFP-tree construction process, it will consume too much memory and will be too expensive for large data sets.

Therefore, we used a different approach by using a modified version of the k-means algorithm. First, we partition the range of the probabilities of items into ϕ parts in equal width, where ϕ is chosen to be significantly larger than k. We store the maximum probability value and the number of distinct item probabilities in each part. After we have all the transactions inserted into the UFP-tree, we then cluster these stored information by k-means.

As mentioned earlier, there are two points in the pattern mining process in which we need to compute the data summarizations. The first relates to the construction of the global UFP-tree. We have discussed the first situation above. The second is the construction of conditional UFP-trees during the mining process. We will discuss this second situation at this point. Suppose we begin to mine the frequent itemsets with prefix item 'g'. By computing the expected support of size 2-itemsets containing 'g' with the method discussed in Section 5.2.0, we could find the locally frequent items. Then, we traverse each path in the global UFP-tree linking the node with a label 'g' to the root to extract the locally frequent items and the corresponding distribution information of the item probabilities stored in each node along the path. This forms a conditional transaction. Here we give such an example of a conditional transaction, which contains three items and corresponds to 30 input transactions: $\{(a, ((0.6, 5), (0.7, 5), (0.8, 20))), (b, ((0.8, 10), (0.9, 20))), (e, ((0.7, 20), (0.88, 10)))\}$. Next, we insert each conditional transaction into the conditional UFP-tree with respect to the item 'g'. Note that the number of probabilities of each item equals 30. This is the number of the probabilities in the node 'g' at the bottom of the corresponding path. This also means that there are 30 input transactions containing 'g' in this path. Notice that we need to merge the clusters after all the transactions are inserted in the conditional UFP-tree in order to keep a limited number of entries in each node. Thus is also done with the k-means clustering algorithm.

Computation of Support Upper Bounds. In order to mine all the frequent itemsets, we first need to mine all the potentially-frequent itemsets using the information stored in each node. As mentioned earlier, the precise support of an itemset cannot be computed directly from the UFP-tree because of the information loss during compression. However, it is possible to compute an upper bound on the support. It is clear that the number of item probabilities in each node along any tree path may vary considerably. Let the number of item probabilities in the last node of a path be denoted by n (namely, the number of transactions containing the path is n). We should take out n largest probabilities in each node along the prefix path from this last node up to the root, and this is an easy task since the item probabilities are summarized in the clusters. For example, suppose all the item probabilities in each tree node are grouped into three clusters (i.e., k=3), and the cluster information in the last node N of a given path P is $\{(c_1{=}0.95, m_1{=}2), (c_2{=}0.9, m_2{=}2), (c_3{=}0.8, m_3{=}1)\}$. That is, the last node contains five item probabilities. Let N' be any node along the path P, and its cluster information be $\{(c_1'{=}0.98, m_1'{=}3), (c_2'{=}0.91, m_2'{=}1), (c_3'{=}0.85, m_3'{=}2)\}$. The five largest item probabilities in N' are 0.98, 0.98, 0.98, 0.91, and 0.85, respectively. The process of computing an upper bound on the expected support of an itemset I with respect to a prefix path P is shown in Algorithm 5.2.0.

Note that an itemset I may be contained in multiple prefix paths, and we can compute an upper bound of the expected support of itemset I with respect to each of these prefix paths. The sum of the upper bounds with respect to these prefix paths must form an upper bound of itemset I. We will prove that the output of Algorithm 5.2.0 is an upper bound on the expected support of the itemset with respect to path P.

LEMMA 15.5 *Given an itemset I, when $|I| = 2$, the support computed according to algorithm 5.2.0 is an upper bound of the expected support of I with respect to path P.*

Proof: *Suppose the number of item probabilities in the last node of path P is n, that is, $\sum_{i=1}^{k} m_{|I|i} = n$. Let us denote the two tree nodes corresponding to the two items in I w.r.t. path P by a and b, the top n largest item probabilities in node a by $a_1 \geq a_2 \geq \ldots \geq a_n$, and the top n largest item probabilities in node b by $b_1 \geq b_2 \geq \ldots \geq b_n$. We will prove the lemma using mathematical induction.*

(1) Let n=2. Since $(a_1 - a_2)(b_1 - b_2) \geq 0$ holds, we have

$$a_1 b_1 + a_2 b_2 \geq a_1 b_2 + a_2 b_1$$

Therefore, the lemma holds when $n = 2$.

Computation of an upper bound on the expected support of an itemset I w.r.t. prefix path P

- **Input:** The cluster information stored in each node along path P corresponding to I, (c_{i1}, m_{i1}), (c_{i2}, m_{i2}), ..., (c_{ik}, m_{ik}), i=1, 2, ..., $|I|$, and $c_{i1}>c_{i2}>\ldots>c_{ik}$ holds.
- **Output:** An upper bound of the expected support of itemset I w.r.t. path P, $E(s(I)|_P)$
- **Initialization:**

  $E(s(I)|_P)$=0;

  $C_1 \leftarrow c_{11}, C_2 \leftarrow c_{21}, \ldots, C_{|I|} \leftarrow c_{|I|1}$;

  $M_1 \leftarrow m_{11}, M_2 \leftarrow m_{21}, \ldots, M_{|I|} \leftarrow m_{|I|1}$;
- **Method:** Repeat the following steps below until no item probability in the last node of the path corresponding to itemset I is left.
  - $E(s(I)|_P)$=$E(s(I)|_P)$+$C_1 \times C_2 \times \ldots \times C_{|I|} \times m$, where $m = min(M_1, M_2, ..., M_{|I|})$;
  - $M_1 \leftarrow M_1 - m, M_2 \leftarrow M_2 - m, \ldots, M_{|I|} \leftarrow M_{|I|} - m$;
  - For $i \in [1, |I|]$ do

    if M_i=0

    Suppose C_i=c_{ij} (where $1 \leq j < k$), then

    $C_i \leftarrow c_{i(j+1)}$ and $M_i \leftarrow m_{i(j+1)}$;

(2) Assume the induction base that when n=k, $\sum_{i=1}^{k} a_i b_i$ is an upper bound. Next, let n=k+1, we will find the maximum sum of products. Let b_{k+1} multiply a_t, $1 \leq t \leq k$, under the assumption above we know that the maximum sum of products of the k numbers left is $(\sum_{i=1}^{t-1} a_i b_i + \sum_{i=t+1}^{k+1} a_i b_{i-1})$. Furthermore,

we have:

$$\sum_{i=1}^{k+1} a_i b_i - (\sum_{i=1}^{t-1} a_i b_i + a_t b_{k+1} + \sum_{i=t+1}^{k+1} a_i b_{i-1})$$
$$= (a_t b_t + a_{t+1} b_{t+1} + a_{t+2} b_{t+2} + \ldots a_{k+1} b_{k+1}) -$$
$$(a_t b_{k+1} + a_{t+1} b_t + a_{t+2} b_{t+1} + \ldots + a_{k+1} b_k)$$
$$= a_{t+1}(b_{t+1} - b_t) + a_{t+2}(b_{t+2} - b_{t+1}) + \ldots$$
$$+ a_{k+1}(b_{k+1} - b_k) + a_t(b_t - b_{k+1})$$
$$\geq a_t(b_{t+1} - b_t) + a_t(b_{t+2} - b_{t+1}) + \ldots$$
$$+ a_t(b_{k+1} - b_k) + a_t(b_t - b_{k+1})$$
$$= a_t(b_{k+1} - b_t) + a_t(b_t - b_{k+1}) = 0$$

then, we derive that when n=k+1, $\sum_{i=1}^{k+1} a_i b_i$ is an upper bound of the sum of the products.
(3) Since when $|I| = 2$ the output of Algorithm 5.2.0 is $\sum_{i=1}^{n} a_i b_i$, the expected support computed following the steps in Algorithm 5.2.0 must be an upper bound.

■

COROLLARY 15.6 *Given two groups of n ($\forall n, n > 0$) floating-point numbers sorted in decreasing order, c_{ij} ($\forall$ i, p, q, $1 \leq i \leq 2$, if $1 \leq p < q \leq n$, then $c_{ip} \geq c_{iq}$ holds), $\sum_{j=1}^{x} \prod_{i=1}^{2} c_{ij}$ is the largest among all the sums of x products, where $1 \leq x \leq n$.*

Proof: *It can be directly derived from the proof of Lemma 15.5 when x=n. For any possible set of x products which are constructed from the two groups of n floating-point numbers, denoted by s_1, s_2,...s_x, we can always find another set of x products which are constructed from the two groups of the first x floating-point numbers, denoted by s_1', s_2', ..., s_x', such that $s_l \leq s_l'$ ($\forall l$, $1 \leq l \leq x$). That is, $(\sum_{j=1}^{x} s_j) \leq (\sum_{j=1}^{x} s_j')$. In addition, according to the proof of Lemma 15.5 we know that $(\sum_{j=1}^{x} \prod_{i=1}^{2} c_{ij}) \geq (\sum_{j=1}^{x} s_j')$ holds. Thus, we have $(\sum_{j=1}^{x} \prod_{i=1}^{2} c_{ij}) \geq (\sum_{j=1}^{x} s_j)$, which means $\sum_{j=1}^{x} \prod_{i=1}^{2} c_{ij}$ is the largest among all the sums of x products, where $1 \leq x \leq n$.* ■

THEOREM 15.7 *Given m groups of n ($\forall n, n > 0$) floating-point numbers sorted in decreasing order, c_{ij} ($\forall$ i, p, q, $1 \leq i \leq m$, if $1 \leq p < q \leq n$, then $c_{ip} \geq c_{iq}$ holds), $\sum_{j=1}^{x} \prod_{i=1}^{m} c_{ij}$ is the largest among all possible sums of x products, where $1 \leq x \leq n$.*

Proof: *We prove the theorem using mathematical induction.*
1. According to Corollary 15.6, we know that it is true when $m = 2$.

2. We assume, when $m = k$, the theorem holds.
3. We will derive from the above assumption that when m=k+1, $\sum_{j=1}^{x}\prod_{i=1}^{m} c_{ij}$ is still the largest among all possible sums of x products, where $1 \leq x \leq n$. Let the $(k+1)$-th group of n floating-point numbers be $c_{(k+1)1} \geq c_{(k+1)2} \geq \ldots \geq c_{(k+1)n}$. As $c_{(k+1)1}$, $c_{(k+1)2}, \ldots,$ and $c_{(k+1)x}$ are among the top x largest values in the $(k+1)$-th group of n floating-point numbers, one of the largest values of the sum of x products constructed from the k+1 groups of n floating numbers must be in the form of $c_{(k+1)1}s_1 + c_{(k+1)2}s_2 + \ldots + c_{(k+1)x}s_x$, where $s_i=\prod_{j=1}^{k} z_{i_j}$, $z_{i_j} \in \{c_{j1}, c_{j2}, \ldots, c_{jn}\}$.
If we use s_y' to denote $\prod_{i=1}^{k} c_{iy}$, we have:

$$\sum_{j=1}^{x}\prod_{i=1}^{m} c_{ij} = c_{(k+1)1}s_1' + c_{(k+1)2}s_2' + \ldots + c_{(k+1)x}s_x'$$

and $s_1' \geq s_2' \geq \ldots \geq s_x'$ must hold.
In addition, we also have:

$$\begin{aligned}
& c_{(k+1)1}s_1' + c_{(k+1)2}s_2' + \ldots + c_{(k+1)x}s_x' - \\
& (c_{(k+1)1}s_1 + c_{(k+1)2}s_2 + \ldots + c_{(k+1)x}s_x) \\
& = (s_1' - s_1)(c_{(k+1)1} - c_{(k+1)2}) + \\
& [(s_1' - s_1) + (s_2' - s_2)](c_{(k+1)2} - c_{(k+1)3}) + \ldots + \\
& [(s_1' - s_1) + (s_2' - s_2) + \ldots + (s_x' - s_x)]c_{(k+1)x} \\
& = (s_1' - s_1)(c_{(k+1)1} - c_{(k+1)2}) + \\
& [(s_1' + s_2') - (s_1 + s_2)](c_{(k+1)2} - c_{(k+1)3}) + \ldots + \\
& [(s_1' + s_2' + \ldots + s_x') - (s_1 + s_2 + \ldots + s_x)]c_{(k+1)x}
\end{aligned}$$

According to our assumption, $\forall l \leq x$, $(\sum_{i=1}^{l} s_i' - \sum_{i=1}^{l} s_i) \geq 0$ holds, and as $(c_{(k+1)l} - c_{(k+1)(l+1)}) \geq 0$ also holds, we get that $(\sum_{j=1}^{x} c_{(k+1)j}s_j' - \sum_{j=1}^{x} c_{(k+1)j}s_j) \geq 0$. Therefore, when m=k+1, $\sum_{j=1}^{x}\prod_{i=1}^{m} c_{ij}$ is still the largest among all possible sums of x products, where $1 \leq x \leq n$. ■

COROLLARY 15.8 *The output of Algorithm 5.2.0 must be an upper bound of the expected support of itemset I ($|I| \geq 2$) w.r.t. prefix path P.*

Proof: *There are $|I|$ nodes in the path P which correspond to the $|I|$ items in I, and each node maintains k clusters. The cluster information of the last node in path P is represented by $c_i(m_i)$, i=1...k, and we let n=$\sum_{j=1}^{k} m_i$. We can then sort the n item probabilities in the last node in descending order.*

For each of the other $|I|-1$ nodes, we can extract its top n largest item probabilities and sort them in descending order. In this way, we get $|I|$ groups of n item probabilities, denoted by z_{ij}, where $1 \leq i \leq |I|$, $1 \leq j \leq n$, and $\forall$ p, q, if $p < q$, $z_{ip} \geq z_{iq}$. According to the computation process of Algorithm 5.2.0 we know that the output of Algorithm 5.2.0 equals $\sum_{j=1}^{n} \prod_{i=1}^{|I|} z_{ij}$. According to Theorem 15.7, we have that it is an upper bound of the expected support of itemset I w.r.t. prefix P. ■

Mining Frequent Patterns with UFP-tree. We used two different approaches for the mining process with UFP-tree. One is the recursive pattern growth approach introduced in [18]. The other is the one described in [31], which constructs a conditional UFP-tree for each frequent item, and then mines frequent itemsets in each conditional tree. In the following, we will explain the two mining methods in detail.

Assume that the frequent item list in support-descending order is $\{e, a, c, d, g\}$. The process of recursively constructing all-level conditional UFP-trees is as follows. First, the algorithm mines frequent itemsets containing g. Second, it mines frequent itemsets containing d but not g. Third, it mines frequent itemsets containing c but neither d nor g. This pattern is repeated until it mines frequent itemsets containing only e. When we are mining frequent itemsets containing item d, we first compute the upper bound of the expected support of each itemset (e.g., $(e, d), (a, d), and(c, d)$) with the method described in Section 5.2.0, and form the locally frequent item list in support descending order (e.g., $\{c, e\}$). Next, the algorithm traverses the UFP-tree by following the node-links of item d again to get the locally frequent itemset information in each path which forms a conditional transaction. We insert the conditional transaction into the conditional UFP-tree with respect to item d. After that, we will repeat the above steps to this conditional UFP-tree of item d, which is the same as the depth-first search in [18].

As we have seen from Observation 2.1, the expected probability of an itemset in a given transaction is defined as the product of the corresponding probabilities. This suggests that the expected support of an itemset decreases quickly when its length increases. While the algorithm proposed in [31] is designed for the deterministic case, we observe that it avoids recursively constructing conditional FP-trees, and can therefore make good use of the *geometric decrease* in the calculations of the expected support of itemsets. This is our rationale for specifically picking the frequent itemset mining algorithm introduced in [31]. It constructs a one-level conditional FP-tree for each frequent item. After we have found the locally frequent items for each conditional FP-tree, we do not reorder the items, but we follow the global frequent item list. The reason for doing so is for the generation of the trie tree. The algorithm in [31] also adopts the popular divide-and-conquer and pattern growth paradigm. Suppose the locally frequent items with respect to the prefix item 'g' are $\{e, a, c, d\}$. The algorithm in [31] computes the itemsets containing 'd' first, then computes those containing 'c' but no 'd', then those containing 'a' but no 'c' nor 'd', and those containing 'e' only at the end. Then, the algorithm proceeds to generate itemsets of increasing size sequentially, until the set of locally frequent items becomes empty.

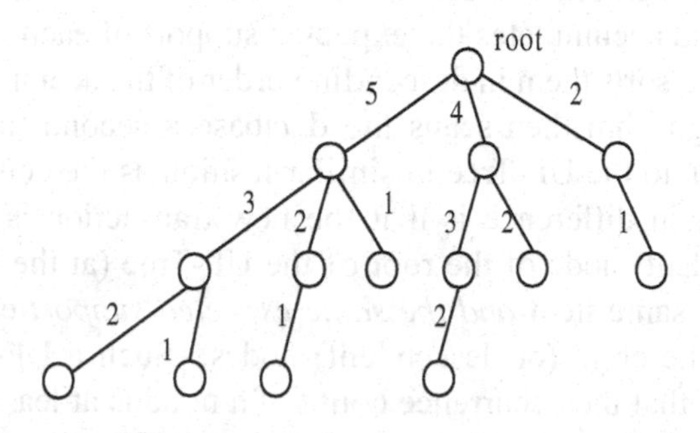

Figure 15.2. An example of a trie tree

Determining Support with a Trie Tree. As mentioned above, the itemsets mined so far are just candidate frequent itemsets and may not be really frequent. In order to determine the real support of each candidate itemset, we store the candidate frequent itemsets in a trie tree structure which is suitable to search and locate an itemset. Figure 15.2 shows an example of a trie tree, which contain a total of 14 nodes including the root node. Each path from the root to a certain node represents an itemset, thus the trie tree in Figure 15.2 stores 13 itemsets. We can also see that along each path from the root to a leaf, the indices of the items are sorted in decreasing order, and the child nodes of each node are also sorted in decreasing order. This arrangement of the items in the trie tree facilitates the search and locating of an itemset.

In order to obtain the exact support of these candidate itemsets stored in the trie tree, we need to read in the transactions one by one, find the candidate itemsets contained in each transaction and calculate the support according to Definition 15.1 and Observation 2.1. Similar to [10], when we deal with a transaction, we maintain two indices, one is for the transaction, the other points to the node of the trie tree. The entire process is an index moving process. The index for the trie tree is moved to find the item pointed by the index for the transaction, and then the index for the transaction is moved to the next item. Once an itemset is found to be contained in a transaction, its expected support is summed according to Definition 15.1 and Observation 2.1. This process continues until the transaction index reaches the last item in the transaction or the trie tree index reaches the last node of the tree.

5.3 Another Variation of the FP-growth Algorithm

Another variation of the FP-tree algorithm (known as UF-Tree) has been proposed in [25]. Each node in the UF-Tree stores the following (i) An item (ii) its expected support, and (iii) the number of occurrences with such expected support for an item. In order to construct the UF-Tree, the algorithm scans the database once, and accumulates the expected support of each item. It finds all frequent items and sorts them in descending order of the accumulated expected support. The algorithm then scans the database a second time, and inserts each transaction into the UF-Tree in similar fashion as the construction of the FP-Tree. The main difference is that the new transaction is merged with a child (or descendent) node of the root of the UF-Tree (at the highest support level) only if the same item *and the same expected support* exist in both the transaction and the child (or descendent) nodes. Such a UF-Tree possesses the nice property that the occurrence count of a node is at least the sum of the occurrence counts of its child (or descendent) nodes. The frequent patterns can be mined from the UF-Tree by keeping track of the expected supports of the itemsets, when forming the projected database for the different itemsets. These expected supports can then be leveraged in order to find the frequent patterns. A number of improvements have also been proposed in [25] in order to improve the memory consumption and performance of the underlying algorithm.

6. A Comparative Study on Challenging Cases

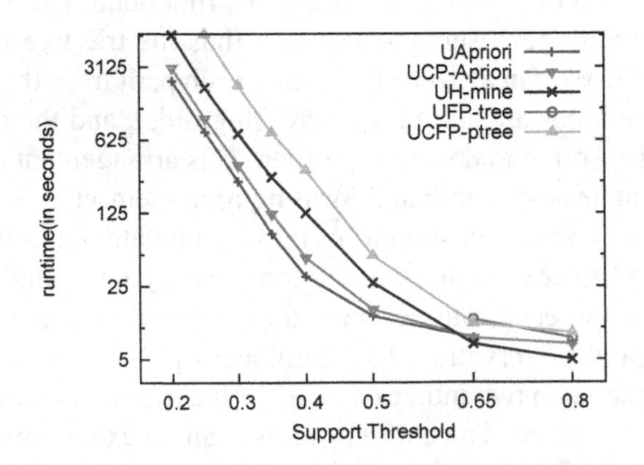

Figure 15.3. Runtime Comparison on Connect4

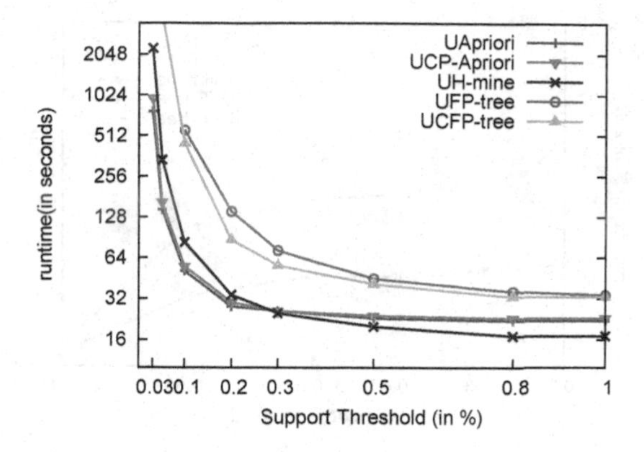

Figure 15.4. Runtime Comparison on kosarak

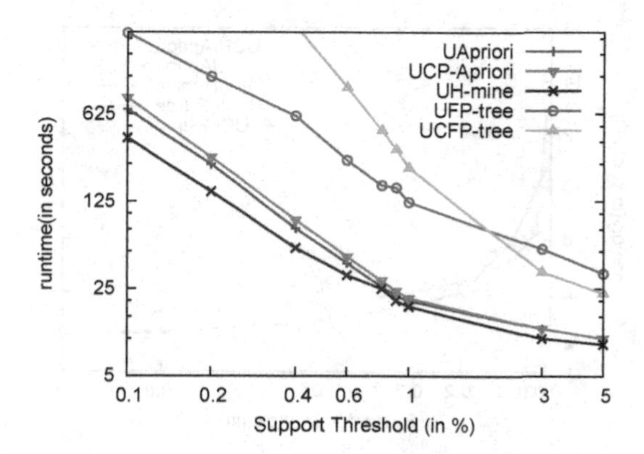

Figure 15.5. Runtime Comparison on T40I10D100K

In this section, we present the performance study for the extended classical frequent pattern mining algorithms of Apriori, H-mine, and FP-growth. The results for the low-existential probability case are well known and presented in [32, 33]. We will study the more difficult case of high-existential probabilities and show that the results are quite different in this case than those presented in [32, 33]. This is because the overhead from the use of pruning techniques is greater in this case than the advantages gained from using the approach. In the following we will denote these revised algorithms by UApriori, UH-mine, and UFP-growth, respectively. We will compare their performance with the state-of-the-art frequent itemset mining algorithm for uncertain data sets, which is the DP approach proposed in chapter [33]. We implemented one of the DP methods proposed in [33] and denote it by UCP-Apriori. The UCP-Apriori

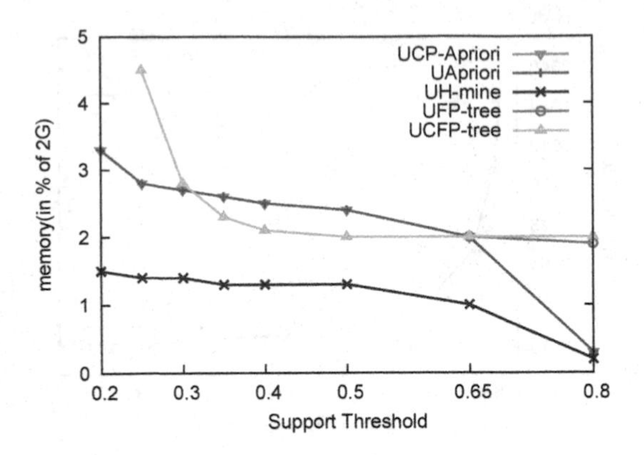

Figure 15.6. Memory Comparison on Connect4

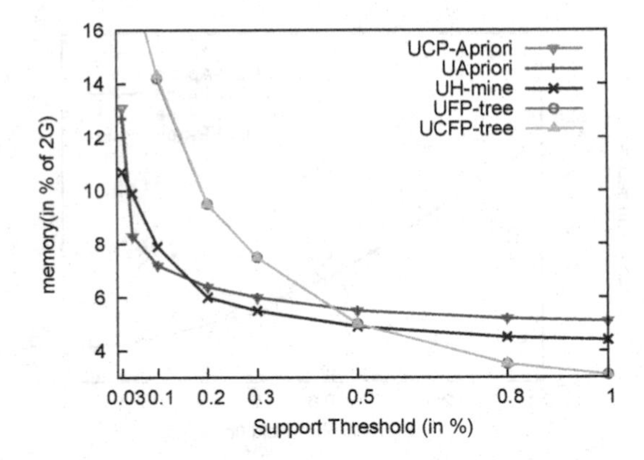

Figure 15.7. Memory Comparison on kosarak

integrates a pruning method called CP with the Apriori frequent itemset mining framework. The experiments were conducted on a machine with 2.66GHz CPU and 2G main memory installed. The operating system is GNU/Linux.

Four data sets were used in the experiments. The first two datasets, *Connect4* and *kosarak*, are real datasets which were downloaded from the FIMI repository.† The *Connect4* data is a very dense data set with 67,000 relatively coherent transactions, and each transaction contains 43 items. *kosarak* is a really sparse data set containing 990,980 transactions. The other two data sets, T40I10D100K and T25I15D320k, were generated using the IBM syn-

† **URL:** *http://fimi.cs.helsinki.fi/data/*

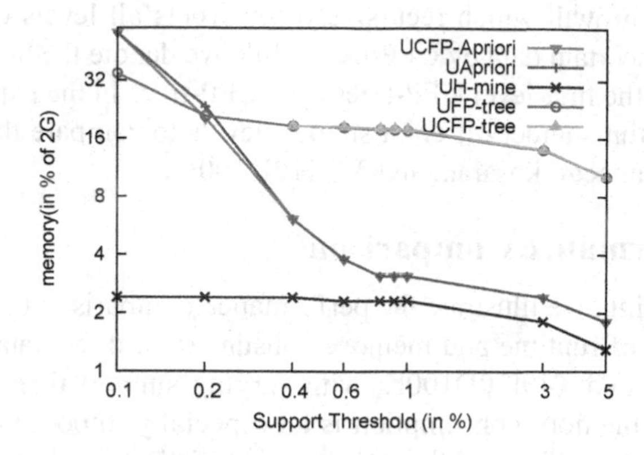

Figure 15.8. Memory Comparison on T40I10D100K

thetic data set generator‡ which was discussed in the Apriori paper [8]. These two data sets contain 100,000 and 320,000 transactions respectively. According to the notation discussed in [8], the parameters used for generating data sets include T (for the average transactions), I (for the maximum potential items per transaction), and D (for the number of transactions in the data set). For example, the data set T25I15D320k has an average transaction size of 25, an average maximal potentially frequent itemset size of 15, and 320K records. In Connect4 and T40I10D100K, there are numerous long frequent itemsets, while in kosarak the short frequent itemsets dominate the majority. As the support threshold goes down, Connect4, kosarak and T40I10D100K all contain a large number of long frequent itemsets as well as some short ones. Thus, the choice of different data sets was designed to test the algorithms in different scenarios.

We note that these are deterministic data sets. In order to obtain uncertain data sets, we introduced the uncertainty to each item in these data sets. We allocated a relatively high probability to each item in the data sets in order to allow the generation of longer itemsets. We assume that the uncertainty of those items follows the normal distribution $N(\mu, \sigma)$. The value of μ was independently and randomly generated in the range of [0.87, 0.99] for each item in each transaction, while the value of σ was generated in the same way but in the range of [1/21, 1/12]. We generated a number between 0 and 1 for every item according to its randomly given distribution.

As discussed in Section 5.2.0, we implemented two variants of the UFP-growth algorithm for uncertain data mining. In the following we denote the

‡**URL:** *http:miles.cnuce.cnr.it/∼palmeri/datam/DCI/datasets.php.*

variant of UFP-growth which recursively constructs all levels of conditional UFP-trees on uncertain data by UFP-tree, while we denote the other one which only constructs the first-level UFP-trees by UCFP-tree. In the experiments, we ran these algorithms under different support levels to compare their efficiency for data sets Connect4, kosarak, and T40I10D100K.

6.1 Performance Comparison

In the following, we illustrate the performance comparison of the five algorithms in terms of runtime and memory consumed on three data sets of Connect4, kosarak, and T40I10D100K, with varying support thresholds. In the uncertain case, memory consumption is an especially important resource because of the additional information about probabilistic behavior which needs to be stored. In resource-constrained hardware, memory-consumption may even decide the range in which a given algorithm may be used. In such cases, memory consumption may be an even more important measure than running time. Therefore, we will test the memory consumption in addition to efficiency. Figures 6 to 15.4 illustrate the runtime comparison results, while Figures 15.5 to 15.7 show the memory usage comparison on these three data sets. We will see that different algorithms provide the best performance with the use of different measures. Our broad observation is that UH-mine is the only algorithm which performs robustly for all measures over all data sets, whereas the variations of candidate generate-and-test also perform quite well, especially in terms of running time. This would suggest that UH-mine is the most practical algorithm to use in a wide variety of scenarios.

Connect4 is a dense data set. Figures 6 and 15.5 show the runtime and memory consumption of UApriori, UCP-Apriori, UH-mine, UFP-tree, and UCFP-tree on this data set. UApriori and UH-mine provide the fastest performance at different support thresholds, whereas UH-mine provides the best memory consumption across all thresholds. Thus, the UH-mine algorithm performs robustly on both measures. UFP-tree and UCFP-tree are the slowest. UCP-Apriori [33] is slower than our version of the Apriori algorithm, which is denoted by UApriori. This is because the method for candidate pruning in UCP-Apriori algorithm is not very efficient and only introduces additional overhead, unless the uncertainty probabilities are set too low as in [33]. However, low uncertainty probabilities are an uninteresting case, since the data will no longer contain long frequent patterns (because of the multiplicative behavior of probabilities and its impact on the expected support), and most algorithms will behave efficiently. It is particularly interesting that the uncertain extension of most deterministic algorithms can perform quite well, whereas the extensions to the well known FP-Tree algorithms do not work well at all. We traced the running process of UFP-tree and UCFP-tree and found that considerable time is spent on the last step of eliminating false positives. Furthermore, in most paths in the UFP-tree, the probabilistic information for thousands of transactions need to be stored, and the concise behavior of the deterministic case is lost. It is this concise behavior which provides the great effectiveness of this technique in the deterministic case, and the loss of this property in the probabilistic case is an important observation from the perspective of algorithmic design. In comparison to the UFP-tree, the UCFP-tree does not need to build all levels of conditional UFP-trees recursively, and it only needs to mine all frequent itemsets in one-level of conditional UFP-tree. Thus, it performs better than UFP-tree.

An important resource in the uncertain case is memory consumption. This is more important in the uncertain case as compared to deterministic data, because of the additional requirements created by probabilistic data. Therefore, in resource constrained hardware, memory-consumption can even dictate the range within which a given algorithm may be used. Figure 15.5 illustrates the comparison of the memory consumption on Connect4. In this case, the behavior of UH-Mine is significantly superior to the other algorithms. As UApriori needs to store a large number of candidate itemsets, UApriori consumes more memory than UH-mine which outputs those mined frequent itemsets on the fly. Connect4 is a relatively coherent data set, and so it is more likely for transactions to share the same prefix path when inserting into the UFP-tree. Thus, it gets the highest compression ratio among all the data sets. However, because the UFP-tree stores uncertainty information, its memory consumption is greater than UApriori. Furthermore, as the support threshold goes down, it generates too many candidate itemsets. This leads to the sharp increase of memory usage.

Data set kosarak is sparse and therefore, the tree like enumeration of the underlying itemsets show a bushy structure. As shown in figure 15.3, both UApriori and UH-mine perform very well on the kosarak data set. In figure 15.3, the Y-axis is in logarithmic scale. UCP-Apriori runs slightly slower than UApriori. UFP-tree and UCFP-tree do not scale well with the decrease of the support threshold. For data set kosarak, the UFP-tree is also bushy and large. When the support is 0.0003, UCFP-tree costs too much time on kosarak for the reason that the one-level conditional UFP-trees are still large. For UFP-trees, too many recursively constructed conditional UFP-trees and the large number of false positives consume too much memory.

Figure 15.6 shows the comparison of memory consumed by these algorithms on kosarak. As the support threshold decreases, the UFP-trees constructed become large and candidate itemsets generated for UFP-tree and UCFP-tree increase quickly, and thus the memory usage increases fast. For UApriori, the memory consumed for storing candidate frequent itemsets increases rapidly and surpasses UH-mine which only holds the H-struct when the support threshold becomes relatively low. The UH-mine maintains its robustness in terms of memory consumption across all datasets, and this is a particularly important property of the algorithm.

Data set T40I10D100K is a synthetic data set containing abundant mixtures of short itemsets and long itemsets. Thus, it is a good data set for testing the behavior of the algorithms when the itemsets cannot be perfectly characterized to show any particular pattern. In this case, the UH-mine is significantly superior to all algorithms both in terms of running time and memory usage. We find that as the support threshold decreases, the gap between UH-mine and UApriori becomes quite large. We note that since the Y-axis in Figure 15.4 is in a logarithm scale of 5, the performance difference between the two algorithms is much greater than might seem visually. As shown in Figure 15.7, the memory cost for UApriori running on T40I10D100K increases dramatically when the support threshold decrease below 0.6%. That is because the number of frequent itemsets increases rapidly with reduction in support. UCP-Apriori is a little slower than UApriori and they consumes similar volume of memory.

According to our experiments on Connect4, kosarak and T40I10D100K data sets, UApriori and UH-mine are both efficient in mining frequent itemsets. Both algorithms run much faster than UFP-tree and UCFP-tree, especially when the support threshold is pretty low. However, with the support level decreases, the number of frequent itemsets increases exponentially, which results in sharp increase of the memory cost. UH-mine is the only algorithm which shows robustness with respect to both efficiency and memory usage. The reason that the FP-growth algorithm is not suitable to be adapted to mine uncertain data sets lies in compressed structure which is not well suited for probabilistic data. UCP-Apriori [33] runs a little slower than UApriori on the three data sets. The memory cost for UCP-Apriori is almost the same as that for UApriori, and therefore UApriori is a more robust algorithm that UCP-Apriori on the whole.

6.2 Scalability Comparison

To test the scalability of UApriori, UH-mine, UFP-tree, and UCFP-tree with respect to the number of transactions, we used the synthetic data set T25I15D320k. It contains 320,000 transactions and a random subset of these is used in order to test scalability. The support threshold is set to 0.5%. The results in terms of running time and memory usage are presented in Figures 15.9 and 15.10 respectively.

Figure 15.9 shows that all these algorithms have linear scalability in terms of running time against the number of transactions varying from 20k to 320k. Among them, UH-mine, UApriori, and UCP-Apriori have much better performance than UFP-tree and UCFP-tree, and among all the algorithms, H-mine has the best performance. For example, when the number of transactions is

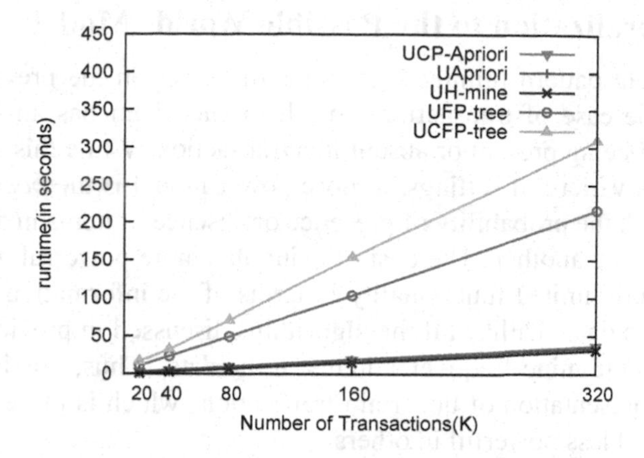

Figure 15.9. Scalability Comparison in terms of runtime

20k, the running times for UH-mine, UApriori, UCP-Apriori, UFP-tree, and UCFP-tree are 1 second, 2.41 seconds, 2.58 seconds, 10.76 seconds, and 16.49 seconds, respectively.

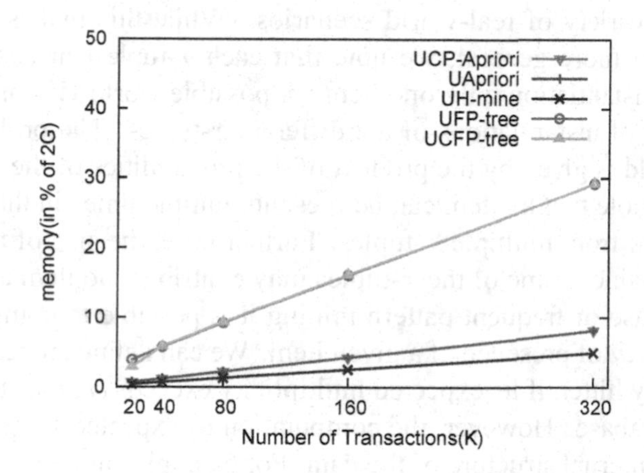

Figure 15.10. Scalability Comparison in terms of Memory

In Figure 15.10, all algorithms shows linear scalability in terms of memory usage. The curves denoted for UFP-tree and UCFP-tree almost coincide, and so do the curves denoted for UApriori and UCP-Apriori. Both UApriori and UH-mine scale much better than UFP-tree and UCFP-tree. UH-mine algorithm scales better than any of the algorithms. Thus, the UH-mine algorithms shows the best scalability both in terms of running time and memory usage.

7. Generalization to the Possible Worlds Model

The frequent pattern mining algorithms discussed in the previous section considered the case of transactions in which each item has an independent probability of being present or absent in a transaction. While this scenario can be useful in a variety of settings, a more power model is the *possible worlds model* in which the probability of presence or absence of items in a given tuple can influence one another. The cost of using this more powerful model is that we have a more limited functionality in terms of the information we can discover from the data. Unlike all the algorithms discussed in previous sections, we can only determine frequent 1-items in the data. Thus, this is a different model for representation of uncertain transactions which is more powerful in some ways and less powerful in others.

In this model, instead of transactions, we have a set of *x-tuples*. Each x-tuple consists of a bag of items with mutually exclusive probabilities of presence. For any item t and x-tuple T, we assume that the probability of presence of t in T is $p(t, T)$. Furthermore, T can contain at most one time. The probability that T contains exactly one item is given by $\sum_{t \in T} p(t, T) \leq 1$. The uncertain data is specified by using a bunch of x-tuples denoted by $T_1 \ldots T_m$. It is assumed that the behavior among different x-tuples is independent of one another. By using different kinds of specifications of $T_1 \ldots T_m$, it is possible to simulate a variety of real-world scenarios. While this makes the specification problem more general, we note that each x-tuple can contain a final probabilistic instantiation only one item. A possible world W corresponds to a combination of instantiations of the different x-tuples. The probability of a particular world is given by the product of the probabilities of the corresponding x-tuples. Note that an item can be present multiple times in the data, since it can be drawn from multiple x-tuples. Furthermore, the size of the database may also vary since some of the x-tuples may contribute no item at all.

As in the case of frequent pattern mining it is possible to compute the expected frequency of presence of a given item. We can define an item to be a ϕ-expected heavy hitter, if its expected multiplicity exceeds ϕ times the expected size of the database. However, the computation of expected frequency seems to ignore the internal structure of the data. For example, an item may not have very high multiplicity, but may be present in the database a certain number of times with very high probability because of how the tuples are structured. In order to provide greater generality to the queries, we can define probabilistic heavy-hitters with the use of a *multiplicity parameter* ϕ as well as a probabilistic parameter τ. The use of two parameters provides greater generality in exploring different scenarios.

DEFINITION 15.9 *An item t is said to be a (ϕ, τ)-heavy hitter, if its multiplicity is at least a fraction ϕ with probability at least τ across all possible worlds.*

The work in [34] presents two different algorithms. The first algorithm is an offline algorithm which runs in polynomial time and uses dynamic programming. The second algorithm works on data streams and uses sampling. The offline algorithm creates the two dimensional table $B^t[i, j]$ for the item t. Thus, the algorithm is designed as a verification algorithm to the binary problem of whether or not the item t is a (ϕ, τ)-heavy hitter. The entry $B^t[i, j]$ denotes the probability that the item t appears i times in the first j x-tuples of the database. We note that $B^t[i, j]$ can be expressed inductively in terms of $B^t[i, j-1]$ and $B^t[i-1, j-1]$ depending upon whether or not the jth item is t. This forms the dynamic programming recursion. Similarly, we can also compute $B^{\bar{t}}[i, j]$ which computes the probability that *any item other than* t occurs i times in the first j x-tuples of the database. Then, for a database containing m x-tuples, the overall probability L can can be computed as follows:

$$L = \sum_{i=1}^{m} B^t[i, m] \left(\sum_{j=1}^{\lfloor (1-\phi)/\phi \rfloor \cdot i} B^{\bar{t}}[j, m] \right) \tag{15.6}$$

We note that the overall algorithm requires $O(m^2)$ space and time. Note that since we may need to use this approach for each item, this may lead to a fairly large running time, when tested across the unverse of n items. The work in [34] also proposes a pruning condition which provides an upper bound to the probability that the item occurs in a fraction ϕ of the database. This allows us to exclude many items as (ϕ, τ)-heavy hitters. This dramatically speeds up the algorithm. We refer the reader to [34] for details of this speedup, as well as the details of the techniques used for streaming algorithms.

8. Discussion and Conclusions

In this chapter, we study the problem of frequent patten mining of uncertain data sets. We present several algorithms for frequent pattern mining proposed by other authors, and also present some novel extensions. We note that the uncertain case has quite different trade-offs from the deterministic case because of the inclusion of probability information. As a result, the algorithms do not show similar relative behavior as their deterministic counterparts. This is especially the case when most of the items of high relative frequencies of presence.

As mentioned in [18], the FP-growth method is efficient and scalable, especially for dense data sets. However, the natural extensions to uncertain data behave quite differently. There are two challenges to the extension of the FP-tree based approach to the uncertain case. First, the compression properties of the FP-Tree are lost in the uncertain case. Second, a large number of false positives are generated, and the elimination of such candidates further affects the efficiency negatively.

As shown in our experiments, UH-mine and UApriori algorithms are efficient and scalable on mining frequent itemsets for uncertain data sets. UH-mine is an algorithm which divides the search space and employs the pattern-growth paradigm, which can avoid generating a large number of candidate itemsets. Both UCP-Apriori [33] and UApriori are extended from the well-known Apriori algorithm. The UCP-Apriori algorithm applies a candidate pruning method during the mining process. According to our performance study, the pruning method proposed for UCP-Apriori results in greater overhead than the efficiency it provides in the most challenging scenarios where uncertainty probabilities are high and long patterns are present. The UH-mine algorithm is especially useful, because it uses the pattern growth paradigm, but does so without using the FP-tree structure which does not extend well to the uncertain case. This also reduces the memory requirements drastically. The UH-mine algorithm proposed in this chapter provides the best trade-offs both in terms of running time and memory usage.

Acknowledgements

Research of the first author was sponsored in part by the US Army Research laboratory and the UK ministry of Defense under Agreement Number W911NF-06-3-0001. The views and conclusions contained in this document are those of the author and should not be interpreted as representing the official policies of the US Government, the US Army Research Laboratory, the UK Ministry of Defense, or the UK Government. The US and UK governments are authorized to reproduce and distribute reprints for Government purposes notwithstanding any copyright notice hereon.

References

[1] R. Agarwal, C. Aggarwal, V. Prasad. Depth First Generation of Long Patterns. *ACM KDD Conference*, 2000.

[2] R. Agarwal, C. Aggarwal, V. Prasad. A Tree Projection Algorithm for Generating Frequent Itemsets. *Journal of Parallel and Distributed Computing*, 2001.

[3] C. C. Aggarwal. On Density Based Transforms for Uncertain Data Mining. *ICDE Conference*, 2007.

[4] C. C. Aggarwal, P. S. Yu. A Framework for Clustering Uncertain Data Streams. *ICDE Conference*, 2008.

[5] C. C. Aggarwal, P. S. Yu. Outlier Detection with Uncertain Data. *SIAM Conference on Data Mining*, 2008.

[6] C. C. Aggarwal, Y. Li, J. Wang, J. Wang. Frequent Pattern Mining with Uncertain Data. *IBM Research Report*, 2008.

[7] C. C. Aggarwal, P. S. Yu. A Survey of Uncertain Data Algorithms and Applications, *IEEE Transactions on Knowledge and Data Engineering*, 2009.

[8] R. Agrawal, R. Srikant. Fast Algorithms for Mining Association Rules in Large Databases. *VLDB Conference*, 1994.

[9] R. J. Bayardo. Efficiently mining long patterns from databases *ACM SIGMOD Conference*, 1998.

[10] Ferenc Bodon. A fast APRIORI implementation.
URL: [http://fimi.cs.helsinki.fi/src/].

[11] D. Burdick, P. Deshpande, T. Jayram, R. Ramakrishnan, S. Vaithyanathan. OLAP Over Uncertain and Imprecise Data. *VLDB Conference*, 2005.

[12] D. Burdick, M. Calimlim, J. Gehrke. MAFIA: A Maximal Frequent Itemset Algorithm. *IEEE Transactions on Knowledge and Data Engineering*, 17(11), pp. 1490–1504, 2005.

[13] A. L. P. Chen, J.-S. Chiu, F. S.-C. Tseng.Evaluating Aggregate Operations over Imprecise Data. *IEEE TKDE*, 8(2), 273–294, 1996.

[14] R. Cheng, Y. Xia, S. Prabhakar, R. Shah, J. Vitter. Efficient Indexing Methods for Probabilistic Threshold Queries over Uncertain Data. *VLDB Conference*, 2004.

[15] N. Dalvi, D. Suciu. Efficient Query Evaluation on Probabilistic Databases. *VLDB Conference*, 2004.

[16] A. Das Sarma, O. Benjelloun, A. Halevy, J. Widom. Working Models for Uncertain Data. *ICDE Conference*, 2006.

[17] A. V. Evfimievski, R. Srikant, R. Agrawal, J. Gehrke. Privacy preserving mining of association rules. *KDD Conference*, 2002.

[18] J. Han, J. Pei, Y. Yin. Mining frequent patterns without candidate generation. *ACM SIGMOD Conference*, 2000.

[19] H. Garcia-Molina, D. Porter. The Management of Probabilistic Data. *IEEE TKDE*, 4:487–501, 1992.

[20] S. R. Jeffery, M. Garofalakis, and M. J. Franklin. Adaptive Cleaning for RFID Data Streams, *VLDB Conference*, 2006.

[21] Sudipto Guha, Nick Koudas, Kyuseok Shim. Approximation and streaming algorithms for histogram construction problems. *ACM Trans. Database Syst.*, 31(1), 396-438, 2006.

[22] H.-P. Kriegel, M. Pfeifle. Density-Based Clustering of Uncertain Data. *ACM KDD Conference*, 2005.

[23] H.-P. Kriegel, M. Pfeifle. Hierarchical Density Based Clustering of Uncertain Data. *ICDM Conference*, 2005.

[24] L. V. S. Lakshmanan, N. Leone, R. Ross, V. S. Subrahmanian. ProbView: A Flexible Probabilistic Database System. *ACM TODS Journal*, 22(3), 419–469, 1997.

[25] C. K.-S. Leung, M. A. F. Mateo, D. A. Brajczuk. A Tree-Based Approach for Frequent Pattern Mining from Uncertain Data, *PAKDD Conference*, 2008.

[26] R. Little, D. Rubin. Statistical Analysis with Missing Data Values. *Wiley Series in Prob. and Stats.*, 1987.

[27] S. I. McClean, B. W. Scotney, M. Shapcott. Aggregation of Imprecise and Uncertain Information in Databases. *IEEE Transactions on Knowledge and Data Engineering*, 13(6), 902–912, 2001.

[28] W. Ngai, B. Kao, C. Chui, R. Cheng, M. Chau, K. Y. Yip. Efficient Clustering of Uncertain Data, *ICDM Conference*, 2006.

[29] J. Pei, J. Han, H. Lu, S. Nishio, S. Tang, D. Yang. H-Mine: Hyper-Struction Mining of Frequent Patterns in Large Databases. *ICDM Conference*, 2001.

[30] R. Rymon. Search through Systematic Set Enumeration. *Third International Conf. Principles of Knowledge Representation and Reasoning*, pp. 539-550, 1992.

[31] Y.G. Sucahyo, R.P. Gopalan. CT-PRO: A Bottom-Up Non Recursive Frequent Itemset Mining Algorithm Using Compressed FP-Tree Data Structure.
URL: [http://fimi.cs.helsinki.fi/src/].

[32] C.-K. Chui, B. Kao, E. Hung. Mining Frequent Itemsets from Uncertain Data. *PAKDD Conference*, 2007.

[33] C.-K. Chui, B. Kao. Decremental Approach for Mining Frequent Itemsets from Uncertain Data. *PAKDD Conference*, 2008.

[34] Q. Zhang, F. Li, and K. Yi. Finding Frequent Items in Probabilistic Data, *ACM SIGMOD Conference*, 2008.

Chapter 16

PROBABILISTIC QUERYING AND MINING OF BIOLOGICAL IMAGES

Vebjorn Ljosa
Broad Institute of MIT and Harvard
Cambridge, MA 02142
ljosa@broad.mit.edu

Ambuj K. Singh
University of California
Santa Barbara, CA 93106-5110
ambuj@cs.ucsb.edu

Abstract Automated sample preparation and image acquisition equipment has enabled new kinds of large-scale biological experiments. Such experiments have led to some notable successes, but their wider application to diverse biological questions requires robust, fully automated image analysis. This chapter examines the role in probabilistic information in managing ambiguity in the different stages of image analysis by presenting techniques for segmenting, measuring, querying, and mining biological images.

Keywords: biological images; probabilistic image segmentation; probabilistic image measurements; index structures for probabilistic data; mining probabilistic spatial objects

1. Introduction

Imaging technology has advanced rapidly in recent years, to the point where automated microscopes can image large numbers of samples with reasonable cost and accuracy. Imaging has long been a prominent tool in biomedical research, but high-throughput imaging has attracted attention because it enables large-scale experiments that lead to insights that have eluded discovery in more narrowly targeted experiments, even when each image has been inspected and measured by an expert. In addition, the greatly increased sample size may reveal trends too subtle to be detectable amidst the individual variation in smaller sets of samples. The huge potential of these large-scale image sets can only be realized with the aid of automated image analysis, not only because manually inspecting hundreds of thousands of images is too labor-intensive, but also because humans have difficulty scoring complex phenotypes consistently and objectively over large sets of images.

Biological images are generally ambiguous in the sense that even an expert cannot accurately identify cells or tissue types and measure properties such as length, count, thickness, and distance. For instance, it may be hard to distinguish between similar cell types, find an accurate boundary between two adjacent cells, or determine how far a thin neurite extends. The entire stack of image informatics software, including image analysis modules and the database, must therefore be able to work with uncertain measurements, i.e., probability distributions over measurement values. Probability distributions are also useful for representing aggregated data, such as the thickness of a tissue layer measured at different spacial locations or the lengths of all microtubules in an image.

This chapter is about modeling and analysing such uncertain measurements. We will explore this aspect of bioimage data through the entire workflow, from image segmentation to data querying and mining. Although the presentation uses retinal images as a running example, the ideas extend to other bioimages at the sub-cellular, cellular, or tissue level.

1.1 An Illustrative Example

To illustrate the probabilistic nature of the computations on bioimage collections, we consider images from the retina [38, 40–42, 48, 54]. The retina contains relatively few classes of cells organized in well-defined layers as shown in Figure 16.1. The electrical signals generated by the photoreceptors in response to light pass through the bipolar cells and ganglion cells, and finally through the optic nerve into the rest of the brain. There are also other types of neurons in the retina; horizontal cells, in particular, are fairly flat neurons located where the photoreceptors and bipolar cells meet [26]. They provide connections that are perpendicular to the main direction of signaling.

Figure 16.2, which shows three horizontal cells, was acquired by staining the retina with anti-neurofilament and anti-calbindin, fluorescent antibodies which together label the entire horizontal cell in cat retinas. Two, three, or even four antibodies can be used together for imaging as long as they fluoresce in response to different wavelengths of light. The measurement of protein distribution within the cells as well as the cells' dendritic branching pattern is important [53].

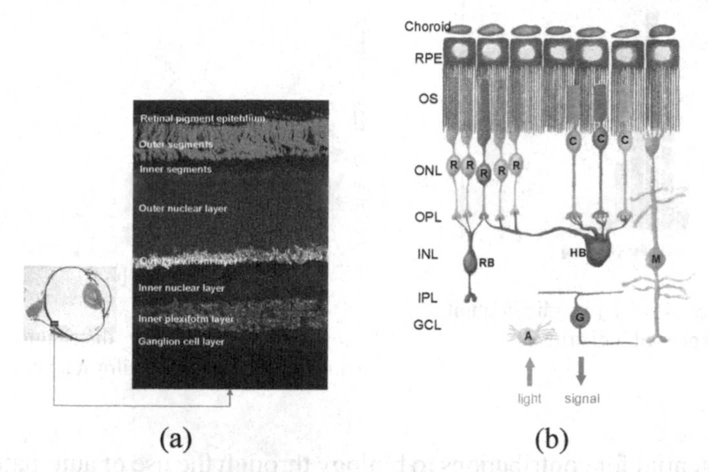

Figure 16.1. Retinal layers. (a) Confocal microscopy image of a vertical section through a cat retina. (b) A diagram of retinal cell types. The synaptic terminals of rod (R) and code (C) photoreceptors end in the OPL. The dendrites of rod bipolar cell (RB) and the axon terminals of the B-type horizontal cell (HB) are postsynaptic to rods. The dendrites of the HB cell are postsynaptic to cones. Axons of rod bipolar cells terminate in the IPL. The dendrites of ganglion cells (G) terminate in the IPL and their axons form the optic nerve. Their cell bodies comprise the GCL, which also contains astrocytes (A). Müller cells (M) are glia; their cell bodies lie in the INL. Illustration by www.blackpixel.net reprinted, with permission, from [27], copyright © 2005 Elsevier.

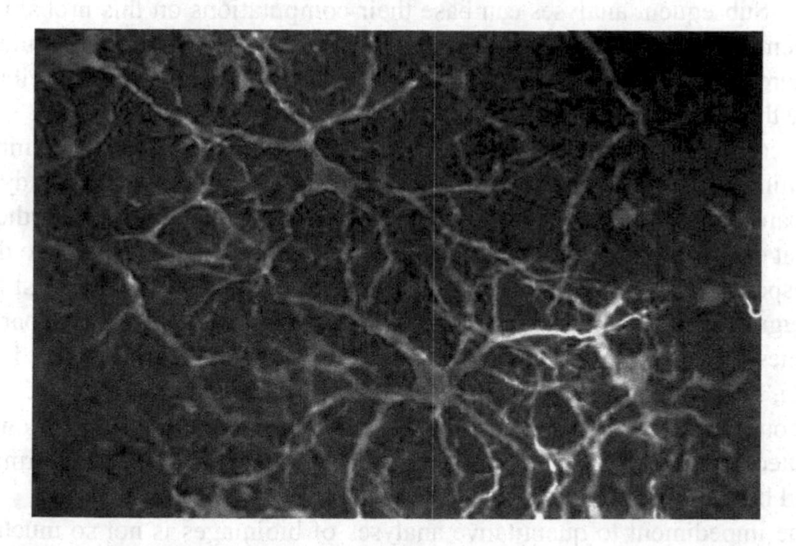

Figure 16.2. Confocal micrograph of three horizontal cells in a detached cat retina, labeled by anti-neurofilament and anti-calbindin. Reprinted with permission from [44], copyright © 2006 IEEE.

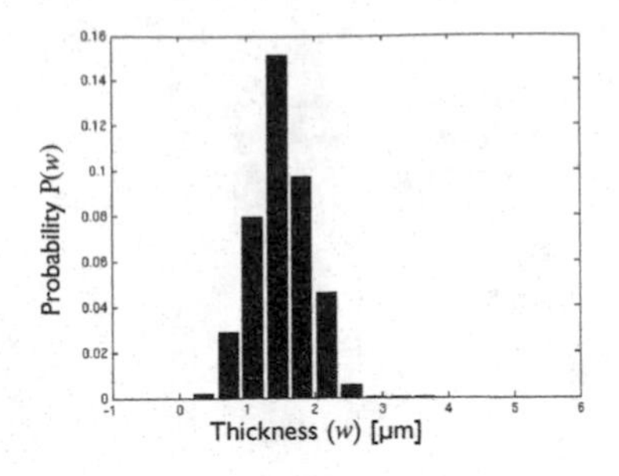

Figure 16.3. Probability distribution of the thickness of a neurite

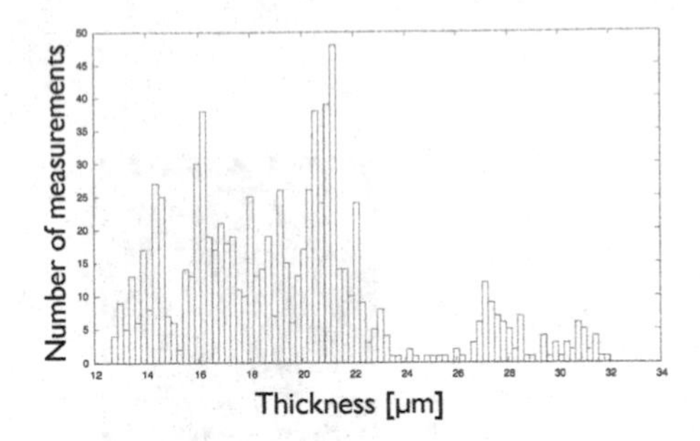

Figure 16.4. Probability distribution of the thickness of the inner plexiform layer in a retina

The potential for contributions to biology through the use of automated analysis techniques is huge. However, there are significant challenges. The first difficulty is that the protein used for labeling is usually present in all cells of a specific type. For instance, neurofilaments are expressed in all horizontal cells in cats, and as their processes intertwine, even an expert cannot reliably determine where one cell ends and another begins. This difficulty in segmenting the cells precludes accurate measurements of dendrite thickness, count, branching patterns, and other important morphological metrics. To alleviate this, the result of the segmentation should be a probabilistic value rather than an exact result. Subsequent analyses can base their computations on this probabilistic segmentation and in turn produce probabilistic measurements. A module that measures neurite thickness, for instance, can produce a probability distribution of the thickness (as in Figure 16.3) rather than a single number.

A second difficulty arises from the individual differences between animals. An animal must be sacrificed in order to image its retina, so in a study that compares normal retinas to retinas at various time points after injury, the images at each time point will be of a different animal. Because there is no direct correspondence between the images, any quantitative comparison must be in the aggregate, by sampling and comparing distributions. One can compare the thicknesses of inner plexiform layer in normal and detached retinas, for instance, by measuring the thickness in numerous places in images of each class and comparing the distributions. Figure 16.4 shows such a distribution, aggregated from over 800 measurements. We see that much of the information would be lost if summarized by just the mean and standard deviation.

The impediment to quantitative analyses of bioimages is not so much that they are tedious to perform manually, but that they are nearly impossible for humans to perform reliably. As a simple example, the bottom cell in Figure 16.2 has at least four neurites (the branching processes from the cell bodies), but it may have as many as eight. For simple analyses such as cell counts, the uncertainty can be controlled by having stringent protocols for how to measure and using multiple humans, but this is infeasible for most interesting analyses.

This chapter examines the aspect of uncertainty from the viewpoints of image analysis and data querying and mining [12, 13, 31, 43]:

- *Probabilistic image analysis techniques.* We will discuss image analysis algorithms that generate probabilistic measurements from biological images. The first such technique is probabilistic segmentation, in which each pixel is assigned a probability of belonging to a cell. We will also investigate measurements of thickness and cell body size, both building on the results of probabilistic segmentation.
- *Querying and mining of probabilistic image data.* Probabilistic measurements need to be stored, accessed, and mined. We discuss an approximation called *APLA* and an index structure called the *APLA-tree*. This index structure can be used for answering range queries and top-k queries on distributions. We also investigate exploratory mining of probabilistic image data, in particular the problem of spatial joins of probabilistic objects. Spatial joins, which find pairs of proximate objects, such as bipolar cell axons that are close to synaptic terminals, are computationally challenging when the objects have uncertain class and extent.

2. Related Work

The field of data management has seen a growing trend in the focus on uncertain data [24, 60]. Many sources inherently provide uncertain data due to technical limitations and acquisition conditions. As accuracy becomes important, we will need to examine this data probabilistically to account for the uncertainty [3].

The accepted model for evaluating probabilistic queries is the "possible-worlds" model [7, 23, 25, 37] in which a set of possible certain worlds is defined for the database objects. The query is evaluated in these certain worlds, and the result is aggregated to return the answer. In general, such evaluation is computationally challenging and makes a straightforward implementation all but infeasible for large databases. A number of strategies have been developed to find subclasses of queries that can be answered efficiently. This includes the isolation of safe query plans [23, 24]. Other approaches combine uncertainty with lineage [8], and consider the decomposition of possible worlds into a manageable set of relations [5]. Aspects of completeness under various models of uncertainty have also been considered [56].

Efficient and interesting techniques exist for answering queries on probability density functions (pdfs) that are Gaussian or uniform. These models can be appropriate for the uncertainty of many individual measurements, but they are too restrictive for pdfs that occur as a result of summarization because the observations being summarized may be generated by different mechanisms. For instance, a pdf of the density of bipolar cells in a detached cat retina will have two peaks, corresponding to parts of the retina that are injured and healthy, respectively. It may be appropriate to fit a model distribution, such as a Gaussian, to data that are well understood, but in a scientific database, the most interesting data to query are precisely the ones that are not well understood.

Researchers have also considered the indexing of uncertain categorical data [59], correlated tuples [58], skyline queries [22], and combination of scores and uncertain values [60]. Monte Carlo simulations and state space searches have been used to answer top-k queries [52, 60]. The Trio system [4], which supports data uncertainty and lineage tracking, is built on top a conventional DBMS. The Orion project [19] is a recent effort aimed at developing an advanced database management system with direct support for uncertain data.

The realization that uncertainty plays an especially important role in biological data has been slow to emerge, yet progress is being made. Louie et al. focus on the integration of uncertain databases specific to the protein annotation domain [47]. Segal and Koller propose providing a model for hierarchical clustering of probabilistic gene expression data [57], and Rattray et al. provide a method for retaining the uncertainty inherent in microarray data analysis [50].

There is a large body of work on image analysis using Markov random fields (MRF) [9–11, 28]. These works introduce the MRF model, explain the theory, and provide an initial set of iterative techniques with which to solve for the *maximum a posteriori* (MAP) configuration. Much of the more recent work in this field focuses on solving for the MAP configuration efficiently using novel methods such as graph cuts [14].

3. Probabilistic Image Analyses

This section is concerned with image analysis techniques that produce probabilistic information. We begin by presenting an algorithm for probabilistic segmentation. Then we discuss how the probabilistic segmentation result can be used to compute probability distributions for neurite thickness and other measurements.

3.1 Probabilistic Segmentation

Our segmentation algorithm can be explained in terms of a simple model of a hypothetical protein that is produced near the center of the cell body and distributed throughout the cell by diffusion [44]. We simulate this process by a Markov random walk [15], starting in the center of the cell. Each step is to one of the eight pixels neighboring the current location in the image, chosen at random, but biased by the relative intensity of the neighbors so that

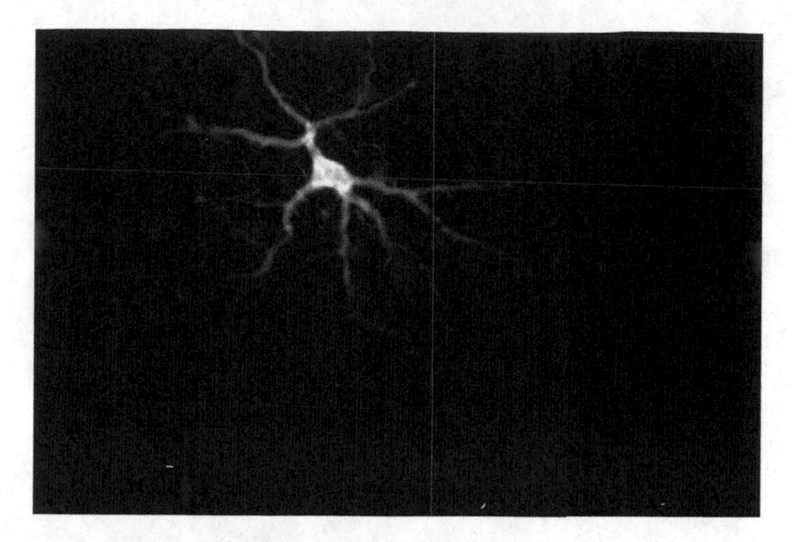

Figure 16.5. One of the cells in Figure 16.2, successfully segmented by the random-walk-with-restarts algorithm (image intensity adjusted for details to be visible in print). Reprinted with permission from [44], copyright © 2006 IEEE.

the step is more likely to be in the direction of a bright neighbor pixel. We count the number of times each pixel has been visited, and use these counts to compute a *probabilistic mask* (*pmask*) that gives the probability that each pixel belongs to the cell. The algorithm described so far works well when the cell is surrounded by unlabeled background, but fails when a neurite touches that of another cell: the walk will eventually stray into the other cell and keep visiting pixels there for a long time. Modifying the algorithm to perform a random walk with *restarts* [17] solves the problem. After each step, the walk returns to its original starting point with a small probability c. Thus, a walk that drifts into the wrong cell will soon return to the cell of interest [44].

Figure 16.5 shows one of the cells in Figure 16.2, segmented by the random-walk-with-restarts algorithm. The intensity of each pixel corresponds to how many times that pixel was visited (called the "visit record"). As expected, pixels farther from the center are generally visited less, and wider neurites are followed more often.

Although a simulation-based implementation of the algorithm, as described above, is sufficiently fast (5 s per cell for a 768-by-512-pixel image), the pmask can also be computed by solving an eigenvalue problem. Each step of the walk can be written as $\vec{x} := (1-c)\boldsymbol{P}\vec{x} + c\vec{s}$, where $\vec{x}$ is the pmask, $\boldsymbol{P}$ is the transition matrix, c is the restart probability, and s is a vector that indicates the center of the cell (the element corresponding to the pixel at the center of the cell is one, the rest are zero).

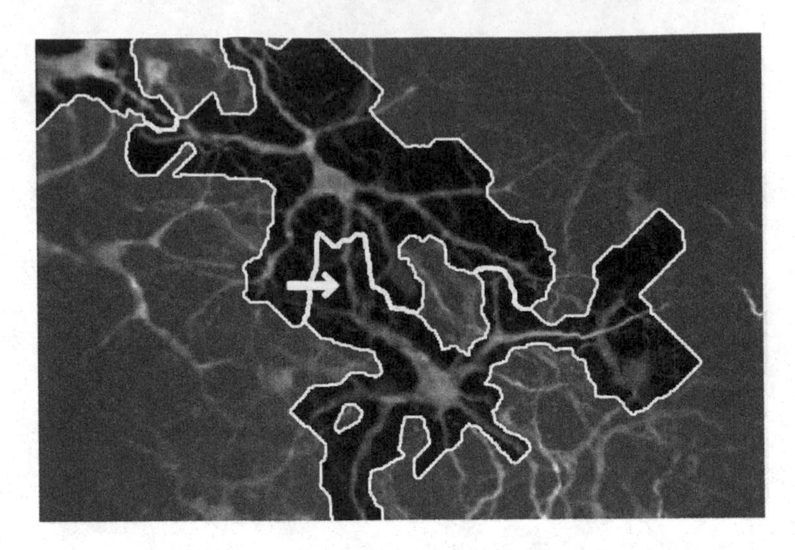

Figure 16.6. The seeded watershed algorithm segments much of the cell well, but makes some large mistakes. For instance, a large piece of a neurite is assigned to the wrong cell (arrow). In contrast, the random-walk-with-restarts algorithm indicates that this piece of the neurite could belong to either cell. Reprinted with permission from [44], copyright © 2006 IEEE.

It is instructive to compare our segmentation to the marker-based (or "seeded") watershed algorithm [64], which is the state of the art for this kind of segmentation problem. The watershed algorithms considers the image as a landscape, where a pixel of intensity x is lowered to a depth x units below the ground. The landscape is adjusted so that local minima occur only at the center of each cell (the foreground markers), and local maxima occur only in areas known to be outside the cell (the background markers). Figure 16.6 shows the three cells from Figure 16.2 segmented by the seeded watershed algorithm, with the original image shown in the background as a frame of reference. We would not expect the algorithm to produce a perfect segmentation, but it is striking that when it does err, the errors are large: long pieces of neurites are completely missing from the lower cell, and most of the neurite indicated by the arrow is misclassified as belonging to the lower cell when it actually belongs to the cell above. In contrast, the random-walk-with-restarts algorithm predicted, through its pmasks, that this part of the neurite could belong to either cell.

Next, we give examples of other image analysis techniques that generate probabilistic information. The first is a technique for measuring the thickness of neurites. The second considers the size of ganglion cells bodies.

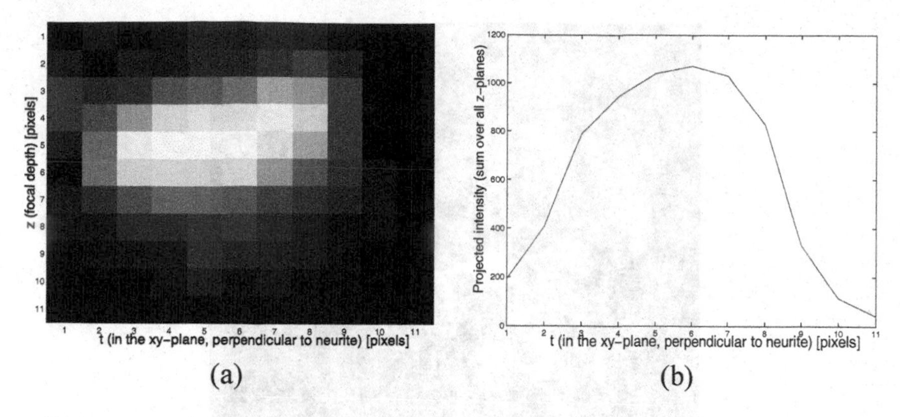

(a) (b)

Figure 16.7. (a) Cross-section through a dendrite. (b) The resulting projected signal.

3.2 Measuring Neurite Thickness

Measuring the thickness or size of an object is useful in understanding biological processes. For instance, we can measure the thickness of neurites to shed light on whether dendrites taper as they extend away from the cell body. Such measurements also serve as a building block for more complex analyses.

Based on the pmask, we can measure the thickness of a neurite as a probabilistic value. We compute $P(w)$, the probability that the neurite is w pixels thick, for $w \in \{1, 2, 3, \ldots\}$. The neurite is w pixels thick if there is a sequence of $w + 2$ pixels (oriented across the neurite) such that only the first and last pixel are not part of the cell. As examples, Figures 16.3 and 16.4 show the probabilistic thicknesses of a neurite and an entire inner plexiform layer, respectively. Such probabilistic measurements are directly useful to biologists because they conveys the variability in the measurement. They are even more useful in automatic analysis and mining because they allow the techniques to deal intelligently with measurements of different accuracy.

When working with 2-D projections of 3-D objects, it is important to develop accurate models for measurements. Figure 16.7(a) shows the cross-section of a neurite. The t axis is in the xy-plane, whereas z is the focal depth. We see that the cross-section is very close to a circular disk. Projected on the t-axis, the cross section yields the signal in Figure 16.7(b). When working with probabilistic masks based on 2-D image, we have assumed that the value of a pixel in the pmask is the probability that the pixel belongs to the cell. From Figure 16.7, however, we see that the intensity in the projected image is lower at the edges of the neurite. The neurite thickness measurement method discussed earlier would therefore be insufficient. Knowing that neurites have circular cross sections and building an appropriate model, however, allows us to make better measurements.

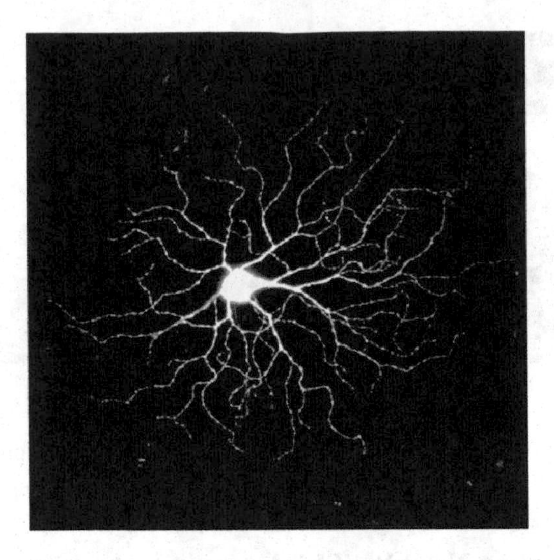

Figure 16.8. An image of a typical ganglion cell (enhanced for visual acuity). The cell body is the circular object at the center, and the dendrites are the arm-like extremities.

3.3 Ganglion Cell Features

Ganglion cells play an important role in the signal pathway from the photoreceptors to the brain in the mammalian eye. We describe a semi-automated method for extracting the size of the soma (cell body). This feature has been useful for classifying ganglion cells into different morphological subtypes [21, 61]. The retinal images considered here are captured on a confocal microscope at a resolution near 0.31 μm/pixel. A typical image is shown in Figure 3.3. Images are acquired by first injecting a fluorescent dye into each cell body. The dye diffuses from the center to the dendrites, and is imaged by a confocal microscope. Dendrites are difficult to identify accurately because they taper as they extend from the cell. Noise from nearby cells and poor diffusion of the dye into the dendrites present additional difficulties. These aspects make it very appropriate to model the information as uncertain.

Calculating the area of a cell body in each *possible world* can be easily accomplished once an outline of the cell body boundary is determined. Fortunately, there is a large body of research in the image processing community on boundary detection of spherical objects. Active contours [34], also known as snakes, are a popular method for detecting object boundaries in an image. Active contours operate by minimizing an energy function composed of two forces, an internal and external force, that move the contour toward object boundaries. Convergence of the snake occurs when the minimum of the energy function has been reached. Once the points along the boundary of the cell body are finalized, the soma size can be determined by computing the convex hull around the snake points.

Once we have computed probabilistic features such as soma size, a group of cells belonging to an experimental condition can be clustered. A technique such as k-medoids [63] can be used for this purpose. The distance measure between two distributions can be based on the earth mover's distance [55]. Comparisons can now be made between two experimental conditions by examining how the clustering patterns change. This can provide biological insights into how cell morphology changes under different conditions.

4. Querying Probabilistic Image Data

As discussed in previous sections, probability distributions are a natural representation for the location and extent of cells and other objects in biomedical images, as well as measurements of them. The following examples show how scientists would like to query such data.

1 *Find all confocal images where the ONL (outer nuclear layer) is 70–90* μm *thick.* Compute the pdf of the ONL thickness of every image, then rank the images by the portion of the pdf in the range 70–90 μm. This is a range query.

2 *Find all images where the ONL is 70–90* μm *thick and the INL (inner nuclear layer) is 50–70* μm *thick.* This is a range query in two dimensions, i.e., the ONL and INL (inner nuclear layer) thickness dimensions.

3 *Find the 10 images with ONL thickness closest to 50* μm. Rank the images by the expected difference between the ONL thickness and the query point of 50 μm. This is a k-NN query.

4 *Find the 10 images with ONL thickness closest to 80* μm *and INL thickness closest to 50* μm. This is an example of a k-NN query whose results are jointly ranked by two distributions.

5 *Find the 10 images where the ONL distribution is most similar to a given model distribution of reattachment and treatment.* A model distribution can be specified by a typical distribution belonging to the class. This is a k-NN query.

6 *Find the 10 images where the ONL distribution is most similar to a given model distribution of reattachment and treatment and the distribution of photoreceptor count is similar to that of a given image.* Here the input consists of two distributions and the best matches using both are returned. This is a k-NN query.

In order to address these types of queries, this section presents *adaptive, piecewise-linear approximations (APLAs)*, which represent arbitrary probability distributions compactly with guaranteed quality [45], as well as an efficient index structure, the APLA-tree.

4.1 Range Queries on Uncertain Data

Let $f_i : \mathbb{R}^d \mapsto [0,1]$ be the joint probability density function (pdf) of d real-valued attributes of an object o_i in the database. For any $\vec{x} = [x_1, \ldots, x_d]$, $f_i(\vec{x})$ is the probability that a random observation of the object will have the value x_1 for o_i's first attribute, the value x_2 for x_i's second attribute, and so on. A probabilistic range query [20, 62] (e.g., queries 1–2 from the list of examples) consists of a query range $R \subseteq \mathbb{R}^d$ and a probability threshold τ. It returns all objects for which the appearance probability (i.e., the probability that a random observation is in the query range) is at least τ.

If the pdf is represented by a histogram, an upper bound of the *appearance probability* can be computed as the sum of the depths of all bins that intersect the query range. The upper bound can be turned into a lower bound by subtracting the bins that intersect the boundary of the query range. Turning the pdf into a cumulative distribution function (cdf) avoids computing the sum; this reduces the amount of computation, but the I/O cost is still unacceptable because detailed multidimensional histograms are bulky and every object's histogram would have to be considered. A different representation is therefore needed. A good representation should (1) be compact in size, (2) give tight lower and upper bounds for the cdf at every point, and (3) be able to summarize the representations of a set of objects, so that the bounds computed from the summary are valid bounds for every object in the set. The third property allows a representation to be used in search trees: during search, entire subtrees can be disregarded because it can be determined from the compact summary stored in an internal node that none of the objects in the subtree satisfies the query.

In previous work, Keogh et al. [35] use piecewise-constant approximations on time series. Korn et al.'s OptimalSplines [36] fit a series of B-splines by maximum likelihood. Tao et al.'s conservative functional boxes (CFBs) [62] treat each dimension separately and envelope the cdf with four lines, found by linear programming. As we will see later, our approximations are more precise and compact.

4.2 k-NN Queries on Uncertain Data

k-NN queries (e.g., queries 3–6 from our list of examples) ask for the nearest neighbors of a given object, but it is not obvious what the semantics of k-NN queries on probability distributions should be. What exactly does it mean for one distribution to be "nearer" than another to a query point $\vec{q}$? Cheng et

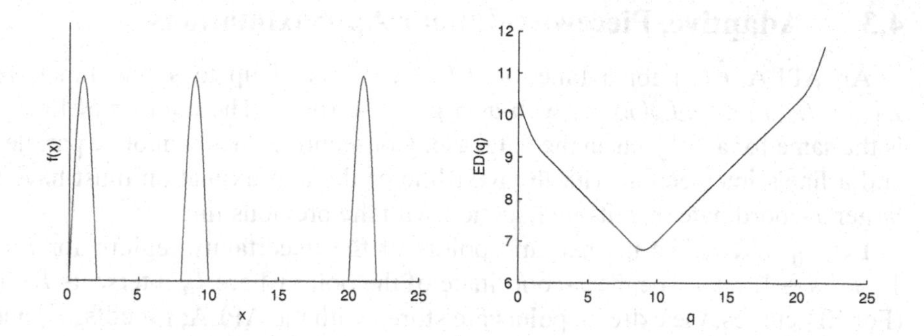

Figure 16.9. An example pdf (mixture of three Gaussians) and its ED-curve. Reprinted with permission from [45], copyright © 2007 IEEE.

al. [18] suggested a definition for 1-NN queries based on the probability p_i that a random observation of an object o_i is closer than random observations from all other objects in the database to $\vec{q}$. The object with the largest p_i is considered $\vec{q}$'s nearest neighbor. They develop efficient solutions for the special cases where the pdf is a uniform distribution constrained to line segments and circles. However, for arbitrary distributions, computing p_i is impractical because p_i depends not only on o_i's pdf, but also on the pdfs of all other objects in the database. In contrast, k-NN queries on deterministic data rank points by their distance to the query point—a quantity that is easily computed for each data point without reference to the rest of the database.

As an alternative, we define probabilistic k-NN queries so they return the k objects that have the smallest *expected distance* (ED) from the query point $\vec{q}$. The expected distance,

$$\mathrm{ED}(\vec{q}, o_i) = \int_{\mathbb{R}^d} d(\vec{q}, \vec{x}) f_i(\vec{x}) \, \mathrm{d}\vec{x},$$

is also known as f_i's first moment about $\vec{q}$, and can be computed solely from the query point $\vec{q}$ and o_i's pdf, f_i. When the distance function d is the L_1 distance (a natural choice for join distributions of attributes), ED has an important property: if $\vec{q}$ is outside the uncertainty region of o_i, then $\mathrm{ED}(\vec{q}, o_i)$ is the distance from $\vec{q}$ to the mean of o_i [46]. The fact that ED has such a simple shape outside the uncertainty region is promising: it means that an indexing scheme need only concentrate on the part of ED that is inside the uncertainty region. Next, we will describe APLA, which can be used to index ED (in order to answer k-NN queries) as well as cdfs (in order to answer range queries).

4.3 Adaptive, Piecewise-Linear Approximations

An APLA $\hat{F}(x)$ for a function $F(x)$ consists of up to s line functions $L_1(x), L_2(x), \ldots, L_s(x)$, as well as a global error ε. The number of lines s is the same for all objects in the database. Consecutive lines cannot be parallel, and a line's intersection with the next line of the approximation must have a larger x-coordinate than its intersection with the previous line.

Let $x_0^{\cap}$ and $x_s^{\cap}$ be the extreme points of the uncertainty region; for $i = 1, \ldots, s-1$, let $x_i^{\cap}$ be the x-coordinate of the point where L_i intersects L_{i+1}. (For ED-curves, the extreme points are stored with the APLA; for cdfs, $x_0^{\cap}$ and $x_s^{\cap}$ can be found by solving $L_1(x_0) = 0$ and $L_s(x_s^{\cap}) = 1$, respectively.) Given an APLA, the function $F(x)$ can be estimated for any value of x.

The global maximal error ε can be defined as $\varepsilon = \max_{i=1}^{s} \varepsilon_i$, where ε_i is the maximal absolute difference between a line L_i and the function F it approximates in the interval $[x_{i-1}^{\cap}, x_i^{\cap}]$, i.e.,

$$\varepsilon_i = \max_{x_{i-1}^{\cap} \leq x \leq x_i^{\cap}} |L_i(x) - F(x)|. \tag{16.1}$$

The APLA provides upper and lower bounds for the function $F(x)$ for every point x.

Assuming that the set of points sampled from $F(x)$ has already been partitioned into subsets $X_1, X_2, \ldots, X_n$, the line $\hat{F}_i$ that approximates a subset X_i while minimizing the error ε_i can be found by linear programming. A dynamic programming algorithm can in turn find the best set of intersection points $\{x_i^{\cap}\}$ for a given function.

Suppose that we have already computed APLAs $\hat{F}_1, \ldots, \hat{F}_n$ for functions $F_1, \ldots, F_n$ (cdfs or ED-curves). Using these APLAs, it is possible to compute an APLA $\hat{G}$ that is a valid approximation of the underlying distributions [46]. This is the key to building hierarchical index structures for APLA, such as the one that will described next.

4.4 Indexing the APLA

The compactness of the APLA speeds up sequential scans. The time to answer queries can be reduced further by the APLA-tree, a dynamic, balanced, paged index structure for the APLA. The APLA-tree is similar to the well-known R-tree [29], but it contains APLAs instead of rectangles. A leaf node contains APLAs of one or more objects as well as the corresponding object identifiers. An internal node consists of a number of entries; each entry has a pointer to a child node and an APLA that summarizes all objects in the subtree rooted at that child.

During range search, an upper bound for the appearance probability is computed for each entry in an internal node. Only if the upper bound is at least τ

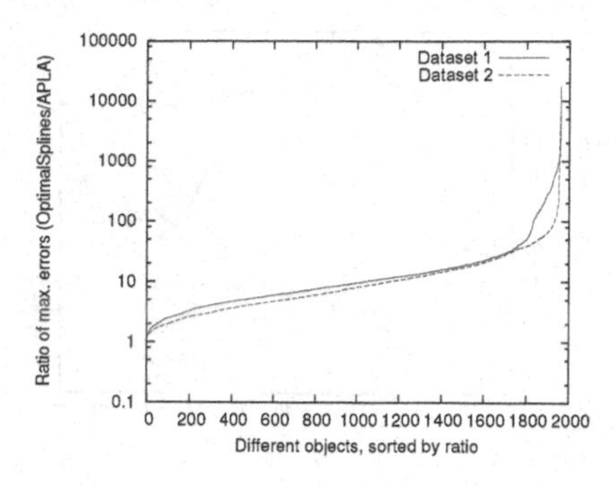

Figure 16.10. Maximal errors for APLA and OptimalSplines. Reprinted with permission from [45], copyright © 2007 IEEE.

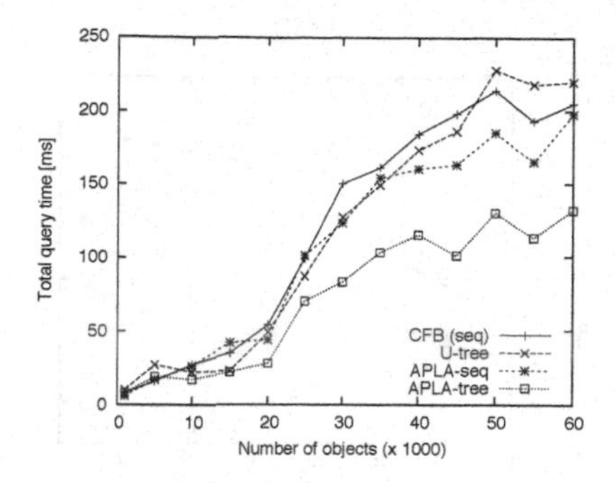

Figure 16.11. Query times for range queries. Reprinted with permission from [45], copyright © 2007 IEEE.

is it necessary to retrieve the subtree rooted in the child node pointed to by the entry. During k-NN search, two priority queues are used [30]: a queue of the best objects seen so far, and a queue that contains nodes the subtrees of which have not yet been searched. The latter is sorted by the ED_{min} of the subtree. Nodes are popped iteratively from the queue of nodes. For internal nodes, the ED_{min} and ED_{max} of each entry are computed. If ED_{min} does not exceed the ED_{max} of the k-th object in the queue of objects, the child node pointed to by the entry is inserted into the queue of nodes.

4.5 Experimental Results

We compared our technique to Tao et al.'s conservative functional boxes (CFBs) and the U-tree index structure based on them [62], as well as Korn et al.'s OptimalSplines [36]. The first set of experiments compares the global errors of APLA and OptimalSplines. Figure 16.10 shows that OptimalSplines had larger error for every object, and for about half of them the error was more than ten times that of APLA. The reason is that OptimalSplines minimizes the expected error instead of the maximal error; the latter is what is important for pruning power. Next, we compared the range query performance of the APLA-tree to that of the U-tree. Figure 16.11 plots the total time to answer range queries as a function of database size for Tao et al.'s techniques (CFB and U-tree) as well as APLA-sequential and APLA-tree. The U-tree is not much faster than sequential scan of CFBs because the precision of CFB is low (42 % compared to 67 % for APLA), so the U-tree is not able to prune enough subtrees to gain an advantage. Sequential scan using APLAs is about 15 % faster, and the APLA-tree is about twice as fast as the other techniques.

Figure 16.12 shows that APLAs can answer a 10-NN query on 30,000 objects in 144 ms by sequential scan or 72 ms using the APLA-tree. The times

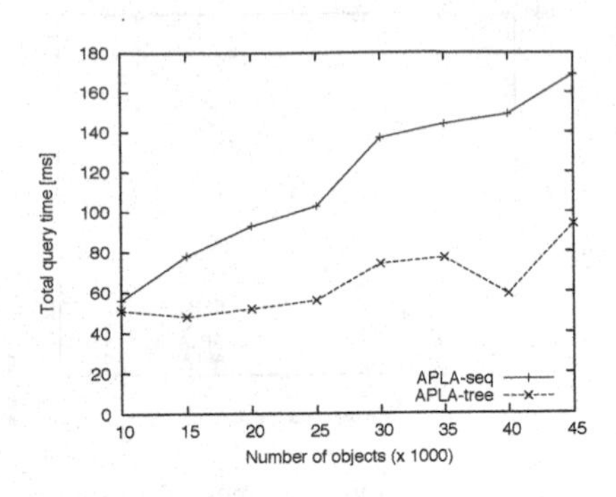

Figure 16.12. Total time to answer a 10-NN query. Reprinted with permission from [45], copyright © 2007 IEEE.

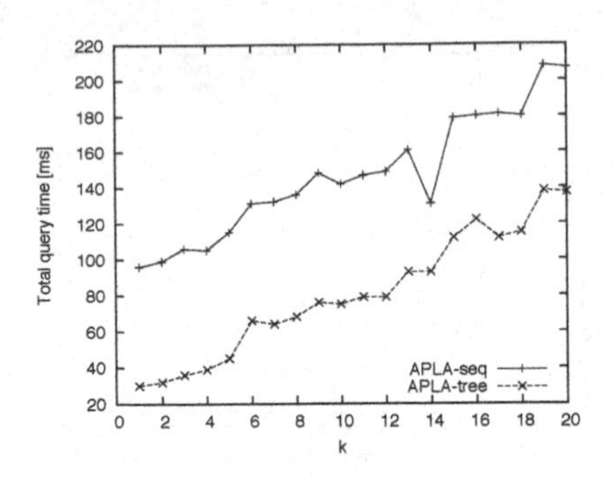

Figure 16.13. The APLA-tree answers k-NN queries twice as fast as APLA-sequential. Reprinted with permission from [45], copyright © 2007 IEEE.

shown are the total times to search the index structure, and then refine the results by retrieving actual objects from disk. Refining is necessary because the APLAs yield a range for the expected distance to each object, so the searches do not return exactly k objects, but a larger set that is guaranteed to contain the k nearest neighbors. Both APLA-sequential and the APLA-tree answer 10-NN queries with more than 80 % precision (in most cases more than 90 %). As a consequence, only 11 or 12 candidate objects need to be refined in most cases. For many applications, this is sufficiently precise, so the refinement step can be skipped. Figure 16.13 plots total query time as a function of k for a 35,000-object dataset. The cost of searching the index structure remains constant for APLA-sequential and APLA-tree, so the linear increase in query time with k is because the number of candidates that must be refined goes up. The APLA-tree is twice as fast as APLA-sequential, regardless of k.

5. Mining Probabilistic Image Data

To illustrate how uncertain information affects high-level analysis, this section discusses algorithms for what is perhaps the simplest form of spatial mining, namely the task of computing spatial joins. In its basic form, a spatial join starts with two sets of points and finds all pairs consisting of one point from each set such that the two points are within a certain distance of each other. Spatial joins are the essence of many biological questions; for instance, a neuroscientist studying the retina may want to find bipolar cell axons that are in close proximity to synaptic terminals [39], or Müller cells that have hypertrophied toward subretinal scars [40]. Spatial joins have been studied extensively for deterministic objects [2, 6, 16, 32, 33, 49, 51], but uncertain data presents algorithmic challenges.

5.1 Defining Probabilistic Spatial Join (PSJ)

We have seen in Section 2 that objects (such as cell, axons, or synaptic terminals) cannot be segmented reliably, so we must work with a probabilistic mask [44]. Even after segmentation, the objects often cannot be identified reliably. For instance, it can be hard to distinguish bipolar cell axons from other parts of the bipolar cells because they express the same protein (used for imaging). This uncertainty in object class constitutes a second source of uncertainty. Consequently, we define a *probabilistic object* a as a pair $\langle M_a, p_a \rangle$, where M_a is a's probabilistic mask and p_a is the confidence value of a's class (e.g., a classifier's estimate for the probability the probability that an object is really a bipolar cell axon). A probabilistic spatial join (PSJ) should take both sources of uncertainty into account when ranking pairs. Because there is often no clear limit for how distant two objects can be and still be of interest, we also incorporate the distance between two objects into our ranking function.

We now derive an appropriate score function for ranking pairs of object, first for point objects and then for the general case. Let a and b be objects with exactly known location, but with probabilities p_a and p_b of belonging to their respective classes. Then, their probabilistic masks have only one point. We define the score s' between two point objects $a = \langle\langle \vec{x}_a, 1\rangle, p_a\rangle$ and $b = \langle\langle \vec{x}_b, 1\rangle, p_b\rangle$ as $s'(p_a, \vec{x}_a, p_b, \vec{x}_b) = p_a\, p_b\, \lambda e^{-\lambda d(\vec{x}_a, \vec{x}_b)}$, where λ is a positive, domain-specific parameter that determines the relative importance of probability and distance, and d is a suitable distance function. Notice that we have adopted an inversely exponential mapping from distance to score. This mapping is appealing because of its simplicity and because it is more sensitive to variations in small distances.

We can now extend the definition from point objects to objects with extent by assuming that the distance that is important now is the *smallest* distance between two pixels that have a high probability of belonging to the objects. To satisfy this requirement, we define the score between two objects as the maximum of all scores between constituent points, weighted by the probability that both points belong to their respective objects: the score $s(a, b)$ of a pair of objects $a = \langle M_a, p_a\rangle$ and $b = \langle M_b, p_b\rangle$ is

$$s(a,b) = \max_{\substack{\langle \vec{x}_a, q_a\rangle \in M_a \\ \langle \vec{x}_b, q_b\rangle \in M_b}} p_a\, p_b\, s'(q_a, \vec{x}_a, q_b, \vec{x}_b) \tag{16.2}$$

$$= p_a\, p_b \max_{\substack{\langle \vec{x}_a, q_a\rangle \in M_a \\ \langle \vec{x}_b, q_b\rangle \in M_b}} q_a\, q_b\, \lambda e^{-\lambda d(\vec{x}_a, \vec{x}_b)}. \tag{16.3}$$

This object-level score function can be used in concert with any technique designed for point-level scores. Variations—e.g., the 90-th percentile of the point-level scores between the objects—are also possible.

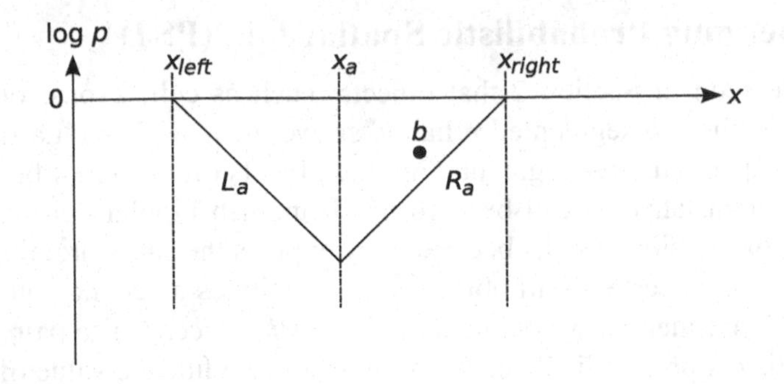

Figure 16.14. Each point *a* defines a triangle. The score between two points *a* and *b* exceeds the threshold τ only if *b* is inside *a*'s triangle. Reprinted with permission from [46], copyright © 2008 IEEE.

We can now proceed to define the *threshold probabilistic spatial join query* ("threshold PSJ query"), a query that, given two sets A and B of probabilistic objects and a score threshold τ, finds all pairs $\langle a, b \rangle \in A \times B$ such that $s(a, b) \geq \tau$. The idea of a *top-k probabilistic spatial join query* ("top-k PSJ query") can be defined similarly.

5.2 Threshold PSJ Query

The naïve solution to a threshold PSJ query is a nested loop join, which computes the score of every pair in $A \times B$ and checks whether it exceeds τ. If A and B contain n and m points, respectively, this takes $O(nm)$ time, which precludes joining datasets of even moderate size. The following algorithm exploits the geometry of the score function to find a more efficient solution.

First, assume 1-D data (i.e., probabilistic points on a line). For a point $\langle x_a, p_a \rangle \in A$, where must a corresponding point $\langle x_b, p_b \rangle$ must be in order to yield a score of at least τ? From the score function, we have that $p_a \, p_b \, \lambda e^{-\lambda |x_a - x_b|} \geq \tau$ and, consequently, that

$$|x_a - x_b| \leq \frac{1}{\lambda} \left(\ln p_a + \ln p_b + \ln \lambda - \ln \tau \right). \tag{16.4}$$

In Figure 16.14, which has x on the first axis and $\ln p$ on the second, Ineq. (16.4) describes an inverted triangle. The left and right side are $|x_a - x_b|$ apart at $\ln p = 0$ (which corresponds to $p = 1$, the highest possible probability), and eventually meet at $x = x_a$.

The algorithm is a plane sweep that (conceptually) slides a vertical line from left to right in Figure 16.14. During execution, it maintains a data structure of all the triangles currently intersecting the sweep line. When the sweep line encounters a point $\langle x_b, p_b \rangle \in B$, we use the data structure to find all triangles in which the point is contained. The time complexity of this lookup operation is critical. In the worst case, the data structure contains n triangles at the same time, and we perform one lookup for each point in B, so the lookup operation must be sublinear in order to avoid $O(nm)$ complexity. The key to achieving this is the slope of the line segments that make up our triangles. From Ineq. (16.4), we see that the left and right edges of the triangle have slopes of $-\lambda$ and λ. They do not depend on p_a or x_a, only on the constant λ. Using this insight, we keep two trees, one for downsloping segments and one for upsloping segments. Then the segments in each tree are parallel and can be indexed according to their unchanging total order. The resulting algorithm uses $O(n+m)$ space, and has a time complexity of $O((n+m)(\log n + k))$— or, if $n = m$, $O(n(\log n + k))$, where k is the size of the answer set. The plane-sweep algorithm can be extended to 2-D data (probabilistic points in the plane) if the distance metric is L_1.

5.3 Top-k PSJ Query

Top-k spatial joins arise from the fact that each pair in the join result has a score, and in many cases only the highest-scoring pairs are of interest. (Of course, if we know an appropriate threshold value, we can use the algorithm for threshold PSJ, but in general, this is not the case.) Our plane-sweep algorithm for threshold queries can be adapted to answer top-k queries as follows. The threshold is initialized to a small positive value (e.g., 10^{-6}), which remains unchanged until k results have been found. Then, after each new result, the threshold is updated to the score of the k-th best result found so far. The performance of this adapted plane-sweep algorithm depends on the order in which results are found, however. We have therefore developed a global scheduling technique for finding good pairs early. The algorithm starts by partitioning each point set into a number of subsets such that each subset has approximately r (a parameter) points that are close both in space and probability. Global scheduling decisions are then made based on the minimum bounding rectangles (MBRs) of these subsets. The algorithm works incrementally, repeatedly deciding which pair of MBRs to join next by finding the pair with the highest likelihood of yielding at least one pair of points that scores above the current threshold.

It is difficult to compute the score distribution for a pair of MBRs for several reasons. First, we do not know how the points are distributed (in space or in probability). Second, even if we assume that the points are distributed uniformly within the MBRs, we cannot find a closed-form solution, but must resort to sampling, which is too time consuming. The fact that a pair of MBRs represents a very large number of point pairs comes to our rescue, however. The highest score between two MBRs is the maximum of many terms, so it can be approximated well by a generalized extreme value distribution (EVD) [65];

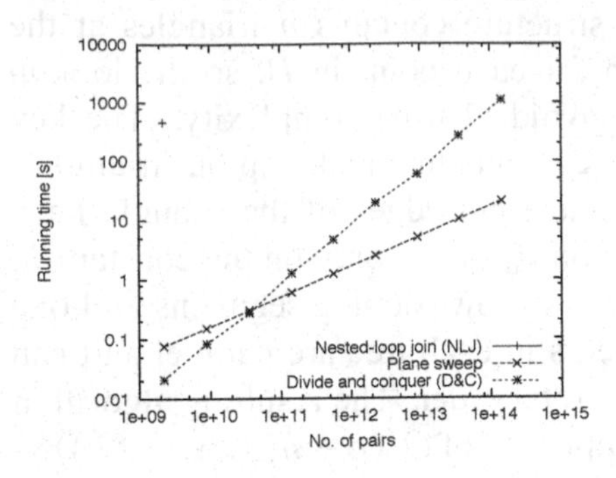

Figure 16.15. Comparative performance of algorithms. Reprinted with permission from [46], copyright © 2008 IEEE.

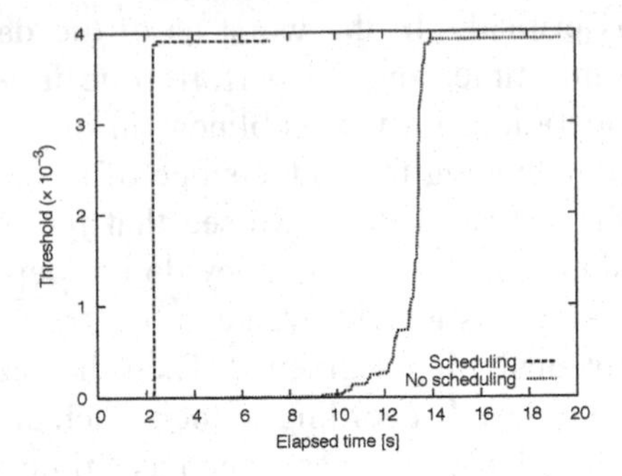

Figure 16.16. Effect of scheduling on development of threshold. Reprinted with permission from [46], copyright © 2008 IEEE.

consequently, the upper bound s_{ub} for the scores between two boxes is a good approximation of the maximum possible score between points belonging to the two MBRs when there are many points in each box. This increasingly becomes the case as the datasets grow—which is precisely when the scheduling needs to be effective. In addition to scheduling, the upper bound is also used for pruning.

We compare our algorithms to a simple divide-and-conquer (D&C) method that can be used both for threshold PSJs and top-k PSJs. The algorithm first checks whether two MBRs A_i and B_j can be pruned based on their s_{ub}. If not, each is split into 2^d quadrants (where d is the number of dimensions, including the probability dimension), and each pair of quadrants is processed recursively. When there are only a few points left, the subdivision stops, and a nested loop finishes the join. Although its worst-case performance is not ideal, the algorithm works for any L_p distance metric, and its performance is good for small datasets.

5.4 Experimental Results

Two datasets of 43k and 52k probabilistic points derived from probabilistic segmentation [44] of wholemount micrographs of horizontal cells were artificially enlarged as needed by randomly copying cells and translating them in space. To assess the performance of the plane-sweep algorithm for threshold PSJ, we varied the number of pairs from 10^9 to 10^{14} and measured the running time of the three join methods (nested loop join, D&C, and plane sweep). The threshold was chosen such that the queries returned 50–100 pairs. The running times are plotted in Figure 16.15 on a log-log scale. We see that the nested loop join did not finish for the larger datasets, and the plane-sweep algorithm is faster than the divide-and-conquer method, except for very small datasets.

For larger datasets, the plane sweep's advantage is significant; for instance, two sets of 10^{14} points can be joined 100 times faster with our plane-sweep algorithm.

Next, we investigated the scheduling algorithm's pruning ability. Figure 16.16 shows how the threshold increased during a run of the top-k algorithms (plane sweep with scheduling and plane sweep without scheduling) on a dataset of 9×10^{12} pairs, with k set to 20. Without scheduling, pairs were examined in an order based solely on the x-coordinates of the points. As a result, it took the algorithm almost 10 s to find good pairs; thereafter, the threshold increased gradually. As long as the threshold was low, the algorithm continued to process candidate pairs with scores too low to affect the final result. In contrast, after investing about 2 s in ordering the pairs of MBRs, the scheduling algorithm was able to find some high-scoring pairs very quickly, and thereby raise the threshold to just below its final value. As a result, most of the remaining pairs of MBRs were pruned.

6. Conclusion

Microscopy is the most important source of data for many biological disciplines. Visual inspection by experts has led to a century of great discoveries, but these are only the beginning, as high-throughput imaging techniques are making new kinds of experiments possible. Automatic analysis is necessary to detect subtle changes and to find patterns in these large image sets, however. The topic of this chapter has been how to manage the uncertainties that arise from analyzing sets of images. We have seen that representing the extent of an object (e.g., a cell) as a probabilistic mask can capture the ambiguity in the image, and that image analysis methods can base their computation on the probabilistic mask in order to produce probability distributions for measurements such as length and thickness. Finally, we have shown that arbitrary probability distributions can be indexed, queried, and mined efficiently. We have proposed a technique that can answer range queries much faster than existing techniques, and which is the first to answer k-nearest-neighbor queries efficiently. We have defined two kinds of spatial joins on objects with probabilistic extent and presented efficient algorithms for these queries. These database techniques make it viable to use probabilistic values in image database systems for large-scale analysis and mining.

Whereas we have presented one specific algorithm for probabilistic segmentation, suitable for horizontal cells and similar objects, the concept of probabilistic segmentation is general, and we hope to see other probabilistic segmentation algorithms emerge for other objects—in particular, the layers of the retina. The most important avenue of future work, however, is developing analyses for specific biological assays and using them to make new scientific discoveries.

Integrating probabilistic techniques such as discussed in this chapter with a database infrastructure is important. In this regard, a data management system called Bisque [1] has been developed at the Center for Bioimage Informatics (CBI) at UCSB. Bisque provides a rich set of biologically relevant image analysis tools to researchers and an integration and testing platform for computer vision and database researchers working at the center and beyond. The system consists of data collection tools, data importers, an image database, web interface, and an analysis suite. The Center is developing new information processing technologies appropriate for extracting a detailed understanding of biological processes from images depicting the distribution of biological molecules within cells or tissues. This is being accomplished by applying pattern recognition and data mining methods to bio-molecular images to fully automate the extraction of information from those images and the construction of statistically-sound models of the processes depicted in them, and by developing a distributed database for large sets of biological images. Efforts are underway to incorporate our probabilistic techniques into Bisque.

Acknowledgements

We would like to thank Geoffrey P. Lewis and Chris Banna from the laboratory of Steven K. Fisher at UCSB for providing the retinal images. Nick Larusso and Brian Ruttenberg contributed to the research on ganglion cells presented in Section 6. This work was supported in part by grant ITR-0331697 from the National Science Foundation.

References

[1] Bisque (Bio-Image Semantic Query User Environment). http://www.bioimage.ucsb.edu/bisque.

[2] D. J. Abel, V. Gaede, R. A. Power, and X. Zhou. Caching strategies for spatial joins. *GeoInformatica*, 3(1):33–59, 1999.

[3] C. Aggarwal. On density based transforms for uncertain data mining. In *Proceedings of the 23nd International Conference on Data Engineering (ICDE)*, pages 866–875, 2007.

[4] P. Agrawal, O. Benjelloun, A. D. Sarma, C. Hayworth, S. U. Nabar, T. Sugihara, and J. Widom. Trio: A system for data, uncertainty, and lineage. In *Proceedings of the 32nd International Conference on Very Large Data Bases (VLDB)*, pages 1151–1154, 2006.

[5] L. Antova, C. Kock, and D. Olteanu. 10^{10^6} worlds and beyond: Efficient representation and processing of incomplete information. In *Proceedings of the 23nd International Conference on Data Engineering (ICDE)*, pages 606–615, 2007.

[6] L. Arge, O. Procopiuc, S. Ramaswamy, T. Suel, and J. S. Vitter. Scalable sweeping-based spatial join. In *Proceedings of the 24rd International Conference on Very Large Data Bases (VLDB)*, pages 570–581, 1998.

[7] D. Barbará, H. Garcia-Molina, and D. Porter. The management of probabilistic data. *IEEE Transactions on Knowledge and Data Engineering*, 4(5):487–502, Oct. 1992.

[8] O. Benjelloun, A. D. Sarma, A. Halevy, and J. Widom. ULDBs: Databases with uncertainty and lineage. In *Proceedings of the 32nd International Conference on Very Large Data Bases (VLDB)*, pages 953–964, 2006.

[9] J. Besag. Spatial interaction and the statistical analysis of lattice systems. *Journal of the Royal Statistical Society, Series B*, 36(2):192–236, 1974.

[10] J. Besag. On the statistical analysis of dirty pictures. *Journal of the Royal Statistical Society B*, 48(3):259–302, 1986.

[11] J. Besag, J. York, A. Mollie, D. Geman, S. Geman, P. Green, S. Mase, Y. Ogata, A. Raftery, J. Banfield, et al. Bayesian image restoration with two applications in spatial statistics. discussion. rejoinder. *Annals of the Institute of Statistical Mathematics*, 43(1):1–59, 1991.

[12] A. Bhattacharya, V. Ljosa, J.-Y. Pan, M. R. Verardo, H. Yang, C. Faloutsos, and A. K. Singh. ViVo: Visual vocabulary construction for mining biomedical images. In *Proceedings of the Fifth IEEE International Conference on Data Mining (ICDM)*, pages 50–57, 2005.

[13] M. V. Boland, M. K. Markey, and R. F. Murphy. Automated recognition of patterns characteristic of subcellular structures in fluorescence microscopy images. *Cytometry*, 3(33):366–375, 1998.

[14] Y. Boykov, O. Veksler, and R. Zabih. Markov random fields with efficient approximations. *Proceedings of the IEEE Computer Society Conference on Computer Vision and Pattern Recognition (CVPR)*, pages 648–655, 1998.

[15] P. Bremaud. *Markov Chains: Gibbs Fields, Monte Carlo Simulation, and Queues*. Springer, 1999.

[16] T. Brinkhoff, H.-P. Kriegel, and B. Seeger. Efficient processing of spatial joins using R-trees. In *Proceedings of the 1993 ACM SIGMOD International Conference on Management of Data*, pages 237–246, 1993.

[17] O. Camoglu, T. Can, and A. K. Singh. Integrating multi-attribute similarity networks for robust representation of the protein space. *Bioinformatics*, 22(13):1585–1592, 2006.

[18] R. Cheng, D. V. Kalashnikov, and S. Prabhakar. Querying imprecise data in moving object environments. *IEEE Transactions on Knowledge Engineering*, 16(9):1112–1127, Sept. 2004.

[19] R. Cheng, C. Mayfield, S. Prabhakar, and S. Singh. Orion: A database system for managing uncertain data. http://orion.cs.purdue.edu/, Mar. 2006.

[20] R. Cheng, Y. Xia, S. Prabhakar, R. Shah, and J. S. Vitter. Efficient indexing methods for probabilistic threshold queries over uncertain data. In *Proceedings of the 30th International Conference on Very Large Data Bases (VLDB)*, 2004.

[21] J. Coombs, D. van der List, G. Wang, and L. Chalupa. Morphological properties of mouse retinal ganglion cells. *Neuroscience*, 140(1):123–136, 2006.

[22] X. Dai, M. Yiu, N. Mamoulis, Y. Tao, and M. Vaitis. Probabilistic spatial queries on existentially uncertain data. In *Advances in Spatial and Temporal Databases*, pages 400–417. Springer, 2005.

[23] N. Dalvi and D. Suciu. Efficient query evaluation on probabilistic databases. In *Proceedings of the 30th International Conference on Very Large Data Bases (VLDB)*, 2004.

[24] N. Dalvi and D. Suciu. Management of probabilistic data: Foundations and challenges. In *Proceedings of the 2007 ACM SIGMOD International Conference on Management of Data*, pages 1–12, 2007.

[25] D. Dey and S. Sarkar. A probabilistic relational model and algebra. *ACM Transactions on Database Systems*, 21(3):339–369, Sept. 1996.

[26] J. E. Dowling. *The Retina: An Approachable Part of the Brain.* Belknap, 1987.

[27] S. K. Fisher, G. P. Lewis, K. A. Linberg, and M. R. Verardo. Cellular remodeling in mammalian retina: Results from studies of experimental retinal detachment. *Progress in Retinal and Eye Research*, 24:395–431, 2005.

[28] S. Geman and D. Geman. Stochastic relaxation, Gibbs distributions, and the Bayesian restoration of images. *IEEE Transactions on Pattern Analysis and Machine Intelligence*, 6(6):721–741, Nov. 1984.

[29] A. Guttman. R-trees: A dynamic index structure for spatial searching. In *Proceedings of the 1984 ACM SIGMOD International Conference on Management of Data*, pages 47–57, 1984.

[30] G. R. Hjaltason and H. Samet. Distance browsing in spatial databases. *ACM Transactions on Database Systems*, 24(2):265–318, 1999.

[31] Y. Hu and R. F. Murphy. Automated interpretation of subcellular patterns from immunofluorescence microscopy. *Journal of Immunological Methods*, 290:93–105, 2004.

[32] Y.-W. Huang, N. Jing, and E. A. Rundensteiner. Spatial joins using R-trees: Breadth-first traversal with global optimizations. In *Proceedings of the 23rd International Conference on Very Large Data Bases (VLDB)*, pages 396–405, 1997.

[33] T. Kahveci, C. A. Lang, and A. K. Singh. Joining massive high-dimensional datasets. In *Proceedings of the 19th International Conference on Data Engineering (ICDE)*, pages 265–276, 2003.

[34] M. Kass, A. Witkin, and D. Terzopoulos. Snakes: Active countour models. *International Journal of Computer Vision*, 1(4):321–331, 1987.

[35] E. Keogh, K. Chakrabarti, S. Mehrotra, and M. Pazzani. Locally adaptive dimensionality reduction for indexing large time series databases. In *Proceedings of the 2001 ACM SIGMOD International Conference on Management of Data*, pages 151–162, 2001.

[36] F. Korn, T. Johnson, and H. Jagadish. Range selectivity estimation for continuous attributes. In *Proceedings of the Eleventh International Conference on Scientific and Statistical Database Management (SSDBM)*, pages 244–253, 1999.

[37] L. V. S. Lakshmanan, N. Leone, R. Ross, and V. S. Subrahmanian. ProbView: A flexible probabilistic database system. *ACM Transactions on Database Systems*, 22(3):419–469, Sept. 1997.

[38] G. Lewis, D. Charteris, C. Sethi, and S. Fisher. Animal models of retinal detachment and reattachment: Identifying cellular events that may affect visual recovery. *Eye*, 16(4):375–387, July 2002.

[39] G. Lewis, K. Linberg, and S. Fisher. Neurite outgrowth from bipolar and horizontal cells after experimental retinal detachment. *Investigative Ophthalmology & Visual Science*, 39(2):424–434, Feb. 1998.

[40] G. P. Lewis and S. K. Fisher. Müller cell outgrowth after retinal detachment: Association with cone photoreceptors. *Investigative Ophthalmology & Visual Science*, 41(6):1542–1545, 2000.

[41] G. P. Lewis, C. S. Sethi, K. A. Linberg, D. G. Charteris, and S. K. Fisher. Experimental retinal detachment: A new perspective. *Molecular Neurobiology*, 28(2):159–175, Oct. 2003.

[42] K. A. Linberg, G. P. Lewis, C. Shaaw, T. S. Rex, and S. K. Fisher. Distribution of S- and M-cones in normal and experimentally detached cat retina. *The Journal of Comparative Neurology*, 430(3):343–356, 2001.

[43] V. Ljosa, A. Bhattacharya, and A. K. Singh. Indexing spatially sensitive distance measures using multi-resolution lower bounds. In *Proceedings of the 10th International Conference on Extending Database Technology (EDBT)*, pages 865–883, 2006.

[44] V. Ljosa and A. K. Singh. Probabilistic segmentation and analysis of horizontal cells. In *Proceedings of the Sixth IEEE International Conference on Data Mining (ICDM)*, pages 980–985, Dec. 2006.

[45] V. Ljosa and A. K. Singh. APLA: Indexing arbitrary probability distributions. In *Proceedings of the 23nd International Conference on Data Engineering (ICDE)*, pages 946–955, 2007.

[46] V. Ljosa and A. K. Singh. Top-k spatial joins of probabilistic objects. In *Proceedings of the 24nd International Conference on Data Engineering (ICDE)*, pages 566–575, 2008.

[47] B. Louie, L. Detwiler, N. Dalvi, R. Shaker, P. Tarczy-Hornoch, and D. Suciu. Incorporating uncertainty metrics into a general-purpose data integration system. *Proceedings of the 19th International Conference on Scientific and Statistical Database Management (SSDBM)*, pages 19–19, 2007.

[48] J. G. Nicholls, A. R. Martin, B. G. Wallace, and P. A. Fuchs. *From Neuron to Brain*. Sinauer, 4th edition, 2001.

[49] J. M. Patel and D. J. DeWitt. Partition based spatial merge join. In *Proceedings of the 1996 ACM SIGMOD International Conference on Management of Data*, pages 259–270, 1996.

[50] M. Rattray, X. Liu, G. Sanguinetti, M. Milo, and N. D. Lawrence. Propagating uncertainty in microarray data analysis. *Briefings in Bioinformatics*, 7(1):37–47, 2006.

[51] S. Ravada, S. Shekhar, C. tien Lu, and S. Chawla. Optimizing join index based spatial join processing: A graph partitioning approach. In *Proceedings of the 17th IEEE Symposium on Reliable Distributed Systems*, pages 302–308, 1998.

[52] C. Ré, N. Dalvi, and D. Suciu. Efficient top-k query evaluation on probabilistic data. In *Proceedings of the 23nd International Conference on Data Engineering (ICDE)*, pages 886–895, 2007.

[53] B. J. Reese, M. A. Raven, and S. B. Stagg. Afferents and homotypic neighbors regulate horizontal cell mor phology, conectivity and retinal coverage. *The Journal of Neuroscience*, 25(9):2167–2175, Mar. 2005.

[54] T. S. Rex, R. N. Fariss, G. P. Lewis, K. A. Linberg, I. Sokal, and S. K. Fisher. A survey of molecular expression by photoreceptors after experimental retinal detachment. *Investigative Ophthalmology & Visual Science*, 43(4):1234–1247, 2002.

[55] Y. Rubner, C. Tomasi, and L. J. Guibas. The earth mover's distance as a metric for image retrieval. *International Journal of Computer Vision*, 40(2):99–121, 2000.

[56] A. D. Sarma, O. Benjelloun, A. Halevy, and J. Widom. Working models for uncertain data. In *Proceedings of the 22nd International Conference on Data Engineering (ICDE)*, 2006.

[57] E. Segal and D. Koller. Probabilistic hierarchical clustering for biological data. In *Proceedings of the Sixth Annual International Conference on Research in Computational Molecular Biology (RECOMB)*, pages 273–280, 2002.

[58] P. Sen and A. Deshpande. Representing and querying correlated tuples in probabilistic databases. In *Proceedings of the 23nd International Conference on Data Engineering (ICDE)*, pages 596–605, 2007.

[59] S. Singh, C. Mayfield, S. Prabhakar, R. Shah, and S. Hambrusch. Indexing uncertain categorical data. In *Proceedings of the 23nd International Conference on Data Engineering (ICDE)*, pages 616–625, 2007.

[60] M. A. Soliman, I. F. Ilyas, and K. C.-C. Chang. Top-k query processing in uncertain databases. In *Proceedings of the 23nd International Conference on Data Engineering (ICDE)*, pages 896–905, 2007.

[61] W. Sun, N. Li, and S. He. Large-scale morphological survey of mouse retinal ganglion cells. *The Journal of Comparative Neurology*, 451(2):115–126, 2002.

[62] Y. Tao, R. Cheng, X. Xiao, W. K. Ngai, B. Kao, and S. Prabhakar. Indexing multi-dimensional uncertain data with arbitrary probability density functions. In *Proceedings of the 31st International Conference on Very Large Data Bases (VLDB)*, 2005.

[63] S. Theodoridis and K. Koutroumbas. *Pattern Recognition.* Academic Press, 3rd edition, 2006.

[64] L. Vincent and P. Soille. Watersheds in digital spaces: An efficient algorithm based on immersion simulations. *IEEE Transactions on Pattern Analysis and Machine Intelligence*, 13:583–598, 1991.

[65] R. E. Walpole and R. H. Myers. *Probability and Statistics for Engineers and Scientists*. Prentice Hall, Englewood Cliffs, New Jersey, 5th edition, 1993.

Index